Statistics and Computing

Series Editors:
J. Chambers
W. Eddy
W. Härdle
S. Sheather
L. Tierney

Springer
New York
Berlin
Heidelberg
Barcelona
Budapest
Hong Kong
London
Milan
Paris
Santa Clara
Singapore
Tokyo

Statistics and Computing

Gentle: Numerical Linear Algebra for Applications in Statistics.

Gentle: Random Number Generation and Monte Carlo Methods.

Härdle/Klinke/Turlach: XploRe: An Interactive Statistical Computing Environment.

Krause/Olson: The Basics of S and S-PLUS.

Ó Ruanaidh/Fitzgerald: Numerical Bayesian Methods Applied to Signal Processing.

Pannatier: VARIOWIN: Software for Spatial Data Analysis in 2D.

Venables/Ripley: Modern Applied Statistics with S-PLUS, 2nd edition.

Lange: Numerical Analysis for Statisticians.

W.N. Venables
B.D. Ripley

Modern Applied Statistics with S-PLUS

Second Edition

With 144 Figures

Springer

W. N. Venables
Department of Statistics
University of Adelaide
Adelaide, South Australia 5005
Australia

B.D. Ripley
Professor of Applied Statistics
University of Oxford
1 South Parks Road
Oxford OX1 3TG
England

Series Editors:

J. Chambers
Bell Labs, Lucent
    Technologies
600 Mountain Ave.
Murray Hill, NJ 07974
USA

W. Eddy
Department of Statistics
Carnegie-Mellon University
Pittsburgh, PA 15213
USA

W. Hårdle
Institut für Statistik und
    Ökonometrie
Humboldt-Universität zu Berlin
Spandauer Str. 1
D-10178 Berlin, Germany

S. Sheather
Australian Graduate School
    of Mathematics
PO Box 1
Kensington
New South Wales 2033
Australia

L. Tierney
School of Statistics
University of Minnesota
Vincent Hall
Minneapolis, MN 55455
USA

Library of Congress Cataloging in Publication Data
Venables, W.N. (William N.)
    Modern applied statistics with S-PLUS / W.N. Venables and B.D.
    Ripley. — 2nd ed.
        p.   cm.
    Includes bibliographical references and index.
    ISBN 0-387-98214-0 (hardcover : alk. paper)
        1. S-PLUS.   2. Statistics—Data processing.   3. Mathematical
    statistics—Data processing.   I. Ripley, Brian D., 1952–
    II. Title.
    QA276.4.V46   1997
    519.5´0285´53—dc 21   97-8922

Printed on acid-free paper.

Production managed by Terry Kornak; manufacturing supervised by Joe Quatela.
Camera-ready copy prepared from the authors' LaTeX files.
Printed and bound by Maple-Vail Book Manufacturing Group, York, PA.
Printed in the United States of America.

9 8 7 6 5 4 3 (Corrected third printing, 1998)

ISBN 0-387-98214-0 Springer-Verlag New York Berlin Heidelberg   SPIN 10695768

To Karen, Bess, Brigid, Gus, Iggy, Lucy, Ben

and Ruth

Preface

S-PLUS is a system for data analysis from the Data Analysis and Products Division of MathSoft (formerly Statistical Sciences), an enhanced version of the S environment for data analysis developed at (the then) AT&T's Bell Laboratories. S-PLUS has become the statistician's calculator for the 1990s, allowing easy access to the computing power and graphical capabilities of modern workstations and now personal computers.

Since the first edition of this book appeared in 1994 S-PLUS has continued to grow in popularity, not least on personal computers as the Windows version now has equivalent capabilities to the versions running on Unix workstations. This edition gives equal weight to the Windows and Unix versions, and covers many of the features that have been added to S-PLUS, for example Trellis and object-oriented editable graphics, linear and non-linear mixed effects models and factor analysis. It also builds on further experience in the classroom and in using S-PLUS.

This edition is designed for users of S-PLUS 3.3 or later. It was written whilst S-PLUS 4.0 for Windows was under development, and covers what we see as the most exciting new features, but not new interfaces to existing functionality. (Some of the simpler statistical operations can be accessed via a menu system or customized buttons, but for the full power of S the command-line interface is still needed, and many users will still prefer it.) The new features are indexed under 'version 4.0'.

Some of the more specialised or rapidly changing functionality is covered in the *on-line complements* (see page 497 for sites) which will be updated frequently. The datasets and S functions that we use are available on-line, and will help greatly in making use of the book. Also on-line are further exercises and answers to selected exercises.

This is not a text in statistical theory, but does cover modern statistical methodology. Each chapter summarizes the methods discussed, in order to set out the notation and the precise method implemented in S. (It will help if the reader has a basic knowledge of the topic of the chapter, but several chapters have been successfully used for specialized courses in statistical methods.) Our aim is rather to show how we analyse datasets using S-PLUS. In doing so we aim to show both how S can be used and how the availability of a powerful and graphical system has altered the way we approach data analysis and allows penetrating analyses to be performed routinely. Once calculation became easy, the statistician's energies could be devoted to understanding his or her dataset.

The core S language is not very large, but it is quite different from most other statistics systems. We describe the language in some detail in the early chapters, but these are probably best skimmed at first reading; Chapter 1 contains the most basic ideas, and each of Chapters 2 and 3 are divided into 'basic' and 'advanced' sections. Once the philosophy of the language is grasped, its consistency and logical design will be appreciated.

The chapters on applying S to statistical problems are largely self-contained, although Chapter 6 describes the language used for linear models, and this is used in several later chapters. We expect that most readers will want to pick and choose amongst the later chapters, although they do provide a number of examples of S programming, and so may be of interest to S programmers not interested in the statistical material of the chapter.

This book is intended both for would-be users of S-PLUS as an introductory guide and for class use. The level of course for which it is suitable differs from country to country, but would generally range from the upper years of an undergraduate course (especially the early chapters) to Masters' level. (For example, almost all the material is covered in the M.Sc. in Applied Statistics at Oxford.) Full exercises are provided only for Chapters 2–5, since otherwise these would detract from the best exercise of all, using S to study datasets with which the reader is familiar. Our library provides many datasets, some of which are not used in the text but are there to provide source material for exercises. (We would stress that it is preferable to use examples from the user's own subject area and experience where this is possible. Further exercises and answers to selected exercises are available from our WWW pages.)

The user community has contributed a wealth of software within S. This book shows its reader how to extend S, and uses this facility to discuss procedures not implemented in S-PLUS, thereby providing fairly extensive examples of how this might be done.

Both authors take responsibility for the whole book, but Bill Venables was the lead author for Chapters 1–4, 6, 7, 9 and 10, and Brian Ripley for Chapters 5, 8 and 11–17. (The ordering of authors reflects the ordering of these chapters.)

The authors may be contacted by electronic mail as

    Bill.Venables@adelaide.edu.au
    ripley@stats.ox.ac.uk

and would appreciate being informed of errors and improvements to the contents of this book.

Acknowledgements:

This book would not be possible without the S environment which has been principally developed by Rick Becker, John Chambers and Allan Wilks, with substantial input from Doug Bates, Bill Cleveland, Trevor Hastie and Daryl Pregibon. The code for survival analysis is the work of Terry Therneau. The S-PLUS code is the work of a much larger team acknowledged in the manuals for that system.

We are grateful to the many people who have read and commented on draft material and who have helped us test the software, as well as to those whose problems have contributed to our understanding and indirectly to examples and exercises. We can not name all, but in particular we would like to thank Adrian Bowman, Bill Dunlap, Sue Clancy, David Cox, Anthony Davison, Peter Diggle, Matthew Eagle, Nils Hjort, Stephen Kaluzny, Francis Marriott, Brett Presnell, Charles Roosen, David Smith and Patty Solomon. We thank MathSoft DAPD, Statistical Sciences (Europe) and CSIRO DMS for early access to versions of S-PLUS and for access to platforms for testing.

Bill Venables
Brian Ripley
January 1997

Contents

Preface   **vii**

Typographical Conventions   **xvii**

1   Introduction   **1**
    1.1   A quick overview of S 2
    1.2   Using S-PLUS under Unix 4
    1.3   Using S-PLUS under Windows 8
    1.4   An introductory session 10
    1.5   What next? 18

2   The S Language   **19**
    2.1   A concise description of S objects 19
    2.2   Calling conventions for functions 29
    2.3   Arithmetical expressions 30
    2.4   Reading data 36
    2.5   Model formulae 41
    2.6   Character vector operations 42
    2.7   Finding S objects 44
    2.8   Indexing vectors, matrices and arrays 47
    2.9   Matrix operations 54
   2.10   Input/Output facilities 61
   2.11   Customizing your S environment 63
   2.12   History and audit trails 66
   2.13   BATCH operation 67
   2.14   Exercises 68

3   Graphical Output   **69**
    3.1   Graphics devices 69
    3.2   Basic plotting functions 73
    3.3   Enhancing plots 78

3.4   Fine control of graphics 82

3.5   Trellis graphics 88

3.6   Object-oriented editable graphics 108

3.7   Exercises 112

4   Programming in S   113

4.1   Control structures 113

4.2   Vectorized calculations and loop avoidance functions 117

4.3   Writing your own functions 125

4.4   Introduction to object orientation 132

4.5   Editing, correcting and documenting functions 139

4.6   Calling the operating system 147

4.7   Recursion and handling vectorization 149

4.8   Frames 153

4.9   Using C and FORTRAN routines 157

4.10 Working within limited memory 157

4.11 Exercises 162

5   Distributions and Data Summaries   163

5.1   Probability distributions 163

5.2   Generating random data 166

5.3   Data summaries 168

5.4   Classical univariate statistics 172

5.5   Density estimation 177

5.6   Bootstrap and permutation methods 186

5.7   Exercises 190

6   Linear Statistical Models   191

6.1   A linear regression example 191

6.2   Model formulae and model matrices 196

6.3   Regression diagnostics 204

6.4   Safe prediction 208

6.5   Factorial designs and designed experiments 209

6.6   An unbalanced four-way layout 214

7   Generalized Linear Models   223

7.1   Functions for generalized linear modelling 227

7.2   Binomial data 230

7.3 Poisson models 238

7.4 A negative binomial family 242

8 Robust Statistics **247**

8.1 Univariate samples 248

8.2 Median polish 254

8.3 Robust regression 256

8.4 Resistant regression 261

8.5 Multivariate location and scale 266

9 Non-linear Models **267**

9.1 Fitting non-linear regression models 268

9.2 Non-linear fitted model objects and method functions 274

9.3 Taking advantage of linear parameters 275

9.4 Confidence intervals for parameters 276

9.5 Assessing the linear approximation 282

9.6 Constrained non-linear regression 284

9.7 General optimization and maximum likelihood estimation 286

9.8 Exercises 295

10 Random and Mixed Effects **297**

10.1 Random effects and variance components 297

10.2 Multistratum models 299

10.3 Linear mixed effects models 304

10.4 Non-linear mixed effects models 314

10.5 Exercises 321

11 Modern Regression **323**

11.1 Additive models and scatterplot smoothers 323

11.2 Projection-pursuit regression 331

11.3 Response transformation models 334

11.4 Neural networks 337

11.5 Conclusions 341

12 Survival Analysis **343**

12.1 Estimators of survivor curves 345

12.2 Parametric models 350

12.3 Cox proportional hazards model 356

12.4 Further examples 363

12.5 Expected survival rates 379

13 Multivariate Analysis   **381**

13.1 Graphical methods 382

13.2 Cluster analysis 389

13.3 Discriminant analysis 395

13.4 An example: *Leptograpsus variegatus* crabs 401

13.5 Factor analysis 405

14 Tree-based Methods   **413**

14.1 Partitioning methods 414

14.2 Cutting trees down to size 425

14.3 Low birth weights revisited 428

15 Time Series   **431**

15.1 Second-order summaries 434

15.2 ARIMA models 443

15.3 Seasonality 449

15.4 Multiple time series 455

15.5 Nottingham temperature data 458

15.6 Regression with autocorrelated errors 463

15.7 Other time-series functions 467

16 Spatial Statistics   **469**

16.1 Spatial interpolation and smoothing 469

16.2 Kriging 475

16.3 Point process analysis 480

17 Classification   **485**

17.1 Theory 485

17.2 A simple example 487

17.3 Forensic glass 493

Appendices

A Datasets and Software   **497**

A.1 Libraries 497

A.2 Caveat 498

B Common S-PLUS Functions   **499**

C Using S-PLUS Libraries   **517**

C.1 Sources of libraries 519

C.2 Creating a library section 520

References   **521**

Index   **533**

Typographical Conventions

Throughout this book S language constructs and commands to the operating system are set in a monospaced typewriter font `like this`. The character ~ may appear as ˜ on your keyboard, screen or printer.

We often use the prompts `$` for the operating system (it is the standard prompt for the Unix Bourne shell) and `>` for S-PLUS. However, we do *not* use prompts for continuation lines, which are indicated by indentation. One reason for this is that the length of line available to use in a book column is less than that of a standard terminal window, so we have had to break lines which were not broken at the terminal.

Some of the S-PLUS output has been edited. Where complete lines are omitted, these are usually indicated by

```
    ....
```

in listings; however most *blank* lines have been silently removed. Much of the S-PLUS output was generated with the options settings

```
    options(width=65, digits=5)
```

in effect, whereas the defaults are `80` and `7`. Not all functions consult these settings, so on occasion we have had to manually reduce the precision to more sensible values.

Chapter 1

Introduction

Statistics is fundamentally concerned with the understanding of structures in data. One of the effects of the information-technology era has been to make it much easier to collect extensive datasets with minimal human intervention. Fortunately the same technological advances allow the users of statistics access to much more powerful 'calculators' to manipulate and display data. This book is about the modern developments in applied statistics which have been made possible by the widespread availability of workstations with high-resolution graphics and computational power equal to a mainframe of a few years ago. Workstations need software, and the S system developed at what was then AT&T's Bell Laboratories (and now at Lucent Technologies) provides a very flexible and powerful environment in which to implement new statistical ideas. Thus this book provides both an introduction to the use of S and a course in modern statistical methods.

S is an integrated suite of software facilities for data analysis and graphical display. Among other things it offers

- an extensive and coherent collection of tools for statistics and data analysis,

- a language for expressing statistical models and tools for using linear and non-linear statistical models,

- graphical facilities for data analysis and display either at a workstation or as hardcopy,

- an effective object-oriented programming language which can easily be extended by the user community.

The term *environment* is intended to characterize it as a planned and coherent system built around a language and a collection of low-level facilities, rather than the perceived 'package' model of an incremental accretion of very specific, high-level and sometimes inflexible tools. Its great strength is that functions implementing new statistical methods can be built on top of the low-level facilities.

Furthermore, most of the environment is open enough that users can explore and, if they wish, change the design decisions made by the original implementors. Suppose you do not like the output given by the regression facility (as we have frequently felt about statistics packages). In S you can write your own summary routine, and the system one can be used as a template from which to start. In many cases sufficiently persistent users can find out the exact algorithm used by listing the S functions invoked.

We have made extensive use of the ability to extend the environment to implement (or re-implement) statistical ideas within S. All the S functions which are used and our datasets are available in machine-readable form; see Appendix A for details of what is available and how to install it.

The name S arose long ago as a compromise name (Becker, 1994), in the spirit of the programming language C (also from Bell Laboratories).

We will assume the reader has S-PLUS, an enhanced version of S supplied in binary form by the Data Analysis Products Division of MathSoft Inc. We refer to the language as S unless we are invoking features unique to S-PLUS.

System dependencies

We have tried as far as is practicable to make our descriptions independent of the computing environment and the exact version of S-PLUS in use. Clearly some of the details must depend on the environment; we used S-PLUS 3.4 on Sun SparcStations under OpenLook to compute the examples, but have also tested them under versions 3.3, 4.0 and 4.5 of S-PLUS for Windows and under S-PLUS 3.3 on Unix. Where timings are given they refer to a laptop PC based on a 133MHz Pentium CPU; a modern workstation (e.g. Sun UltraSparc 1/170) will be at least twice as fast.

S-PLUS is normally used at a workstation running a graphical user interface (GUI) such as Motif or Microsoft Windows, or at an X-terminal. We will assume such a graphical environment. The ability to cut and paste is very convenient in writing S code; the examples in this book were developed by cut and paste between the LATEX source and a window running an S-PLUS session.

S-PLUS 4.x for Windows has a new user interface. Differences in future Unix versions will be dealt with in the on-line complements.

Reference manuals

The basic references are Becker, Chambers & Wilks (1988) for the basic environment and Chambers & Hastie (1992) for the statistical modelling, object-oriented programming and data structuring features; these should be supplemented by checking the on-line help pages for changes and corrections as S-PLUS has evolved considerably since these books were written. Our aim is not to be comprehensive nor to replace these manuals, but rather to explore much further the use of S-PLUS to perform statistical analyses.

1.1   A quick overview of S

Most things done in S are permanent; in particular, data, results and functions are all stored in operating system files. These are referred to as *objects*. On most Unix systems these files are stored in the .Data subdirectory of the current directory under the same name, so that a variable stock is stored (in an internal format) in the file .Data/stock. (The Windows version of S-PLUS uses a _Data subdirectory and translates names to conform to the limitations of the MS-DOS file system.)

Variables can be used as scalars, matrices or arrays, and S provides extensive matrix manipulation facilities. Further, objects can be made up of collections (known as *lists*) of such variables, allowing complex objects such as the result of a regression calculation. This means that the result of a statistical procedure can be saved for further analysis in a future session. Typically the calculation is separated from the output of results, so one can perform a regression and then print various summaries and compute residuals and leverage plots from the saved regression object.

Technically S is a function language. Elementary commands consist of either *expressions* or *assignments*. If an expression is given as a command, it is evaluated, printed, and the value is discarded. An assignment evaluates an expression and passes the value to a variable but the result is not printed automatically. An expression can be as simple as 2 + 3 or a complex function call. Assignments are indicated by the *assignment operator* <-. For example

```
> 2 + 3
[1] 5
> sqrt(3/4)/(1/3 - 2/pi^2)
[1] 6.6265
> mean(hstart)
[1] 137.99
> m <- mean(hstart); v <- var(hstart)/length(hstart)
> m/sqrt(v)
[1] 32.985
```

Here > is the S-PLUS prompt, and the [1] states that the answer is starting at the first element of a vector.

More complex objects will have printed a short summary instead of full details. This is achieved by the object-oriented programming mechanism; complex objects have *classes* assigned to them which determine how they are printed, summarized and plotted.

S-PLUS can be extended by writing new functions, which then can be used in the same way as built-in functions (and can even replace them). This is very easy; for example to define functions to compute the standard deviation and the two-tailed p–value of a t statistic we can write

```
std.dev <- function(x) sqrt(var(x))
t.test.p <- function(x, mu=0) {
    n <- length(x)
    t <- sqrt(n) * (mean(x) - mu) / std.dev(x)
    2 * (1 - pt(abs(t), n - 1))
}
```

It would be useful to give both the t statistic and its p-value, and the most common way of doing this is by returning a list; for example we could use

```
t.stat <- function(x, mu=0) {
    n <- length(x)
    t <- sqrt(n) * (mean(x) - mu) / std.dev(x)
    list(t = t, p = 2 * (1 - pt(abs(t), n - 1)))
}
z <- rnorm(300, 1, 2)  # generate 300 N(1, 4) variables.
t.stat(z)
$t:
[1] 8.2906
$p:
[1] 3.9968e-15

unlist(t.stat(z, 1))  # test mu=1, compact result
        t       p
-0.56308 0.5738
```

The first call to `t.stat` prints the result as a list; the second tests the non-default hypothesis $\mu = 1$ and using `unlist` prints the result as a numeric vector with named components.

Linear statistical models can be specified by a version of the commonly-used notation of Wilkinson & Rogers (1973), so that

```
time ~ dist + climb
time ~ transplant/year + age + prior.surgery
```

refer to a regression of `time` on both `dist` and `climb`, and of `time` on year within each transplant group and on age, with a different intercept for each type of prior surgery. This notation has been extended in many ways, for example to survival and tree models and to allow smooth non-linear terms.

1.2   Using S-PLUS under Unix

One system dependency is the mapping of mouse buttons. We refer to button 1, usually the left button, and button 2, the middle button on Unix interfaces (or perhaps both together of two).

Getting started

S-PLUS makes use of the file system and for each project we strongly recommend that you have a separate working directory to hold the files for that project.

The suggested procedure for the first occasion on which you use S-PLUS is:

1. Create a separate directory, say `SwS`, for this project, which we suppose is 'Statistics with S-PLUS', and make it your working directory.

```
$ mkdir SwS
$ cd SwS
```

Copy any data files you need to use with S-PLUS to this directory.

2. Create a subdirectory of SwS called .Data by

```
$ mkdir .Data
```

This subdirectory is for use by S-PLUS itself and hence has a Unix 'dot name' to be hidden from casual inspection.

3. Start the system with

```
$ Splus
```

4. At this point S commands may be issued (see later). The prompt is > unless the command is incomplete, when it is +. To use our software library issue

```
library(MASS, first=T)
```

5. To quit the S program the command is

```
>  q()
$
```

If you do not create a working directory .Data, S-PLUS will use that subdirectory of your home directory, or create such a directory for you (with a warning). To keep projects separate, we strongly recommend that you *do* create a working directory.

For subsequent sessions the procedure is simpler: make SwS the working directory and start the program as before:

```
$ cd SwS
$ Splus
```

issue S commands, terminating with the command

```
>  q()
$
```

On the other hand, to start a new project start at step 1.

Bailing out

One of the first things we like to know with a new program is how to get out of trouble. S-PLUS is generally very tolerant, and can be interrupted by Ctrl-C. (This means hold down the key marked Control or Cntrl and hit the second key.) This will interrupt the current operation, back out gracefully (so, with rare exceptions, it is as if it had not been started) and return to the prompt.

Sometimes this is not sufficient, for example if a long sequence of operations has been pasted into a terminal window. C-shell users can always suspend the process (with Ctrl-Z) and then kill it using `jobs` and then `kill %n` for the appropriate *n*. Again, because S-PLUS only commits changes at the end of an operation, this is almost always harmless.

Sometimes it is necessary to get to the operating system. On a multi-window system we suggest you use another window! If that is not possible, C-shell users can suspend the S-PLUS process. However, it is also possible to issue commands to the operating system by starting the line with ! , for example

```
> !date
Thu Feb 17 22:23:03 GMT 1994
```

Note: this starts a new copy of the shell or command interpreter, which inherits its environment from that used to run S-PLUS.

Getting help with functions and features

There are two ways to access the help system.

(1)   A help facility similar to the `man` facility. This can be invoked from the command line. For example, to get information on the function `var` the command is

```
> help(var)
```

which will put up a pager in the terminal window running S-PLUS to view the help file. If you prefer, a separate help window (which can be left up) can be obtained by

```
> help(var, window=T)
```

A faster alternative (to type) is

```
> ?var
```

The help page can be printed by

```
> help(var, offline=T)
```

provided the system manager has set this facility up when the system was installed.

For a feature specified by special characters and in a few other cases (one is `"function"`), the argument must be enclosed in double or single quotes, making it an entity known in S as a character string. For example two alternative ways of getting help on the list component extraction function, `[[`, are

```
> help("[[")
> ?"[["
```

Many S commands have additional help for *name*.`object` describing their result: for example `lm` also has a help page for `lm.object`.

(2)   Using S-PLUS running under a Unix windowing system there is another way to interact with the help system. For example, under Motif we can use

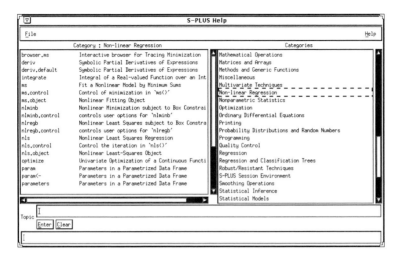

Figure 1.1: A view of the help system started by `help.start` on Unix.

```
> help.start(gui = "motif")
```

to bring up a window (see Figure 1.1) listing the help topics available. Items may be selected interactively from a series of menus, and the selection process causes other windows to appear with the help information. These may be scanned at the screen and then either dismissed or sent to a printer. (Similar facilities are available under OpenLook by setting `gui="openlook"`.) This help system is shut down with `help.off()` and *not* by quitting the help window.

If `help` or `?` are used when this help system is running, the requests are sent to its window.

Command line editing

Working in a windowing system such as X-windows under Unix usually allows you to modify and re-submit previous commands by a cut-and-paste method using the mouse. This is not available when working with a simple terminal, and even with a windowing system some users prefer a keyboard-based recall and edit mechanism.

Unix versions of S-PLUS have a command line editor which is made available by invoking S-PLUS with the `-e` flag:

```
$ Splus -e
```

To avoid forgetting to include the flag a handy alias for C-shell users is to include in their `.cshrc` file a line like

```
alias S+ 'Splus -e'
```

There are two conventions available for the line editor, either emacs or vi style, set according to the shell environment variable S_CLEDITOR. In a `csh` shell, to get the emacs conventions use

```
$ setenv S_CLEDITOR emacs
```

and for the vi conventions, use vi instead of emacs. (The default is vi, so setting S_CLEDITOR is not usually necessary in that case.) For a list of the commands use

```
?Command.edit
```

For users of GNU emacs there is the independently developed ESS package available from

```
http://franz.stat.wisc.edu/pub/ESS/
```

which provides a comprehensive working environment for S programming.

1.3   Using S-PLUS under Windows

Versions 3.2 and later of S-PLUS for Windows have almost exactly the same capabilities as the Unix versions. (The only significant drawbacks come in the use of compiled C and FORTRAN code, as specific compilers from Watcom are needed unless DLLs are constructed.)

One system dependency is the mapping of mouse buttons. We refer to button 1, usually the left button, and button 2, the right button (not as in Unix the middle button, even on a three-button mouse).

Getting started

The methods for setting up an S-PLUS project differ by Windows interface.

Windows 3.x *and* NT3.51,   S-PLUS 3.x

1. Create a separate directory, say SWS, for this project, and a subdirectory _Data.

2. Copy any data files you need to use with S-PLUS to this directory.

3. Within Windows, open the S-PLUS for Windows Group from the Program Manager window.

4. Click on the S-PLUS for Windows icon. Select Copy from the File menu, and select the S-PLUS for Windows Group from the To Group list box.

5. Click on the copy icon, choose Properties from the File menu, and edit the working directory field to give the project's directory, and the description field to reflect the project.

6. Double-click on the project's S-PLUS icon.

Windows 95 *and* NT4.0

1. Create a new folder, say SWS, for this project, then (3.x) create a new folder _Data within that folder.

2. Copy any data files you need to use with S-PLUS to the folder SWS.

3. From the Start menu, select Settings, Taskbar, the Start Menu Programs page and click on the Advanced button.

4. Open the S-PLUS folder under Programs.

5. Create a duplicate copy of the S-PLUS for Windows icon for example, using Copy from the Edit menu. Change the name of this icon to reflect the project it will be used for.

6. Right-click on the new icon, and select Properties from the pop-up menu.

7. On the page labelled Shortcut,
   (3.x) type the complete path to your directory as the field Start in, or
   (4.0) add at the end of the Target field S_PROJ= followed by the complete path to your directory.

8. Select the project's S-PLUS icon from the Start menu tree.

9. (4.0) You will be asked if _Data and _Prefs directories should be created. Click on OK. When the program has initialized, click on the Commands Window button.

At this point S commands may be issued (see later). The prompt is > unless the command is incomplete, when it is +. The usual Windows command-line recall and editing are available. To use our software library issue

```
library(MASS, first=T)
```

You can exit the program from the File menu, and entering the command q() in the Commands window will also exit the program.

To keep projects separate, we strongly recommend that you *do* create a working directory.

For subsequent sessions the procedure is much simpler; just launch S-PLUS by double-clicking on the project's icon. On the other hand, to start a new project start at step 1.

Bailing out

One of the first things we like to know with a new program is how to get out of trouble. S-PLUS is generally very tolerant, and can be interrupted by Esc (4.0) or Ctrl-C. (3.x; this means hold down the key marked Control or Cntrl and hit the second key.) This will interrupt the current operation, back out gracefully (so, with rare exceptions, it is as if it had not been started) and return to the S-PLUS prompt.

Using the interrupt key will not *always* interrupt the Windows version of S-PLUS; it is possible to write code that when running never calls the check for an interrupt, although this is unlikely to happen with useful code. If interruption does not work, the only alternative is to kill the S-PLUS task via the Ctrl-Alt-Del key sequence. (In Windows 3.x this will reboot the machine.)

In S-PLUS 3.3 the Tools menu has an Execute item that will allow users to run Windows and MS-DOS programs from within the S-PLUS session.

Getting help with functions and features

The Windows version of S-PLUS has a standard Windows help system accessible from the Help item on the main menu, and this can give a tutorial to its use. It is also possible to use the commands help or ? , for example by

```
> help(var)
> ?var
```

For inbuilt functions these send the request to the Windows help system. For most user-written functions the help file is displayed in a pager (default Notepad) window.

Many S commands have additional help for *name*.object describing their result: for example lm also has a help page for lm.object .

Under S-PLUS 3.x, for some libraries (including ours) it is necessary to specify the library, for example

```
> help(lda, library="MASS")
```

which uses a Windows help file specific to the library. To make help and ? behave in the same way for functions in libraries as for inbuilt functions use

```
library(helpfix, first=T)
```

1.4   An introductory session

The best way to learn S-PLUS is by using it. We invite readers to work through the following familiarization session and see what happens. First-time users may not yet understand every detail, but the best plan is to type what you see and observe what happens as a result.

As explained above you should first create a special directory and make it your working directory for the session, then start S-PLUS.

The whole session takes most first-time users one to two hours at the appropriate leisurely pace. The left column gives commands; the right column gives brief explanations and suggestions.

`library(MASS)`	A command to make available our datasets. Your local advisor can tell you the correct form for your system.
`help.start(gui="motif")`	Start the help facility.
	This step only applies if you are working in a windowing system in S-PLUS under Unix. Otherwise use the Windows help system. Users of Open Windows may prefer `gui="openlook"`.
Explore the help windows using the mouse.	
Close the help window	Under Unix close it to an icon; do not quit the window.
`trellis.device()`	Turn on the graphics window. You may need to re-position and re-size to make it convenient to work with both windows. Try not to change the aspect ratio of the graphics window when you re-size it.
On a monochrome Unix system use `trellis.device(color=F)`	

`x <- rnorm(50)` `y <- rnorm(50)`	Generate two pseudo-random normal vectors of x and y coordinates, and assign them to x and y.
`h <- chull(x, y)`	Find their convex hull in the plane.

`plot(x, y)`
`polygon(x[h], y[h], dens=15,`
`    angle=30)`
	Plot the points in the plane, and mark in their convex hull, filling it with lines at the rate of 15 to the inch, at an angle of 30 degrees to the horizontal. The result should be similar to Figure 1.2(a).

`objects()`	List the S objects which are now in the .Data directory.
`rm(x,y,h)`	Remove objects no longer needed. (The clean up phase.)
`x <- rnorm(10000)` `y <- rnorm(10000)`	Generate 10000 pairs of normal variates
`truehist(c(x,y+2), nbins=25)`	Histogram of a mixture of normal distributions. Experiment with the number of bins (25) and the shift (2) of the second component.

```
dd <- con2tr(
    hist2d(x,y,,,15,15))
contourplot(z ~ x + y,
    data=dd, aspect=1)
```

2D histogram on an 15×15 grid. We convert its output into a form suitable for the Trellis visualization routines.

```
wireframe(z ~ x + y,
    data=dd, drape=T)
```

Perspective plot with superimposed coloured levels.

```
levelplot(z ~ x + y,
    data=dd, aspect=1)
```

Greyscale or pseudo-colour plot.

```
x <- seq(1, 20, 0.5)
x
```

Make $x = (1, 1.5, 2, \ldots, 19.5, 20)$ and list it.

```
w <- 1 + x/2
y <- x + w*rnorm(x)
```

w will be used as a 'weight' vector and to give the standard deviations of the errors.

```
dum <- data.frame(x, y, w)
dum
rm(x, y, w)
```

Make a *data frame* of three columns named x, y and w, and look at it. Remove the original x, y and w.

```
fm <- lm(y ~ x, data=dum)
summary(fm)
```

Fit a simple linear regression of y on x and look at the analysis.

```
fm1 <- lm(y ~ x, data=dum,
    weight=1/w^2)
summary(fm1)
```

Since we know the standard deviations, we can do a weighted regression.

```
lrf <- loess(y ~ x, dum)
```

Fit a smooth regression curve using a modern regression function.

```
attach(dum)
```

Make the columns in the data frame visible as variables.

```
plot(x, y)
```

Make a standard scatterplot. To this plot we will add the three regression lines (or curves) as well as the known true line.

```
lines(spline(x, fitted(lrf)))
```

First add in the local regression curve using a spline interpolation between the calculated points.

```
abline(0, 1, lty=3)
```

Add in the true regression line (intercept 0, slope 1) with a different line type.

```
abline(fm)
```

Add in the unweighted regression line. abline() is able to extract the information it needs from the fitted regression object.

```
abline(fm1, lty=4)
```
Finally add in the weighted regression line, in line type 4. This one should be the most accurate estimate, but may not be, of course. One such outcome is shown in Figure 1.2(b).

You may be able to make a hardcopy of the graphics window by selecting the Print option from a menu.

```
plot(fitted(fm), resid(fm),
    xlab="Fitted Values",
    ylab="Residuals")
```
A standard regression diagnostic plot to check for heteroscedasticity, that is, for unequal variances. The data are generated from a heteroscedastic process, so can you see this from this plot?

```
qqnorm(resid(fm))
qqline(resid(fm))
```
A normal scores plot to check for skewness, kurtosis and outliers. (Note that the heteroscedasticity may show as apparent non-normality.) Diagnostic plots are shown in Figure 1.3.

```
detach()
rm(fm,fm1,lrf,dum)
```
Remove the data frame from the search path and clean up again.

We look next at a set of data on record times of Scottish hill races against distance and total height climbed.

```
hills
```
List the data

```
splom(~ hills)
```
Show a matrix of pairwise scatterplots (Figure 1.4).

```
brush(as.matrix(hills))
```

Click on the Quit button in the graphics window to continue.

Try highlighting points and see how they are linked in the scatterplots (Figure 1.5). Also try rotating the points in 3D.

```
attach(hills)
```
Make columns available by name.

```
plot(dist, time)
identify(dist, time,
    row.names(hills))
```
Use mouse button 1 to identify outlying points, and button 2 to quit. Their row numbers are returned.

```
abline(lm(time ~ dist))
```
Show least squares regression line.

```
abline(ltsreg(dist, time),
    lty=3)
```
Fit a very resistant line. See Figure 1.6.

```
detach()
```
Clean up again.

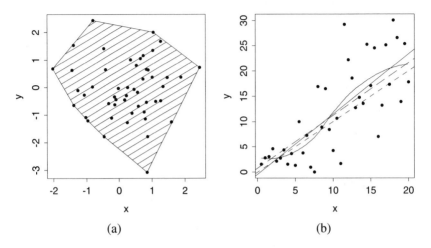

Figure 1.2: A convex hull and an heteroscedastic regression plot.

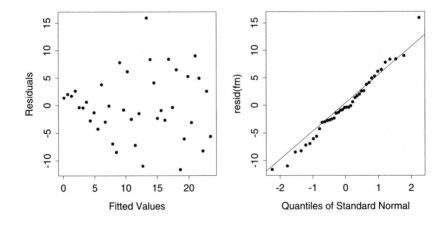

Figure 1.3: Two residual plots for the artificial heteroscedastic regression data.

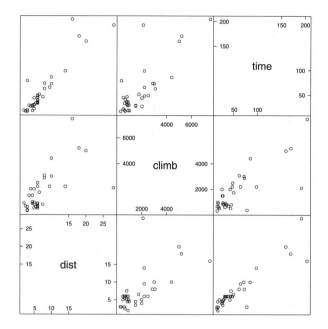

Figure 1.4: Pairs plot for data on Scottish hill races.

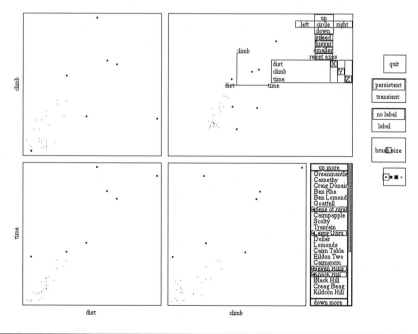

Figure 1.5: Screendump of a `brush` plot of dataset `hills` (from openlook).

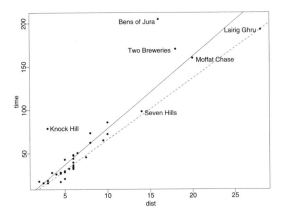

Figure 1.6: Annotated plot of time vs distance for `hills` with regression line and resistant line (dashed).

We can explore further the effect of outliers on a linear regression by designing our own examples interactively. Try this several times.

```
plot(c(0,1), c(0,1), type="n")
xy <- locator(type="p")
```
Make our own dataset by clicking with button 1, then with button 2 to finish.

```
abline(lm(y ~ x, xy), col=4)
abline(rreg(xy$x, xy$y),
    lty=3, col=3)
abline(ltsreg(xy$x, xy$y),
    lty=2, col=2)
```
Fit least-squares and robust regression lines, and a resistant line. (Omit the col= on a mono display.) Repeat to try the effect of outliers, both vertically and horizontally.

```
rm(xy)
```
Clean up again.

We now look at data from the 1879 experiment of Michelson to measure the speed of light. There are five experiments (column `Expt`) and each has 20 runs (column `Run`) and `Speed` is the recorded speed of light, in km/sec, less 299000. (The currently accepted value on this scale is 734.5.)

```
attach(michelson)
```
Make the columns visible by name.

```
search()
```
The *search path* is a sequence of places, either directories or data frames, where S-PLUS looks for objects required for calculations.

```
plot(Expt, Speed,
    main="Speed of Light Data",
    xlab="Experiment No.")
```
Compare the five experiments with simple boxplots. The result is shown in Figure 1.7.

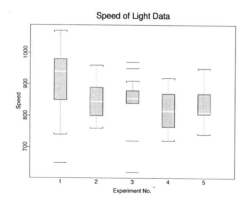

Figure 1.7: Boxplots for the speed of light data.

```
fm <- aov(Speed ~ Run + Expt)        Analyse as a randomized block design ,
summary(fm)                          with runs and experiments as factors.
```

	Df	Sum of Sq	Mean Sq	F Value	Pr(F)
Run	19	113344	5965	1.1053	0.36321
Expt	4	94514	23629	4.3781	0.00307
Residuals	76	410166	5397		

```
fm0 <- update(fm, . ~ .-Run)         Fit the sub-model omitting the nonsense
anova(fm0, fm)                       factor, runs, and compare using a formal
                                     analysis of variance.
```

```
Analysis of Variance Table
Response: Speed
```

	Terms	Resid. Df	RSS	Test	Df	Sum of Sq	F Value	Pr(F)
1	Expt	95	523510					
2	Run + Expt	76	410166	+Run	19	113344	1.1053	0.36321

```
detach()                             Clean up before moving on.
rm(fm, fm0)
```

---

type in the listing in Figure˜1.8

```
butterfly                            Note how naming the function object
butterfly()                          butterfly without the parentheses
                                     gives the function definition. Adding
                                     the parentheses causes it to be evalu-
                                     ated, which should give you a picture of
                                     a rather mathematical-looking butterfly
                                     in your graphics window, Figure 1.9.
```

```
usa()                                A map, which can be used for geograph-
                                     ical display, Figure 1.9.
```

```
q()                                  Quit S-PLUS.
```

```
butterfly <- function()
{
    theta <- seq(0, 24 * pi, len = 2000)
    radius <- exp(cos(theta)) - 2 * cos(4 * theta) +
                            sin(theta/12)^5
    plot(radius * sin(theta), - radius * cos(theta),
        type = "l", axes = F, xlab = "", ylab = "")
}
```

Figure 1.8: The source of the butterfly function. The indentation is unimportant, but be careful where lines are broken.

Figure 1.9: A mathematical butterfly and an S-PLUS map

1.5 What next?

We hope that you now have a flavour of **S-PLUS** and are inspired to delve more deeply. We suggest that you read Chapter 2, perhaps cursorily, and the Sections 3.1–3 and 3.5. Thereafter, tackle the statistical topics that are of interest to you. Chapters 5 to 17 are fairly independent, and contain cross-references where they do interact. Chapters 7 to 10 build on Chapter 6, especially its first two sections.

Chapter 4 comes early, because it is about S not about statistics, but will be most useful to advanced users and to those writing extensions to S. Many users are surprised to find themselves in this category; the philosophy of S encourages writing re-usable functions, friends get to hear about them, ...

Chapter 2

The S Language

S is a language for the manipulation of objects. It aims to be both an interactive language (like, for example, a Unix shell language) as well as a complete programming language with some convenient object-oriented features. In this chapter we shall be concerned with the interactive language, and hence certain language constructs used mainly in programming will be postponed to Chapter 4.

This chapter divides into two parts. The first seven sections give an informal overview of the language which will suffice at first reading. Later sections are more formal, and useful for answering "why did it do that?" questions. Appendix B gives brief synopses of commonly used S functions, with references to their description in the main text.

2.1 A concise description of S objects

In this section we discuss the most important types of S objects and the simplest ways to manipulate them,

Naming conventions

Standard S names for objects are made up from the upper- and lower-case roman letters, the digits, 0–9, in any non-initial position and also the period, ' . ', which behaves as a letter except in names such as .37 where it acts as a decimal point. Some important conventions make use of a period in object names and users will become aware of these in due course. For example 'three dots', ..., is a reserved identifier, only used in the context of defining functions (see pages 29 and 129). Other reserved identifiers include break, for, function, if, in, next, repeat, return and while.

Non-standard names are allowed using any collection of characters, but special steps have to be taken to use them. They are needed in some contexts but not in ordinary use and we suggest that they be avoided where possible. For example so-called *replacement functions* must have names that end in the two characters "<-", but they are not used explicitly.

In some (but not all) cases non-standard names may be handled by placing them within quotes. In particular this is true of function names (when used for function calls), object names to which a value is being assigned[1], and argument or component names. Some examples are

```
> "style<-" <- function(p, value)    # a replacement function
        structure(p, style = value)
> Ttest(x1, x2)$"t-stat"  # one component of a result
[1] -3.6303
```

Note that S is *case sensitive*, so Alfred and alfred are distinct S names and that the underscore, _ , is *not* available as a letter in the S standard name alphabet. (Periods are often used to separate words in names.)

Avoid using system names for your own objects; in particular avoid c, q, s, t, C, D, F, I, T, diff, mean, pi, range, rank, tree and var.

Language layout

Commands to S are either expressions or assignments. Commands are separated by either a semi-colon, ; , or a newline. The # symbol marks the rest of the line as comments.

The S prompt is > unless the command is syntactically incomplete, when the prompt changes to +.[2] The only way to extend a command over more than one line is by ensuring that *it is* syntactically incomplete until the final line.

In this book we will often omit both prompts and indicate continuation by simple indenting. When we do include prompts it is usually to separate input from output within a display.

An expression command is evaluated and (normally) printed. For example

```
> 1 - pi + exp(1.7)
[1] 3.332355
```

This rule allows any object to be printed by giving its name. Note that pi is the value of π. Giving the name of an object will normally print it or a short summary; this can be done explicitly using the function print, and summary will often give a full description.

An assignment command evaluates an expression and passes the value to a variable but the result is not printed. The recommended assignment symbol is the combination, <- , so an assignment in S looks like

```
a <- 6
```

which gives the object a the value 6. To improve readability of your code we strongly recommend that you put at least one space before and after binary operators, especially the assignment symbol. (We regard the use of _ for assignments as unreadable, but it is allowed.) Assignments using the right-pointing combination

---

[1] When S writes objects onto an external file in assignment form (using dump) the left-hand side of each top level assignment is always placed in quotes, whether the name is standard or not.

[2] These prompts can be altered: see Section 2.11.

"->" are also allowed to make assignments in the opposite direction, but these are never needed and are little used in practice.

An assignment is a special case of an expression with value equal to the value assigned. When this value is itself passed by assignment to another object the result is a multiple assignment, as in

```
b <- a <- 6
```

Multiple assignments are evaluated from right to left, so in this example the value 6 is first passed to a and then to b.

It is useful to remember that the most recently evaluated non-assignment expression in the session is stored as the variable .Last.value [3] and so may be kept as a permanent object by a following assignment such as

```
keep <- .Last.value
```

This is also useful if the result of an expression is unexpectedly not printed; just print(.Last.value).

Vectors and matrices

The objects in S that most users deal with directly are vectors, functions or lists. Note that there are no scalars; vectors of length one are used instead. Vectors consist of numeric or logical values or character strings (and may not mix them). Normally it is unnecessary to be aware if the numeric values are integer, real or even complex, and whether they are stored to single or double precision; S handles all the possibilities in a unified way.

The simplest way to create a vector is to specify its elements by the function c (for concatenate):

```
> mydata <- c(2.9, 3.4, 3.4, 3.7, 3.7, 2.8, 2.8, 2.5, 2.4, 2.4)
> colours <- c("red", "green", "blue", "white", "black")
> x1 <- 25:30
> x1
[1] 25 26 27 28 29 30
> mydata[7]
[1] 2.8
> colours[3]
[1] "blue"
```

and the colon expression specifies a range of integers. The [1] indicates that the printout of the vector starts at element one. Individual elements of a vector are accessed by specifying them by number in square brackets, as in these examples. Character strings may be entered with either double or single quotes (in matching pairs), but will always be printed with double quotes.

Logical values are represented as T and F:

---

[3] If the expression consists of a simple name such as x, only, the .Last.value object is not changed.

```
> mydata > 3
 [1] F T T T T F F F F F
```

and may also be entered as TRUE and FALSE. Logical operators apply element-by-element to vectors, as do arithmetical expressions.

The elements of a vector can also be named and accessed by name:

```
> names(mydata) <- c('a','b','c','d','e','f','g','h','i','j')
> mydata
  a   b   c   d   e   f   g   h   i   j
2.9 3.4 3.4 3.7 3.7 2.8 2.8 2.5 2.4 2.4
> names(mydata)
 [1] "a" "b" "c" "d" "e" "f" "g" "h" "i" "j"
> mydata["e"]
  e
3.7
```

An assignment with the left-hand side something other than a simple identifier, as in the first of these, will be called a *replacement*. They are disguised calls to a special kind of function called a *replacement function* that re-constructs the entire modified object. The expanded call in this case has the form

```
> mydata <- "names<-"(mydata,
             c('a','b','c','d','e','f','g','h','i','j'))
```

More generally, several elements of a vector can be selected by giving a vector of element names or numbers or a logical vector:

```
> letters[1:5]
 [1] "a" "b" "c" "d" "e"
> mydata[letters[1:5]]
  a   b   c   d   e
2.9 3.4 3.4 3.7 3.7
> mydata[mydata > 3]
  b   c   d   e
3.4 3.4 3.7 3.7
```

(The object letters is an S vector containing the 26 lower-case English letters; the object LETTERS contains the upper-case equivalents.) We will refer to taking *subsets* of vectors (although to mathematicians they are subsequences).

Elements of a vector can be <u>omitted</u> by giving negative indices, for example,

```
> mydata[-c(3:5)]
  a   b   f   g   h   i   j
2.9 3.4 2.8 2.8 2.5 2.4 2.4
```

but in this case the names can not be used.

There are many special uses of vectors. For example, they can appear to be matrices, arrays or factors. This is handled by giving the vectors *attributes*. All objects have two attributes, their mode and their length:

```
> mode(mydata)
[1] "numeric"
> mode(letters)
[1] "character"
> mode(sin)
[1] "function"
> length(mydata)
[1] 10
> length(letters)
[1] 26
> length(sin)
[1] 2
```

(The length of a function is one plus the number of arguments.) The components of some objects can be given names, which are then the `names` attribute of the object.

Giving a vector a `dim` attribute allows, and sometimes causes, it to be treated as a matrix or array. For example

```
> dim(mydata) <- c(2, 5)
> mydata
     [,1] [,2] [,3] [,4] [,5]
[1,]  2.9  3.4  3.7  2.8  2.4
[2,]  3.4  3.7  2.8  2.5  2.4
attr(, "names"):
 [1] "a" "b" "c" "d" "e" "f" "g" "h" "i" "j"
> dim(mydata) <- NULL
```

Notice how the names are retained as an attribute, and how the matrix is filled down columns rather than across rows. The final assignment removes the `dim` attribute and restores `mydata` to a named vector. A simpler way to create a matrix from a vector is to use `matrix`

```
> matrix(mydata, 2, 5)
     [,1] [,2] [,3] [,4] [,5]
[1,]  2.9  3.4  3.7  2.8  2.4
[2,]  3.4  3.7  2.8  2.5  2.4
```

by which the names are discarded. The function `matrix` can also fill matrices by row,

```
> matrix(mydata, 2, 5, byrow = T)
     [,1] [,2] [,3] [,4] [,5]
[1,]  2.9  3.4  3.4  3.7  3.7
[2,]  2.8  2.8  2.5  2.4  2.4
```

As these displays suggest, matrix elements can be accessed as `mat[m,n]`, and whole rows and columns by `mat[m,]` and `mat[,n]` respectively.

Arrays are multi-way extensions of matrices, formed by giving a `dim` attribute of length three or more. They can be accessed by `A[r, s, t]` and so on. Sections 2.8 and 2.9 discuss matrices and arrays in more detail.

Lists

A list is used to collect together items of different types. For example, an employee record might be created by

```
Empl <- list(employee="Anna", spouse="Fred", children=3,
             child.ages=c(4,7,9))
```

As this example shows, the elements of a list do not have to be of the same length. The components of a list are always numbered and may always be referred to as such. Thus Empl is a list of length 4, and the individual components may be referred to as Empl[[1]], Empl[[2]], Empl[[3]] and Empl[[4]]. Further, since Empl[[4]] is a vector, Empl[[4]][1] is its first entry. However, it is more convenient to refer to the components explicitly by name, in the form

```
> Empl$employee
[1] "Anna"
> Empl$child.ages[2]
[1] 7
```

Names of components may be abbreviated to the minimum number of letters needed to identify them uniquely. Thus Empl$employee may be minimally specified as Empl$e since it is the only component whose name begins with the letter 'e', but Empl$children must be specified as at least Empl$childr because of the presence of another component called Empl$child.ages.

In many respects lists are like vectors. For example, the vector of component names is simply a names attribute of the list like any other object and may be treated as such; to change the component names of Empl to a, b, c and d we can use the replacement

```
> names(Empl) <- letters[1:4]
> Empl[3:4]
$c:
[1] 3
$d:
[1] 4 7 9
```

Notice that we select *components* as if this were a vector (and not [[3:4]] as might have been expected). (The distinction is that [ ] returns a list with the selected component(s) so Empl[3] is a list with one component, whereas Empl[[3]] extracts that component from the list.)

The concatenate function, c, can also be used to concatenate lists or to add components, so

```
Empl <- c(Empl, service = 8)
```

would add a component for years of service.

The function unlist converts a list to a vector:

```
> unlist(Empl)
 employee  spouse children child.ages1 child.ages2 child.ages3
  "Anna"    "Fred"   "3"       "4"         "7"         "9"
> unlist(Empl, use.names=F)
[1] "Anna" "Fred" "3"       "4"       "7"       "9"
```

which can be useful for a compact printout (as here). (Mixed types will all be converted to character, giving a character vector.)

The function c has named argument recursive; if this is T the list arguments are unlisted before being joined together. Thus

```
c(list(x = 1:3, a = 3:6), list(y = 8:23, b = c(3, 8, 39)))
```

is a list with four (vector) components, but adding recursive=T gives a vector of length 26.

Factors

A factor is a special type of vector, normally used to hold a categorical variable, for example

```
> citizen <- factor(c("uk","us","no","au","uk","us","us"))
> citizen
[1] uk us no au uk us us
```

Although this is entered as a character vector, it is printed without quotes. Appearances here are deceptive, and a special print method is used. Internally the factor is stored as a set of codes, and an attribute giving the *levels*:

```
> print.default(citizen)
[1] 3 4 1 2 3 4 4
attr(, "levels"):
[1] "au" "no" "uk" "us"
attr(, "class"):
[1] "factor"
> unclass(citizen)
[1] 3 4 1 2 3 4 4
```

Why might we want to use this rather strange form? Using a factor indicates to many of the statistical functions that this is a categorical variable (rather than just a list of labels), and so it is treated specially. As just one example, using a factor as the response variable signals to the function tree that a classification rather than regression tree is required (see Chapter 14).

By default the levels are sorted into alphabetical order, and the codes assigned accordingly. Some of the statistical functions give the first level a special status, so it may be necessary to specify the levels explicitly:

```
> citizen <- factor(c("uk","us","no","au","uk","us","us"),
      levels = c("us", "fr", "no", "au", "uk"))
> citizen
[1] uk us no au uk us us
Levels:
[1] "us" "fr" "no" "au" "uk"
```

Note that levels which do not occur can be specified, in which case the levels *are* printed. This often occurs when subsetting factors.[4]

A frequency table for the categories defined by a factor is given by the function `table`

```
> table(citizen)
 us fr no au uk
  3  0  1  1  2
```

If `table` is given several factor arguments the result is a multi-way frequency array, as discussed more fully on pages 120ff.

Sometimes the levels of a categorical variable are naturally ordered, as in

```
> income <- ordered(c("Mid","Hi","Lo","Mid","Lo","Hi","Lo"))
> income
[1] Mid Hi  Lo  Mid Lo  Hi  Lo

 Hi < Lo < Mid
> as.numeric(income)
[1] 3 1 2 3 2 1 2
```

Again the effect of alphabetic ordering is not what is required, and we need to set the levels explicitly, or use an `ordered` replacement:

```
> inc <- ordered(c("Mid","Hi","Lo","Mid","Lo","Hi","Lo"),
      levels = c("Lo", "Mid", "Hi"))
> inc
[1] Mid Hi  Lo  Mid Lo  Hi  Lo

 Lo < Mid < Hi
> ordered(income) <- c("Lo", "Mid", "Hi")
> income
[1] Mid Hi  Lo  Mid Lo  Hi  Lo

 Lo < Mid < Hi
> as.numeric(income)
[1] 2 3 1 2 1 3 1
```

---

[4] In S-PLUS 3.4 and later an extra argument may be included when subsetting factors, as in `f[i, drop=T]`, to ensure that the levels are pruned to include only those which occur in the subset. Under the default, `drop=F`, the levels are not changed.

Note that whereas an `ordered` replacement permutes the current levels, a `levels` replacement merely changes the names of the levels and leaves the codes unchanged.

Ordered factors are a special case of factors that some functions (including `print`) treat in a special way.

The function `cut` can be used to create ordered factors by sectioning continuous variables into discrete class intervals. For example

```
> erupt <- cut(faithful$eruptions, breaks = 1:6)
> ordered(erupt) <- levels(erupt)
> erupt
   [1] 3+ thru 4 1+ thru 2 3+ thru 4 2+ thru 3 4+ thru 5
   [6] 2+ thru 3 4+ thru 5 3+ thru 4 1+ thru 2 4+ thru 5
    ....
  1+ thru 2 < 2+ thru 3 < 3+ thru 4 < 4+ thru 5 < 5+ thru 6
```

Note that the intervals are of the form $(n, n + 1]$, so an eruption of 4 minutes is put in category `3+ thru 4`.

Data frames

A data frame is the type of object normally used in S to store a data matrix. It should be thought of as a list of variables of the same length, but possibly of different types (numeric, character or logical). Consider our data frame `painters`:

```
> painters
             Composition Drawing Colour Expression School
   Da Udine           10       8     16          3      A
   Da Vinci           15      16      4         14      A
   Del Piombo          8      13     16          7      A
   Del Sarto          12      16      9          8      A
   Fr. Penni           0      15      8          0      A
    ....
```

which has four numerical variables and one character variable. Since it is a data frame, it is printed in a special way. The components are printed as columns (rather than as rows as vector components of lists are) and there is a set of names, the `row.names`, common to all variables.

```
> row.names(painters)
   [1] "Da Udine"      "Da Vinci"         "Del Piombo"
   [4] "Del Sarto"     "Fr. Penni"        "Guilio Romano"
   [7] "Michelangelo"  "Perino del Vaga"  "Perugino"
    ....
```

Further, neither the row names nor the values of the character variable appear in quotes.

Data frames can be indexed in the same way as matrices:

```
> painters[1:5, c(2, 4)]
            Drawing Expression
   Da Udine       8          3
   Da Vinci      16         14
Del Piombo       13          7
 Del Sarto       16          8
  Fr. Penni      15          0
```

But they may also be indexed as lists, which they are. Note that a single index behaves as it would for a list, so `painters[c(2,4)]` gives a data frame of the second and fourth variables which is the same as `painters[, c(2,4)]`.

Variables which satisfy suitable restrictions (having the same length, and the same names, if any) can be collected into a data frame by the function `data.frame`, which resembles `list`:

```
mydat <- data.frame(MPG, Dist, Climb, Day = day)
```

although data frames are most commonly created by reading a file (see `read.table` on page 37).

There is a side effect of `data.frame` which needs to be considered; all character and logical columns are converted to factors unless their names are included in `I()`, so for example

```
mydat <- data.frame(MPG, Dist, Climb, Day = I(day))
```

preserves `day` as a character vector, `Day`.

Compatible data frames can be joined by `cbind`, which adds columns of the same length, and `rbind`, which stacks data frames vertically. The result is a data frame with appropriate names and row names.

It is also possible to include matrices and lists within data frames. If a matrix is supplied to `data.frame`, it is as if its columns were supplied individually; suitable labels are concocted. If a list is supplied, it is treated as if its components had been supplied individually.

Coercion

There are a series of functions named `as.xxx` which convert to the specified type in the best way possible. For example, `as.matrix` will convert a numerical data frame to a numerical matrix, and a data frame with any character or factor columns to a character matrix. The function `as.character` is often useful to generate names and other labels.

Functions `is.xxx` test if their argument is of the required type. Note that these do not always behave as one might guess; for example `is.vector(mydata)` will be *false* as this tests for a 'pure' vector without any attributes such as names. Similarly, `as.vector` has the (often useful) side effect of discarding all attributes.

2.2 Calling conventions for functions

Functions may have their arguments *specified* or *unspecified* when the function is defined. (We saw how to write simple functions on page 3.)

When the arguments are unspecified there may be an arbitrary number of them. They are shown as ... when the function is defined or printed. Examples of functions with unspecified arguments include the concatenation function, `c(...)`, and the parallel maximum and minimum functions `pmax(...)` and `pmin(...)`.

Where the arguments are specified there are two conventions for supplying values for the arguments when the function is called:

1. arguments may be specified in the same order in which they occur in the function definition, in which case the values are supplied in order, and

2. arguments may be specified as `name=value`, when the order in which the arguments appear is irrelevant. The name may be abbreviated providing it partially matches just one named argument.

It is important to note that these two conventions may be mixed. A call to a function may begin with specifying the arguments in positional form but specify some later arguments in the named form. For example the two calls

```
t.test(x1, y1, var.equal = F, conf.level = 0.99)
t.test(conf.level = 0.99, var.equal = F, x1, y1)
```

are equivalent. (The precise rules for argument matching are discussed on page 130.)

Functions with named arguments also have the option of specifying *default values* for those arguments, in which case if a value is not specified when the function is called the default value is used. For example, the function `t.test` has an argument list defined as

```
t.test <- function(x, y = NULL, alternative = "two.sided",
        mu = 0, paired = F, var.equal = T, conf.level = 0.95)
```

so that our previous calls can also be specified as

```
t.test(x1, y1, , , , F, 0.99)
```

and in all cases the default values for `alternative`, `mu` and `paired` are used. Using the positional form and omitting values, as in this last example, is rather prone to error, so the named form is preferred except for the first couple of arguments.

Some functions (for example `paste`) have both unspecified and specified arguments, in which case the specified arguments occurring after the ... argument on the definition must be named exactly if they are to be matched at all.

The argument names and any default values for an S function can be found from the on-line help, by printing the function itself or succinctly using the `args` function. For example

```
> args(hist)
function(x, nclass, breaks, plot = TRUE, probability = FALSE,
        include.lowest = T, ..., xlab = deparse(substitute(x)))
NULL
```

shows the arguments, their order and those default values which are specified for
hist function for plotting histograms. (The return value from args always ends
with NULL.) Note that even when no default value is specified the argument itself
may not need to be specified. If no value is given for nclass or breaks when
the hist function is called, default values are calculated within the function.
Unspecified arguments are passed on to a plotting function called from within
hist.

S-PLUS 3.3 for Windows has a function arg.dialog which shows the
arguments of a function in a dialog box, allowing them to be entered as needed
and the call executed. This can be useful if the function has many arguments.

Functions are considered in much greater detail in Chapter 4.

2.3   Arithmetical expressions

We have seen that a basic unit in S is a vector. Arithmetical operations are
performed on vectors, element by element. The standard operators + - * / ^
are available, where ^ is the power (or exponentiation) operator (giving x^y).

Vectors may be empty. The expression numeric(0) is both the expression to
create an empty numeric vector and the way it is represented when printed. It has
length zero. It may be described as "a vector such that if there were any elements
in it, they would be numbers"!

Vectors can be complex, and almost all the rules for arithmetical expressions
apply equally to complex quantities. A complex number is entered in the form
3.1 + 2.7i, with no space before the i. Functions Re and Im return the real
and imaginary parts. Note that complex arithmetic is not used unless explicitly
requested, so sqrt(x) for x real and negative produces an error. If the complex
square root is desired use sqrt(as.complex(x)) or sqrt(x + 0i).

The recycling rule

The expression y + 2 is a syntactically natural way to add 2 to each element of
the vector y, but 2 is a vector of length 1 and y may be a vector of any length.
A convention is needed to handle vectors occurring in the same expression but
not all of the same length. The value of the expression is a vector with the same
length as that of the longest vector occurring in the expression. Shorter vectors
are *recycled* as often as need be, perhaps fractionally, until they match the length
of the longest vector. In particular a single number is repeated the appropriate
number of times. Hence

```
x <- c(10.4, 5.6, 3.1, 6.4, 21.7)
y <- c(x, 0, x)
v <- 2 * x + y + 1
```

generates a new vector v of length 11 constructed by

1. repeating the number 2 five times to match the length of the vector x and multiplying element by element, and

2. adding together, element by element, 2*x repeated 2.2 times, y as it stands, and 1 repeated eleven times.

The example is very artificial. Fractional recycling is usually a sign of an error and when it takes place like this S issues a warning that this has happened:

```
> v <- 2 * x + y + 1
Warning messages:
  Length of longer object is not a multiple of the
         length of the shorter object in: 2 * x + y
```

Some standard S functions

Some examples of standard functions follow.

1. There are several functions to convert to integers; round will normally be preferred, and rounds to the nearest integer. (It can also round to any number of digits in the form round(x, 3). Using a negative number rounds to a power of ten, so that round(x,-3) rounds to thousands.) Each of trunc, floor and ceiling round in a fixed direction, towards zero, down and up respectively.

2. Other arithmetical operators are %/% for integer divide and %% for modulo reduction.[5]

3. The common functions are available, including abs, sign, log, log10, sqrt, exp, sin, cos, tan, acos, asin, atan, cosh, sinh and tanh with their usual meanings. Note that the value of each of these is a vector of the same length as their argument. The function log has a second argument, the base of the logarithms, defaulting to e.

   Less common functions are gamma and lgamma ($\log_e \Gamma(x)$).

4. There are functions sum and prod to form the sum and product of a whole vector, as well as cumulative versions cumsum and cumprod.

5. The functions max(x) and min(x) select the largest and smallest elements of a vector x. The functions cummax and cummin give cumulative maxima and minima.

---

[5] The result of e1 %/% e2 is floor(e1/e2) if e2!=0 and 0 if e2==0. The result of e1 %% e2 is e1-floor(e1/e2)*e2 if e2!=0 and e1 otherwise (see Knuth, 1968, §1.2.4). Thus %/% and %% always satisfy e1==(e1%/%e2)*e2+e1%%e2.

6. The functions pmax(x1, x2, ...) and pmin(x1, x2, ...) take an arbitrary number of vector arguments and return the element-by-element maximum or minimum values respectively. Thus the result is a vector of length that of the longest argument and the recycling rule is used for shorter arguments. For example

```
xtrunc <- pmax(0, pmin(1,x))
```

is a vector like x but with negative elements replaced by 0 and elements larger than 1 replaced by 1.

7. The function range(x) returns c(min(x), max(x)). If range, max or min is given several arguments these are first concatenated into a single vector.

8. Two useful statistical functions are mean(x) which calculates the sample mean, which is the same as sum(x)/length(x), and var(x) which gives the sample variance, namely sum((x-mean(x))^2)/(length(x)-1).[6]

9. sort returns a vector of the same size as x with the elements arranged in increasing order. See page 52 for further details.

10. The function rev arranges the components of a vector or list in reverse order. duplicated produces a logical vector with value T only where a value in its vector argument has occurred previously and unique removes such duplicated values.

11. Set operations may be done with the functions union, intersect and setdiff, which enact the set operations $A \cup B$, $A \cap B$ and $A \cap \overline{B}$ respectively. Their arguments (and hence values) may be vectors of any mode but, like true sets, they should contain no duplicated values.

Operator precedence

The formal precedence of operators is given in Table 2.1. However, as usual it is better to use parentheses to group expressions rather than rely on remembering these rules. They can be found on-line from help(Syntax).

Generating regular sequences

There are several ways in S to generate sequences of numbers. For example, 1:30 is the vector c(1, 2, ..., 29, 30). The colon operator has a high precedence within an expression, so 2*1:15 is the vector c(2, 4, 6, ..., 28, 30). Put n <- 10 and compare the sequences 1:n-1 and 1:(n-1).

A construction such as 10:1 may be used to generate a sequence in reverse order.

---

[6] If the argument to var is an $n \times p$ matrix the value is a $p \times p$ sample covariance matrix obtained by regarding the rows as sample vectors.

Table 2.1: Precedence of operators, from highest to lowest.

$	for list extraction
[, [[	vector and list element extraction
^	exponentiation
−	unary minus
:	sequence generation
%%, %/%, %*%	and other special operators %...%
* /	multiply and divide
+ −	addition and subtraction
< > <= >= == !=	comparison operators
!	logical negation
& | && ||	logical operators
~	formula
<<−	assignment within a function (see page 155)
<−, _, −>	assignment

The function `seq` is a more general facility for generating sequences. It has five arguments, only some of which may be specified in any one call. The first two arguments, named `from` and `to`, if given specify the beginning and end of the sequence, and if these are the only two arguments the result is the same as the colon operator. That is `seq(2,10)` and `seq(from=2, to=10)` give the same vector as `2:10`.

The third and fourth arguments to `seq` are named `by` and `length`, and specify a step size and a length for the sequence. If `by` is not given, the default `by=1` is used. For example,

```
s3 <- seq(-5, 5, by=0.2)
s4 <- seq(length=51, from=-5, by=0.2)
```

generate in both `s3` and `s4` the vector $(-5.0, -4.8, -4.6, \ldots, 4.6, 4.8, 5.0)$.

The fifth argument is named `along` and has a vector as its value. If it is the only argument given it creates a sequence 1, 2, ..., `length`(*vector*), or the empty sequence if the value is empty. (This makes `seq(along=x)` preferable to `1:length(x)` in most circumstances.) If specified rather than `to` or `length` its length determines the length of the result.

A companion function is `rep` which can be used to repeat an object in various ways. The simplest form is

```
s5 <- rep(x, times=5)
```

which will put five copies of `x` end-to-end in `s5`.

A `times=v` argument may specify a vector of the same length as the first argument, `x`. In this case the elements of `v` must be non-negative integers, and the result is a vector obtained by repeating each element in `x` a number of times as specified by the corresponding element of v. Some examples will make the process clear:

```
x <- 1:4          # puts c(1,2,3,4)                    into x
i <- rep(2, 4)    # puts c(2,2,2,2)                    into i
y <- rep(x, 2)    # puts c(1,2,3,4,1,2,3,4)            into y
z <- rep(x, i)    # puts c(1,1,2,2,3,3,4,4)            into z
w <- rep(x, x)    # puts c(1,2,2,3,3,3,4,4,4,4) into w
```

As a more useful example, consider a two-way experimental layout with 4 row classes, 3 column classes and 2 observations in each of the 12 cells. The observations themselves are held in a vector y of length 24 with column classes stacked above each other, and row classes in sequence within each column class. Our problem is to generate two indicator vectors of length 24 which will give the row and column class, respectively, of each observation. Since the 3 column classes are the first, middle and last 8 observations each, the column indicator is easy. The row indicator requires three calls to rep:

```
> colc <- rep(1:3,rep(8,3));  colc
> [1] 1 1 1 1 1 1 1 1 2 2 2 2 2 2 2 2 3 3 3 3 3 3 3 3
> rowc <- rep(rep(1:4, rep(2,4)), 3); rowc
> [1] 1 1 2 2 3 3 4 4 1 1 2 2 3 3 4 4 1 1 2 2 3 3 4 4
```

These can also be generated arithmetically using the ceiling function

```
> 1 + (ceiling(1:24/8) - 1) %% 3 -> colc; colc
> [1] 1 1 1 1 1 1 1 1 2 2 2 2 2 2 2 2 3 3 3 3 3 3 3 3
> 1 + (ceiling(1:24/2) - 1) %% 4 -> rowc; rowc
> [1] 1 1 2 2 3 3 4 4 1 1 2 2 3 3 4 4 1 1 2 2 3 3 4 4
```

In general the expression 1 + (ceiling(1:n/r) - 1) %% m generates a sequence of length n consisting of the numbers 1, 2, ..., m each repeated r times. This often a useful idiom.

Logical expressions

Logical vectors are most often generated by *conditions*. The logical operators are <, <=, >, >= (which have self-evident meanings), == for exact equality and != for exact inequality. If c1 and c2 are vector valued logical expressions, c1 & c2 is their intersection ('and'), c1 | c2 is their union ('or') and !c1 is the negation of c1. These operations are performed separately on each component with the recycling rule applying for short arguments.

Logical vectors may be used in ordinary arithmetic. They are *coerced* into numeric vectors, F becoming 0 and T becoming 1. For example, assuming the value or values in sd are positive

```
N.extreme <- sum(y < ybar - 3*sd | y > ybar + 3*sd)
```

would count the number of elements in y that were further than 3*sd from ybar on either side. The right-hand side can be expressed more concisely as sum(abs(y-ybar) > 3*sd).

The function xor computes (element-wise) the exclusive or of its two arguments.

The functions any and all are useful to collapse a logical vector. The function all.equal provides a way to test for equality up to a tolerance if appropriate.

The missing value marker, NA

Not all the elements of a vector may be known. When an element or value is 'not available' or a 'missing value', a place within a vector may be reserved for it by assigning the special value NA.

In general any operation on an NA becomes an NA. The motivation for this rule is simply that if the specification of an operation is incomplete, the result cannot be known and hence is not available.

The function is.na(x) gives a logical vector of the same length as x with values which are true if and only if the corresponding element in x is NA.

```
ind <- is.na(z)
```

Notice that the logical expression x == NA is not equivalent to is.na(x). Since NA is really not a value but a marker for a quantity that is not available, the first expression is incomplete. Thus x == NA is a vector of the same length as x *all* of whose values are NA irrespective of the elements of x.

S functions differ markedly in their policy for handling missing values. Many will omit rows that contain a missing value after reducing the data matrix to only those columns needed for a calculation. For statistical functions the na.action argument to may allow other possibilities. The default na.action is usually na.fail which causes the procedure to stop; the alternative na.omit implements the row-omission policy. Another possibility, na.gam.replace, replaces missing numerical values in a data frame by the mean of the non-missing values in the column, but adds a level, NA, to factors and ordered factors (resulting in an unordered factor).

Missing values are output as NA, and can be input as NA or by ensuring that a value is missing (for example, a field is blank). When character strings are coerced to mode numeric the result is NA unless the string parses as a number.

Elementary matrix operations

We have seen that a matrix is merely a data vector with a dim attribute specifying a double index. However, S contains many operators and functions for matrices; for example t(X) is the transpose function. The functions nrow(A) and ncol(A) give the number of rows and columns in the matrix A.

The operator %*% is used for matrix multiplication. Vectors which occur in matrix multiplications are if possible promoted either to row or to column vectors, whichever is multiplicatively coherent. (This may be ambiguous, as we shall see.) Note carefully that if A and B are square matrices of the same size, then A * B is the matrix of element-by-element products whereas A %*% B is the matrix product. If x is a vector, then

```
x %*% A %*% x
```

is a quadratic form $x^T A x$, where x is the column vector and T denotes transpose.

Note that x %*% x seems to be ambiguous, as it could mean either $x^T x$ or $x x^T$. A more precise definition of %*% is that of an inner product rather than a matrix product, so in this case $x^T x$ is the result. (For $x x^T$ use x %o% x; see page 54.)

Matrices can be built up from other vectors and matrices by the functions cbind and rbind. Informally cbind forms matrices by binding together vectors or matrices column-wise, and rbind binds row-wise. The arguments to cbind must be either vectors of any length, or matrices with the same column size, that is the same number of rows. The result is a matrix with the concatenated arguments forming the columns. If some of the arguments to cbind are vectors they may be shorter than the column size of any matrices present, in which case they are cyclically extended to match the matrix column size (or the length of the longest vector if no matrices are given). The function rbind performs the corresponding operation for rows. In this case any vector arguments, possibly cyclically extended, are taken as rows. Note that we have already seen cbind and rbind operating on data frames.

Further matrix operations are discussed in Sections 2.8 and 2.9.

2.4   Reading data

Large data objects will usually be read as values from external files rather than entered during an S session at the keyboard. The S input facilities are simple and their requirements are fairly strict and rather inflexible. There is a clear presumption by the designers of S that you will be able to modify your input files using other tools, such as file editors and the Unix utilities sed and awk, to fit in with the requirements of S. There are some tools within S that can cater for non-standard situations, as we discuss below.

If variables are to be held in data frames, as we strongly suggest they should be, an entire data frame can be read directly with the read.table function. There is also a more general input function, scan, that is useful in special circumstances.

The read.table function

In order to be read into a data frame, an external file should have a standard form:

1. The first line of the file should have a *name* for each variable in the data frame. (Header lines can be ignored by setting the argument `skip` to skip an appropriate number of lines.)

2. Each additional line of the file has as its first item a *row name* and the values for each variable, separated by spaces, tabs or both. Character strings containing blanks must be contained within quotes, which are otherwise optional.

The first few lines of a file to be read as a data frame are shown in Figure 2.1. By default numeric items (except row names) are read as numeric variables and non-numeric variables, such as `Cent.heat` in the example, as factors. (This can be changed if necessary via the argument `as.is`.)

	Price	Floor	Area	Rooms	Age	Cent.heat
01	52.00	111.0	830	5	6.2	no
02	54.75	128.0	710	5	7.5	no
03	57.50	101.0	1000	5	4.2	no
04	57.50	131.0	690	6	8.8	no
05	59.75	93.0	900	5	1.9	yes

....

Figure 2.1: Input file form with names and row names.

The function `read.table` is used to read in the data frame

```
HPrice <- read.table("houses.dat")
```

It is often convenient to generate row names within S-PLUS. In this case the file should omit the row name column (as in Figure 2.2), and we must specify `header=T` in the call to `read.table`. The row names may be specified as an argument, `row.names`, to `read.table`, but to ensure the default row names, namely the row numbers, are generated we must specify `row.names=NULL`, as in

```
HPrice <- read.table("houses1.dat", header=T, row.names=NULL)
```

If the `row.names` argument is omitted, the first non-numeric column with all components different will be used as the row names, if one exists. This can be very puzzling if it is not anticipated.

The argument `na.strings` can be used to specify a character vector of input strings to map to `NA`. The `skip` argument may be used to specify a number of lines in the file to be omitted before reading data.

Note that data frames *must* have both row and column names so `read.table` supplies default values where necessary. The default names can be surprisingly unhelpful, so we recommend always supplying at least column names.

S-PLUS for Windows 3.3 has a menu item Import ASCII on its File menu that provides a dialog-box interface to `read.table`.

Price	Floor	Area	Rooms	Age	Cent.heat
52.00	111.0	830	5	6.2	no
54.75	128.0	710	5	7.5	no
57.50	101.0	1000	5	4.2	no
57.50	131.0	690	6	8.8	no
59.75	93.0	900	5	1.9	yes

....

Figure 2.2: Input file form without row names.

Data file names

Unix users can use any legal file name for data files. So can Windows users, but they have to take care how they specify them, as backslashes (\) within names have to be doubled inside S, for example as

```
"c:\\mywork\\splus\\sws\\file.dat"
```

For functions expecting the name of a file as a particular argument this can (usually) also be specified with slashes as

```
"c:/mywork/splus/sws/file.dat"
```

which can be easier to use, especially for users conversant with Unix. The filename `clipboard` may be used in Windows to refer to the Windows clipboard for input or output (but the file length is limited to 32 Kb).

The function `scan`

The simplest use of the `scan` function is to input a single vector from the keyboard:

```
> x <- scan()
1: 23.4 45.6 77.8 12.9
5: 20 10 11
8: 33
9:
> x
[1] 23.4 45.6 77.8 12.9 20.0 10.0 11.0 33.0
```

The default arguments specify that input is to come from the keyboard. Data items are entered as the prompt changes to `n:` where `n` is the index of the next item to be read. Reading is terminated by an empty input line (only from the keyboard) or by Ctrl-D.

Data may be read as a vector from an external file using `scan` with the file name as argument. Suppose, for example, the file `mat.dat` contains

```
12     5     4     3
 5    17     2     1
 6     4    19     0
 4     5     1    21
```

where each line is intended to become a row of an S matrix M. To read the matrix we use

```
M <- matrix(scan("mat.dat"), ncol=4, byrow=T)
```

It is not necessary to specify how many rows the matrix has since this is deduced from the number of columns and the total number of items read. The argument byrow=T indicates that the vector is to fill the matrix by rows rather than by columns.

Like read.table, scan also has an argument skip that allows a number of lines at the top of the input file to be omitted before reading data.

Another common way to use scan is similar to read.table but with more flexibility. Suppose the data vectors are of equal length and are to be read in parallel from a data file input.dat. Suppose that there are three vectors, the first of mode character and the remaining two of mode numeric. The first step is to use scan to read in the three vectors as a list, as follows

```
indat <- scan("input.dat", list(id="", x=0, y=0))
```

The second argument is a template list that establishes the mode of the three vectors to be read and the structure of the result, which is a list. If we wish to access the variables separately they may either be re-assigned to variables:

```
label <- indat$id; x <- indat$x; y <- indat$y
```

or the list itself may be attached to the search path, by

```
attach(indat)
```

As a final example, suppose we need to read in a large file with 50 numeric variables, and each case occupies 5 lines of the file big.dat. We are willing to to label the variables X1, X2, ..., X50. The first step is to construct the template list:[7]

```
inlist <- as.list(numeric(50))          # a list of 50 zeros.
names(inlist) <- paste("X", 1:50, sep="")
```

To read the file we must also specify that each case will occupy more than one line of the data file:

```
datlist <- scan("big.dat", what=inlist, multi.line=T)
```

This is only possible using the multi.line=T argument with scan; multi-line files cannot yet be read with read.table.

There is a function count.fields that will count the number of fields on each line of a file, which can be useful in locating faulty lines in a large file.

---

[7] paste is discussed on page 42.

Reading non-standard data files

Occasionally it is helpful to use a different field separator from the default, which is any consecutive sequence of blanks, tabs or newlines (known as "whitespace"). For example, some spreadsheet programs' output files are written with fields separated by commas. The argument `sep=","` of `read.table` and `scan` will allow this. (In this case empty and blank fields will be read as missing values `NA`.) Using `sep="\t"` specifies that the fields are to be separated by a single tab character, thus allowing strings to be read that themselves contain blanks, and `sep="\n"` allows only the newline as a field separator, so every line will be read as a single field.

If input data are to be read from a file where data items occupy specific columns without guaranteed whitespace separators between items, the first option to consider is to modify the file so that there is separating whitespace. This makes the data easier to read and simpler to check directly from the data file if necessary, but is not an option if the items themselves may contain embedded blanks and tab characters. If it is necessary to use a file with fixed-width fields there is an argument, `widths`, to `scan` that allows fixed width input by specifying an integer vector of field widths. There is also a function `make.fields` that can be used to convert a file with fixed width, non-separated, input fields into a file with separated fields. It should be noted that the `widths` argument of `scan` uses the `make.fields` function itself and a temporary file, so its use with very large files may pose a problem.

Another solution to handling complex or non-standard input is to read in each line of the file as a character string vector and extract the data items later. For example, suppose we wish to read 40 variables from each line of a file, `v40.dat`. Each variable occupies two columns, making 80 columns of data in all. Missing values are denoted by a blank entry. We may use

```
chdata <- scan("v40.dat", what="", sep="\n")
```

Specifying `sep="\n"` ensures that data items are separated only by the newline, and `what=""` establishes that the data to be read in are to be of mode `character` and so `chdata` is a character vector with each complete line of the data file as its elements.

To extract each column in numeric form we need to use a simple loop

```
dat <- list()              # initially empty list
for(i in 1:40)
      dat[[i]] <- as.numeric(substring(chdata, 2*i-1, 2*i))
```

Coercion to numeric of a fully blank string results in a missing value, as required. (This also happens in the case of a string that does not parse correctly as a number.) To make the data more easily accessible it is a good idea to convert it to a data frame:

```
digits <- substring(1000+seq(along=dat), 3, 4)
names(dat) <- paste("V", digits, sep="")
dat <- as.data.frame(dat)
```

This device ensures the names are of constant length, so the first variable is V01 rather than V1. In turn this ensures that consecutive variables remain consecutive when the names are sorted. The function substring is discussed on page 43.

One fairly common problem in reading files is a failure to ensure that the end-of-file characters have been converted whilst transferring the data to the computer on which S-PLUS is used. All versions of S-PLUS will accept either LF (the Unix norm) or CRLF (the MS-DOS norm), but files transferred from Macintosh computers must be converted.

2.5 Model formulae

Model formulae were introduced into S as a compact way to specify linear models, but have since been adopted for so many diverse purposes in S-PLUS that they are now best regarded as an integral part of the S language. The various uses of model formulae all have individual features which are treated in the appropriate chapter, based on the common features described here.

A formula is of the general form

```
response ~ expression
```

where the left-hand side, response, may sometimes be absent and the right-hand side, expression, is a collection of terms joined by operators usually resembling an arithmetical expression. The meaning of the right-hand side is context dependent. For example, in non-linear regression it is an arithmetical expression and all operators have their usual arithmetical meaning. In linear and generalized linear modelling it specifies the form of the model matrix and the operators have a different meaning. In Trellis graphics it is used to specify the abscissa variable for a plot, but a vertical bar, |, operator is allowed to indicate conditioning variables.

Some functions such as lme and nlme for mixed effects models may require several formulae as arguments.

It is conventional (but not quite universal) that a function which interprets a formula also has arguments data, subset and na.action. Then the formula is interpreted in the *context* of the argument data which is usually a data frame or a list; the objects named on either side of the formula are looked for first in data and then searched for in the usual way (described in detail in the next section). The subset argument is also interpreted in the context of the data frame.

We have seen a few formulae in Chapter 1, all for linear models, where the response is the dependent variable and the right-hand side specifies the explanatory variables. We had

```
fm <- lm(y ~ x,  data=dum)
abline(lm(time ~ dist))
fm <- aov(Speed ~ Run + Expt)
fm0 <- update(fm, . ~ . - Run)
```

Notice that in these cases + indicates inclusion, not addition, and – exclusion.

In most cases we had already attached the data frame, so did not specify it via a data argument. The function `update` is a very useful way to change the call to functions using model formulae; it reissues the call having updated the formula (and any other arguments specified when it is called). The formula term " . " has a special meaning in a call to `update`; it means 'what is there already' and may be used on either side of the ~.

It is implicit in this description that the objects referred to in the formula are of the same length, or constants that can be replicated to that length; they should be thought of as all being measured on the same set of units. The other two arguments allow that set of units to be altered: `subset` is an expression evaluated in the the context of `data` which should evaluate to a valid indexing vector (of types 1, 2 or 4 on page 47). The `na.action` argument specifies what is to be done when missing values are found by specifying a function to be applied to the data frame of all the data needed to process the formula. The default action is usually `na.fail`, which reports an error and stops, but some functions have more accommodating defaults.

Further details of model formulae are given for Trellis graphics (page 91), linear models (Section 6.2), non-linear models (page 269), mixed models (pages 301, 306 and 314), survival analysis (pages 344, 345, 357 and 379), multivariate analysis (page 384) and tree functions (page 420). Many of these involve special handling of factors and functions appearing on the right-hand side of a formula.

2.6 Character vector operations

The form of character vectors can be unexpected and should be carefully appreciated. Unlike say C, they are vectors of character strings, not of characters, and most operations are performed separately on each component.

Note that `""` is a legal character string with no characters in it, known as the empty string. This should be contrasted with `character(0)` which is empty character vector. As vectors, `""` has length 1 and `character(0)` has length 0.

Character vectors may be created by assignment and may be concatenated by the `c` function. They may also be used in logical expressions, such as `"ann" < "belinda"`, in which case lexicographic ordering applies using the ASCII collating sequence.

There are several functions for operating on character vectors. The function `nchar(text)` gives (as a vector) the number of characters in each element of its character vector argument. The function `paste` takes an arbitrary number of arguments, coerces them to strings or character vectors if necessary and joins them together, element by element, as character vectors. For example

```
> paste(c("X","Y"), 1:4)
[1] "X 1" "Y 2" "X 3" "Y 4"
```

Any short arguments are re-cycled in the usual way. By default the joined elements are separated by a blank; this may be changed by using the argument, sep=string, often the empty string:

```
> paste(c("X","Y"), 1:4, sep="")
[1] "X1" "Y2" "X3" "Y4"
```

Another argument, collapse, allows the result to be concatenated into a single long string. It prescribes another character string to be inserted between the components during concatenation. If it is NULL, the default, or character(0), no such global concatenation takes place. For example

```
> paste(c("X","Y"), 1:4, sep="", collapse=" + ")
[1] "X1 + Y2 + X3 + Y4"
```

Substrings of the strings of a character vector may be extracted (element-by-element) using the substring function. It has three arguments

```
substring(text, first, last = 1000000)
```

where text is the character vector, first is a vector of first character positions to be selected and last is a vector of character positions for the last character to be selected. If first or last are shorter vectors than text they are re-cycled in the usual way.

The first argument is coerced to character if necessary, so one way to generate number labels padded with leading zeros to a constant width is to use a construction like the example on page 40.

For another example, the dataset state.name is a character vector of length 50 containing the names of the states of the United States of America in alphabetic order. To extract the first four letters in the names of the last seven states:

```
> substring(state.name[44:50], 1, 4)
[1] "Utah" "Verm" "Virg" "Wash" "West" "Wisc" "Wyom"
```

Note the use of the index vector, [44:50], to select the last seven states.

The function abbreviate provides a more general mechanism for generating abbreviations. In this example it gives

```
> as.vector(abbreviate(state.name[44:50]))
[1] "Utah" "Vrmn" "Vrgn" "Wshn" "WsVr" "Wscn" "Wymn"
> as.vector(abbreviate(state.name[44:50], use.classes=F))
[1] "Utah" "Verm" "Virg" "Wash" "WVir" "Wisc" "Wyom"
```

We used as.vector to suppress the names attribute which contains the unabbreviated names!

The function grep searches for patterns in a vector of character strings, and returns the indices of the strings in which a match was found. On Unix systems it is based on the function egrep. There are important differences between the pattern matching protocols used by grep on Unix and Windows so the details should be noted; an example occurs on page 45. There is also a command regexpr in versions 4.x which behaves like grep on Unix.

Simpler forms of matching are done by the functions `match`, `pmatch` and and `charmatch`. Each tries to match each element of its first argument against the elements of its second argument. The function `match` seeks the first exact match (equality) whereas the other two look for partial matches (the search string starts the character string.) All return the value of their `nomatch` argument (which defaults to `NA`) if there is no match. The function `charmatch` returns the index of a unique match, and 0 if there is more than one match. With argument `duplicates.ok=F` (the default), `pmatch` returns `nomatch` for duplicate matches, whereas with `duplicates.ok=T` it returns the index of the first match.

2.7   Finding S objects

It is important to understand where S keeps its objects and where it looks for objects on which to operate. The objects that S creates at the interactive level during a session it (usually) stores permanently, as *files* in the `.Data` subdirectory of your working directory. (Subdirectory `_Data` under Windows.) On the other hand, objects created at a higher level, such as within a function, are kept in what is known as a *local frame* which is only temporary and such objects are deleted when the function is finished.[8]

Each permanent object is held as a file using a *mapped* file name. In Unix the mapped file name is, except in exceedingly rare cases, the same as the object name in S-PLUS. In Windows, however, it is usually different unless the object has a lower-case name that as a character string makes a legal MS-DOS file name.[9] What is important to note is that when S is restarted at a later time, objects created in previous sessions are still available.

This explains why we recommend that you should use separate working directories for different jobs. For some users common names for objects are single letter names like `x`, `y` and so on, and if two projects share the same `.Data` subdirectory their objects easily become mixed up. However, data frames provide another convenient way of encapsulating and partitioning collections of related objects within the same `.Data` directory.

When S looks for an object, it searches through a sequence of places known as the *search path*. Usually the first entry in the search path is the `.Data` subdirectory of the current working directory. The names of the places currently making up the search path are given by invoking the function

```
search()
```

To get a list of objects currently held in the first place on the search path, use the command

```
objects()
```

---

[8] For precise details see page 153.

[9] These object files should normally not be manipulated except through S-PLUS, although it is sometimes convenient to transfer them directly to another compatible computer.

The names of the objects held in any place in the search path can be displayed by
giving the `objects` function an argument. For example,

```
objects(2)
```

lists the contents of the entity at position 2 of the search path. It is also possible
to list selectively, using the `pattern` argument of `objects`. This restricts the
listing to objects matching the pattern (in the sense of `grep` regular expressions,
under Unix those of the `egrep` command). For example, with our library MASS
attached at position 8,

```
> objects(8, pattern=".*lda")
[1] "lda"          "predict.lda"
> objects(8, pattern="lda$")
[1] "lda"          "predict.lda"
```

both list all objects whose names end with `lda`. (These examples were done on
Unix.) As they show, the regular expressions used by the `pattern` argument are
not the same as the 'wildcard' matching expressions of Unix shells; `.` matches
any character (use `\.` to match `.`) and `.*` matches zero or more occurrences of
any character, that is any character string.

The function `objects` uses another function, `grep`, for pattern matching and
this is currently different for Unix and Windows. In Windows `*` *is* a wildcard
character matching zero or more characters and `?` matches precisely one character.
Windows 3.3 users also have an Object Manager under their Tools menu which
can be used to list all objects anywhere on the search path, to subset by type or
pattern, and to view, summarize or edit the object.

Conversely the function `find`(*object*) discovers where an object appears on
the search path, perhaps more than once. For example (on one of our Unix systems
under S-PLUS 3.3 after `library(treefix, first=T)`)

```
> search()
[1] ".Data"
[2] "/usr/local/splus/library/treefix/.Data"
[3] "/usr/local/splus/splus/.Functions"
[4] "/usr/local/splus/stat/.Functions"
[5] "/usr/local/splus/s/.Functions"
[6] "/usr/local/splus/s/.Datasets"
[7] "/usr/local/splus/stat/.Datasets"
[8] "/usr/local/splus/splus/.Datasets"
> find(prune.tree)
[1] "/usr/local/splus/library/treefix/.Data"
[2] "/usr/local/splus/splus/.Functions"
> as.vector(find("prune.tree", numeric.=T))
[1] 2 3
```

The first argument to `find` may be an object name or a character string giving
the object name. If the object name is nonstandard, as in `find("+")` it must be
given as a character string. More subtly, if the object being sought *is* a character

string vector, such as `letters`, the name must also be given in quotes, since otherwise the search will be made not for the object itself but for an object whose name is given by the first component, usually resulting in a puzzling error. The safest policy is to quote the argument when in any doubt.

A second argument to `find` allows the result to be returned as a numeric vector giving the positions on the search path where the object is located, but as it has the data directory names as a `names` attribute it is often difficult to read unless the names are removed.

In the example above the object named `prune.tree` occurs in two places on the search path. Since the our version from the `treefix` library occurs in position 2 before the S-PLUS version in position 3, ours is the one that will be found and used. By this mechanism changes may be made to S-PLUS functions without affecting the original version.

The 'places' on the search path can be of two main types.[10] As well as data directories of (S created) files, they can also be S lists, usually data frames. In the S literature any entity that can be placed on the search path is sometimes referred to as a *dictionary*, or as a *database*. The directory at position 1 (normally `.Data`) is called the *working database*. A library section is a specialized use of a directory, which is discussed in Appendix C.

As we have noted, if several different objects with the same name occur on the search path all but the first will be masked and normally unreachable. To bypass this, the function `get` may be used to select an object from any given position on the search path. For example,

```
s.prune.tree <- get("prune.tree", where = 3)
```

copies the object called `prune,tree` from position 3 on the search path to an object called `s.prune.tree` in the working database.[11] A function, `masked`, is available to check if system objects are being masked by user-defined objects, whether intentionally or not.

Extra directories, lists or data frames can be added to this list with the `attach` function and removed with the `detach` function. Examples were included in the introductory session. Normally a new entity is attached at position 2, and `detach()` removes the entity at position 2, normally the result of the last `attach`. All the higher-numbered databases are moved up or down accordingly. Note that lists can be attached by `attach(alist)` but must be detached by `detach("alist")`. If a list is attached, a copy is used, so any subsequent changes to the original list will not be reflected in the attached copy. When a list is detached the copy is normally discarded, but if any changes have been made to that database it will be saved unless the argument `save=F` is set. The name used for the saved list is of the form `.Save.alist.2` (and is reported by `detach`) unless `save` is a character string when the database will be saved under that name.

---

[10] There are further types of possible database which we have never had occasion to use.

[11] Note that the object name as the argument to `get` must be given in quotes.

The value returned by `attach` is `NULL` and of `detach` is a character string vector listing the previous entries on the search path. Automatic printing is suppressed in both cases, but the `.Last.value` variable is affected.

To remove objects permanently from the working database the function `rm` is used, whose arguments are the names of the objects to be discarded, as in

```
rm(x, y, z, ink, junk)
```

If the names of objects to be removed are held in a character vector it may be specified as a named argument. An equivalent form of the preceding command is

```
junk <- c("x", "y", "z", "ink", "junk")
rm(list = junk)
```

The function `remove` can be used to remove objects with non-standard names or from data directories other than position 1 of the search path. The objects to be removed must be specified as a character vector. A further equivalent to the preceding command is

```
remove(junk, where = 1)
```

A considerable degree of *caching* of databases is performed, so their view may differ inside and outside the S session. To avoid this, use the `synchronize` function. With no argument, it writes out objects to the current working database. With a numerical vector argument, it re-reads the specified directories on the search path, which can be necessary if some other process has altered the database.

This is the end of the informal part of Chapter 2. You might like to take a rest, start skimming, or move on to Chapter 3 at this point.

2.8   Indexing vectors, matrices and arrays

We have already seen how subsets of the elements of a vector (or an expression evaluating to a vector) may be selected by appending to the name of the vector an *index vector* in square brackets. We now consider indexing more formally. For vector objects, index vectors can be any of four distinct types:

1. **A logical vector.**   The index vector must be of the same length as the vector from which elements are to be selected. Values corresponding to `T` in the index vector are selected and those corresponding to `F` omitted. For example,

   ```
   y <- x[!is.na(x)]
   ```

   creates an object `y` which will contain the non-missing values of `x`, in the same order as they originally occurred. Note that if `x` has any missing values, `y` will be shorter than `x`. Another example is

```
z <- (x+y)[!is.na(x) & x > 0]
```

which creates an object z and places in it the values of the vector x+y
for which the corresponding value in x was positive (and non-missing).
The system function unique uses the function duplicated to give
x[!duplicated(x)].

2. **A vector of positive integers.**   In this case the values in the index vector must
lie in the the the set { 1, 2, ..., length(x) }. The corresponding elements
of the vector are selected and concatenated, in that order, in the result. The
index vector can be of any length and the result is of the same length as the
index vector. For example x[6] is the sixth component of x and x[1:10]
selects the first 10 elements of x (assuming length(x) ⩾ 10, otherwise
there will be an error). For another example we use the dataset letters,
a character vector of length 26 containing the lower-case letters:

```
> letters[1:3]
[1] "a" "b" "c"
> letters[1:3][c(1:3,3:1)]
[1] "a" "b" "c" "c" "b" "a"
```

3. **A vector of negative integers.**   This specifies the values to be *excluded* rather
than included. Thus

```
> y <- x[-(1:5)]
```

drops the first five elements of x.

4. **A vector of character strings.**   This possibility only applies where an object
has a names attribute to identify its components. In that case a subvector
of the names vector may be used in the same way as the positive integers in
case **2**. For example,

```
> longitude <- state.center[["x"]]
> names(longitude) <- state.name
> longitude[c("Hawaii", "Alaska")]
  Hawaii  Alaska
 -126.25 -127.25
```

finds the longitude of the geographic centre of two most western states of
the USA. The names attribute is retained in the result.

A vector with an index expression attached can also appear on the left-hand side
of an assignment, making the operation a *replacement*. In this case the assignment
operation appears to be performed only on those elements of the vector implied
by the index. For example,

```
x[is.na(x)] <- 0
```

replaces any missing values in x by zeros. Note that this is really a disguised
call to the function "[<-", which copies the entire object. Replacing even one
element of a large object can be a memory expensive operation. For another
example note that

```
y[y < 0] <- -y[y < 0]
```

has the same effect as y <- abs(y).

The case of a zero index falls outside these rules. A zero index in a vector of an expression being assigned passes nothing, and a zero index in a vector to which something is being assigned accepts nothing. For example

```
> a <- 1:4
> a[0]
numeric(0)
> a[0] <- 10
> a
[1] 1 2 3 4
```

Zero indices may be included with otherwise negative indices or with otherwise positive indices but not with both positive and negative.

Another case to be considered is if the absolute value of an index falls outside the range 1, ..., length(x). In an expression this gives NA if positive and imposes no restriction if negative. On the left-hand side of a replacement, a positive index greater than length(x) extends the vector, assigning NAs to any gap, and a negative index less than -length(x) is ignored.

The functions replace and append are convenience functions; replace(x, pos, values) returns a copy of x with x[pos] <- values without affecting the original. The function append(x, values, after=length(x)) appends values to x in the specified place, and returns a copy without changing x.

Array indices

An *array* can be considered as a multiply indexed collection of data entries. Any array with just two indices is called a *matrix*, and this special case is perhaps the most important.

A *dimension vector* is a vector of positive integers of length at least 2. If its length is k then the array is k-dimensional, or as we prefer to say, k-indexed. The values in the dimension vector give the upper limits for each of the k indices. Index ranges for S objects always start at 1 (unlike those of C which start at 0). Note that singly subscripted arrays do not exist other than as vectors.

A vector can be used by S as an array only if it has a dimension vector as its dim attribute. Suppose, for example, a is a vector of 150 elements. Either of the assignments

```
a <- array(a, dim=c(3,5,10))  # make a a 3x5x10 array
dim(a) <- c(3,5,10)           # alternative direct form
```

gives it the dim attribute that allows it to be treated as a $3 \times 5 \times 10$ array. The elements of a may now be referred to either with one index, as before, as in a[!is.na(z)] *or* with three, comma separated indices, as in a[2,1,5].

To create a matrix the function array may be used, or the simpler function matrix. For example, to create a 10×10 matrix of zeros we could use

```
Zmat <- matrix(0, nrow=10, ncol=10)
```

Note that a vector used to define an array or matrix is recycled if necessary, so the 0 here is repeated 100 times.

It is important to know how the two indexing conventions correspond; which element of the vector is a[2,1,5] ? S arrays use 'column major order' (as used by FORTRAN but not C). This means the first index moves fastest, and the last slowest. So the correspondence is

$$
\begin{aligned}
\texttt{a[1]} &\longleftrightarrow \texttt{a[1,1,1]}, & \texttt{a[5]} &\longleftrightarrow \texttt{a[2,2,1]}, \\
\texttt{a[2]} &\longleftrightarrow \texttt{a[2,1,1]}, & \ldots &\longleftrightarrow \ldots, \\
\texttt{a[3]} &\longleftrightarrow \texttt{a[3,1,1]}, & \texttt{a[149]} &\longleftrightarrow \texttt{a[3,5,9]}, \\
\texttt{a[4]} &\longleftrightarrow \texttt{a[1,2,1]}, & \texttt{a[150]} &\longleftrightarrow \texttt{a[3,5,10]}
\end{aligned}
$$

A formal definition of column major order can be given by saying if the dimension vector is $(d_1, d_2, \ldots, d_m)$ then the element with multiple index $i_1, i_2, \ldots, i_m$ corresponds to the element with single index

$$
i_1 + \sum_{j=2}^{m} \left\{ (i_j - 1) \prod_{k=1}^{j-1} d_k \right\}
$$

The function matrix has an additional argument byrow which if set to true allows a matrix to be generated from a vector in row-major order.

Each of the dimensions can be given a set of names, just as for a vector. The names are stored in the dimnames attribute which is a list of (possibly NULL) vectors of character strings. For example we can name the first two dimensions of a by

```
dimnames(a) <- list(letters[1:3],
                    c("i", "ii", "iii", "iv", "v"), NULL)
```

For a k-fold indexed array any of the four forms of indexing is allowed in each index position. There are two additional possibilities:

5. **Any array index position may be empty.** In this case the index range implied is the entire range allowed for that index.

6. **An array may be indexed by a matrix.** In this case if the array is k-indexed the index matrix must be an $m \times k$ matrix with integer entries and each row of the index matrix is used as an index vector specifying one element of the array. Thus the matrix specifies m elements of the array to be extracted or replaced.

So if a is a $3 \times 5 \times 10$ array, then a[1:2,,] is the $2 \times 5 \times 10$ array obtained by omitting the last level of the first index. The same sub-array could be specified in this case by a[-3,,].

To give a simple example of a matrix index, consider extracting the diagonal elements of a square matrix X, that is, X[1,1], X[2,2], ..., X[n,n] in a vector, say Xii, and then zeroing the diagonal.[12]

---

[12] This example is artificial as there is a function diag that can be used for both purposes.

```
n <- dim(X)[1]              # same as nrow(X)
i.i <- matrix(1:n, n, 2)    # using the recycling rule
Xii <- X[i.i]               # extract the diagonal
X[i.i] <- 0                 # replace each by zero.
```

Note that by default `a[2,,]` is not a $1 \times 5 \times 10$ array but a 5×10 array. Also `a[2,,1]` and `Xii` in the previous example are vectors and not arrays. In general if any index range reduces to a single value the corresponding element of the dimension vector is removed in the result. This default convention is sometimes helpful and sometimes not. To override it a named argument `drop=F` can be given in the array reference:

```
sua <- a[2,,]           # a   5x10 matrix
sub <- a[2,,, drop=F]   # a 1x5x10 array
```

Note that the drop convention also applies to columns (but not rows) of data frames, so if subscripting leaves just one column, a vector is returned.

However, this convention does not apply to matrix operations. Thus

```
X <- matrix(1:30, 10, 3)
X[,1]
X %*% c(1,3,5)
```

result in a ten-element vector and a 10×1 matrix respectively, which appears inconsistent. (One can argue for either convention, as reducing $1 \times n$ matrices to vectors is often undesirable.) The function `drop` forces dropping, so `drop(X %*% c(1,3,5))` returns a vector and `X[,1, drop=F]` returns a matrix. (Function writers have often overlooked these rules, which can result in puzzling or incorrect behaviour when just one observation or variable meets some selection criterion.)

The function `dim` can be used to find the dimensions of an array or matrix. For matrices two convenience functions `nrow` and `ncol` are provided[13] to access the number of rows and columns respectively. There are further convenience functions `row` and `col` that can be applied to matrices to produce a matrix of the same size filled with the row or column number. Thus to produce the upper triangle of a square matrix `A` we can use

```
A[col(A) >= row(A)]
```

This is a logical vector index, and so returns the upper triangle in column-major order. For the lower triangle we can use `<=` or the function `lower.tri`. A few S functions want the lower triangle of a symmetric matrix in row-major order: note that this is the upper triangle in column-major order.

More generally there is a function `slice.index(A, k)` which generates an array like `A` with entries the value of the kth index. Thus if `M` is a matrix `row(M)`, for example, is the same as `slice.index(M, 1)`.

---

[13] Note that the names are singular: it is all too easy to write `nrows`!

Array arithmetic

Arrays may be used in ordinary arithmetic expressions and the result is an array formed by element-by-element operations on the data vector. The `dim` attributes of operands generally need to be the same, and this becomes the dimension vector of the result. So if `A`, `B` and `C` are all arrays of the same dimensions

```
D <- 2*A*B + C + 1
```

makes `D` a similar array with data vector the result of the evident element-by-element operations. However the precise rule concerning mixed array and vector calculations has to be considered a little more carefully. From experience we have found the following to be a reliable guide, although it is to our knowledge undocumented and hence liable to change.

- The expression is scanned from left to right.
- Any short vector operands are extended by recycling their values until they match the size of any other operands.
- As long as short vectors and arrays, only, are encountered, the arrays must all have the same `dim` attribute or an error results.
- Any vector operand longer than some previous array immediately converts the calculation to one in which all operands are coerced to vectors. A diagnostic message is issued if the size of the long vector is not a multiple of the (common) size of all previous arrays.
- If array structures are present and no error or coercion to vector has been precipitated, the result is an array structure with the common `dim` attribute of its array operands.

Sorting

The S function `sort` at its simplest takes one vector argument and returns a vector of sorted values. The vector to be sorted may be numeric or character, and if there is a names attribute the correspondence of names is preserved. As a simple example consider sorting `mydata`:

```
> mydata
   a   b   c   d   e   f   g   h   i   j
 2.9 3.4 3.4 3.7 3.7 2.8 2.8 2.5 2.4 2.4
> sort(mydata)
   i   j   h   f   g   a   b   c   d   e
 2.4 2.4 2.5 2.8 2.8 2.9 3.4 3.4 3.7 3.7
```

Note that the ordering of tied values is preserved.

The second argument to `sort` allows partial sorting. The argument specifies an index or set of indices which represent the order statistics guaranteed to be correct in the result. The values between these reference indices will all be intervening values, but may not be in sorted order. For example, consider finding the median of a large sample from the $N(0, 1)$ distribution directly:

```
> x <- rnorm(100001)
> sort(x, partial=50001)[50001]
[1] 0.0028992
```

This is most often used to find quantiles, for which the functions `median` and `quantile` are provided.

A more flexible sorting tool is `sort.list`[14], which produces an index vector which will arrange its argument in increasing order. Thus `x[sort.list(x)]` returns the same value as `sort(x)` and `x[sort.list(-x)]` is the vector `x` arranged in *decreasing* order. For example, to arrange the states of the USA from west to east according to their geographical centres we could use

```
> latitude <- state.center[["y"]]
> names(latitude) <- state.name
> i <- sort.list(longitude)
> cbind(latitude = latitude[i], longitude = longitude[i])
               latitude longitude
       Alaska   49.250  -127.250
       Hawaii   31.750  -126.250
....
 Rhode Island   41.593   -71.124
        Maine   45.623   -68.980
```

Notice how under `cbind` the `names` attribute of the vector can become part of the `dimnames` attribute of the matrix.

A further sorting function is `order`. It takes an arbitrary number of arguments and returns the index vector which would arrange the first in increasing order, with ties broken by the second, and so on. For example, to arrange employees by age, by salary within age, and by employment number within salary, and print them together with their number of dependents, we might use:

```
m <- order(Age, Salary, No)
cbind(Age[m], Salary[m], No[m], Depndts[m])
```

The function `rank` is related to sorting: it computes the ranks of the elements of a vector. This is not quite the same as the inverse of `sort.list`, as ties are averaged.

```
> shoes$B
[1] 14.0  8.8 11.2 14.2 11.8  6.4  9.8 11.3  9.3 13.6
> rank(shoes$B)
[1]  9  2  5 10  7  1  4  6  3  8
> rank(round(shoes$B))
[1] 9.0 2.5 5.5 9.0 7.0 1.0 4.0 5.5 2.5 9.0
> sort.list(sort.list(round(shoes$B)))
[1]  8  2  5  9  7  1  4  6  3 10
```

All four functions have an argument `na.last` that determines the handling of missing values. With `na.last=NA` (the default for `sort`) missing values are deleted; with `na.last=T` (the default for `sort.list`, `order` and `rank`) they are put last, and with `na.last=F` they are put first.

---

[14] The use of "list" is unfortunate here since it has nothing to do with S list objects.

2.9   Matrix operations

The standard matrix operators and indexing of matrices have already been covered in Sections 2.3 and 2.8. In this section we cover more specialized calculations involving matrices. Many (but not all) of these will work with numerical data frames, which can always be coerced by `as.matrix(dataframe)` or `data.matrix(dataframe)` (see page 58).

The function `crossprod` forms "crossproducts", meaning that

```
XT.y <- crossprod(X, y)
```

calculates $X^T y$. This matrix could be calculated as `t(X) %*% y` but using `crossprod` is more efficient. If the second argument is omitted it is taken to be the same as the first. Thus `crossprod(X)` calculates the matrix $X^T X$.

An important operation on arrays is the *outer product*. If a and b are two numeric arrays, their outer product is an array whose dimension vector is obtained by concatenating their two dimension vectors (order is important), and whose data vector is obtained by forming all possible products of elements of the data vector of a with those of b. The outer product is formed by the operator `%o%`:

```
ab <- a %o% b
```

or by the function `outer`:

```
ab <- outer(a, b, "*")
ab <- outer(a, b)          # as "*" is the default.
```

The multiplication function may be replaced by an arbitrary function of two variables (or its name as a character string). For example if we wished to evaluate the function

$$f(x, y) = \frac{\cos(y)}{1 + x^2}$$

over a regular grid of values with $x-$ and $y-$coordinates defined by the S vectors x and y respectively, we could use

```
f <- function(x, y) cos(y)/(1 + x^2) # define the function
z <- outer(x, y, f)                  # use it.
```

If the function is not required elsewhere we could do the whole operation in one step using an *anonymous* function as the third argument, as in

```
z <- outer(x, y, function(x, y) cos(y)/(1 + x^2))
```

The function `diag` either creates a diagonal matrix from a vector argument, or extracts as a vector the diagonal of a matrix argument. Used on the assignment side of an expression it allows the diagonal of a matrix to be replaced[15]. For example, to form a covariance matrix in multinomial fitting we could use

---

[15] This is one of the few places where the recycling rule is disabled: the replacement must be a scalar or of the correct length.

```
> p <- dbinom(0:4, size=4, prob=1/3)  # an example prob vector
> CC <- -(p %o% p)
> diag(CC) <- p + diag(CC)
> structure(3^8 * CC, dimnames=list(0:4, 0:4))  # convenience
      0     1     2     3    4
0  1040  -512  -384  -128  -16
1  -512  1568  -768  -256  -32
2  -384  -768  1368  -192  -24
3  -128  -256  -192   584   -8
4   -16   -32   -24    -8   80
```

In addition `diag(n)` for a positive integer n generates an $n \times n$ identity matrix. This is an exception to the behaviour for vector arguments; `diag(x,length(x))` will give a diagonal matrix with diagonal x for a vector of any length, even one.

The function `vecnorm` gives the p-norm of a vector, by default for $p = 2$.

Functions operating on matrices

The standard operations of linear algebra are either available as functions or can easily be programmed, making S a flexible matrix manipulation language (if rather slower than specialized matrix languages).

The function `solve` inverts matrices and solves systems of linear equations; `solve(A)` inverts A and `solve(A, b)` solves A `%*%` x = b. (If the system is over-determined, the least-squares fit is found, but matrices of less than full rank give an error.)

The function `chol` returns the Choleski decomposition $A = U^T U$ of a non-negative definite symmetric matrix. (Note that this is an unusual convention; more commonly the lower-triangular form $A = LL^T$ with $L = U^T$ is used.) Function `backsolve` solves upper triangular systems of matrices, and is often used in conjunction with `chol`. (There is an almost identical function `solve.upper`, but no analogue for lower-triangular matrices.)

Eigenvalues and eigenvectors

The function `eigen` calculates the eigenvalues and eigenvectors of a square matrix. The result is a list of two components, `values` and `vectors`. If we only need the eigenvalues we can use:

```
eigen(Sm, only.values=T)$values
```

Real symmetric matrices have real eigenvalues, and the calculation for this case can be much simpler and more stable. A further named argument, `symmetric`, may be used to specify whether or not a matrix is (to be regarded as) symmetric. The default value is T if the matrix exactly equals its transpose, otherwise F.

Singular value decomposition and generalized inverses

An $n \times p$ matrix X has a *singular value decomposition* (SVD) of the form

$$X = U \Lambda V^T$$

where U and V are $n \times \min(n, p)$ and $p \times \min(n, p)$ matrices of orthonormal columns, and Λ is a diagonal matrix. Conventionally the diagonal elements of Λ are ordered in decreasing order; the number of non-zero elements is the rank of X. A proof of its existence can be found in Golub & Van Loan (1989), and a discussion of the *statistical* value of SVDs in Thisted (1988).

The function svd takes a matrix argument, M, and calculates the singular value decomposition. The components of the result are u and v, the orthonormal matrices and d, a vector of singular values. If either U or V is not required its calculation can be avoided by the argument nu=0 or nv=0.

In some applications a generalized inverse of a matrix, X, is required. This is defined as any matrix X^- such that $XX^-X = X$. One choice is known as the Moore-Penrose or spectral generalized inverse, usually written as X^+ and defined as

$$X^+ = V \Lambda^- U^T$$

where Λ^- is a diagonal matrix like Λ with entries λ_i^{-1} if $\lambda_i > 0$ or 0 if $\lambda_i = 0$. Deciding when a singular value is small enough to be considered zero requires some care and the following function for the generalized inverse uses a protected reciprocal based on the relative sizes of the singular values.

```
ginv <- function(X, tol = sqrt(.Machine$double.eps))
    s <- svd(X)
    nz <- s$d > tol * s$d[1]
    if(any(nz)) s$v[, nz] %*% (t(s$u[, nz])/s$d[nz]) else X * 0
}
```

(An enhanced version[16] is in library MASS.) The generalized inverse may be used to find a particular solution to a system of consistent linear equations of less than full rank, or a least squares solution to an overdetermined system, not necessarily of maximum rank. Useful references include Rao & Mitra (1971), Pringle & Rayner (1971) and Dodge (1985).

The QR decomposition

A faster decomposition to calculate than the SVD is the QR decomposition, defined as

$$M = QR$$

where, if M is $n \times p$, Q is an $n \times n$ matrix of orthonormal columns (that is, an orthogonal matrix) and R is an $n \times p$ matrix with zero elements apart from the first p rows that form an upper triangular matrix. (See Golub & Van Loan, 1989, §5.2). The function qr(M) implements the algorithm detailed in Golub & Van Loan, and hence returns the result in a somewhat inconvenient form to use directly. The result is a list that can be used by other tools. For example

---

[16] S-PLUS 4.0 has a similar function ginverse.

```
M.qr <- qr(M)              # QR decomposition
Q <- qr.Q(M.qr)            # Extract  a Q (n x p) matrix
R <- qr.R(M.qr)            # Extract an R (p x p) matrix
y.res <- qr.resid(M.qr, y) # Project onto error space
```

The last command finds the residual vector after projecting the vector y onto the column space of M. Other tools that use the result of qr include qr.fitted for fitted values and qr.coef for regression coefficients.

Note that by default qr.R only extracts the first p rows of the matrix R and qr.Q only the first p columns of Q, which form an orthonormal basis for the column space of M if M is of maximal rank. To find the complete form of Q we need to call qr.Q with an extra argument complete=T. The columns of Q beyond the rth, where $r \leqslant p$ is the rank of M, form an orthonormal basis for the null space or *kernel* of M, which is occasionally useful for computations. A simple function to extract it is

```
Null <- function(M) {
  tmp <- qr(M)
  set <- if(tmp$rank == 0) 1:ncol(M) else -(1:tmp$rank)
  qr.Q(tmp, complete=T)[, set, drop=F]
}
```

Determinant and trace

The only function provided for finding the determinant of a square matrix, M, is in the Matrix library (see page 60). For older systems lacking the Matrix library there are several ways to write determinant functions. Often it is known in advance that a determinant will be non-negative or that its sign is not needed, in which case methods to calculate the absolute value of the determinant suffice. In this case it may be calculated as the product of the singular values, or, slightly faster but possibly less accurately from the QR-decomposition as

```
absdet <- function(M) abs(prod(diag(qr(M)$qr)))
```

If the sign is unknown and important, the determinant may be calculated as the product of the eigenvalues. These will in general be complex and the result may have complex roundoff error even though the exact result is known to be real, so a simple function to perform the calculation is

```
det <- function(M) Re(prod(eigen(M, only.values=T)$values))
```

The trace function below is so simple the only reason for having it might be to make code using it more readable.

```
tr <- function(M) sum(diag(M))
```

Converting data frames to and from matrices

The coercion function `as.data.frame` takes a matrix argument (of mode numerical, character or logical) and produces a data frame from it using the columns as variables. If the matrix has a `dimnames` attribute it is retained in the names and row names of the data frame. If not, names are constructed in a default manner with the variable (column) names incorporating the original matrix name.

The constructor function, `data.frame`, may also be used to convert a matrix to a data frame in a similar way, but in this case by default any column names that are nonstandard variable names are standardized by changing the offending characters to periods.

There are two functions which may be used for converting a data frame into a matrix, and they behave slightly differently.

1. `as.matrix(dataframe)` produces a matrix from the argument data frame of mode numeric if all constituent variables in the data frame are numeric, and of mode character otherwise. So any factor will cause the entire matrix result to be of mode character. Note that it is not possible to have mixed mode matrices.

2. `data.matrix(dataframe)` produces a numeric matrix from the argument data frame in all circumstances. Any factors among the variables are first coerced to numeric, and the levels information of the factors is retained in the attribute `column.levels` of the outcome.

The `Matrix` library

S-PLUS 3.3 introduced a completely distinct set of matrix operations based on the LAPACK library (Anderson *et al.*, 1995) of FORTRAN linear algebra routines. These are made available by

```
library(Matrix)
```

Note the 'M'.

The objects on which this library of routines operates are known as Matrices rather than matrices, and are objects of *class* `"Matrix"` [17]. They are created by the `Matrix` function, for example

```
> Matrix(mydata, 2, 5)
     [,1] [,2] [,3] [,4] [,5]
[1,]  2.9  3.4  3.7  2.8  2.4
[2,]  3.4  3.7  2.8  2.5  2.4
attr(, "class"):
[1] "Matrix"
```

Only the `class` attribute reveals any difference. Elementwise operations and matrix multiplication work just as for matrices. For subscripting there are a few differences:

---

[17] The concept of class is discussed in Section 4.4 but the details are not required to understand the present section.

1. Dimensions are not dropped (page 51) and the `drop=F` argument is not needed (nor allowed).

2. Logical subscripts must be of the correct length: the recycling rule is disabled in this context.

3. A Matrix may be given a single index which is itself a two-column matrix.[18] The rows of this index matrix specify the rows and columns of the original Matrix to be used. The index matrix may be numeric or character, in the latter case referring to the dimnames. In this way an operation may be carried out on an arbitrary set of elements of the Matrix. It appears that a two-column index Matrix may also be used.

However, Matrices are not matrices, and you should not expect functions necessarily to handle them correctly.

The benefits of the Matrix library come from the existence of specialized forms of Matrices which can be stored more compactly and operated on more efficiently. There are for example identity, diagonal, symmetric, Hermitian, lower triangular, upper triangular, orthonormal and permutation forms. These are created by class functions, for example

```
> Identity(3)
[1] 3
attr(, "class"):
[1] "Identity" "Matrix"
> Diagonal(c(7,5,3))
[1] 7 5 3
attr(, "class"):
[1] "Diagonal" "Matrix"
> A <- Matrix(c(7,1,1,7), 2, 2)
> class(A) <- Matrix.class(A)
> A
     [,1] [,2]
[1,]   7    1
[2,]   1    7
attr(, "class"):
[1] "Hermitian" "Matrix"
> A[lower.tri(A)] <- 0
> class(A) <- Matrix.class(A)
> A
     [,1] [,2]
[1,]   7    1
[2,]   0    7
attr(, "class"):
[1] "UpperTriangular" "Matrix"
> RowPermutation(c(3,1,2))
[1] 3 1 2
attr(, "class"):
```

---

[18] There is a bug in both 3.3 and 3.4 which prevents matrix indices being used on an assignment.

```
[1] "RowPermutation" "Matrix"
> unpack(RowPermutation(c(3,1,2)))
     [,1] [,2] [,3]
[1,]    0    0    1
[2,]    1    0    0
[3,]    0    1    0
attr(, "class"):
[1] "Orthonormal" "Matrix"
```

Note how the function `Matrix.class` computes the appropriate specializations for its argument. The function `unpack` unpacks the sparse representations, although they can be used directly.

There is a set of functions to operate on Matrices, making use of specialized computations where appropriate. There are methods for the functions `t`, `cbind`, `rbind` (page 36), `solve` (page 55), `eigen`, `qr` and `svd` (page 56). The function `det` computes the log and sign of the determinant. The function `lu` computes the LU decomposition; that is a row permutation matrix P and lower and upper triangular matrices L and U such that $PA = LU$. (The matrix L has ones on its diagonal. For symmetric matrices it gives $PAP^T = LDL^T$ for a block-diagonal matrix D.) The function `facmul` multiplies by factors from the LU decomposition (or extracts specific factors) and `expand` returns them as full Matrix objects.

There is no special method for the Choleski decomposition, but this can be computed from the LU decomposition of a symmetric matrix. For the $A = LL^T$ form of the decomposition we can use

```
Choleski <- function(x)
{
    x <- as.Matrix(x)
    class(x) <- Matrix.class(x)
    if(!inherits(x, "Hermitian")) stop("x must be symmetric")
    LU <- expand(lu(x))
    B <- LU$block.diagonal
    if(!inherits(B, "Diagonal") || any(diag(B) <= 0)
       || !inherits(LU$permutation, "Identity"))
          stop("x is not positive definite")
    x <- LU$triangular %*% sqrt(B)
    class(x) <- Matrix.class(x)
    x
}
```

This function is rather conservative, in that it assumes that `x` is (numerically) strictly positive definite.

The function `backsolve` is replaced by the methods of `solve` for triangular matrices (lower as well as upper).

The more specialized functions `norm` (matrix norms), `rcond` (condition numbers) and `schur` (Schur decomposition) are most likely to be of interest to numerical analysts.

2.10   Input/Output facilities

In this section we cover a miscellany of topics connected with input and output.

Writing data to a file

A character representation of any S vector may be written on an output device
(including the session window) by the `write` function

```
write(x, file="outdata")
```

where the session window is specified by `file=""`. Little is allowed by way
of format control, but the number of columns can be specified by the argument
`ncolumns` (default 5 for numeric vectors, 1 for character vectors). This is
sometimes useful to print out vectors or matrices in a fixed layout. For data
frames, `write.table` is provided:

```
> write.table(dataframe, file="", sep=",")
> write.table(painters, "", sep="\t", dimnames.write="col")
Composition     Drawing Colour  Expression      School
10       8      16       3      A
15      16       4      14      A
 8      13      16       7      A
12      16       9       8      A
    . . . .
```

Omitting `dimnames.write="col"` writes out both row and column labels (if
any) and setting it to F omits both.

   We have found the following simple function useful to print out matrices or
data frames:

```
write.matrix <- function(x, file="", sep=" ")
{
        x <- as.matrix(x)
        p <- ncol(x)
        cat(dimnames(x)[[2]],format(t(x)), file=file,
            sep=c(rep(sep, p-1), "\n"))
}
```

This produces a neatly formatted layout, and starts with the column labels (if any)
ready for reading in by `read.table` with `header=T`.

Executing commands from, or diverting output to, a file

If commands are stored on an external file, say `commands.q` in the current
directory, they may be executed at any time in an S session with the command

```
source("commands.q")
```

It is often useful to use options(echo=T) before a source file, to have the commands echoed. (If used in the file itself, it takes effect *after* the file is completed.) Similarly

```
sink("record.lis")
```

will divert all subsequent output from the session window to an external file record.lis. The command

```
sink()
```

restores output to the window once more.

Two functions are useful for externally manipulating or transmitting S objects. The function dump takes as its first argument a character vector giving the names of S objects to be written on an external file, by default the file dumpdata. These are written in *assignment form* so that executing the file dumpdata with the source command will re-create the objects.

```
dump(c("a", "x", "ink"), file="outdata") # dump objects
    ....
source("outdata") # re-create a, x and ink
```

The dumped objects are in a form which is fairly easy to read and to edit, but reading such a file with source can be slow, particularly for large numeric objects.

The function data.dump does a similar job to dump but the dumped objects may only be re-created using the companion function data.restore to read the file, and this operation is relatively fast. These two functions are intended for transmission of S objects between remote or incompatible computers, and unlike dump are guaranteed to preserve the storage mode (integer, single or double precision) of S objects.

General printing

The function cat is similar to paste with argument collapse="" in that it coerces its arguments to character strings and concatenates them. However instead of returning a character string result it prints out the result in the session window or optionally on an external file. For example to print out today's date on our Unix system:

```
> d <- date()
> cat("Today's date is:", substring(d,1,10),
            substring(d,25,28), "\n")
Today's date is: Mon Jan  6 1997
```

which is needed occasionally for dating output from within a function. Note that an explicit newline ("\n") is needed.

Other arguments to cat allow the output to be broken into lines of specified length, and optionally labelled:

```
> cat(1,2,3,4,5,6, fill=8, labels=letters)
a 1 2
c 3 4
e 5 6
```

and `fill=T` fills to the current output width.

The function `format` provides the most general way to prepare data for output. It coerces data to character strings in a common format, which can then be used by `cat`. For example, the print function `print.summary.lm` for summaries of linear regressions contains the lines

```
cat("\nCoefficients:\n")
print(format(round(x$coef, digits = digits)), quote = F)
cat("\nResidual standard error:",
    format(signif(x$sigma, digits)), "on", rdf,
    "degrees of freedom\n")
cat("Multiple R-Squared:", format(signif(x$r.squared, digits)),
    "\n")
cat("F-statistic:", format(signif(x$fstatistic[1], digits)),
    "on", x$fstatistic[2], "and", x$fstatistic[3],
    "degrees of freedom, the p-value is", format(signif(1 -
    pf(x$fstatistic[1], x$fstatistic[2], x$fstatistic[3]),
    digits)), "\n")
```

Note the use of `signif` and `round` to specify the accuracy required. (For `round` the number of digits are specified, whereas for `signif` it is the number of significant digits.) To see the effect of `format` *vs* `write`, consider:

```
> write(iris[,1,1], "", 15)
5.1 4.9 4.7 4.6 5 5.4 4.6 5 4.4 4.9 5.4 4.8 4.8 4.3 5.8
5.7 5.4 5.1 5.7 5.1 5.4 5.1 4.6 5.1 4.8 5 5 5.2 5.2 4.7
4.8 5.4 5.2 5.5 4.9 5 5.5 4.9 4.4 5.1 5 4.5 4.4 5 5.1
4.8 5.1 4.6 5.3 5
> cat(format(iris[,1,1]), fill=60)
5.1 4.9 4.7 4.6 5.0 5.4 4.6 5.0 4.4 4.9 5.4 4.8 4.8 4.3 5.8
5.7 5.4 5.1 5.7 5.1 5.4 5.1 4.6 5.1 4.8 5.0 5.0 5.2 5.2 4.7
4.8 5.4 5.2 5.5 4.9 5.0 5.5 4.9 4.4 5.1 5.0 4.5 4.4 5.0 5.1
4.8 5.1 4.6 5.3 5.0
```

There is a tendency to output values such as `0.6870000000000001`, even after rounding to (here) 3 digits. (Not from `print`, but from `write`, `cat`, `paste`, `as.character` and so on.) Use `format` to avoid this.

By default the accuracy of printed and converted values is controlled by the `options` parameter `digits`, which defaults to 7.

2.11   Customizing your S environment

The S environment can be customized in many ways, down to replacing system functions by your own versions. For multi-user systems, all that is normally desirable is to use the `options` command to change the defaults of some variables, and if it is appropriate to change these for every session of a project, to use `.First` to set them (see the next subsection).

The function `options` accesses or changes the dataset `.Options`, which can also be manipulated directly. Its exact contents will differ between operating systems and S-PLUS releases, but one example is

```
> unlist(options())
      echo prompt continue width length         keep check digits
   "FALSE" "> "     "+ "      "80"   "48"   "function" "0"   "7"
         memory object.size audit.size          error    show
   "2147483647" "5000000"   "500000"   "dump.calls" "TRUE"
    compact scrap free warn editor expressions reference
   "100000" "500" "1"  "0"   "vi"     "256"       "1"
   contrasts.factor contrasts.ordered  ts.eps  pager
   "contr.helmert"   "contr.poly"        "1e-05" "less"
```

(The function `unlist` has converted all entries to character strings.) Other options (such as `gui`) are by default unset. Calling `options` with no argument or a character vector argument returns a list of the current settings of all options, or those specified. Calling it with one or more `name=value` pairs resets the values of component `name`, or sets it if it was unset. For example

```
> options("width")
$width:
[1] 80
> options("width"=65)
> options(c("length", "width"))
$length:
[1] 48
$width:
[1] 65
```

The meaning of the more commonly used options are given in Table 2.2.

There is a similar command, `ps.options`, to customize the actions of the `postscript` graphics driver in the Unix versions of S-PLUS.

S-PLUS 3.3 for Windows provides a dialog-box interface to the `options` function via the menu item Session options on the Options menu.

Session startup and finishing functions

If the `.Data` subdirectory of the working directory contains a function `.First` this function is executed silently at the start of any S session. This allows some automatic customization of the session which may be particular to that working directory. A typical `.First` function might include commands such as

Table 2.2: Commonly used options to customize the S environment.

`width`	The page width, in characters. Not always respected.
`length`	The page length, in lines. Used to split up listings of large objects, repeating column headings. Set to a very large value to suppress this.
`digits`	number of significant digits to use in printing. Set this to 17 for full precision.
`gui`	The preferred graphical user interface style.
`echo`	Logical for whether expressions are echoed before being evaluated. Useful when reading commands from a file.
`prompt`	The primary prompt.
`continue`	The command continuation prompt.
`editor`	The default text editor for `ed` and `fix`.
`error`	Function called to handle errors. See Section 4.5.
`warn`	The level of strictness in handling warnings. The default, 0, collects them together; 1 reports them immediately and 2 makes any warning an error condition.
`memory`	The maximum memory (in bytes) that can be allocated.
`object.size`	The maximum size (in bytes) of any object.

```
.First <- function()
{
  options(prompt = "> ", continue = "+   ", digits = 5,
     length = 99999, gui = "motif", editor="vi")
  ps.options(paper = "a4", font = 3, pointsize = 10,
     horizontal = F)
  library(MASS, first=T)
}
```

If it were known that the project would always use the same windowing system it might be appropriate to open a graphics window and also a help window with statements such as:

```
motif("-geometry 600x500-0+0")
help.start()
```

If there is a function `.Last` it is executed when the session is terminated. A typical `.Last` function on a Unix system where disc space was always in short supply might be:

```
.Last <- function()
{
  unix("rm -f ps.out.*.ps") # remove unwanted PostScript files
  cat("Adios.\n")
}
```

2.12   History and audit trails

Unless specifically disabled, S keeps a compact record of all commands in a file `.Audit` (`_Audit` in Windows) in the current working database. This can be accessed in two ways under Unix. The `history` command will retrieve commands (by default the last 10) and allow them to be edited and re-submitted, and `again` does so for the last command. Both can search for commands matching a pattern, their first argument. For example

```
history(max=50, rev=F, evaluate=F)
```

will list the last up to 50 commands in chronological order, and

```
again("lda", ed=T)
```

allows the last command containing the string "lda" to be edited and then submitted.

The command `Splus AUDIT` used from the Unix operating system prompt allows a much more detailed investigation of the audit trail. It has a cryptic command language detailed in its help page. For details of the power of this facility, see Becker & Chambers (1988).

The audit trail can grow large. The option `audit.size`, default 0.5Mb, if reached triggers a warning at the beginning of a session. Run (under Unix)

```
Splus TRUNC_AUDIT n
```

at any time (outside S) to truncate the file to about n bytes (default 100000). The Windows equivalent is `TRUNC_AU n`.

An audit file may be created under Windows, but there are no equivalents of `history`, `again` and `Splus AUDIT`. It is common to disable auditing by setting the environment variable `S_NOAUDIT=X`. (This is set in `Splus.ini` by default under 3.x.)

S-PLUS 3.3 for Windows has an alternative command-history mechanism. There is a Commands History item on the Tools menu. This brings up a window within which commands can be reviewed, selected, edited and executed. These commands can be saved in a file from that window's File menu. Setting the environment variable

```
S_CMDFILE=+history.q
```

causes the command history to automatically be appended to the file `history.q` in the current directory, and this file will be used to initialize the Commands History window in subsequent sessions. If auditing is turned on, `S_CMDFILE=+_Audit` will use the audit file to initialize the commands history.

S-PLUS 4.0 for Windows will have an identical command-history mechanism for its command window. It also has a History window which will show both S commands and operations in the GUI.

2.13   BATCH operation

It is sometimes desirable to run an S-PLUS job non-interactively. On a Unix machine under csh we might use

```
Splus < infile >& outfile &
```

but unless options(echo=T) has been set the input commands will not be echoed to the output file. (We use >& to redirect prompts and errors to the output file.) The Splus BATCH command automates this process, so

```
Splus BATCH infile outfile
```

sets options(echo=T) and then reads infile and redirects all output to outfile. The command is implemented using nohup, the precise effect of which varies with Unix platform; on BSD systems (such as SunOS4.1.3) it reduces the priority of the process somewhat, and on all systems it blocks some software signals to ensure that the S-PLUS job continues to run when the shell terminates.

Although the input is echoed, ends-of-lines are often changed to semi-colons within braced expressions, so the outfile can be hard to read.

Windows

S-PLUS 4.x provides a command-line interface program sqpe.exe whose input and output can be redirected and which can be used from an MS-DOS window in Windows 95 or in other shells. Both prompts and error messages are written to the output stream, but commands are not echoed unless options(echo=T) is set. Before using sqpe ensure that the environment variable SHOME is set: an example of its use (in an MS-DOS window) is

```
set SHOME=C:\Program Files\splus4
%SHOME%\cmd\sqpe < infile > outfile
```

All versions have a BATCH switch. Under 3.x the syntax is

```
splus BATCH infile [outfile [errfile] ]
```

where the last two filenames are optional, and each file can be specified as stdin, stdout or stderr respectively to use the command window as usual. This cannot be used directly in a MS-DOS window in Windows 95 but a line such as

```
start /w splus.exe BATCH infile outfile errfile
```

can be used. Under 4.x the BATCH syntax is

```
splus [S_PROJ=dir] /BATCH infile [outfile [errfile]]
```

where error messages are sent to outfile if errfile is not supplied. Setting S_PROJ is needed to change the working directory _Data from its default. The input commands are not echoed (use options(echo=T)) and the prompts will not appear in the output file. The files are specified relative to the working directory.

2.14 Exercises

The exercises in Chapter 4 also exercise the material of this chapter, but are rather harder.

2.1.   How would you find the index(es) of specified values within a vector? For example, where is the hill race (in `hills`) with a climb of 2100 feet?

2.2.   The column `ftv` in data frame `birthwt` counts the number of visits. Reduce this to a factor with levels 0, 1 and '2 or more'. [Hint: manipulate the `levels`, or investigate functions `cut` and `merge.levels`.]

2.3.   Write a simple function to compute the median absolute deviation (used in robust statistics; see Chapter 8) median$|x - \mu|$ with default μ the sample median. Compare your answer with the S-PLUS function `mad`.

2.4.   Suppose `x` is an object with named components and `out` is a character string vector. How would you make a new object obtained from `x` by *excluding* any components whose names are in `out`?

Chapter 3

Graphical Output

S-PLUS provides comprehensive graphics facilities, from simple facilities for producing common diagnostic plots by plot (*object*) to fine control over publication-quality graphs. In consequence, the number of graphics parameters is huge. In this chapter, we build up the complexity gradually. Most readers will not need the material in Section 3.4, and indeed the material there is not used elsewhere in this book. However, we *have* needed to make use of it, especially in matching existing graphical styles.

Some graphical ideas are best explored in their statistical context, so that, for example, histograms are covered in Chapter 5, survival curves in Chapter 12, biplots in Chapter 13 and time-series graphics in Chapter 15. Table 3.1 gives an overview of the high-level graphics commands with page references; many of the call sequences can be found in Appendix B.

There are many books on graphical design. Cleveland (1993) discusses most of the methods of this chapter and detailed design choices (such as aspect ratios of plots and the presence of grids) which can affect the perception of graphical displays. As these are to some extent a matter of personal preference and this is also a guide to S-PLUS, we have kept to the default choices.

Trellis graphics are a recent addition to S-PLUS with a somewhat different style and philosophy to the basic plotting functions. We describe the basic functions first, then the Trellis functions in Section 3.5. S-PLUS 4.0 for Windows introduced a radically different style of object-oriented editable graphics which we cover in Section 3.6.

3.1 Graphics devices

Before any plotting commands can be used, a graphics device must be opened to receive graphical output. Most commonly this is a window on the screen of a workstation or a plotter or printer. A list of supported devices on the current hardware with some indication of their capabilities is available from the on-line help system by

```
?Devices
```

Table 3.1: High-level plotting functions. Page references are given to the most complete description in the text: see also Appendix B. Those marked by † are superseded by Trellis functions.

	page	
abline	75	Adds lines to the current plot in slope-intercept form.
axis	80	Adds an axis to the plot.
barplot	73	Bar graphs.
biplot	388	Represent rows and columns of a data matrix.
brush spin	382	Dynamic graphics.
contour †	77	Contour plot. contourplot is preferred.
dotchart		Produce a dot chart.
eqscplot	76	Plots with geometrically equal scales (our library).
faces		Chernoff's faces plot of multivariate data.
frame	78	Advance to next figure region.
hist †	168	Histograms using barplot.
hist2d	184	Two-dimensional histogram calculations.
identify locator	80	Interact with an existing plot.
image † image.legend	77	High-density image plot functions. levelplot is preferred.
interaction.plot		Interaction plot for a two-factor experiment.
legend	81	Add a legend to the current plot.
matplot	87	Multiple plots specified by the columns of a matrix.
mtext	81	Add text in the margins.
pairs †	76	All pairwise plots between multiple variables. splom is preferred.
par	82	Set or ask about graphics parameters.
persp † perspp persp.setup	77	Three-dimensional perspective plot functions. Partially superseded by wireframe and cloud.
pie		Produce a pie diagram.
plot		Generic plotting function.
polygon		Add a polygon to the present plot, possibly filled.
points lines	75	Adds points or lines to the current plot.
qqplot † qqnorm †	165	Quantile-quantile and normal Q-Q plots. Trellis functions qq and qqmath are preferred.
scatter.smooth	326	Scatterplot with a smooth curve.
segments arrows	88	Draw line segments or arrows on the current plot.
stars		Star plots of multivariate data.
symbols		Draw variable sized symbols on a plot.
text	75	Add text symbols to the current plot.
title	80	Add title(s).

Table 3.2: Some of the graphical devices available for various versions of S-PLUS.

`motif`	X11–windows systems.
`openlook`	X11–windows systems.
`iris4d`	Silicon Graphics Iris workstations.
`win.graph`	Windows.
`graphsheet`	S-PLUS 4.0—see Section 3.6
`postscript`	POSTSCRIPT printers.[1]
`hplj`	Hewlett-Packard LaserJet printers.
`hpgl`	Hewlett-Packard HP-GL plotters.
`win.printer`	Use Windows printing.
`pdf.graph`	Adobe's PDF format (S-PLUS 4.0).

(Note the capital letter.) A number of graphics terminals are supported, but as these are now unlikely to be encountered they are not considered here.

A graphics device is opened by giving the command in Table 3.2, possibly with parameters giving the size and position of the window, for example

```
motif("-geometry 600x400-0+0")
```

opens a small graphics window initially positioned in the top right-hand corner of the screen. (Note that `motif` may be used on any X11 display, not just one running the Motif window manager, although `openlook` may be preferred for OpenLook window managers.) Many screen devices support the argument `ask`, which if true will request permission to clear the screen and start the next plot. (This can be set later by `par(ask=T)` or using `dev.ask`.)

Users of S-PLUS for Windows 3.3 may find it convenient to open or close a `win.graph` or `win.printer` device from the Graphics Device item on their Tools menu.

Graphics devices may be closed using

```
graphics.off()
```

This will close all the graphics devices currently open and perform any further wrap-up actions that need to be taken, such as sending trailer information to printer devices. Note that quitting with `q()` will automatically perform a `graphics.off` operation.

It is possible to have several graphical devices open at once. By default the most recently opened one is used, but `dev.set` can be used to change the current device (by number). The function `dev.list` lists currently active devices, and `dev.off` closes the current device, or one specified by number. There are also commands `dev.cur`, `dev.next` and `dev.prev` which return the number of the current, next or previous device on the list.

---

[2] `postscript` exists on both Unix and Windows systems, but the driver and the function arguments are completely different.

The `motif` and `openlook` devices have a Copy option on their Graph menu which allows a (smaller) copy of the current plot to be copied to a new window, perhaps for comparison with latter plots. (The copy window can be dismissed by the Delete or Destroy item from its Graph menu.) There is also a `dev.copy` function which copies the current plot to the specified device (by default the next device on the list).

Many of the graphics devices on windowing systems have menus of choices, for example to make hardcopies and to alter the colour scheme in use. (For S-PLUS 3.3 for Windows the colour scheme is selected for all graphics windows from the Options menu, and also by the function `win.colorscheme`.)

Graphical hardcopy

There are several ways to produce a hardcopy of a plot on a suitable printer or plotter. The `motif` and `openlook` devices have a Print item on their Graph menu which will send a full-page copy of the current plot to the default printer. A little more control (the orientation and the print command) is available from the Printing item on the Options or Properties menu.

Users of the Windows version of S-PLUS can use the Print option on their File menu or (in S-PLUS 4.0) the printer icon on the toolbar. This prints the window with focus, so bring the desired graphics window to the top first.

The function `printgraph` (Unix) will copy the current plot to a POSTSCRIPT or LASERJET printer, and allows the size and orientation to be selected, as well as paper size and printer resolution where appropriate.

The function `dev.print` will copy the current plot to a printer device (default `postscript` under Unix and `win.printer` under Windows) and allow its size, orientation, pointsize of text and so on to be set.

Finally, it is normally possible to open an appropriate printer device and repeat the plot commands, although this does preclude interacting with the plot on-screen.

A warning: none of these methods appear to work correctly for `brush` and `spin` plots. The plots in this book were produced by screen dumps using X11 facilities. Windows users can copy to the clipboard and then paste into other applications.

Hardcopy to a file

It is very useful to be able to copy a plot to a file for inclusion in a paper or book (as here). Since each of the hardcopy methods allows the printer output to be re-directed to a file, there are many possibilities.

Users of S-PLUS 3.x under Windows can use the Print option on the File menu to select printing to a file and may also check the Setup box to include the header in the file. The graphics window can then be printed in the usual way. Under S-PLUS 4.0 it is preferable to use the Export Graph... item on the File menu, which can save in a variety of graphics formats including Windows metafile and Encapsulated PostScript (with a preview image if desired).

If the plot file will be rescaled subsequently, the simplest way under Unix is to edit the print command in the `Printing` item of a `motif` or `openlook` device to be

```
cat > plot_file_name <
```

or to make a Bourne-shell command file `rmv` by

```
$ cat > rmv
mv $2 $1
^D
$ chmod +x rmv
```

place this in your path and use `rmv plot_file_name`. (This avoids leaving around temporary files with names like `ps.out.0001.ps`.) Then click on Print to produce the plot.

POSTSCRIPT users will probably want Encapsulated PostScript (EPSF) format files. These are produced automatically by the procedure in the last paragraph, and also by setting both `onefile=F` and `print.it=F` as arguments to the `postscript` device under Unix. Note that these are *not* EPSI files and include neither a preview image nor the PICT resource that MACINTOSH EPSF files require (but can often be used as a TEXT file on a MACINTOSH). It would be unusual to want an EPSF file in landscape format; select 'Portrait' on the `Printing` menu item, or use `horizontal=F` as argument to the `postscript` device under Unix. The default pointsize (14) is set for a full-page landscape plot, and 10 or 11 are often more appropriate for a portrait plot. Set this in the call to `postscript` or use `ps.options` (*before* the `motif` or `openlook` device is opened, or use `ps.options.send` to alter the print settings of the current device).

For Windows users the `placeable metafile` format may be the most useful for plot files, as it is easily incorporated into other Windows applications whilst retaining full resolution. This is automatically used if the graphics device window is copied to the clipboard, and may also be generated by `win.printer`.

3.2 Basic plotting functions

The function `plot` is a generic function, which when applied to many S objects will give a plot. (However, in many cases the plot does not seem appropriate to us.) Many of the plots appropriate to univariate data such as boxplots and histograms are considered in Chapter 5.

Barcharts

The function to display barcharts is `barplot`. This has many options (described in the on-line help), but some simple uses are shown in Figure 3.1. (Many of the details are covered in Section 3.3.)

UK deaths from lung disease

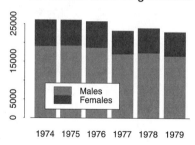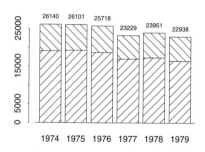

Figure 3.1: Two different styles of barchart showing the annual UK deaths from certain lung diseases. In each case the lower block is for males, the upper block for females.

```
lung.deaths <- aggregate(ts.union(mdeaths, fdeaths), 1)
barplot(t(lung.deaths), names=dimnames(lung.deaths)[[1]],
    main="UK deaths from lung disease")
legend(locator(1), c("Males", "Females"), fill=c(2,3))
loc <- barplot(t(lung.deaths), names=dimnames(lung.deaths)[[1]],
    style="old", dbangle=c(45,135), density=10)
total <- apply(lung.deaths, 1, sum)
text(loc, total + par("cxy")[2], total, cex=0.7)
```

Line and scatterplots

The default plot function takes arguments x and y, vectors of the same length, or a matrix with two columns, or a list (or data frame) with components x and y and produces a simple scatterplot. The axes, scales, titles and plotting symbols are all chosen automatically, but can be overridden with additional graphical parameters which can be included as named arguments in the call. The most commonly used ones are:

type="c"	Type of plot desired. Values for c are: p for points only (the default), l for lines only, b for both points and lines, (the lines miss the points), s, S for step functions (s specifies the level of the step at the left end, S at the right end), o for overlaid points and lines, h for high density vertical line plotting, and n for no plotting (but axes are still found and set).
axes=L	If F all axes are suppressed. Default T, axes are automatically constructed.
xlab="string" ylab="string"	Give labels for the x– and/or y–axes (default: the names, including suffices, of the x and y coordinate vectors).
sub="string" main="string"	sub specifies a title to appear under the x–axis label and main a title for the top of the plot in larger letters. (default: both empty).

`xlim=c(lo ,hi)`	Approximate minimum and maximum values for x– and/or y–axis
`ylim=c(lo, hi)`	settings. These values are normally automatically rounded to make
	them 'pretty' for axis labelling.

The functions `points`, `lines`, `text` and `abline` can be used to add to a plot, possibly one created with `type="n"`. Brief summaries are:

`points(x,y,...)`	Add points to an existing plot (possibly using a different plotting character). The plotting character is set by `pch=` and the size of the character by `cex=` or `mkh=`.
`lines(x,y,...)`	Add lines to an existing plot. The line type is set by `lty=` and width by `lwd=`. The `type` options may be used.
`text(x,y,labels,...)`	Add text to a plot at points given by `x,y`. `labels` is an integer or character vector; `labels[i]` is plotted at point `(x[i],y[i])`. The default is `seq(along=x)`. The character size is set by `cex=`.
`abline(a,b,...)` `abline(h=c,...)` `abline(v=c,...)` `abline(lmobject,...)`	Draw a line in intercept and slope form, (a,b), across an existing plot. `h=c` may be used to specify y–coordinates for the heights of horizontal lines to go across a plot, and `v=c` similarly for the x–coordinates for vertical lines. The coefficients of a suitable *lmobject* are used.

These are the most commonly used graphics functions; we have seen examples of their use in Chapter 1, and will see many more. (There are also functions `arrows` used in Figure 13.14 and `symbols` described in Appendix B that we do not use in this book.) The plotting characters available for `plot` and `points` can be characters of the form `pch="o"` or numbered from 0 to 18 (and up to 27 on a graphsheet in S-PLUS 4.0), which uses the marks shown in Figure 3.2.

Figure 3.2: Plotting symbols or marks, specified by `pch=n`.

Size of text and symbols

Confusingly, the size of plotting characters is selected in one of two very different ways. For plotting characters (by `pch="o"`) or text (by `text`), the parameter `cex` (for 'character expansion') is used. This defaults to the global setting (which defaults to 1), and rescales the character by that factor. For a mark set by `pch=n`, the size is controlled by the `mkh` parameter which gives the height of the symbol *in inches*. (This will be clear for printers; for screen devices the default device region is about 8in × 6in and this is not changed by re-sizing the window.) However if `mkh=0` (the default) the size is then controlled by `cex` and the default size of each symbol is approximately that of `0`. Care is needed in changing `cex` on a call to `plot`, as this will also change the size of the axis labels. It is better to use, for example,

```
plot(x, y, type="n")                    # axes only
points(x, y, pch=4, mkh=0, cex=0.7)    # add the points
```

If `cex` is used to change the size of all the text on a plot, it will normally be desirable to set `mex` to the same value to change the interline spacing. An alternative to specifying `cex` is `csi`, which specifies the absolute size of a character (in inches). (There is no `msi`.)

The default text size can be changed for some graphics devices, for example by argument `pointsize` for the `postscript` and `win.printer` devices.

Equally scaled plots

There are many plots, for example in multivariate analysis, which represent distances in the plane and for which it is essential to have a scaling of the axes which is geometrically accurate. This can be done in many ways, but most easily by our function `eqscplot` which behaves as the default plot function but shrinks the scale on one axis until geometrical accuracy is attained.

Caveat: When screen devices (except `graphsheet` s) are resized the S-PLUS process is not informed, so `eqscplot` can only work for the original window shape.

Multivariate plots

The plots we have seen so far deal with one or two variables. To view more we have several possibilities. A *scatterplot matrix* or *pairs* plot shows a matrix of scatterplots for each pair of variables, as we saw in Figure 1.4, which was produced by `splom(~ hills)`. Enhanced versions of such plots are a *forte* of Trellis graphics, so we will not discuss how to make them in the base graphics system.

Dynamic graphics

The function `brush` allows interaction with the (lower half) of a scatterplot matrix. Examples are shown in Figure 1.5 on page 15 and Figure 13.1 on page 382. As it is much easier to understand these by using them, we suggest you try

```
brush(as.matrix(hills))
```

and experiment.

Points can be highlighted (marked with a symbol) by moving the brush (a rectangular window) over them with button 1 held down. When a point is highlighted, it is shown highlighted in all the displays. Highlighting is removed by brushing with button 2 held down. It is also possible to add or remove points by clicking with button 1 in the scrolling list of row names.

One of four possible (device-dependent) marking symbols can be selected by clicking button 1 on the appropriate one in the display box on the right. The marking is by default persistent, but this can be changed to 'transient' in which only points under the brush are labelled (and button 1 is held down). It is also possible to select marking by row label as well as symbol.

The brush size can be altered under Unix by picking up a corner of the brush in the `brush size` box with the mouse button 1 and dragging to the required size. Under Windows, move the brush to the background of the main `brush` window, hold down the left mouse button and drag the brush to the required size.

The plot produced by `brush` will also show a three-dimensional plot (unless `spin=F`), and this can be produced on its own by `spin`. Clicking with mouse button 1 will select three of the variables for the x, y and z axes. The plot can be spun in several directions and re-sized by clicking in the appropriate box. The `speed` box contains a vertical line or slider indicating the current position.

Plots from `brush` and `spin` can only be terminated by clicking with the mouse button 1 on the `quit` box or button.

Obviously `brush` and `spin` are only available on suitable screen devices, including `motif`, `openlook`, `graphsheet` and `win.graph`. Hardcopy is only possible by directly printing the window used, not by copying the plot to a printer graphics device.

Plots of surfaces

The functions `contour`, `persp` and `image` allow the display of a function defined on a two-dimensional regular grid. Their Trellis equivalents give more elegant output, so we will not discuss them in detail. The function `contour` allows more control than `contourplot`. We anticipate an example from Chapter 16 of plotting a smooth topographic surface for Figure 3.3, and contrast it with `contourplot`.

```
topo.loess <- loess(z ~ x * y, topo, degree=2, span = 0.25)
topo.mar <- list(x = seq(0, 6.5, 0.2), y=seq(0, 6.5, 0.2))
topo.lo <- predict(topo.loess, expand.grid(topo.mar))
par(pty="s")        # square plot
contour(topo.mar$x, topo.mar$y, topo.lo, xlab="", ylab="",
    levels = seq(700,1000,25), cex=0.7)
points(topo)
par(pty="m")
contourplot(z ~ x * y, mat2tr(topo.lo), aspect=1,
    at = seq(700, 1000, 25), xlab="", ylab="",
    panel = function(x, y, subscripts, ...) {
        panel.contourplot(x, y, subscripts, ...)
        panel.xyplot(topo$x,topo$y, cex=0.5)
    }
)
```

This generates values of the surface on a regular 33×33 grid generated by `expand.grid`. We provide the functions `con2tr` and `mat2tr` to convert objects designed for input to `contour` and matrices as produced by `predict.loess` into data frames suitable for the Trellis 3D plotting routines.

There are examples of the output of `persp` on pages 333 and 334 produced by `plot.gam`.

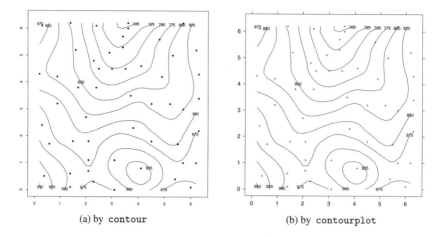

(a) by contour   (b) by contourplot

Figure 3.3: Contour plots of loess smoothing of the topo dataset.

3.3 Enhancing plots

In this section we cover a number of ways that are commonly used to enhance plots, without reaching the level of detail of Section 3.4.

Some plots (such as Figure 1.2) have been square, whereas others are rectangular. Which is selected by the graphics parameter pty. Setting par(pty="s") selects a square plotting region, whereas par(pty="m") selects a maximally sized (and therefore usually non-square) region.

Multiple figures on one plot

We have already seen several examples of plotting two or more figures on a single device surface, apart from scatterplot matrices. The graphics parameters mfrow and mfcol subdivide the plotting region into an array of figure regions. They differ in the order in which the regions are filled. Thus

```
par(mfrow=c(2,3))
par(mfcol=c(2,3))
```

both select a 2×3 array of figures, but with the first they are filled along rows, and with the second along columns. A new figure region is selected for each new plot, and figure regions can be skipped by using frame.

All but two of the multi-figure plots in this book were produced with mfrow. Most of the side-by-side plots were produced with par(mfrow=c(2,2)), but using only the first two figure regions.

The split.screen function provides an alternative and more flexible way of generating multiple displays on a graphics device. An initial call such as

```
split.screen(figs=c(3,2))
```

subdivides the current device surface into a 3×2 array of *screens*. The screens created in this example are numbered 1 to 6, *by rows*, and the original device surface is known as screen 0. The current screen is then screen 1 in the upper left corner, and plotting output will fill the screen as it would a figure. Unlike multi-figure displays, the next plot will use the same screen unless another is specified using the `screen` function. For example the command

```
screen(3)
```

causes screen 3 to become the next current screen.

On screen devices the function `prompt.screen` may be used to define a screen layout interactively. The command

```
split.screen(prompt.screen())
```

allows the user to define a screen layout by clicking mouse button 1 on diagonally opposite corners. In our experience this requires a steady hand, although there is a `delta` argument to `prompt.screen` that can be used to help in aligning screen edges. Alternatively, if the `figs` argument to `split.screen` is specified as an N by 4 matrix, this divides the plot into N screens (possibly overlapping) whose corners are specified by giving (xl, xu, yl, yl) as the row of the matrix (where the whole region is $(0, 1, 0, 1)$).

The `split.screen` function may be used to subdivide the current screen recursively, thus leading to non-regular arrangements. In this case the screen numbering sequence continues from where it had reached.

Split-screen mode is terminated by a call to `close.screen(all=T)`; individual screens can be shut by `close.screen(n)`.

The function `subplot` provides a third way to subdivide the device surface. This has call

```
subplot(fun, x, y, size=c(1, 1), vadj=.5, hadj=.5, pars=NULL)
```

which adds the graphics output of `fun` to an existing plot. The size and position can be determined in may ways (see the on-line help); if all but the first argument is missing a call to `locator` is used to ask the user to click on any two opposite corners of the plot region[2].

Use of the `fig` parameter to `par` provides an even more flexible way to subdivide a plot; see Section 3.4.

With multiple figures it is normally necessary to reduce the size of the text. If either the number of rows or columns set by `mfrow` or `mfcol` is three or more, the text size is halved by setting `cex=0.5` (and `mex=0.5`; see Section 3.4). This may produce characters that are too small and some resetting may be appropriate. (On the other hand, for a 2×2 layout the characters will usually be too large.) For all other methods of subdividing the plot surface the user will have to make an appropriate adjustment to `cex` and `mex` or to the default text size (for example by changing `pointsize` on the `postscript` and `win.printer` devices).

---

[2] not the figure region; Figure 3.4 on page 81 shows the distinction.

Adding information

The basic plots produced by `plot` often need additional information added to give context, particularly if they are not going to be used with a caption. We have already seen the use of `xlab`, `ylab`, `main` and `sub` with scatterplots. These arguments can all be used with the function `title` to add titles to existing plots. The first argument is `main`, so

```
title("A Useful Plot?")
```

adds a main title to the current plot.

Further points and lines are added by the `points` and `lines` functions. We have seen how plot symbols can be selected with `pch=`. The line type is selected by `lty=`. This is device-specific, but usually includes solid lines (1) and a variety of dotted, dashed and dash-dot lines. Line width is selected by `lwd=`, with standard width being 1, and the effect being device-dependent. On colour devices it is also possible to select a colour number by parameter `col=`, again device-specific. (Colours are often rendered as greylevels on printer devices.)

The `type="s"`, `"S"` argument to `lines` allows 'staircase' lines. It is better to use `stepfun` with dashed lines, as the dash sequence is not restarted at each change of angle (at least on the `postscript` device).

Identifying points interactively

The function `identify` has a similar calling sequence to `text`. The first two arguments give the $x-$ and $y-$coordinates of points on a plot and the third argument gives a vector of labels for each point. (The first two arguments may be replaced by a single list argument with two of its components named x and y, or by a two-column matrix.) The labels may be a character string vector or a numeric vector (which is coerced to character). Then clicking with mouse button 1 near a point on the plot causes its label to be plotted; labelling all points or clicking anywhere in the plot with button 2 terminates the process. (The precise position of the click determines the label position, in particular to left or right of the point.) We saw an example in Figure 1.6 on page 16. The function returns a vector of index numbers of the points which were labelled.

In Chapter 1 we used the `locator` function to add new points to a plot. This function is most often used in the form `locator(1)` to return the (x, y) coordinates of a single button click to place a label or legend, but can also be used to return the coordinates of a series of points, terminated by clicking with mouse button 2.

Adding further axes and grids

It is sometimes useful to add further axis scales to a plot, as in Figure 9.1 on page 268 which has scales for both kilograms and pounds. This is done by the function `axis`. There we used

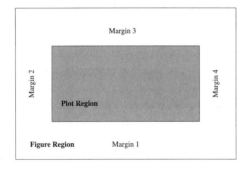

Figure 3.4: Anatomy of a graphics figure.

```
attach(wtloss)
oldpar <- par()
# alter margin 4; others are default
par(mar=c(5.1, 4.1, 4.1, 4.1))
plot(Days, Weight, type="p",ylab="Weight (kg)")
Wt.lbs <- pretty(range(Weight*2.205))
axis(side=4, at=Wt.lbs/2.205, lab=Wt.lbs, srt=90)
mtext("Weight (lb)", side=4, line=3)
detach()
par(oldpar)
```

This adds an axis on side 4 (labelled clockwise from the bottom; see Figure 3.4) with labels rotated by $90°$ (`srt=90`) and then uses `mtext` to add a label 'underneath' that axis. Other parameters are explained in Section 3.4. Please read the on-line documentation very carefully to determine which graphics parameters are used in which circumstances. (For example, why did we need `srt=90` for `axis` but not `mtext` ?)

Grids can be added by using `axis` with long tick marks, setting parameter `tck=1` (yes, obviously). For example, a dotted grid is created by

```
axis(1, tck=1, lty=2)
axis(2, tck=1, lty=2)
```

and the location of the grid lines can be specified using `at=`.

Adding legends

Legends are added by the function `legend`. Since it can label many types of variation such as line type and width, plot symbol, colour, and fill type, its description is very complex. All calls are of the form

```
legend(x, y, legend, ...)
```

where x and y give either the upper left corner of the legend box or both upper left and lower right corners. These are often most conveniently specified on-screen by using `locator(1)` or `locator(2)` . Argument `legend` is a character vector giving the labels for each variation. The remaining arguments are vectors of the same length as `legend` giving the appropriate coding for each variation, by `lty=`, `lwd=`, `pch=`, `col=`, `fill=`, `angle=` and `density=`.

By default the legend is contained in a box; the drawing of this box can be suppressed by argument `bty="n"`.

The Trellis function `key` provides a more flexible approach to constructing legends, and can be used with basic plots. (See page 107 for further details.)

Mathematics in labels

Users frequently wish to include the odd subscript, superscript and mathematical symbol in labels. There is no general solution, but for the Unix `postscript` driver Alan Zaslavsky's package `postscriptfonts` adds these features. It is available from `statlib`: see Appendix C.

3.4   Fine control of graphics

The graphics process is controlled by *graphics parameters*, which are set for each graphics device. Each time a new device is opened these parameters for that device are reset to their default values. Graphics parameters may be set, or their current values queried, using the `par` function. If the arguments to `par` are of the `name=value` form the graphics parameter `name` is set to `value`, if possible, and other graphics parameters may be reset to ensure consistency. The value returned is a list giving the previous parameter settings. Instead of supplying the arguments as `name=value` pairs, `par` may also be given a single list argument with named components.

If the arguments to `par` are quoted character strings, `"name"`, the current value of graphics parameter `name` is returned. If more than one quoted string is supplied the value is a list of the requested parameter values, with named components. The call `par()` with no arguments returns a list of all the graphics parameters.

Some of the many graphics parameters are given in Tables 3.3 and 3.4 (on page 85). They have short names, usually of three characters. Those in Table 3.4 can also be supplied as arguments to high-level plot functions, when they apply just to the figure produced by that call. (The layout parameters are ignored by the high-level plot functions.)

The figure region and layout parameters

When a device is opened it makes available a rectangular surface on which one or more plots may appear. Each plot occupies a rectangular section of the device surface called a *figure*. A figure consists of a rectangular *plot region* surrounded by a *margin* on each side. The margins or sides are numbered 1 to 4, clockwise starting from the bottom. The plot region and margins together make up the *figure*

Table 3.3: Some graphics layout parameters with example settings.

`din, fin, pin`	Absolute device size, figure size and plot region size in inches. `fin=c(6,4)`
`fig`	Define the figure region as a fraction of the device surface. `fig=c(0,0.5,0,1)`
`font`	Small positive integer determining a text font for characters and hence an interline spacing. `font=3`
`mai, mar`	The four margin sizes, in inches (`mai`), or in text line units (`mar`, that is, *relative* to the current font size). Note that `mar` need not be an integer. `mar=c(3,3,1,1)+0.1`
`mex`	Number of text lines per interline spacing. `mex=0.7`
`mfg`	Defines a position within a specified multi-figure display. `mfg=c(2,2,3,2)`
`mfrow, mfcol`	Defines a multi-figure display. `mfrow=c(2,2)`
`new`	Logical value indicating whether the current figure has been used or not. `new=T`
`oma, omi, omd`	Defines outer margins in text lines or inches, or by defining the size of the array of figures as a fraction of the device surface. `oma=c(0,0,4,0)`
`plt`	Define the plot region as a fraction of the figure region. `plt=c(0.1,0.9,0.1,0.9)`
`pty`	Plot type, or shape of plotting region, `"s"` or `"m"`
`uin`	Returns inches per user coordinate for x and y.
`usr`	Limits for the plot region in user coordinates. `usr=c(0.5, 1.5, 0.75, 10.25)`

---

region, as in Figure 3.4. The device surface, figure region and plot region have their vertical sides parallel and hence their horizontal sides also parallel.

The size and position of figure and plot regions on a device surface are controlled by *layout parameters*, most of which are listed in Table 3.3. Lengths may be set in either absolute or relative units. Absolute lengths are in *inches*, whereas relative lengths are in *text lines* (so relative to the current font size).

Margin sizes are set using `mar` for text lines or `mai` for inches. These are 4–component vectors giving the sizes of the lower, left, upper and right margins in the appropriate units. Changing one causes a consistent change in the other; changing `mex` will change `mai` but not `mar`.

Positions may be specified in relative units using the unit square as a coordinate system for which some enclosing region, such as the device surface or the figure region, is the unit square. The `fig` parameter is a vector of length 4 specifying the current figure as a fraction of the device surface. The first two components give the lower and upper x–limits and the second two give the y–limits. Thus to put a point plot in the left-hand side of the display and a Q-Q plot on the right-hand side we could use:

```
postscript(file="twoplot.ps")    # open a postscript device
par(fig=c(0, 2/3, 0, 1))         # set a figure on the left
plot(x,y)                        # point plot
par(fig=c(2/3, 1, 0, 1))         # set a figure on the right
qqnorm(resid(obj))               # diagnostic plot
dev.off()
```

The left hand figure occupies $2/3$ of the device surface and the right hand figure $1/3$. For regular arrays of figures it is simpler to use mfrow or split.screen.

Positions in the plot region may also be specified in absolute *user coordinates*. Initially user coordinates and relative coordinates coincide, but any high-level plotting function changes the user coordinates so that the x– and y–coordinates range from their minimum to maximum values as given by the plot axes. The graphics parameter usr is a vector of length 4 giving the lower and upper x– and y–limits for the user coordinate system. Initially its setting is usr=c(0,1,0,1). Consider another simple example:

```
> motif()                # open a device
> par("usr")             # usr coordinates
[1] 0 1 0 1
> x <- 1:20
> y <- x + rnorm(x)      # generate some data
> plot(x, y)             # produce a scatterplot
> par("usr")             # user coordinates now match the plot
[1]   0.2400 20.7600   1.2146 21.9235
```

Any attempt to plot outside the user coordinate limits causes a warning message unless the general graphics parameter xpd is set to T.

Figure 3.5 shows some of the layout parameters for a multi-figure layout. Such an array of figures may occupy the entire device surface, or it may have *outer margins*, which are useful for annotations that refer to the entire array. Outer margins are set with the parameter oma (in text lines) or omi (in inches). Alternatively omd may be used to set the region containing the array of figures in a similar way to which fig is used to set one figure. This implicitly determines the outer margins as the complementary region. In contrast to what happens with the margin parameters mar and mai, a change to mex will leave the outer margin size, omi, constant but adjust the number of text lines, oma.

Text may be put in the outer margins by using mtext with parameter outer=T.

Common axes for figures

There are at least two ways to ensure that several plots share a common axis or axes.

1. Use the same xlim or ylim (or both) setting on each plot and ensure that the parameters governing the way axes are formed, such as lab, las, xaxs and allies, do not change.

Table 3.4: Some of the more commonly used general and high-level graphics parameters with example settings.

Text:

adj	Text justification. 0=left justify, 1=right justify, 0.5=centre.
cex	Character expansion. cex=2
csi	Height of font (inches). csi=0.11
font	Font number: device-dependent.
srt	String rotation in degrees. srt=90
cin cxy	Character width and height in inches and usr coordinates (for information, not settable).

Symbols:

col	Colour for symbol, line or region. col=2
lty	Line type: solid, dashed, dotted, etc. lty=2
lwd	Line width, usually as a multiple of default width. lwd=2
mkh	Mark height (inches). mkh=0.05
pch	Plotting character or mark. pch="*" or pch=4 for marks. (See page 75.)

Axes:

bty	Box type, as "o", "l", "7", "c", "n".
exp	Notation for exponential labels. exp=1
lab	Tick marks and labels. lab=c(3,7,4)
las	Label orientation. 0=parallel to axis, 1=horizontal, 2=vertical.
log	Controls log axis scales. log="y"
mgp	Axis location. mgp=c(3,1,0)
tck	Tick mark length as signed fraction of the plot region dimension. tck=-0.01
xaxp yaxp	Tick mark limits and frequency. xaxp=c(2, 10, 4)
xaxs yaxs	Style of axis limits. xaxs="i"
xaxt yaxt	Axis type. "n" (null), "s" (standard), "t" (time) or "l" (log)

High Level:

ask	Prompt before going on to next plot? ask=F
axes	Print axes? axes=F
main	Main title. main="Figure 1"
sub	Subtitle. sub="23-Jun-1994"
type	Type of plot. type="n"
xlab ylab	Axis labels. ylab="Speed in km/sec"
xlim ylim	Axis limits. xlim=c(0,25)
xpd	May points or lines go outside the plot region? xpd=T

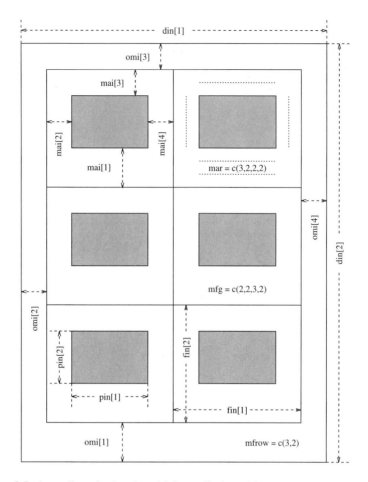

Figure 3.5: An outline of a 3×2 multi-figure display with outer margins showing some graphics parameters. The current figure is at position $(2, 2)$ and the display is being filled by rows. In this figure "fin[1]" is used as a shorthand for par("fin")[1], and so on.

2. Set up the desired axis system with the first plot and then set the low-level parameter xaxs="d", yaxs="d" or both as appropriate. This ensures that the axis or axes are not changed by further high-level plot commands on the same device.

An example: A Q-Q normal plot with envelope

In Chapter 5 we recommend assessing distributional form by quantile-quantile plots. A simple way to do this is to plot the sorted values against quantile approximations to the expected normal order statistics and draw a line through the 25% and 75% percentiles to guide the eye, performed for the variable Infant.Mortality of the Swiss provinces data (on fertility and socio-economic factors on Swiss provinces in about 1888) by

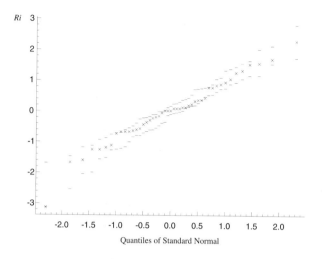

Figure 3.6: The Swiss fertility data. A Q-Q normal plot with envelope for infant mortality.

```
swiss.df <- data.frame(Fertility=swiss.fertility, swiss.x)
attach(swiss.df)
qqnorm(Infant.Mortality)
qqline(Infant.Mortality)
```

The reader should check the result and compare it with the style of Figure 3.6.

Another suggestion to assess departures is to compare the sample Q-Q plot with the envelope obtained from a number of other Q-Q plots from generated normal samples. This is discussed in Atkinson (1985, §4.2) and is based on an idea of Ripley (see Ripley, 1981, Chapter 8). The idea is simple. We generate a number of other samples of the same size from a normal distribution and scale all samples to mean 0 and variance 1 to remove dependence on location and scale parameters. Each sample is then sorted. For each order statistic the maximum and minimum values for the generated samples form the upper and lower envelopes. The envelopes are plotted on the Q-Q plot of the scaled original sample and form a guide to what constitutes serious deviations from the expected behaviour under normality. Following Atkinson our calculation uses 19 generated normal samples.

We begin by calculating the envelope and the x–points for the Q-Q plot.

```
samp <- cbind(Infant.Mortality, matrix(rnorm(47*19), 47, 19))
samp <- apply(scale(samp), 2, sort)
rs <- samp[,1]
xs <- qqnorm(rs, plot=F)$x
env <- t(apply(samp[,-1], 1, range))
```

As an exercise in building a plot with specific requirements we will now present the envelope and Q-Q plot in a style very similar to Atkinson's. To ensure that the Q-Q plot has a y–axis large enough to take the envelope we could calculate the y–limits as before, or alternatively use a matrix plot with type=n for the envelope at this stage. The axes are also suppressed for the present:

```
matplot(xs, cbind(rs,env), type="pnn", pch=4, mkh=0.06, axes=F)
```

The argument setting `type="pnn"` specifies that the first column (`rs`) is to produce a point plot and the remaining two (`env`) no plot at all, but the axes will allow for them. Setting `pch=4` specifies a 'cross' style plotting symbol (see Figure 3.2) similar to Atkinson's, and `mkh=0.06` establishes a suitable size for the plotting symbol.

Atkinson uses small horizontal bars to represent the envelope. We can now calculate a half length for these bars so that they do not overlap and do not extend beyond the plot region. Then we can add the envelope bars using `segments`:

```
xyul <- par("usr")
smidge <- min(diff(c(xyul[1], xs, xyul[2])))/2
segments(xs-smidge, env[,1], xs+smidge, env[,1])
segments(xs-smidge, env[,2], xs+smidge, env[,2])
```

Atkinson's axis style differs from the default S style in several ways. There are many more tick intervals; the ticks are inside the plot region rather than outside; there are more labelled ticks, and the labelled ticks are longer than the unlabelled. From experience ticks along the x–axis at 0.1 intervals with labelled ticks at 0.5 intervals seems about right but this is usually too close on the y–axis. The axes require four calls to the `axis` function:

```
xul <- trunc(10*xyul[1:2])/10
axis(1, at=seq(xul[1], xul[2], by=0.1), labels=F, tck=0.01)
xi <- trunc(xyul[1:2])
axis(1, at=seq(xi[1], xi[2], by=0.5), tck=0.02)
yul <- trunc(5*xyul[3:4])/5
axis(2, at=seq(yul[1], yul[2], by=0.2), labels=F, tck=0.01)
yi <- trunc(xyul[3:4])
axis(2, at=yi[1]:yi[2], tck=0.02)
```

Finally we add the L–box, put the x–axis title at the centre and the y–axis title at the top:

```
box(bty="l")            # lower case "L"
ps.options()$fonts
mtext("Quantiles of Standard Normal", side=1, line=2.5, font=3)
mtext("Ri", side=2, line=2, at=yul[2], font=10)
```

where fonts 3 and 10 are Times-Roman and Times-Italic on the device used (`postscript` under Unix), found from the list given by `ps.options`.

The final plot is shown in Figure 3.6.

3.5   Trellis graphics

Trellis graphics were developed at the former AT&T's Bell Laboratories to provide a consistent graphical 'style' and to extend conditioning plots. The 'style' is a development of that used in Cleveland (1993). There are two major versions of Trellis in use, with several minor versions. Version 2.1, which we describe here, is part of S-PLUS 3.4 (all Unix platforms). Version 2.0 is part of the later versions (Windows, SGI, DEC Alpha) of S-PLUS 3.3, and a slightly revised version 2.0 is part of the S+SPATIALSTATS module for the platforms of the earlier versions (Sun, HP, IBM) of S-PLUS 3.3.[3] Version 1 was available as a separate product, and is incorporated in the earlier versions of S-PLUS 3.3 (Sun, HP, IBM). One way to tell which version of Trellis you have is that `contourplot` only exists in version 2.x and `contour.formula` only exists in version 1. We describe Trellis version 2.x here. Many of the functions we use do not exist in version 1.x, and where they do the syntax may be very different. The changes from 2.0 to 2.1 are mainly bug fixes. S-PLUS 4.0 has a later version of Trellis 2.

Trellis is very prescriptive, and changing the display style is not always an easy matter.

It may be helpful to understand that Trellis is written entirely in the S language, as calls to the basic plotting routines. Two consequences are that it can be slow and memory-intensive, and that it takes over many of the graphics parameters for its own purposes. (Global settings of graphics parameters are usually not used, the outer margin parameters `omi` being a notable exception.) Computation of a Trellis plot is done in two passes, once when a Trellis object is produced, and once when that object is printed (producing the actual plot).

The `trellis` library contains a large number of examples: use

```
?trellis.examples
```

to obtain an up-to-date list. These are all functions which can be called to plot the example, and listed to see how the effect was achieved.

Trellis graphical devices

The `trellis.device` graphical device is provided by the `trellis` library. It is perhaps more accurate to call it a meta-device, for it uses one of the underlying graphical devices (currently `motif`, `openlook`, `iris4d`, `postscript`, `graphsheet`, `win.graph` and `win.printer` where these are available), but customizes the parameters of the device to use the Trellis style, and in particular its colour schemes.

Trellis devices by default use colour for screen windows and greylevels for printer devices. The settings for a particular device can be seen by running the command `show.settings()` (see Figure 3.7). These settings are not the same for all colour screens, nor for all printer devices. Trellis colour schemes have a mid-grey background on colour screens (but not colour printers). It is important to select `color=F` for black and white screens, since most of the default plotting colours *and* the background will be mapped to white if the Trellis colour scheme is used.

---

[3] for which it needs to be invoked as `library(ntrellis)`

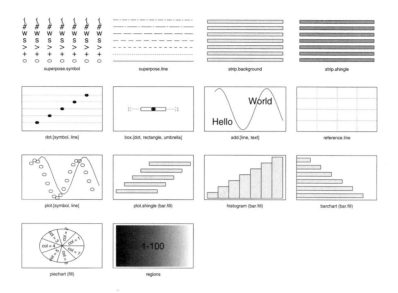

Figure 3.7: The settings of the greylevel PostScript `trellis.device`.

If a Trellis plot is used without a graphics device already in use, a Trellis device is started. Beware that this may not be what you want, as a colour device is started on a black and white screen, so that most plotted points and lines are then invisible!

Hardcopy from Trellis plots

The basic paradigm of a Trellis plot is to produce an object that the device 'prints', that is plots. Thus the simplest way to produce a hardcopy of a Trellis plot is to switch to a printer device, 'print' the object again, and switch back. For example (under Unix)

```
trellis.device()
p1 <- histogram(faithful$eruptions)
p1      # plots it on screen
trellis.device("postscript", file="hist.eps",
    onefile=F, print.it=F)
p1      # print the plot
dev.off()
```

However, it can be difficult to obtain precisely the same layout in this way (since this depends on the aspect ratio and size parameters), and it is impossible to interact with such a graph (for example by using `identify`). Fortunately, the methods for hardcopy described on page 72 can still be used. It is important to set the options for the `postscript` device to match the colour schemes in use. For example, on Unix with hardcopy via the `postscript` device we can use

```
ps.options(colors=colorps.trellis[, -1])
```

before the Trellis device is started. Then the rmv method and dev.print will use the Trellis colour scheme and produce colour PostScript output. Conversely, if greylevel PostScript output is required (for example, for figures in a book or article) we can use (for a motif device; xcm.trellis.openlook.bw for openlook)

```
ps.options(colors=bwps.trellis, pointsize=11)
trellis.settings.motif.bw <- trellis.settings.bwps
xcm.trellis.motif.bw <- xcm.trellis.motif.grey
trellis.device(color=F)
```

using xcm objects in our library. This sets up a color screen device to mimic the 'black and white' (actually greylevel) PostScript device.

Trellis model formulae

Trellis graphics functions make use of the language for model formulae described in Section 2.5. The Trellis code for handling model formulae to produce a data matrix from a data frame (specified by the data argument) allows the argument subset to select a subset of the rows of the data frame, as one of the first three forms of indexing vector described on page 47. (Character vector indices are not allowed.)

There are a number of inconsistencies in the use of the formula language. There is no na.action argument, and missing values are handled inconsistently; generally rows with NAs are omitted, but splom fails if there are missing values. Surprisingly, splom uses a formula, but does not accept a data argument.

Trellis uses an extension to the model formula language, the operator | which can be read as 'given'. Thus if a is a factor, lhs ~ rhs | a will produce a plot for each level of a of the subset of the data for which a has that level (so estimating the conditional distribution given a). Conditioning on two or more factors gives a plot for each combination of the factors, and is specified by an interaction, for example | a*b. For the extension of conditioning to continuous variates via what are known as shingles, see page 103.

Trellis plot objects can be kept, and update can be used to change them, for example to add a title or change the axis labels.

Basic Trellis plots

As Table 3.5 and Figure 3.7 show, the basic styles of plot exist in Trellis, but with different names and different default styles. Their usage is best seen by considering how to produce some figures in the Trellis style.

Figure 1.4 (page 15) was produced by splom(~ hills). Trellis plots of scatterplot matrices read from bottom to top (as do all multi-panel Trellis displays, like graphs rather than matrices, despite the meaning of the name splom). By default the panels in a splom plot are square. The axes are omitted in Trellis 2.0.

Figure 3.8 is a Trellis version of Figure 1.6. Note that the y axis numbering is horizontal by default (equivalent to the option par(las=1)), and that points are

Table 3.5: Trellis plotting functions. Page references are given to the most complete description in the text: see also Appendix B.

	page	
xyplot	96	Scatterplots.
bwplot	93	Boxplots.
stripplot	99	Display univariate data against a numerical variable.
dotplot	100	ditto in another style
histogram	169	'Histogram', acrually frequency plot.
densityplot	179	Kernel density estimates.
barchart	93	Horizontal bar plots.
piechart		Pie chart.
splom	91	Scatterplot matrices.
contourplot	96	Contour plot of a surface on a regular grid.
levelplot	96	Pseudo-colour plot of a surface on a regular grid.
wireframe	96	Perspective plot of a surface evaluated on a regular grid.
cloud	106	A perspective plot of a cloud of points.
key	107	Add a legend.
color.key	97	Add a color key (as used by levelplot).
trellis.par.get	94	Save Trellis parameters.
trellis.par.set	94	Reset Trellis parameters.
equal.count	103	Compute a shingle.

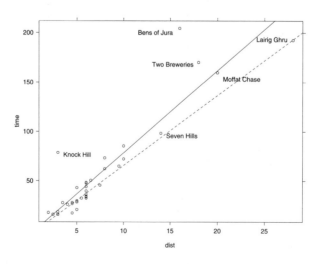

Figure 3.8: A Trellis version of Figure 1.6 (page 16).

plotted by open circles rather than filled circles or stars. It is not possible to add
to a Trellis plot (as the user coordinate system is not retained), so the Trellis call
has to include all the desired elements. This is done by writing a *panel function,*
in this case

```
xyplot(time ~ dist, data = hills,
    panel = function(x, y, ...) {
        panel.xyplot(x, y, ...)
        panel.lmline(x, y, type="l")
        panel.abline(ltsreg(x, y), lty=3)
        identify(x, y, row.names(hills))
    }
)
```

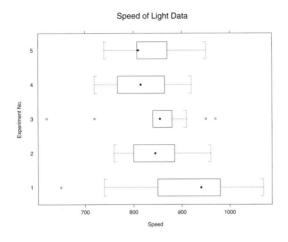

Figure 3.9: A Trellis version of Figure 1.7 (page 17).

Figure 3.9 is a Trellis version of Figure 1.7. Boxplots are known as box-and-
whisker plots, and are displayed horizontally. This figure was produced by

```
bwplot(Expt ~ Speed, data=michelson, ylab="Experiment No.")
title("Speed of Light Data")
```

Note the counter-intuitive way the formula is used. This plot corresponds to a
one-way layout splitting Speed by experiment, so it is tempting to use Speed as
the response[4]. It may help to remember that the formula is of the y ~ x form for
the x and y axes of the plot. (The same ordering is used for all the univariate
plot functions.)

Figure 3.10 is a Trellis version of Figure 3.1. This is the first example we have
seen of a plot with more than one panel.

---

[4] This was the meaning in Trellis 1.x in boxplot.formula and stripplot!

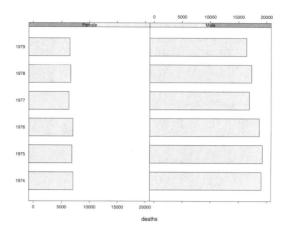

Figure 3.10: A Trellis version of Figure 3.1 (page 74).

```
lung.deaths.df <- data.frame(year = rep(1974:1979, 2),
    deaths = c(lung.deaths[, 1], lung.deaths[ ,2]),
    sex = rep(c("Male", "Female"), rep(6,2)))
barchart(year ~ deaths | sex, lung.deaths.df, xlim=c(0, 20000))
```

Figure 3.11 is an enhanced scatterplot matrix, again using a panel function to add to the basic display. Now we see the power of panel functions, as the basic plot commands can easily be applied to multi-panel displays. The `aspect="fill"` command allows the array of plots to fill the space: by default the panels are square as in Figure 1.4.

```
splom(~ swiss.df, aspect="fill",
    panel = function(x, y, ...) {
        panel.xyplot(x, y, ...); panel.loess(x, y, ...)
    }
)
```

Most Trellis graphics functions have a `groups` parameter, which we can illustrate on the `stormer` data used in Section 9.4 (see Figure 3.12).

```
sps <- trellis.par.get("superpose.symbol")
sps$pch <- 1:7
trellis.par.set("superpose.symbol", sps)
xyplot(Time ~ Viscosity, stormer, groups = Wt,
    panel = panel.superpose, type = "b",
    key = list(columns = 3,
        text = list(paste(c("Weight:     ", "", ""),
                          unique(stormer$Wt), "gms")),
        points = Rows(sps, 1:3)
        )
)
```

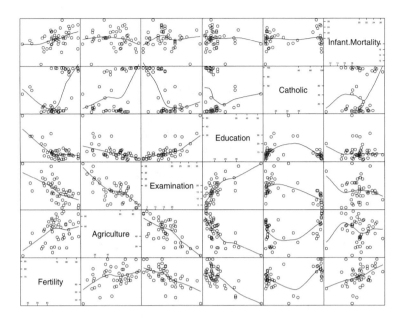

Figure 3.11: A Trellis scatterplot matrix display of the Swiss provinces data.

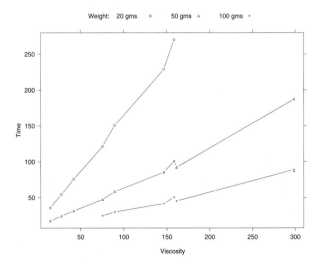

Figure 3.12: A Trellis plot of the `stormer` data.

Here we have changed the default plotting symbols (which differ by device) to the first seven `pch` characters shown in Figure 3.2 on page 75. (We could just use the argument `pch=1:7` to `xyplot`, but then specifying the key becomes much more complicated.)

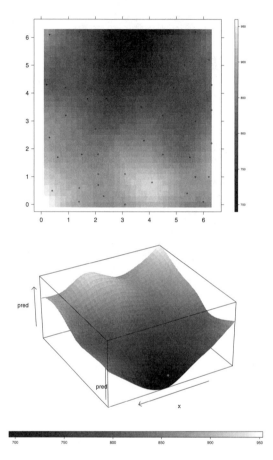

Figure 3.13: Trellis `levelplot` and `wireframe` plots of a `loess` smoothing of the `topo` dataset.

Figure 3.13 shows further Trellis plots of the smooth surface shown in Figure 3.3. Once again panel functions are needed to add the points. The grid has been reduced in extent, since whereas `contourplot` copes with the `NA`s produced by `predict.loess` for Figure 3.3, prior to S-PLUS 3.4 `levelplot` produced incorrect results (without any warning). The `aspect=1` parameter ensures a square plot. The `drape=T` parameter to `wireframe` is optional, producing the superimposed greylevel (or pseudo-colour) plot.

```
topo.mar <- list(x = seq(0.2, 6.3, 0.2), y=seq(0, 6.2, 0.2))
topo.plt <- expand.grid(topo.mar)
topo.plt$pred <- as.vector(predict(topo.loess, topo.plt))
```

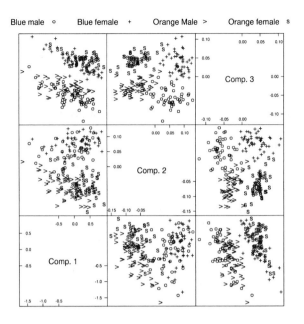

Figure 3.14: A scatterplot matrix of the first three principal components of the crabs data.

```
levelplot(pred ~ x * y, topo.plt, aspect=1,
    at = seq(690, 960, 10), xlab="", ylab="",
    panel = function(x, y, subscripts, ...) {
        panel.levelplot(x, y, subscripts, ...)
        panel.xyplot(topo$x,topo$y, cex=0.5, col=1)
    }
)
wireframe(pred ~ x * y, topo.plt, aspect=c(1, 0.5), drape=T,
    screen = list(z = -150, x = -60),
    colorkey=list(space="bottom", y=0.1))
```

(We would have liked to change the size of the `colorkey`, but this gave a faulty plot. The arguments given by `colorkey` refer to the `color.key` function.) We can see no way to add the points to the perspective display.

Trellises of plots

In multivariate analysis we necessarily look at several variables at once, and we explore here several ways to do so. We can produce a scatterplot matrix of the first three principal components of the `crabs` data (Section 13.4) by

```
lcrabs.pc <- predict(princomp(log(crabs[,4:8])))
crabs.grp <- c("B", "b", "O", "o")[rep(1:4, rep(50,4))]
splom(~ lcrabs.pc[, 1:3], groups = crabs.grp,
    panel = panel.superpose,
```

```
key = list(text = list(c("Blue male", "Blue female",
                         "Orange Male", "Orange female")),
           points = Rows(trellis.par.get("superpose.symbol"), 1:4),
           columns = 4)
)
```

A 'black and white' version of this plot is shown in Figure 3.14. On a 'colour' device the groups are distinguished by colour and are all plotted with the same symbol (o).

However, it might be clearer to display these results as a trellis of splom plots, by

```
sex <- crabs$sex; levels(sex) <- c("Female", "Male")
sp <- crabs$sp; levels(sp) <- c("Blue", "Orange")
splom(~ lcrabs.pc[, 1:3] | sp*sex, cex=0.5, pscales=0)
```

as shown in Figure 3.15[5]. Notice how this is the easiest method to code. It is at the core of the paradigm of Trellis, which is to display many plots of subsets of the data in some meaningful layout.

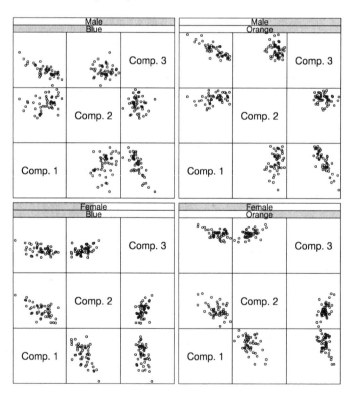

Figure 3.15: A multi-panel version of Figure 3.14.

---

[5] Users of S-PLUS 3.3 should omit `pscales=0`.

Now consider data from a multi-factor study, Quine's data on school absences discussed in Sections 6.6 and 7.4. It will help to set up more informative factor labels, as the factor names are not given (by default) in trellises of plots.

```
Quine <- quine
levels(Quine$Eth) <- c("Aboriginal", "Non-aboriginal")
levels(Quine$Sex) <- c("Female", "Male")
levels(Quine$Age) <- c("primary", "first form",
                        "second form", "third form")
levels(Quine$Lrn) <- c("Average learner", "Slow learner")
bwplot(Age ~ Days | Sex*Lrn*Eth, data=Quine)
```

This gives an array of 8 boxplots, which by default takes up two pages. On a screen device there will be no pause between the pages unless the argument ask=T is set for par. It is more convenient to see all the panels on one page, which we can do by asking for a different layout (Figure 3.16).

```
bwplot(Age ~ Days | Sex*Lrn*Eth, data=Quine, layout=c(4,2))
```

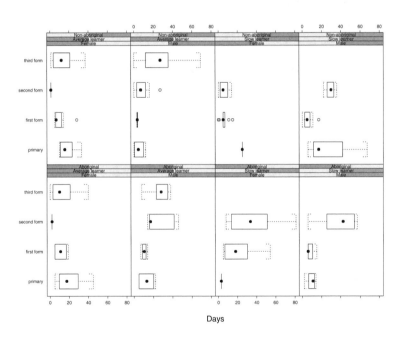

Figure 3.16: A multi-panel boxplot of Quine's school attendance data.

A stripplot allows us to look at the actual data. We jitter the points slightly to avoid over-plotting.

```
stripplot(Age ~ Days | Sex*Lrn*Eth, data=Quine,
    jitter = T, layout = c(4,2))

stripplot(Age ~ Days | Eth*Sex, data=Quine,
```

```
        groups = Lrn, jitter=T,
        panel = function(x, y, subscripts, jitter.data=F, ...) {
            if(jitter.data)  y <- jitter(y)
            panel.superpose(x, y, subscripts, ...)
        },
        xlab = "Days of absence",
        between = list(y=1), par.strip.text = list(cex=1.2),
        key = list(columns = 2, text = list(levels(Quine$Lrn)),
            points = Rows(trellis.par.get("superpose.symbol"), 1:2)
            ),
        strip = function(...)
                   strip.default(..., strip.names=c(T, T), style=1)
   )
```

The second form of plot, shown in Figure 3.17, uses different symbols to distinguish one of the factors.

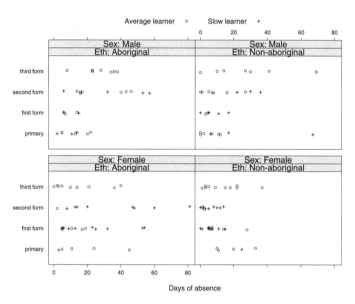

Figure 3.17: A `stripplot` of Quine's school attendance data.

It is possible to include the factor name in the strip labels, by using a custom `strip` function as in Figure 3.17. The parameter `par.strip.text` controls the size, font and colour of the strip labels. The alternative `styles` can be useful with panels labelled by factors; there are currently six.

The Trellis function `dotplot` is very similar to `stripplot`; its panel function includes horizontal lines at each level. `stripplot` uses the styles of `xyplot` whereas `dotplot` has its own set of defaults; for example the default plotting symbol is a filled rather than open circle.

As a third example, consider our dataset `fgl`. This has 10 measurements on 214 fragments of glass from forensic testing, the measurements being of the

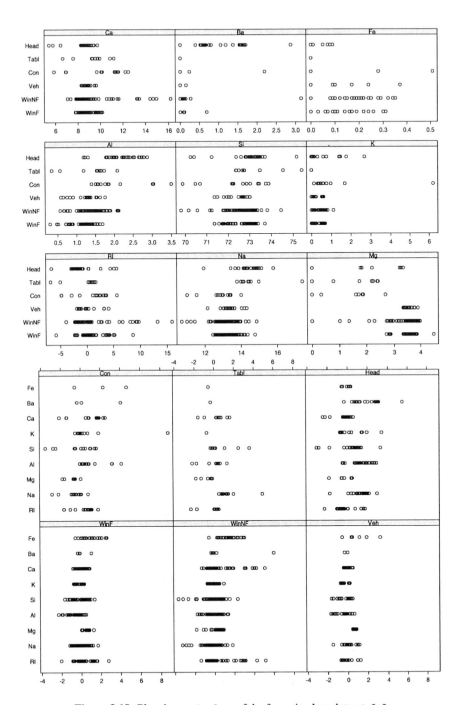

Figure 3.18: Plots by stripplot of the forensic glass dataset fgl.

refractive index and composition (percent weight of oxides of Na, Mg, Al, Si, K, Ca, Ba and Fe). The fragments have been classified by six sources. We can look at the types for each measurement by

```
fgl0 <- fgl[ ,-10] # omit type.
fgl.df <- data.frame(type = rep(fgl$type, 9),
    y = as.vector(as.matrix(fgl0)),
    meas = factor(rep(1:9, rep(214,9)), labels=names(fgl0)))
stripplot(type ~ y | meas, data=fgl.df, scales=list(x="free"),
    strip=function(...) strip.default(style=1, ...), xlab="")
```

and by measurement for each type by

```
fgl.df2 <- data.frame(type = rep(fgl$type, 9),
    y = as.vector(scale(as.matrix(fgl0), T, T)),
    meas = factor(rep(1:9, rep(214,9)), labels = names(fgl0)))
stripplot(meas ~ y | type, data=fgl.df2,
    strip=function(...) strip.default(style=1, ...), xlab="")
```

The first version uses separate scales for each measurement; the second rescales each to zero mean and unit variance.

Layout of a trellis

A trellis of plots is generated as a sequence of plots which are then arranged in rows, columns and pages. The sequence is determined by the order in which the conditioning factors are given, the first varying fastest. The order of the levels of the factor is that of its `levels` attribute.

How the sequence of plots is displayed on the page(s) is controlled by an algorithm which tries to optimize the use of the space available, but it can be controlled by the `layout` parameters. A specification layout = $c(c, r, p)$ asks for c columns, r rows and p pages. (Note the unusual ordering.) Using $c = 0$ allows the algorithm to choose the number of columns; p is used to produce only the first p pages of a many-page trellis.

If the number of levels of a factor is large and not easily divisible (for example 7), we may find a better layout by leaving some of the cells of the trellis empty. For example we might use (extending an example on the next page)

```
Cath <- equal.count(swiss.df$Catholic, number=2, overlap=0)
Agr5 <- equal.count(swiss.df$Agric, number=5, overlap=0.25)
xyplot(Fertility ~ Education | Agr5 * Cath, data=swiss.df,
        layout=c(2,3), skip = c(F,F,F,F,F,T))
```

to use 5 panels on each page of a 3×2 layout.

The `between` parameter can be used to specify gaps in the trellis layout, as in Figure 3.17. It is a list with x and y components, numeric vectors which specify the gaps in units of character height. The `page` parameter can be used to invoke a function (with argument n, the page number) to label each page. The default page function does nothing.

It sometimes helps to re-order the levels of a factor to emphasis trends in the data. The trellis function reorder.factor provides one way to do this. For example, we could use

```
reorder.factor(Quine$Age, Quine$Days, median)
```

to reorder the forms by the median number of days absence. (This returns an ordered factor.)

Conditioning plots and shingles

The idea of a trellis of plots conditioning on combinations of one or more factors can be extended to conditioning on real-valued variables, in what are known as conditioning plots or *coplots*. Two variables are plotted against each other in a series of plots with the values of further variable(s) restricted to a series of possibly overlapping ranges. This needs an extension of the concept of a factor known as a *shingle*[6].

Suppose we wished to examine the relationship between Fertility and Education in the Swiss fertility data as the variable Catholic ranges from predominantly non-Catholic to mainly Catholic provinces. We add a smooth fit to each panel.

```
Cath <- equal.count(swiss.df$Catholic, number=6, overlap=0.25)
xyplot(Fertility ~ Education | Cath, data=swiss.df,
   panel = function(x, y) {
      panel.xyplot(x, y); panel.loess(x, y)
   }
)
```

The result is shown in Figure 3.19, with the strips continuing to show the (now overlapping) coverage for each panel. Fertility generally falls as education rises and rises as the proportion of Catholics in the population rises. Note that the level of education is lower in predominantly Catholic provinces.

The function equal.count is used to construct a shingle with suitable ranges for the conditioning intervals.

Conditioning plots may also have more than one conditioning variables. Let us condition on Catholic and agriculture simultaneously. Since the dataset is small it seems prudent to limit the number of panels to 6 in all.

```
Cath <- equal.count(swiss.df$Catholic, number=2, overlap=0)
Agr <- equal.count(swiss.df$Agric, number=3, overlap=0.25)
xyplot(Fertility ~ Education | Cath * Agr, data=swiss.df,
   panel = function(x, y) {
      panel.xyplot(x, y); panel.loess(x, y)
   }
)
```

---

[6] which in American usage is a rectangular wooden tile laid partially overlapping on roofs or walls.

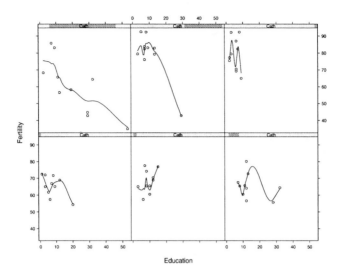

Figure 3.19: A conditioning plot for the Swiss provinces data.

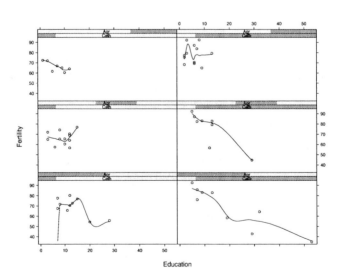

Figure 3.20: Another conditioning plot with two conditioning shingles.

The result is shown in Figure 3.20. In general the fertility rises with the proportion of Catholics and agriculture and falls with education. There is no convincing evidence of substantial interaction.

Shingles have levels, and can be printed and plotted:

```
> Cath <- equal.count(swiss.df$Cath, number=6, overlap=0.25)
> Cath
Data:
 [1] 10.0  84.8  93.4  33.8   5.2  90.6  92.9  97.2  97.7  91.4
   ....
Intervals:
  min   max count
  2.2   4.5    10
  4.2   7.7    10
   ....
Overlap between adjacent intervals:
[1] 3 2 3 2 3
> levels(Cath)
  min   max
  2.2   4.5
   ....
> plot(Cath, aspect = 0.3)
```

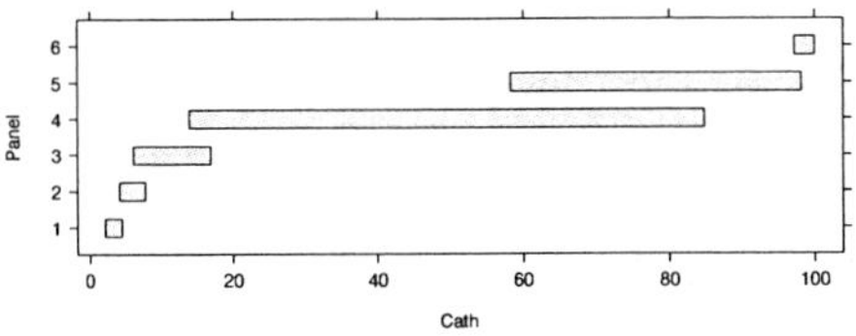

Figure 3.21: A plot of the shingle Cath.

Multiple displays per page

Recall that a Trellis object is plotted by printing it. The method print.trellis has optional arguments position, split and more. The argument more should be set to T for all but the last part of a figure. The position of individual parts on the device surface can be set by either split or position. A split argument is of the form c(x, y, nx, ny) for four integers. The second pair give a division into a nx × ny layout, just like the mfrow and mfcol arguments to par. The first pair give the rectangle to be used within that layout, with origin at the bottom left.

A position argument is of the form c(*xmin, ymin, xmax, ymax*) giving the corners of the rectangle within which to plot the object. (This is a different order from split.screen.) The coordinate system for this rectangle is [0, 1] for both axes, but the limits can be chosen outside this range. It will often be necessary to experiment with slightly overlapping subregions to avoid white space (as shown in Figure 5.7 on page 180).

The print.trellis works by manipulating the graphics parameter omi, so the outer margin settings are preserved. However, none of the basic plot methods of subdividing the device surface will work, and if a trellis print fails omi is not reset. (Using par(omi=rep(0,4), new=F) will reset the usual defaults.)

Fine control

Detailed control of Trellis plots may be accomplished by a series of arguments described in the help page for trellis.args, with variants for the wireframe and cloud perspective plots under trellis.3d.args.

We have seen some of the uses of panel functions. Some care is needed with computations inside panel functions that use any data (or user-defined objects or functions) other than their arguments. First, the computations will occur inside a deeply nested set of function calls, so care is needed to ensure that the data is visible (see page 153ff). Second, those computations will be done at the time the result is printed (that is, plotted) and so the data need to be in the desired state at plot time.

If non-default panel functions are used, we may want these to help control the coordinate system of the plots, for example to use a fitted curve to decide the aspect ratio of the panels. This is the function of the prepanel argument, and there are prepanel functions corresponding to the densityplot, lmline, loess, qq, qqmath and qqmathline panel functions. These will ensure that the whole of the fitted curve is visible, and they may affect the choice of aspect ratio.

The parameter aspect controls the aspect ratio of the panels. A numerical value (most usefully one) sets the ratio, "fill" adjusts the aspect ratio to fill the space available and "xy" attempts to bank the fitted curves to $\pm 45°$. (See exercise 3.4.)

The scales argument determines how the x and y axes are drawn. It is a list of components of name=value form, and components x and y may themselves be lists. The default relation="same" ensures that the axes on each panel are identical. With relation="sliced" the same number of data units are used, but the origin may vary by panel, whereas with relation="free" the axes are drawn to accommodate just the data for that panel. One can also specify most of the parameters of the axis function, and also log=T to obtain a $\log_{10}$ scale[7] or even log=2 for a $\log_2$ scale.

The function splom has an argument varnames which sets the names of the variables plotted on the diagonal. In Trellis version 2.1 the argument pscales determines how the axes are plotted: set pscales=0 to omit them.

---

[7] This seems undocumented in S-PLUS 3.3, and unreliable.

Keys

The function `key` is a replacement for `legend`, and can also be used as an argument to Trellis functions. If used in this way, the Trellis routines allocate space for the key, and repeat it on each page if the trellis extends to multiple pages.

The call of key specifies the location of the key by the arguments `x`, `y` and `corner`. By default `corner=c(0,1)`, when the coordinate (x, y) specifies the upper left corner of the key. Any other coordinate of the key can be specified by setting `corner`, but the size of the key is computed from its contents. (If the argument `plot=F`, the function returns a two-element vector of the computed width and height, which can be used to allocate space.) When `key` is used as an argument to a Trellis function, the position is normally specified not by `x` and `y` but by the argument `space` which defaults to `"top"`.

Most of the remaining arguments to `key` will specify the contents of the key. The (optional) arguments `points`, `lines`, `text` and `rectangles` (for `barchart`) will each specify a column of the key in the order in which they appear. Each argument must be a *list* giving the graphics parameters to be used (and for `text`, the first argument must be the character vector to be plotted). (The function `trellis.par.get` is useful to retrieve the actual settings used for graphics parameters.)

The third group of arguments to `key` fine-tunes its appearance—should it be transparent (`transparent=T`), the presence of a border (specified by giving the border colour as argument `border`), the spacing between columns (`between.columns` in units of character width), the background colour, the font(s) used, the existence of a title and so on. Consult the on-line help for the current details. The argument `columns` specifies the number of columns in the key—we used this in Figures 3.12 and 3.14.

Perspective plots

The argument `aspect` is a vector of two values for the perspective plots, giving the ratio of the y and z sizes to the x size; its effect can be seen in Figure 3.13.

The arguments `distance`, `perspective` and `screen` control the perspective view used. If `perspective=T` (the default), the `distance` argument (default 0.2) controls the extent of the perspective, although not on a physical distance scale as 1 corresponds to viewing from infinity. The `screen` argument (default `list(z = 40, x = -60)`) is a list giving the rotations (in degrees) to be applied to the specified axis in turn. The initial coordinate system has x pointing right, z up and y into the page.

The argument `zoom` (default 1) may be used to scale the final plot, and the argument `par.box` controls how the lines forming the enclosing box are plotted.

Auxiliary functions

Trellis graphics relies heavily on data frames, and so provides some 'helper' functions to create suitable data frames.

The function make.groups takes a number of data vectors and forms a data frame with components data (the concatenated vectors) and which (a category specifying the source vector number, with attribute levels giving the names of the vectors, taken from a name=vector form if appropriate). Thus

```
deaths.bygrp <- make.groups(male = mdeaths, female = fdeaths)
bwplot(which ~ data, data = deaths.bygrp)
```

gives parallel box-and-whisker plots of the two series. As these are in fact time series, we could also use

```
deaths.ts <- as.data.frame.ts(male =. mdeaths, female = fdeaths)
bwplot(which ~ series, data = deaths.ts)
xyplot(series ~ time | which, data = deaths.ts, type = "l",
    scale = list(y="free"),
    main = list("Deaths from Lung Diseases", cex=2))
```

This creates a data frame with columns series, which, time and cycle.

The function as.data.frame.array turns multidimensional arrays into data frames, for example

```
iris.df <- as.data.frame.array(iris, col.dims = 2)
names(iris.df)[5:6] <- c("specimen", "species")
```

3.6 Object-oriented editable graphics

S-PLUS 4.0 for Windows introduced a completely separate style of graphics based on menus and toolbars with simple command-line equivalents. The style of the interface is designed for intuitive exploration by experienced Windows users; most of the options are set from dialog boxes brought up by selecting and double-clicking or right-clicking elements of the plot. There are separate dialog boxes for different plot elements: at least the background, plotted objects, axes and any annotations.

This graphical system works in a new graphical device called a *Graph Sheet*. A graphsheet device can be launched from the command window by the function graphsheet(), but it will normally be opened from a plot palette button or from the new document icon on the toolbar. Graphsheets can be used as graphics devices with both base and Trellis command-line graphics, and provide limited editing facilities, for example to edit the text of labels and the colour and width of lines, if the 'Object-oriented Graphs' button has been selected. To produce a hardcopy of a graphsheet to a printer the normal Windows printing facilities can be used: it is also possible to export the graph(s) to a file via the Export Graph... item on the File menu. Finally, graphsheets can be saved (as S-PLUS graph files with extension .sgr) and re-imported for editing or additions. Their structure can be browsed in the graphsheets view of the object browser (another part of the new interface).

2D plots

The 2D plot palette on the toolbar contains many buttons for different plot types plus six which control the placement of axes and three the production of conditioning plots. The basic types of plots are scatter and line plots (of types `"l"`, `"s"` and `"S"`), smoothed plots, and various fitted lines. There are also barplots, histograms, and pie charts.

Most of these plots work on two columns of data[8]. As they are not specified as arguments to a function, they need to be specified in some other way, and there are many possibilities. The most convenient way will often be to double-click a data frame in the object browser, select the desired columns (in the order x, y, z if needed, ...) and either drag-and-drop onto the appropriate button in the 2D plot palette, or just click on the button. The variables may be specified or changed using the either of the dialog boxes brought up by double-clicking the plotted points or the background.

All plots can be conditioned, in the Trellis style but with a little less flexibility. The default is to use four panels in a 2×2 layout. The conditioning variable(s) can be set in the Multipanel tab of the plot background dialog box or by drag-and-drop. For the latter, select the desired variable(s) in an object browser of an open view of the data frame, then click and hold over the columns (not their headers) and drag to the graphsheet window. A rectangular drop target will appear in the graph window above the plot area: 'drop' the columns there. (Dropping the variables in the plot area will replace the plotted variables.) Alternatively, additional columns can be selected and the Conditioning button selected on the toolbar before generating the plot.

Buttons on the 2D plot palette can select no conditioning or a 3×3 layout as well as the default 2×2 layout. Considerably greater control over the conditioning is available from the Multipanel tab. This allows the number and layout of panels to be changed, the partitioning of continuous variables to be set (with overlap as in shingles if desired), and strips to be plotted or not. (If they are, they can be selected and their properties edited.)

3D plots

The distinction between 2D and 3D plots is by appearance not data; contour, level and filled contour plots are on the 2D plot palette even though they require *three* columns to be selected. Like Trellis plots, the contour and surface plots require a z coordinate evaluated at a rectangular grid of x and y coordinates, but unlike Trellis they will interpolate irregular data to an automatically chosen grid (controlled from the Gridding tab of the plot dialog box).

The contour and levels plots are analogues of `contourplot` and `levelplot`; the filled contour plots are contour plots with the regions between contours filled in different colours. The details of the contours, levels and colours can be altered

---

[8] If only one column is selected this is used for y and the index is used for x.

from the dialog box brought up by double-clicking or right-clicking inside the plot region.

The analogue of the Trellis function `cloud` is a 3D scatterplot from the 3D plot palette. A cloud of points can be represented as points, connected by lines (with or without highlighting the points) or as a 'dropped line scatter' plot where each point is represented by a line segment from (x, y, z) to $(0, 0, z)$. There are also 3D bar charts (sometimes known as Manhattan diagrams) for data on a rectangular grid.

Surfaces can be represented in many ways:

(a) As a wireframe surface, plotted at the grid spacing or at every other grid point.

(b) As a wireframe interpolated to a finer grid (default half the spacing) by splines.

(c) Filled versions of (a) and (b), in which the surface is shown in a solid colour (selected from the Fills tab of the dialog box selected from the surface).

(d) A draped surface, with levels represented by 8, 16 or 32 colours. (Many representations intermediate between (c) and (d) can be selected from the Lines and Fills tabs.)

(e) Contours or filled contours, in which contour levels are plotted at the appropriate height as horizontal sections of the surface.

To change the representation of a surface, select the surface by clicking on it, then the appropriate button in the 3D plot palette.

All of these 3D plots can be rotated. First select the plot area by clicking within the plot region delimited by the axes, but not on the surface. Four circles and a triangle will appear. Any of these can be dragged to rotate the plot: the circles give horizontal rotation (about the vertical axis) and the triangle rotates the vertical axis. It may be a good idea to change to a simple view of the surface (such as a coarse wireframe grid) if rotation proves to be slow.

The software also allows multiple (2, 4 or 6) views of the plot from different (equispaced) angles, and these can be rotated simultaneously by rotating one of the panels. Multiple views are selected by buttons on the 3D plot palette. As these are a form of conditioning, a single view is selected again by clicking on the 'no conditioning' button in the palette.

Plots can also be conditioned on the x, y or z variables and shown as a series of 'exploded' views.

All the plots from the 3D palette can be conditioned on additional variables, selecting 2×2 or 2×3 layouts from a palette button, with fine-tuning from the background dialog box.

The shape of the enclosing cuboid (the `aspect` parameter in 3D Trellis) can be set from the 3D Workbox tab of the plot's dialog box. Figure 3.22 shows a 32-colour draped surface plot of the `topo.plt` object fitted by `loess`. We adjusted the aspect ratio to scale the x and y axes equally.

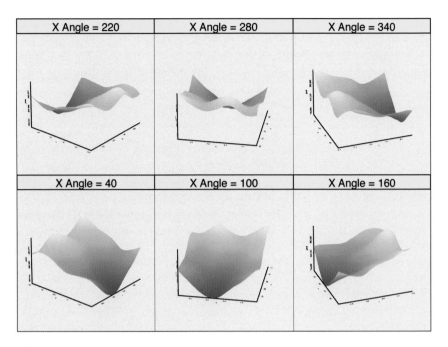

Figure 3.22: Views of the loess smoothing of the topo dataset from the Axum-based graphics of S-PLUS 4.0 for Windows.

Editing plots

Many of the properties of a graph can be altered from dialog boxes. To select a part of the graph (such as an axis or fitted line or label) (left-)click on it. Clicking on any of the data points in a 2D plot will select both the points and the fitted curve. Then either double-clicking or right-clicking will bring up or a tabbed dialog box or a shortcut menu to the tabs from which the properties of that part can be selected.

The plot region or the whole graph can be selected. The dialog box has at least four tabs, Plot Summary (including the data frame used), Position/Size (which includes aspect ratio, with a welcome option for 'proportional units'), Fill/Border (colours, patterns, . . .) and Multipanel (for conditioning). The 3D plots add the 3D Workbox tab. When the whole graph is selected it can be resized by dragging the handles, or moved by dragging a point outside the plot region, and similarly for the plot region.

Once an axis label or title is selected, clicking on the text brings up an 'in-place' edit box for replacement text. Double-clicking on the surrounding box enables properties such as font and colour to be altered: these can also be altered from the toolbar when the text is selected.

Legends and titles can be added from the Insert menu; showing a legend can also be toggled from a toolbar button.

There is an *annotation* palette which has tools to label points (as in `identify`) and to add text, a date stamp or various symbols to a graph. This palette is selected from its toolbar button or from the Toolbars item on the Views menu. Its effect is to provide a simple drawing package with which to enhance graphs.

3.7 Exercises

3.1.   The data frame `survey` contains the results of a survey of 237 first-year Statistics students at Adelaide University. For a graphical summary of all the variables, use `plot(survey)`. Note that this produces a dotchart for factor variables, and a normal scores plot for the numeric variables.

One component of this data frame, `Exer`, is a factor object containing the responses to a question asking how often the students exercised. Produce a barchart of these responses. Use `table` and `pie` or `piechart` to create a pie chart of the responses. Do you like this better than the bar plot? Which is more informative? Which gives a better picture of exercise habits of students? The `pie` function takes an argument `names` which can be used to put labels on each pie slice. Redraw the pie chart with labels. Alternatively, you could add a legend to identify the slices.

You might like to try the same things with the `Smoke` variable, which records responses to the question "How often do you smoke?" Note that `table` and `levels` ignore missing values; if you wish to include non-respondents in your chart use `summary` to generate the values, and `names` on the summary object to generate the labels.

3.2.   Make a plot of petal width *vs* petal length of the `iris` data for a partially sighted audience, identifying the three species. You will need to double the annotation size, thicken the lines and change the layout to allow larger margins for the larger annotation.

3.3.   Plot $\sin(x)$ against x, using 200 values of x between $-\pi$ and π, but do not plot any axes yet (use parameter `axes=F` in the call to `plot`.) Add a y axis passing through the origin using the 'extended' style and horizontal labels. Add an x axis with tick-marks from $-\pi$ to π in increments of $\pi/4$, twice the usual length.

3.4.   Cleveland (1993) recommends that the aspect ratio of line plots is chosen so that lines are 'banked' at $45°$. By this he means that the averaged absolute value of the slope should be around $\pm45°$. Write a function to achieve this for a time-series plot, and try it out on the `sunspots` dataset. (Average along the arc length of the curve. See the function `banking` for Cleveland's solution.)

3.5.   The Trellis function `splom` produces a complete matrix of scatterplots, as does the basic plotting functions `pairs`, but in earlier versions of S-PLUS `pairs` only plotted the lower triangle of the matrix. Write a function to emulate the earlier behaviour. (HINT: look at `pairs.default`. The graphics parameter `mfg` may be useful.)

Chapter 4

Programming in S

The S language is both an interactive language and a language for adding new functions to the S-PLUS system. It is a complete programming language with control structures, recursion and a useful variety of data types. The S-PLUS environment provides many functions to handle standard operations, but most users need occasionally to write new functions. This chapter is concerned with designing, writing, testing and correcting your own S functions.

Much of the system software is itself written in the S language. This code may be inspected, providing a wealth of examples. A user who understands S programming can often clarify points of detail from the code itself in cases where the documentation is unclear.

Writing clear, correct programs in any language always presents a challenge and S is no exception. The first four sections of this chapter are relevant to the casual writer of functions; the rest of the chapter is also important, but especially for those intending to make extensions to S, particularly if these are to be used by others.

4.1 Control structures

Control structures are the commands that make decisions or execute loops. These statements are most often used inside functions rather than interactively.

Conditional execution of statements

Conditional execution uses either the `if` statement or the `switch` function. The `if` statement has the form

```
if (condition)  true.branch   else   false.branch
```

First the expression `condition` is evaluated. If the result is `T` (or non-zero) the value of the `if` statement is that of the expression `true.branch`, otherwise that of the expression `false.branch`. The `else` part is optional and omitting it is equivalent to using "`else NULL`". If `condition` has a vector value only the first component is used and a warning is issued. The `if` function can be extended over several lines, and the statements may be compound statements enclosed in braces { }.

Two additional logical operators, && and || , are useful with if statements. Unlike & and | , which operate componentwise on vectors, these operate on scalar logical expressions. With && the right-hand expression is only evaluated if the left-hand one is true, and with || only if it is false. This conditional evaluation property can be used as a safety feature, as in

```
if (is.numeric(x) && min(x) > 0)  sx <- sqrt(x)
else  stop("x must be numeric and all components positive")
```

The expression min(x) > 0 is invalid for non-numeric x.

The functions any and all are often useful in defining scalar logicals. They evaluate respectively the logical 'or' and 'and' of all components of their argument. For example the usual test for exact symmetry of a matrix is all(X == t(X)); a less strict version would be all(abs(X-t(X)) < eps) where eps is some appropriate tolerance. Alternatively, we could use all.equal(X, t(X)), which uses a tolerance.

The if statement should be distinguished from the ifelse function, which is its vector counterpart. Its form is

```
ifelse(test, true.value, false.value)
```

where all arguments are vectors, and the recycling rule applies if any are short. All arguments are evaluated and test is coerced to logical if necessary. In those component positions where the value is T the corresponding component of true.value is the result and elsewhere it is that of false.value. Since ifelse operates on vectors, it is fast and should be used if possible.

Note that ifelse can generate NA warning messages even in cases where the result contains none. In an assignment such as

```
y.logy <- ifelse(y <= 0, 0, y*log(y))
```

all three arguments to ifelse are evaluated, so warning messages will be generated if y has any negative components even though the final result will have no NA components. Two ways of producing the desired result for *frequency* vectors y are

```
y.logy <- y * log(y + (y==0))
y.logy <- y * log(pmax(1, y))  # alternative
```

The switch function provides a more readable alternative to nested if statements. For example if you wished to allow the user a choice of tests for equality of variances the conditional statements

```
result <- if (test == "Levene")  levene(y, f)
          else
             if (test == "Cochran") cochran(y, f)
             else bartlett(y, f)
```

would allow three possibilities, Levene's, Cochran's or Bartlett's test (assuming the three functions were available). This sequence of checks can be replaced by the construction

```
result <- switch(test,
            Levene = levene(y, f),
            Cochran = cochran(y, f),
            bartlett(y, f))
```

If the first argument to `switch` evaluates to a character string the value of the expression is that of the matching named argument, or, if none does match exactly, that of a final unnamed argument, if any. If the actual argument for the matching string is missing, `switch` uses the next available argument. For example one could use

```
result <- switch(test,
    Levene =, levene =, "Levene's test" = levene(y, f),
    Cochran =, cochran =, "Cochran's test" = cochran(y, f),
    Bartlett =, bartlett =, "Bartlett's test" =,
    bartlett(y, f))
```

to allow alternative possibilities for specifying the test to be used, still retaining Bartlett's test as the 'catch all'. Note that non-standard names may be used if quoted.

Abbreviated names are not matched but we can allow them by using the function pmatch (see page 44).

```
result <- switch(
    pmatch(test, c("Levene","levene", "Cochran", "cochran"),
            nomatch = ""),
    "1" =, "2" = levene(y, f),
    "3" =, "4" = cochran(y, f),
    bartlett(y, f))
```

Note that the result is coerced to character mode by the `nomatch` argument.

The first argument to `switch` may also evaluate to a number, which is coerced to an integer. Argument names are then ignored and the appropriate numbered argument among those remaining, if there is one, is selected. There is no default final argument; if the number is outside the range 1 to nargs()-1 the result is NULL. If a selected argument position is present but the argument itself is vacant there is no 'drop through' convention. In this case `switch` returns no result, which can result in a puzzling error.

Loops: the `for`, `while` and `repeat` statements

A `for` loop allows a statement to be iterated as a variable assumes values in a specified sequence. The statement has the form

```
for(variable in sequence) statement
```

where in is a keyword, variable is the loop variable and sequence is the vector of values it assumes as the loop proceeds. This is often of the form 1:10 or seq(along=x) but it may be a list, in which case variable assumes the value of each component in turn. The statement part will often be a grouped statement and hence enclosed within braces, { }.

The while and repeat loops do not make use of a loop variable. Their forms are

        while (condition) statement

and

        repeat statement

In both cases the commands in the body of the loop are repeated. For a while loop the normal exit occurs when condition becomes F; the repeat statement continues indefinitely unless exited by a break statement.

The next statement within the body of a for, while or repeat loop causes a jump to the beginning of the next iteration. The break statement causes an immediate exit from the loop.

There are also For loops, which are similar to for loops but 'unroll' the loop and use a separate S-PLUS process to perform the calculations. We defer the details to page 159.

A single-parameter maximum-likelihood example

For a simple example with a statistical context we will estimate the parameter λ of the zero-truncated Poisson distribution by maximum likelihood. We will use artificial data but the same problem sometimes occurs in practice.

The probability distribution is specified by

$$\Pr(Y = y) = \frac{e^{-\lambda} \lambda^y}{(1 - e^{-\lambda}) \, y!} \qquad y = 1, 2, \dots$$

and corresponds to observing only non-zero values of a Poisson count. The mean is

$$E(Y) = \frac{\lambda}{1 - e^{-\lambda}}$$

The maximum likelihood estimate $\hat{\lambda}$ is found by equating the sample mean to its expectation

$$\bar{y} = \frac{\hat{\lambda}}{1 - e^{-\hat{\lambda}}}$$

If this equation is written as $\hat{\lambda} = \bar{y}(1 - e^{-\hat{\lambda}})$, Newton's method leads to the iteration scheme

$$\hat{\lambda}_{m+1} = \hat{\lambda}_m - \frac{\hat{\lambda}_m - \bar{y}(1 - e^{-\hat{\lambda}_m})}{1 - \bar{y} \, e^{-\hat{\lambda}_m}}$$

which we will now implement. This is a natural situation in which a while loop might be used in S-PLUS. First we generate our artificial sample from a distribution with $\lambda = 1$.

```
> yp <- rpois(50, lam=1)    # full Poisson sample of size 50
> table(yp)
  0  1  2 3 5
 21 12 14 2 1
> y <- yp[yp > 0]           # truncate the zeros; n = 29
```

We will have a termination condition based both on convergence of the process and an iteration count limit just for safety. An obvious starting value is $\hat{\lambda}_0 = \bar{y}$.

```
> ybar <- mean(y); ybar
> [1] 1.7586
> lam <- ybar
> it <- 0                   # iteration count
> del <- 1                  # iterative adjustment
> while (abs(del) > 0.0001 && (it <- it + 1) < 10) {
    del <- (lam - ybar*(1 - exp(-lam)))/(1 - ybar*exp(-lam))
    lam <- lam - del
    cat(it, lam, "\n")}
1 1.32394312696735
2 1.26142504977282
3 1.25956434178259
4 1.25956261931933
```

To generate output from a loop in progress an explicit call to a function such as print or cat has to be used. For tracing output cat is usually convenient since it can combine several items. Numbers are coerced to character in full precision; using format(lam) in place of lam is the simplest way to reduce the number of significant digits to the options default.

4.2 Vectorized calculations and loop avoidance functions

Programmers coming to S from other languages are often slow to take advantage of the power of S to do vectorized calculations, that is calculations that operate on entire vectors rather than on individual components in sequence. This often leads to unnecessary loops. For example consider calculating the Pearson chi-squared statistic for testing independence in a two-way contingency table. This is defined as

$$X_P^2 = \sum_{i=1}^{r} \sum_{j=1}^{s} \frac{(f_{ij} - e_{ij})^2}{e_{ij}}$$

where $e_{ij} = f_{i.} f_{.j}/f_{..}$ are the expected frequencies. Two nested for loops may seem to be necessary, but in fact no explicit loops are needed. Assuming the frequencies f_{ij} are held as a matrix the most efficient calculation in S uses matrix operations:

```
fi. <- f %*% rep(1, ncol(f))
f.j <- rep(1, nrow(f)) %*% f
e <- (fi. %*% f.j)/sum(fi.)
X2p <- sum((f-e)^2/e)
```

Explicit loops in S should be regarded as potentially expensive in time and memory use and ways of avoiding them should be considered. (See page 158. Note that this will be impossible with genuinely iterative calculations such as our Newton scheme.)

The functions `apply`, `tapply`, `sapply` and `lapply` discussed below offer a way around explicit loops. Further comments and an example are given on page 159.

The functions `apply` and `sweep`

The function `apply` allows functions to operate on an array using sections successively. For example, consider the dataset `iris` which is a $50 \times 4 \times 3$ array of four observations on 50 specimens of each of three species. Suppose we want the means for each variable by species; we can use `apply`.

The arguments of `apply` are

1. The name of the array, `X`.
2. An integer vector, `MARGIN`, giving the indices defining the sections of the array to which the function is to be separately applied. It is helpful to note that if the function applied has a scalar result, the result of `apply` is an array with `dim(X)[MARGIN]` as its dimension vector.
3. The function, or the name of the function, `FUN`, to be applied separately to each section.
4. Any additional arguments needed by the function as it is applied to each section.

Thus we need to use

```
> apply(iris, c(2,3), mean)
          Setosa Versicolor Virginica
Sepal L.   5.006     5.936     6.588
Sepal W.   3.428     2.770     2.974
Petal L.   1.462     4.260     5.552
Petal W.   0.246     1.326     2.026
> apply(iris, c(2,3), mean, trim=0.1)
          Setosa Versicolor Virginica
Sepal L.  5.0025    5.9375    6.5725
Sepal W.  3.4150    2.7800    2.9625
Petal L.  1.4600    4.2925    5.5100
Petal W.  0.2375    1.3250    2.0325
```

where we also show how arguments can be passed to the function, in this case to give a trimmed mean. If we want the overall means we can use

```
> apply(iris, 2, mean)
 Sepal L. Sepal W. Petal L. Petal W.
   5.8433    3.0573     3.758    1.1993
```

Note how dimensions have been dropped to give a vector. If the result of FUN is itself a vector of length d, say, then the result of apply is an array with dimension vector c(d, dim(X)[MARGIN]), with single-element dimensions dropped. Note that matrix results are reduced to vectors; if we ask for the covariance matrix for each species,

```
ir.var <- apply(iris, 3, var)
```

we get a 16×3 matrix. We can add back the correct dimensions, but in so doing we lose the dimnames. We can restore both by

```
ir.var <- array(ir.var, dim = dim(iris)[c(2,2,3)],
                dimnames = dimnames(iris)[c(2,2,3)])
```

The function apply is often useful to replace explicit loops. Earlier versions of the function used loops internally, but since S-PLUS 3.2 these have been replaced by a call to an internal function. Note too that for *linear* computations it is rather inefficient. We can form the means by matrix multiplication:

```
> matrix(rep(1/50, 50) %*% matrix(iris, nrow = 50), nrow = 4,
         dimnames = dimnames(iris)[-1])
         Setosa Versicolor Virginica
Sepal L.  5.006      5.936     6.588
Sepal W.  3.428      2.770     2.974
Petal L.  1.462      4.260     5.552
Petal W.  0.246      1.326     2.026
```

which will be very much faster on larger examples, but is much less transparent.

The function aperm is often useful with array/matrix arithmetic of this sort. It permutes the indices, so that aperm(iris, c(2,3,1)) is a $4 \times 3 \times 50$ array. Note that the matrix transpose operation is a special case.

Functions sweep *and* scale

The functions apply and sweep are often used together. For example, having found the means of the iris data, we may want to remove them by subtraction or perhaps division. We can use sweep in each case:

```
ir.means <- apply(iris, c(2,3), mean)
sweep(iris, c(2,3), ir.means)
log(sweep(iris, c(2,3), ir.means, "/"))
```

Of course, we could have subtracted the log means in the second case.

The function scale takes a matrix argument and operates on each column in turn. By default it mean-corrects the column and re-scales to unit variance. If the argument center is F no centring is done, and if it is a vector of length the number of columns, that vector is subtracted from each row. Similarly, the *argument* scale controls the scaling of each column. Thus for a matrix A, scale(A, T, F) is equivalent to sweep(A, apply(A,2,mean), "-"), but more efficient.

Functions operating on factors and lists

Factors are used to define groups in vectors that have the same length. Each level of the factor defines a group, possibly empty.

In this section we will use the `quine` data frame from the `MASS` library giving data from a survey of school children. The study giving rise to the data set is described more fully on page 214; in brief the data frame has four factors, `Sex`, `Eth` (ethnicity—two levels), `Age` (four levels) and `Lrn` (Learner group—two levels) and a quantitative response, `Days`, the number of days the child in the sample was away from school in a year.

The function `table`

It is often necessary to tabulate factors to find frequencies. The main function for this purpose is `table` which returns a frequency cross-tabulation as an array. For example:

```
> attach(quine)
> table(Age)
 F0 F1 F2 F3
 27 46 40 33
> table(Sex, Age)
   F0 F1 F2 F3
 F 10 32 19 19
 M 17 14 21 14
```

Note that the factor levels become the appropriate `names` or `dimnames` attribute for the frequency array. If the arguments given to `table` are not factors they are effectively coerced to factors.

The function `crosstabs` may also be used. It takes a formula and data frame as its first two arguments, so the data frame need not be attached. A call to `crosstabs` for the same table is

```
> tab <- crosstabs(~Sex + Age, quine)
> print.default(tab)
   F0 F1 F2 F3
 F 10 32 19 19
 M 17 14 21 14
 ....
```

The print method for objects of class `crosstabs` gives extra information of no interest to us here, but the object behaves in calculations as a frequency table.

Ragged arrays and the function `tapply`

The combination of a vector and a labelling factor or factors is an example of what is called a *ragged array*, since the group sizes can be irregular. (When the group sizes are all equal the indexing may be done implicitly and more efficiently using arrays, as was done with the `iris` data.)

To calculate the average number of days absent for each age group we can use the ragged array function `tapply`.

```
> tapply(Days, Age, mean)
     F0      F1    F2      F3
  14.852 11.152 21.05 19.606
```

The first argument is the vector for which functions on the groups are required, the second argument, INDICES, is the factor defining the groups and the third argument, FUN, is the function to be evaluated on each group. If the function requires more arguments they may be included as additional arguments to the function call, as in

```
> tapply(Days, Age, mean, trim = 0.1)
     F0       F1     F2     F3
  12.565 9.0789 18.406 18.37
```

for 10% trimmed means.

If the second argument is a list of factors the function is applied to each group of the cross-classification. Thus to find the average days absent for age by sex classes we could use

```
> tapply(Days, list(Sex, Age), mean)
       F0      F1     F2     F3
  F 18.700 12.969 18.421 14.000
  M 12.588  7.000 23.429 27.214
```

To find the standard errors of these we could use an anonymous function as the third argument, as in

```
> tapply(Days, list(Sex, Age),
           function(x) sqrt(var(x)/length(x)))
       F0     F1     F2     F3
  F 4.2086 2.3299 5.3000 2.9409
  M 3.7682 1.4181 3.7661 4.5696
```

As with table coercion to factor takes place where necessary.

Operating along structures: the functions lapply, sapply *and* split

The functions lapply and sapply are the analogues of tapply and apply for operations on the individual components of lists or vectors. The two functions are called with exactly the same arguments but whereas sapply will simplify the result to a vector or an array if possible, lapply will always return a list.

The function split takes as arguments a data vector and a vector defining groups within it. The value is a list of vectors, one component for each group.

For ragged arrays defined by a single factor the calls tapply(vec, fac, fun) and sapply(split(vec, fac), fun) often give the same result. (The only time they will differ is when sapply simplifies a list to an array and tapply leaves it as a list.) Even though tapply actually uses lapply with split for the critical computations, calling them directly can sometimes be much faster. For example consider the problem of finding the *convolution* of two numeric vectors. If they have components a_i and b_j respectively, their convolution has components $c_k = \sum_{i+j=k} a_i b_j$. This is the same problem as finding the coefficient vector for the product of two polynomials. A simple function to do this is

```
> conv1 <- function(a, b) {
    ab <- outer(a, b)
    unlist(lapply(split(ab, row(ab) + col(ab)), sum))
}
```

A second version, `conv2`, replaces the last line with

```
tapply(ab, row(ab) + col(ab), sum)
```

We can compare times (on a PC):

```
> a <- runif(1000)
> b <- runif(100)
> dos.time(conv1(a, b))
[1] 2.98
> dos.time(conv2(a, b))
[1] 4.38
```

The timing functions are described on pages 148 and 149.

Notice that with `sapply`, `lapply` and `split` if the relevant argument is not
a factor or list of factors, once again the appropriate coercions to factor take place.

The first argument to `lapply` or `sapply` may be any vector, in which case
the function is applied to each component. For example one way to discover
which single-letter names are currently visible on the search path is

```
> Letters <- c(LETTERS, letters)
> Letters[sapply(Letters, function(xx) exists(xx))]
[1] "C" "D" "F" "I" "T" "c" "q" "s" "t"
```

This example can be used to expose some anomalies. First, if we had used x
as an argument instead of xx it would have shown up as an object in the result
although the object is not globally visible.

```
> Letters[sapply(Letters, function(x) exists(x))]
[1] "C" "D" "F" "I" "T" "c" "q" "s" "t" "x"
```

Second, if we had used the function `exists` by name, another phantom variable,
X, would be found, again because of the temporary visibility of internal coding.

```
> Letters[sapply(Letters, exists)]
[1] "C" "D" "F" "I" "T" "X" "c" "q" "s" "t"
```

These problems occur because `exists` is a generic function and the relevant
method, `exists.default`, contains a reference to the function `.Internal`. In
general any function with an `.Internal` reference may give problems if used as
an argument to `lapply` or `sapply`, particularly as a simple name. For example
`dim` is such a function

```
> dim
function(x)
.Internal(dim(x), "S_extract", T, 10)
> sapply(list(quine, quine), dim)   # does NOT work!
[[1]]:
NULL
[[2]]:
NULL
> sapply(list(quine, quine), function(x) dim(x))
      [,1] [,2]
[1,]  146  146
[2,]    5    5
```

The second call should be equivalent to the first, but in this case the second gives correct results and the first does not.

Frequency tables as data frames

Returning to the `quine` data frame, we may find which of the variables are factors using

```
> sapply(quine, is.factor)
  Eth Sex Age Lrn Days
    T   T   T   T    F
```

since a data frame is a list.

Consider the general problem of taking a set of n factors and constructing the complete n-way frequency table as a data frame, that is, as a frequency vector and a set of n classifying factors. The four factors in the `quine` data will serve as an example, but we will work in a way that generalizes immediately to any number of factors.

First we remove any non-factor components from the data frame.

```
quineFO <- quine[sapply(quine, is.factor)]
```

To calculate the frequencies we could `attach` the data frame `quineFO` and call `table` with all factors, but this would be tedious if the number of factors were large, and more importantly this method could not be used in a general function.

The function `do.call` takes two arguments: the name of a function (as a character string) and a list. The result is a call to that function with the list supplying the arguments. List names become argument names. Hence we may find the frequency table using

```
tab <- do.call("table", quineFO)
```

The result is a multi-way array of frequencies. We next find the classifying factors, which correspond to the indices of this array. One way to do this is to use a `for` loop and `slice.index(tab, k)` to generate the factor levels, but a more convenient way is to use `expand.grid`. This function takes any number of vectors and generates a data frame consisting of all possible combinations of values, in the correct order to match the elements of a multi-way array with the lengths of the vectors as the index sizes. Argument names become component names in the data frame. Alternatively expand.grid may take a single list whose components are used as individual vector arguments.

Hence to find the index vectors for our data frame we may use

```
QuineF <- expand.grid(lapply(quineF0, levels))
```

If expand.grid did not accept the single list form of argument we could have used

```
QuineF <- do.call("expand.grid", lapply(quineF0, levels))
```

Finally we put together the frequency vector and classifying factors.

```
> QuineF$Freq <- as.vector(tab)
> QuineF
   Eth Sex Age Lrn Freq
1   A   F  F0  AL    4
2   N   F  F0  AL    4
3   A   M  F0  AL    5
....
30  N   F  F3  SL    0
31  A   M  F3  SL    0
32  N   M  F3  SL    0
```

We use as.vector to remove all attributes since some might conflict with the data frame properties.

Functions operating on data frames

Three functions, merge, by and aggregate [1], operate on data frames in a way which mimics common database operations in other languages. (There is already a meaning for aggregate with time series discussed on page 434, which is unaffected.)

Both by and aggregate are closely related to tapply. Using aggregate is equivalent to using tapply on each column of a data frame. Instead of an INDICES argument, the grouping factors are specified by an argument called by, which is most usefully a named list. For example

```
> aggregate(crabs[, 4:8], list(sp=crabs$sp, sex=crabs$sex),
            median)
  sp sex    FL    RW    CL    CW    BD
1  B   F 13.15 12.20 27.90 32.35 11.60
2  O   F 18.00 14.65 34.70 39.55 15.65
3  B   M 15.10 11.70 32.45 37.10 13.60
4  O   M 16.70 12.10 33.35 36.30 15.00
```

---

[1] All three functions were introduced in S-PLUS 3.3.

It is important to ensure that the function used is applicable to each column, which is why we must omit the non-metric columns here.

The function `by` takes a data frame and splits it by the second argument, `INDICES`, passing each data frame in turn to its `FUN` argument. This is equivalent to using `tapply` on the columns in parallel rather than one at a time.

```
> by(crabs[,4:8], list(crabs$sp, crabs$sex), summary)
crabs$sp:B
crabs$sex:F
        FL            RW            CL            CW
 Min.   : 7.2  Min.   : 6.5  Min.   :14.7  Min.   :17.1
 1st Qu.:11.5  1st Qu.:10.6  1st Qu.:23.9  1st Qu.:27.9
 Median :13.1  Median :12.2  Median :27.9  Median :32.4
 Mean   :13.3  Mean   :12.1  Mean   :28.1  Mean   :32.6
 3rd Qu.:15.3  3rd Qu.:13.9  3rd Qu.:32.8  3rd Qu.:37.8
 Max.   :19.2  Max.   :16.9  Max.   :40.9  Max.   :47.9
  ....
```

The function `merge` provides a *join* of two data frames as databases. That is, it combines pairs of rows which have common values in specified columns to a row with all the information contained in either data frame. See the on-line help for full details.

As an example, consider combining the `animals` and `mammals` data frames, selecting the entries with the same row names in each. We could do this using

```
> merge(animals, mammals, by="row.names")
           Row.names    body.x  brain.x    body.y brain.y
 1 African elephant  6654.000   5712.0  6654.000  5712.0
 2    Asian elephant  2547.000   4603.0  2547.000  4603.0
 ....
 20    Rhesus monkey     6.800    179.0     6.800   179.0
 21            Sheep    55.500    175.0    55.500   175.0
```

Note how the original row names are discarded.

4.3 Writing your own functions

Functions in S are created by assignment using the keyword `function`:

```
fname <- function(arg1, arg2, etc.) function.body
```

where `arg1`, `arg2`, ... [2] are arguments to be satisfied on the call. The statement, `function.body` defining the body of the function can be any S statement, but is usually a grouped statement and so enclosed within braces, `{ }`.

A function is called by giving its name with an argument sequence in parentheses

---

[2] Note that we are using `etc.` to denote an indefinite number of similar items, as " ... " is an allowable literal argument with a special meaning. We always use *four* dots when indicating an omission.

```
fname(val1, val2, etc.)
```

Several protocols are available for specifying both the arguments in the function definition and their values on a call.

The value returned by the function is the value of the statement, which is usually an unassigned final expression. Alternatively the action of a function may be terminated at any stage by calling the `return` function where the argument specifies the value to be returned. For example if some argument x is empty it might be appropriate to stop the function and return the missing value marker immediately.

```
    . . . .
if(length(x) == 0) return(NA)
    . . . .
```

If `return` is given several, possibly named, arguments the value returned is a list of components with names as supplied, which seems to be an undocumented feature.

As a simple example consider a function to perform a two-sample t-test the way it is often taught in elementary courses.

```
ttest <- function(y1, y2, test = "two-sided", alpha = 0.05)
{
  n1 <- length(y1)
  n2 <- length(y2)
  ndf <- n1 + n2 - 2
  s2 <- ((n1 - 1) * var(y1) + (n2 - 1) * var(y2))/ndf
  tstat <- (mean(y1) - mean(y2))/sqrt(s2 * (1/n1 + 1/n2))
  tail.area <- switch(test,
    "two-sided" = 2 * (1 - pt(abs(tstat), ndf)),
    lower = pt(tstat, ndf),
    upper = 1 - pt(tstat, ndf),
    {
      warning("test must be 'two-sided', 'lower' or 'upper'")
      NULL
    }
  )
  list(tstat = tstat, df = ndf,
       reject = if(!is.null(tail.area)) tail.area < alpha,
       tail.area = tail.area)
}
```

This function requires two arguments to be specified, y1 and y2, the two sample vectors. It also allows two more, `test` and `alpha`, but if no values for these are given on the call, the default values are used. Thus if `test` is not specified a two-sided test is assumed, and if `alpha` is not specified a significance level of 5% is used.

The result is a list of components giving information about the test result. This is conventional, although giving a result as a primary object with additional information carried as attributes is sometimes appropriate.

It is easy to test the function using simulated data.

```
> x1 <- round(rnorm(10), 1); x1
>  [1]  0.5 -1.4  1.5  0.5 -0.9 -0.9 -0.1 -0.5 -1.2  0.1
> x2 <- round(rnorm(10) + 1,1); x2
>  [1]  0.4 -0.3  0.9  2.2  1.8  0.7  2.7  1.0  1.1  2.9
> ttest(x1, x2)
$tstat:
[1] -3.6303
$df:
[1] 18
$reject:
[1] T
$tail.area:
[1] 0.0019138
```

A more compact way to print the result is to coerce it to a numeric vector using `unlist`:

```
> unlist(ttest(x1, x2))
   tstat df reject tail.area
 -3.6303 18      1 0.0019138
```

S-PLUS has similar function, `t.test`, which we can use to check ours.

```
> t.test(x1, x2)

            Standard Two-Sample t-Test

data:  x1 and x2
t = -3.6303, df = 18, p-value = 0.0019
    ....
```

Before extending[3] our function to make it more flexible and hence more useful, we discuss some other aspects of functions, and facilities available for writing them.

Lazy evaluation

When an S function is called the argument expressions are parsed but not evaluated. Inside a function when a formal argument is required the actual argument expression (or the default if it is not supplied) is then evaluated and its value is used by the function. In particular if an argument not used by the function on a particular call, it is never evaluated and it could involve variables that do not even exist. It must merely parse correctly. This is in striking contrast to many compiled languages where the actual arguments are evaluated before the call to the function is made.

---

[3] Readers wishing to work through this example at the keyboard as it develops over the rest of this section may wish to review briefly the section on editing functions, Section 4.5 on pages 139f.

It is important to note that actual argument expressions are evaluated in the context of the calling function, known as the *parent frame*, whereas default values, if needed, are evaluated in the context of the function itself, known as the *local frame*. (Frames are considered more formally in Section 4.7.)

The protocol is called *lazy evaluation* and it has some important consequences for S programming, including

1. Default expressions for formal arguments may involve not only other arguments to the function but also variables in the search path, or variables local to the function itself. The rule is that the expression must be capable of evaluation when it is needed; in particular any local variables involved must have values at that point. Thus it is possible to use a call such as

   ```
   glm.obj <- glm(y ~ x, binomial(link = probit))
   ```

   even though there is no object with name `probit` visible in the search path.

2. Within a function it is possible to retrieve the unevaluated argument expressions using the function `substitute`. The function `deparse` can be used to coerce such an object to mode character, which is essentially the reverse of parsing. Hence if x is a formal argument `deparse(substitute(x))` gives whatever expression was used for argument x (or the default if none was used) and returns it as a character string. (The `binomial` function uses `as.character(substitute(link))`.)

The second possibility is often used to generate labels on graphs. For example

```
myplot <- function(x, y, lab = deparse(substitute(y))) {
    ....
    title(main = paste("A plot of", lab))
    ....
}
```

The functions `warning`, `stop` and `missing`

The functions `warning` and `stop` are used inside functions to handle unexpected situations; `warning` arranges for a warning to be issued when control returns to the session level but the action of the function continues. For example if we call our `ttest` function with an invalid character string for `test` a warning message is issued and the default test performed. We use `unlist` again for a compact display:

```
> unlist(ttest(x1, x2, test="left"))
   tstat df
 -3.6303 18
Warning messages:
  test must be 'two-sided', 'lower' or 'upper' in: \
       switch(test, ....
```

The function `stop` terminates the action of the function, issues an error message and returns control to the session level immediately. It does not terminate the S-PLUS session as `q()` does. As we shall see in Section 4.5 it can also be made to precipitate a dump of information on the state of the calculation that can often help in tracing errors.

The function `missing` may be used within a function to check if a value was specified for some argument. For example we could modify our `ttest` function to include a check for `y1` and `y2`:

```
ttest <- function(y1, y2, test = "two-sided", alpha = 0.05)
{
    if(missing(y1) || missing(y2))
        stop("two samples needed")
    ....

> ttest(y2 = x2)
Error in ttest(y2 = x2): two samples needed
Dumped
```

It is often tempting to use `missing` to specify default values of arguments, but the standard default mechanism is usually better.

A common idiom is to use `stop("some message")` as a default value when some argument for a function must be supplied. For example the function `rpois` for generating artificial Poisson samples requires that the mean parameter be specified:

```
> args(rpois)
function(n, lambda = stop("no lambda arg"))
NULL
```

The special argument "..."

The 'three dots' argument, " ... ", is special in that any number of named arguments may be specified for it on the call. Inside the function, normally it is simply passed on as an argument to some other function, in which case all corresponding arguments are passed on as they were specified in the original call. One standard way to process " ... " arguments is to evaluate them by passing them to `list`.

Continuing our t-test example, we allow the option of producing a boxplot. To customize the boxplot the user has to be able to pass on additional arguments to `plot`. We need to make two changes to our function. The first is to add two more arguments to the definition, `boxp` and ... :

```
ttest <- function(y1, y2, test = "two-sided", alpha = 0.05,
                boxp = T, ...)
{
    ....
```

A better default value for `boxp` would be `T` if a graphics device other than the null device is currently open, otherwise `F`. This is easy to do within **S-PLUS** by setting the default `boxp = dev.cur()>1`. We also need to do the boxplot. A convenient place for this is just before the return expression:

```
....
if(boxp) {
    sam <- paste("Sample", 1:2)
    f <- factor(rep(sam, c(length(y1), length(y2))))
    plot(f, c(y1, y2), ...)
}
list(tstat = ....
```

Any additional arguments given to `ttest` will now be passed on to the `plot` function. For example, it might be a good idea to add a more informative label to the vertical axis.

```
ttest(x1, x2, ylab="Water Quality")
```

Since `ylab` does not match any other argument name it is included in the " ... " argument and passed on to `plot`.

Argument matching

When arguments are given in the `name=value` form, the `name` may be abbreviated in a similar way to component names of lists. For example, in our `ttest` function the argument names are `y1`, `y2`, `test` and `alpha`. The last two arguments may be abbreviated to the single letter `t` and `a` respectively.

```
> unlist(ttest(y1=x2, y2=x1, a=0.99, t="upper"))
   tstat df reject   tail.area
  3.6303 18       1 0.00095688
```

The complete rules by which formal and actual arguments are matched are

1. First, any actual arguments specified in the `name=value` form where the name *exactly* matches the name of a formal argument are matched.

    In the case of formal arguments that occur *after* a ... argument, this is the only way they can be matched.

2. Next, arguments specified in the `name=value` form for which the `name` is a unique partial match for a formal argument name are matched, provided the formal argument does not occur after a ... argument.

3. Any unnamed actual arguments are then matched to the remaining formal arguments one by one in sequence as far as possible.

4. All remaining unmatched actual arguments become part of the ... formal argument, if there is one, or else an error occurs.

With character arguments such as `test` it is possible to provide a set of allowable values from which the user may select by partial matching. The function `match.arg(arg, choices)` uses the character string `arg` to select that value from the character string vector `choices` for which a unique partial match occurs. No or more than one partial match results in an error.

If the call to `match.arg` is made from within a function and `arg` is an argument then `choices` may be omitted. The default value for `arg` must then be a character string vector and this default value is used as the `choices` argument. If the function is called and no value for `arg` is supplied, the first entry in the default character string vector is the value used. To see this in action, consider the changes we could make to `ttest`.

```
ttest <- function(y1, y2,
        test = c("two-sided", "lower", "upper"), alpha = 0.05)
{
    test <- match.arg(test)
    ....
    tail.area <- switch(test,
        "two-sided" = 2 * (1 - pt(abs(tstat), ndf)),
        lower = pt(tstat, ndf),  upper = 1 - pt(tstat, ndf))
    list(tstat=tstat, df=ndf, reject=tail.area < alpha,
        tail.area = tail.area)
}
```

The call `ttest(x1, x2)` will perform a two-sided t-test. To specify a lower-tailed one-sided t-test we could now use

```
ttest(x1, x2, tes="low")
```

Now `test="lower"` is selected by partial matching. An invalid choice for the argument `test` now precipitates an error from `match.arg`, as in

```
> ttest(x1, x2, tes="left")
Error in call to "ttest1": Argument "test" should be one of:\
        "two-sided", "lower", "upper"
Dumped
```

The function `match.arg` should be distinguished from the related functions `match`, `pmatch` and `charmatch` discussed on page 44.

The `on.exit` function for exit actions

A common use of the " ... " argument is to allow the user to specify temporary changes to the graphical parameters for some plot done within a function. There is a conventional way to do this, namely

1. The " ... " argument is included, but only arguments to `par` may be substituted for it on the call.

2. Before any plotting is done the following statements are executed:

```
oldpar <- par(...)
on.exit(par(oldpar))
```

The first statement records the state of the graphics parameters. The second statement using `on.exit` arranges for the previously current settings to be re-set when the action of the function is terminated (whether normally or by an error exit).

The `on.exit` function can be used in a similar way to allow temporary changes to be made to the `options` settings.

An argument `add` to the `on.exit` function allows additional actions to be added to those already specified; this is good practice. Calling `on.exit()` with no arguments cancels any actions previously requested.

Saving partial results and restarting

One important application for `on.exit` is to save partial results if a (long) S-PLUS session should be interrupted. A typical construction is

```
on.exit(assign(".partial.res", res[1:(iter-1)],
               where=1, immediate=T))
res <- numeric(lots)
for(iter in 1:lots) res[iter] <- myfun(iter)
```

(The function `assign` is discussed on page 155.)

For repetitive calculations that involve applying the same steps to different data sets, some of which may cause an error, there is a function called `restart` that allows the user to over-ride error exits and continue with the calculation by re-calling the present function. *This function should only be used with extreme caution by experienced* S *programmers.*

4.4   Introduction to object orientation

The primary purpose of the S programming environment is to construct and manipulate objects. These objects may be fairly simple, such as numeric vectors, factors, arrays or data frames, or reasonably complex such as an object conveying the results of a model fitting process. The manipulations fall naturally into broad categories such as plotting, printing, summarising and so forth. They may also involve several objects such as performing an arithmetic operation on two objects to construct a third.

Since S is intended to be an extensible environment new kinds of object are designed by users to fill new needs, but it will usually be convenient to manipulate them using familiar functions such as `plot`, `print` and `summary`. For the new kind of object the standard manipulations will usually have an obvious purpose, even though the precise action required differs at least in detail from any previous action in this category.

Consider the `print` function. It is normally invoked implicitly simply by evaluating an expression, or giving a name, at the session level, as in

```
> x <- 1:5
> x
[1] 1 2 3 4 5
```

If a user designs a new kind of object called, say, a "newfit" object, it will often be useful to make available a method of printing such objects so that the important features are easy to appreciate and the less important details are suppressed.

One way of doing this is to write a new function, say print.newfit, which could be used to perform the particular kind of printing action needed. If myobj is a particular newfit object it could then be printed using

```
> print.newfit(myobj)
```

We could also write functions with names plot.newfit, summary.newfit, residuals.newfit, coefficients.newfit for the particular actions appropriate for newfit objects for plotting, summarising, extracting residuals and extracting coefficients, and so on.

In the case of printing, however, it would be much more convenient for the user if simply naming the object at the session level

```
> myobj
```

were enough to cause the special printing action to happen. Similarly it would be useful to be able to use the standard functions such as plot(myobj) and summary(myobj) and have the S evaluator select the appropriate action for the class of object presented to it.

This is one important idea behind *object-oriented programming*. To make it work we need to have a standard method by which the evaluator may recognise the different classes of object being presented to it. In S this is done by giving the object a class attribute, which at its simplest is a character string naming the class. There is a class replacement function available for making this attribute assignment, which would normally be done when the object was created. For example

```
> class(myobj) <- "newfit"
```

In many cases the class attribute is a character vector of several components; the object is then said to *belong* to the class specified by the first and to *inherit* from classes specified by the second, third, and so on, in turn.

Functions like print, plot and summary are called *generic* functions. They have the property of adapting their action to match the class of object presented to them. Typically their definition is very short, often only one line:

```
> print
function(x, ...)
UseMethod("print")
```

The UseMethod function is treated specially by the evaluator. The first step taken is to evaluate the principal argument, (here x), and examine its class attribute, if any. If the object x belongs to class newfit, and if there is a function print.newfit visible on the search path, then the remaining body of the generic function (usually, as here, all of it) is replaced by a call to print.newfit of the form print.newfit(x, ...), but using the existing frame rather than generating a new frame and leaving the arguments unevaluated. Objects created in the frame before the call to UseMethod will remain visible unless masked by an argument of the same name. In addition the frame at this point will always contain a few extra objects named .Class, .Method, .Group and .Generic. These are character strings giving information on the process underway. Any code in the generic function that would normally be executed after the call to UseMethod is (silently) discarded.

Functions such as print.newfit are called *method* functions for the print generic function. The entire process is called *method dispatch*.

If no function print.newfit is visible the class vector of x is examined to see if there is a print method function available for one of the classes from which the object inherits. If so, the first one found is used. If there is no primary or inherited print method function available, or if the object x has no class attribute, then the default method, here print.default, is used. If no method function can be found for some generic function and no default method is supplied an error results.

The advantages of this approach include being able to add methods to existing generic functions for new classes without any need to change existing code, being able to manipulate objects using a consistent and familiar suite of generic functions, and through inheritance being able to capitalise on existing code when some existing method caters for objects in the new class sufficiently well. The scheme offers most advantage to the programmer if the classes can be built incrementally so that maximum use may be made of the inheritance mechanism to re-use existing code. One disadvantage is that the principal argument will be evaluated twice, once by the generic function and again by the method. It pays to present the principal argument as an already evaluated object rather than as an expression requiring substantial evaluation effort.

The function unclass removes the class(es) of its argument.

A generic *t*-test function

We illustrate some of the ideas of object-oriented programming by revisiting the *t*-test problem. We will allow the two samples to be specified in a number of different ways, namely

1. as two separate numeric vectors, as in the previous case,
2. as a two-column matrix,
3. as a list of two components, or
4. as a two-level factor and a single numeric vector.

The first argument will determine the particular method dispatched, but since numeric vectors, matrices and lists normally will have no class attribute, we need to assign one to them so that the method dispatch will be appropriate. The function data.class infers a class from attributes of the object. If a class attribute exists the inferred class is the first component of the class vector. In other cases inferred attributes often coincide with the mode of the object, such as "numeric" or "list" but for matrices the inferred class is "matrix". For our purposes it is important to know that our four cases will be distinguished.

As usual the generic function is very simple:

```
Ttest <- function(z, ...) {
  if(is.null(class(z))) class(z) <- data.class(z)
  UseMethod("Ttest")
}
```

Since data.class does not preserve inheritance it can be important not to use it when a class for the object already exists.

Next, consider a default method. This is almost the same as the previous ttest function with the plot option removed and with the original samples and arguments returned as part of the result.

```
Ttest.default <- function(y1, y2, ..., alpha = 1/20,
          test = c("two-sided", "lower tailed", "upper tailed"))
{
  n1 <- length(y1)
  n2 <- length(y2)
  ndf <- n1 + n2 - 2
  s2 <- ((n1 - 1) * var(y1) + (n2 - 1) * var(y2))/ndf
  tstat <- (mean(y1) - mean(y2))/sqrt(s2 * (1/n1 + 1/n2))
  tail.area <- switch(test <- match.arg(test),
    "two-sided" = 2 * (1 - pt(abs(tstat), ndf)),
    "lower tailed" = pt(tstat, ndf),
    "upper tailed" = 1 - pt(tstat, ndf),
    {
      warning("test must be 'two-sided', 'lower' or 'upper'")
      NULL
    })
  structure(list("t-stat"=tstat, d.f.=ndf, y1=y1, y2=y2,
                test=test, tail.area=tail.area,
                reject=tail.area < alpha, alpha=alpha),
            class = "my.t.test")
}
```

The arguments alpha and test now occur after a ... argument which means they must be named, in full, if they are supplied by the user. This is a minor disadvantage but we shall see reasons for it. The function structure used for the result returns an object like its first argument with attributes given by any remaining arguments, so the result of Ttest.default is still a list, but with class attribute "my.t.test".

Consider now a print method for objects of this class. The key components are the first two, so a succinct print method simply prints those out.

```
print.my.t.test <- function(x, ...) {
  y <- x
  x <- unlist(x[1:2])
  NextMethod("print")
  invisible(y)
}
```

The function NextMethod is, like UseMethod, somewhat special. Used as here it performs a call like the original to the generic function named (as a character string as its first argument) but for the next inherited method. In this case it will be to the default print method, but in principle that could change. Unlike UseMethod which discards following code in the generic function, control passes back to the original method function after the NextMethod call is complete.

By convention print methods return the value of their principal argument invisibly. This is why it is saved as a temporary variable, y, not involved on the call constructed by NextMethod, but only used for the return value. The invisible function turns off automatic printing, thus preventing an infinite recursion when printing is done implicitly at the session level.

We can now test the function with the two samples generated previously and shown on page 127:

```
> Ttest(x1, x2)
  t-stat d.f.
  -3.6303    18
```

If on some occasion we were only interested in the result and not the statistic, that component can still be accessed directly, as in

```
if(Ttest(x1, x2, test="upper")$reject) cat("x2 is better\n")
```

For most purposes the standard print method will be sufficient output, but if a more extensive display is required it can be done with a summary method. Most summary methods merely enhance the object with further components found by additional calculations, provide a new class and the display is done by a new print method. In this case the calculations are done already, so we merely change the class attribute and write a print method for the new class:

```
summary.my.t.test <- function(object)
  structure(object, class = c("sum.my.t.test", class(object)))

print.sum.my.t.test <- function(x, ...) {
  n1 <- length(x$y1)
  n2 <- length(x$y2)
  cat("\n T-test of  samples: \n")
  cat("Sample 1:", format(x$y1[1:min(n1, 5)], ...))
  cat(if(n1 <= 5) "" else " ...", "\n")
  cat("Sample 2:", format(x$y2[1:min(n2, 5)], ...))
```

```
      cat(if(n2 <= 5) "" else " ...", "\n")
      cat("\nt =", format(x$"t-stat", ...), "on",
          x$d.f., "d.f.\n")
      cat("Test:", x$test, "\n")
      cat("Tail area:", format(x$tail.area, ...), "\n")
      cat("Level:", format(x$alpha, ...), "\t")
      cat("Null hypothesis",
          if(x$reject) "rejected." else "retained.", "\n")
      invisible(x)
    }
```

The result is much more detail, including the first few sample members:

```
> tst <- Ttest(x1, x2, test = "lower")
> summary(tst)

 T-test of  samples:
Sample 1:  0.5 -1.4  1.5  0.5 -0.9 ...
Sample 2:  0.4 -0.3  0.9  2.2  1.8 ...

t = -3.6303 on 18 d.f.
Test: lower tailed
Tail area: 0.00095688
Level: 0.05     Null hypothesis rejected.
```

The boxplots option was removed because a better strategy is to supply a `plot` method for `my.t.test` objects rather than include it as yet another argument to remember in the original call. The idea is very simple. We simply extract the sample components from the object and call `boxplot`:

```
> plot.my.t.test <- function(t.tst, ...)
      invisible(boxplot(tst[c("y1", "y2")], ...))
```

Any customisation needed may be supplied as additional arguments which are passed on to `boxplot`. The actual call is to `plot`, as in

```
tst <- Ttest(x1, x2)
plot(tst)
```

Finally consider the alternative ways we wish to allow for calling `Ttest`. These are handled by very short method functions that take the initial argument or arguments, manufacture arguments that suit `Ttest.default` and effectively call that function.

```
Ttest.matrix <- function(y, ...)
    NextMethod("Ttest", y[, 1], y[, 2])
Ttest.list <- function(lis, ...)
    NextMethod("Ttest", lis[[1]], lis[[2]])
Ttest.factor <- function(f, y, ...) {
    lev <- levels(f)
    NextMethod("Ttest", y[f == lev[1]], y[f == lev[2]])
  }
```

Notice that NextMethod function call has additional arguments. In the constructed call to Test.default these arguments will appear first, and so provide values for y1 and y2 in the default method. All arguments supplied on the original call to the function containing the NextMethod call are placed next. In this case they will be soaked up by the ... argument in Ttest.default and discarded, unless their names are precisely test or alpha, when they match the appropriate argument. (In such instances an explicit call to Ttest.default would probably be less confusing.)

A test of the factor method is

```
> x12 <- c(x1, x2)
> sam <- factor(rep(0:1, c(length(x1), length(x2))))
> Ttest(sam, x12)
  t-stat d.f.
 -3.6303    18
```

Some care is needed in using NextMethod, especially when as here new arguments are specified. The arguments in the call that NextMethod constructs consist of any that are specified as arguments to NextMethod (apart from the first), followed by the arguments to the original call to the function calling NextMethod which are not missing. In our example these original arguments are unnecessary and the ... argument in the default method absorbs them rather than let them supply spurious values for genuinely useful arguments to the next method. In many cases the argument sequence will not require a change and in these cases NextMethod may be called with only the first argument supplied. Even that may be omitted if the next method is for the same generic function as the current one, but it is regarded as good programming style to specify it, particularly if the name of the generic function contains a period.

Our libraries contain many examples of the use of the object-oriented programming paradigm. Most often we just add method functions for generic functions such as print, summary and predict, as in classes "lda", "qda" and "nnet". Similarly objects of our class "negbin" inherit from class "glm" but need their own method functions for anova, summary and family to cater for the few special needs of negative binomial models.

The robust linear regression functions discussed in Chapter 8 define a new class "rlm", and also inherit from the linear regression class "lm". Both print and summary functions are provided. However, summary.rlm produces a result of class "summary.lm", so printing the summary invokes print.summary.lm. We could have used inheritance for the print method too, with a function something like

```
print.rlm <- function(x, ...) {
   NextMethod("print")
   cat("Scale estimate:", format(signif(x$s, 3)), "\n")
   if(x$converged)
       cat("Converged in", length(x$conv), "iterations\n")
   else cat("Ran", length(x$conv),
```

```
                     "iterations without convergence\n")
        invisible(x)
   }
```

In general inheritance works well only when the classes are designed in parallel with inheritance in mind. For example, it might appear that we could use inheritance from `summary.lm` to build `summary.rlm`, but scale estimation by the root mean square of the residuals is so embedded in `summary.lm` that this proves to be impossible.

Group methods

A generic function is one that either has a call to `UseMethod` or to `.Internal`. The latter are part of the basic software of S (and not user-written) but methods for them may be written in exactly the same way as for the first type. Many of the most basic S functions are `.Internal` generic functions, such as the coercion functions

```
> as.numeric
function(x)   .Internal(as.numeric(x), "As_vector", T, 4)
```

the extraction and replacement functions

```
> get("[[")
function(x, ..., drop = T)
.Internal(x[[..., drop = drop]], "S_extract", T, 3)
```

and the arithmetic and logical operators

```
> get("+")
function(e1, e2)   .Internal(e1 + e2, "do_op", T, 5)
```

Many of these generic functions occur in related groups, and if methods for, say, the arithmetic operators are needed there are some short cuts allowed so that instead of writing separate method functions for each operator, a single *group method* function may be written to handle them all.

Examples of group method functions are given in the on-line complements to the book. In particular we discuss methods for manipulating polynomial objects in S using the ordinary arithmetic operators.

4.5   Editing, correcting and documenting functions

Many functions are permanent objects used again and again by the programmer and other users. It is important they be written well, free of errors and clearly documented. Removing errors is often a process requiring several phases.

Editing S functions and objects

An S object may be modified in several ways, but in all cases the object is given a text representation in an external file or window, the file is edited in some way, and the object re-created within S-PLUS by assignment. Commonly used methods are as follows.

1. Small adjustments can often be made by cut-and-paste between the S-PLUS session and an editable window. Sometimes it is helpful to save the function definition in a file (from the editor or with dump("obj", "obj.q")) and use source("obj.q") to read it into S-PLUS again. Note that .q is the preferred filename extension, as .s is used for assembler code in Unix.

2. The function fix takes as argument an object to edit:

```
> fix(obj)
```

This initiates an editing session with a text version of obj available for correction. On completion of the editing session the corrected version is assigned to the object and the S-PLUS session is resumed. The editor used is a system default. Under Unix this is usually vi and in Windows notepad, but a different editor can be specified as an options argument:

```
> options(editor = "emacs -nw")
```

The most recently edited object can be re-edited by invoking fix().

3. The function ed can achieve the same outcome by explicit assignment. It takes an object to edit as its principal argument and the name of an editor to use as another optional argument:

```
> obj <- ed(obj, editor = "emacs -nw")
```

If vi is the editor chosen we can use obj <- vi(obj). Again, the function can be re-edited by invoking ed or vi without specifying an object.

Locating and correcting errors

When an abnormal exit occurs from a function it is often useful to investigate the state of the variables when and where the error occurred. There are two conceptually separate ways of going about this, one using an inbuilt interactive symbolic debugger called inspect, and another more traditional method using a suite of tools including debugger, browser and trace. Both are simple and effective but since inspect is possibly the more powerful we will concentrate on it and only refer briefly to the more traditional methods later.

Error location using `inspect`

As an example of error location and remedy consider the initial `ttest` function on page 126. Suppose our data comes in the form of a factor `sam` and data vector `x12`, as constructed on page 138, but here repeated for convenience. We might mistakenly call the function as follows:

```
> x12 <- c(x1, x2)
> sam <- factor(rep(0:1, c(length(x1), length(x2))))
> ttest(x12[sam == 1], x12[sam == 2])
Error in dim<-: Invalid value for dimension 1: c(0, ..)
Dumped
```

The error message is unenlightening, but when an error leads to the Dumped message as above, the first thing to do is to consult the dump to see some details of the state of things when the error occurred:

```
> traceback()
Message: Invalid value for dimension 1
4: dim(xm) <- c(n, 1)
3: var(y2)
2: ttest(x12[sam == 1], x12[sam == 2])
1:
```

This is already quite informative. It tells us that at the time of the error the function `ttest` was calling `var`, which in turn was executing a statement `dim(xm) <- c(n, 1)`. These tiers of evaluation are called *frames* which we consider in more detail later, but for now we can think of them as the local environments in which functions operate. They contain the arguments to functions under their formal names and the names and values of any local variables currently defined.

In many cases a call to `traceback` will trigger a realisation of the error and lead immediately to a fix, but in more subtle or stubborn situations some more active investigation is needed. This is most easily done using `inspect` which allows the expression to be evaluated in a controlled way. The programmer can step through the task as finely as one expression or function call at a time if need be, move up and down the frames currently in existence and inspect variables or even change their values as the operation proceeds.

We start the controlled evaluation by giving `inspect` the expression causing problems:

```
> inspect(ttest(x12[sam == 1], x12[sam == 2]))
entering function ttest
stopped in ttest (frame 3), at:
        n1 <- length(y1)
d>
```

Execution is temporarily halted at the beginning of the first function being called. The special prompt

Table 4.1: Main options for the function `inspect`.

Individual help entries are available for the following.

Advance evaluation:
  step – walk through expressions
  do – do expressions atomically
  complete – a loop or function
  resume – continue to next mark
  enter – descend into a function call
  quit – abandon evaluation

Display:
  where – current calls and expr'n
  objects – local frame objects
  show – installed tracks and marks
  find – location of an S-PLUS object
  return.value – function return value
  on.exit – scheduled on.exit expr'ns
  fundef – function definition

Halt evaluation; track functions:
  mark, unmark – arrange to stop in
    a function or at an expression
  track, untrack – install or change
    function call reporting

Examine other current frames:
  up, down

Miscellaneous:
  eval – S-PLUS expressions
  help – syntax and description
  debug.options – option settings

Informational help entries:
  names – when to quote
  keywords – list of reserved words

```
d>
```

indicates that `inspect` is waiting for instructions on what to do next. Notice that this is now frame 3 whereas in the traceback it was frame 2. This is because we have enclosed our expression within `inspect`, which itself is now occupying frame 2.

Instructions to `inspect` have the form of a keyword optionally followed by arguments of an appropriate kind. For example the instruction

```
d> help
```

will produce a table of keywords and brief descriptions, as shown in Table 4.1. If `help` is followed by another keyword, more complete information for that specific keyword is displayed, for example:

```
d> help fundef
syntax: fundef [name]
    Prints the (original) function definition for 'name'.  The
    default is the local function.
    ....
d>
```

To remind ourselves of the original function we can ask to have it displayed:

```
d> fundef ttest
ttest :
function(y1, y2, test = "two-sided", alpha = 0.05)
```

```
{
  n1 <- length(y1)
  n2 <- length(y2)
  ndf <- n1 + n2 - 2
  s2 <- ((n1 - 1) * var(y1) + (n2 - 1) * var(y2))/ndf
       . . . .
  return(list(tstat = tstat, df = ndf,
          reject = if(!is.null(tail.area)) tail.area < alpha,
          tail.area = tail.area))
}
d>
```

This is not quite the same as our original definition as the return value is now explicitly marked with the `return` statement.

The keywords come in groups. Those of the first group containing `step`, `do`, ..., `quit` control the course of execution. Before proceeding, however, it may be prudent to arrange in advance for other places to stop. We know that `var` may be causing a problem, so we can mark it as a place to stop.

```
d> mark var
entry mark  set for var
exit mark(s) set for var
```

There are functions called *primitives* which cannot be marked, for example `dim` and `dim<-`, the access and replacement functions for the `dim` attribute:

```
d> mark dim
dim is a primitive; cannot mark
d> mark dim<-
dim<- is a primitive; cannot mark
```

If we step through the process marks are not necessary, but with marks in place even if we resume normal execution there will be a temporary halt wherever the process enters or leaves the `var` function.

Let us proceed. First one step at a time:

```
d> step
stopped in ttest (frame 3), at:
        length(y1)
```

So from the first assignment we are now about to do the first step and evaluate the right-hand side.

The `resume` keyword allows execution to proceed unless halted by a mark or an error (or by the user using Ctrl-C).

```
d> resume
entering function var
stopped in var (frame 4), at:
        if(length(which.na(x)))
                stop("missing values in x not allowed")
```

The mark has halted evaluation at the top of the var function. We know from the initial traceback that the problem was with y2 rather than y1 so we will step over this one and proceed to the next call to var. Entering resume or complete a few more times takes us over our marks to the point where we enter var again, this time with the argument y2 to var.

We can now remind ourselves of the formal arguments to var.

```
d> eval args(var)
function(x, y = x)
NULL
d> eval x
numeric(0)
d> eval y
numeric(0)
```

So var may have two arguments, but if the second is omitted it is the same as the first. Moreover on this occasion the first is empty, which is clearly a problem we must fix.

We can now look to see precisely where we are:

```
d> where
Frame numbers and calls:

5: debug.tracer(what = TR.GENERIC, index = c(3, 1)) from 4
4: var(y2) from 3
3: ttest(x12[sam == 1], x12[sam == 2]) from 1
2: inspect(ttest(x12[sam == 1], x12[sam == 2])) from 1
1:    from 1
--------------------
stopped in var (frame 4), at:
        if(length(which.na(x)))
                stop("missing values in x not allowed")
```

We are actually in frame 4, as the inspect engine running the process occupies frame 5. Let us go up to the frame of the function calling this one, frame 4, and look around. (This is called the frame "above", although it has a lower number and is always printed below!)

```
d> up
ttest (frame 3)
```

We can now see what objects are in the local environment there and look at some of them:

```
d> objects
 [1] ".Auto.print" ".entered."   ".name."      "alpha"
 [5] "n1"          "n2"          "ndf"         "test"
 [9] "y1"          "y1"          "y2"          "y2"
d> eval y1
 [1]   0.4 -0.3  0.9  2.2  1.8  0.7  2.7  1.0  1.1  2.9
d> eval y2
numeric(0)
```

Some of the objects visible have been placed there by inspect, but most look familiar. We now note that y1 looks like what should be y2 and y2 is empty! Clearly the problem lies with what we used for indices in the original call. (The error was to use a non-existent level for the factor leading to an empty sample. This is a very common mistake.) At this stage we might want to quit and fix the problem, or we might want to resume execution until the point where the error occurs.

If an error occurs during inspect it triggers a display of frames similar to that obtained by traceback. A final quit will terminate the debug session.

Some more traditional debugging facilities

When an expression has an error leading to the message Dumped it creates an object called last.dump in the first position of the search path, usually the primary permanent database. Normally this contains little more information than that shown by traceback, but by setting

```
options(error = dump.frames)
```

the last.dump object will include all information on all frames in existence when the error caused the exit. This can be an extremely large object, so when the facility is not needed the option should be returned to the default by

```
options(error = dump.calls)
```

After an error exit, the dump object can be investigated in detail using the debugger function. This allows the variables to be interactively inspected in a manner very similar to that allowed by inspect and the error located by *post mortem* analysis.

The traditional way to locate errors during execution is to insert calls to the function browser, with no arguments, inside the code and to re-evaluate the expression. When the expression browser() is encountered execution halts and a secondary interactive session commences within the frame of the function. Variables may be inspected or changed in a way again like inspect, but it is not possible to step through the code. Unlike inspect, browser does not use keywords, so the secondary session requires only the usual S expressions.

Calls to either debugger and browser are terminated by evaluating the expression 0 (zero).

Table 4.2 summarises most of the debugging tools available in S-PLUS, and the reader is invited to look up the help information for more details.

Creating a help document

Functions or data frames that may be used by more than one user, or even by one user over a long period of time, should be documented. This is easy to do in a machine readable form, but the forms for Unix and Windows platforms are different.

Table 4.2: Some debugging facilities in S-PLUS.

`print, cat`	Printing key quantities from within a function may be all that is needed to locate an error.
`traceback`	Prints the calls in process of evaluation at the time of any error that causes a dump.
`options(warn=2)`	Changes a warning into an error, precipitating a dump.
`options(error=FUN)`	Specifies the dump action. The default `FUN` is `dump.calls` which dumps the details of calls, only, but `dump.frames` can be used to cause a dump of all evaluation frames in existence at the time of the error.
`last.dump`	The object in the `.Data` directory that contains a list of calls or frames after a dump.
`debugger`	Function to investigate `last.dump` after an error.
`browser`	Function that may be inserted to interrupt the action and allow variables in the frame to be investigated before the error occurs.
`trace`	Place tracing information at the head of, or inside, functions. May be used to insert calls to `browser` at specified positions.
`tprint`	Produces a numbered listing of the body of a function for use with the `at` argument of `trace`.
`untrace`	Turns some or all tracing off.
`inspect`	An interactive debugger.

Unix

The first step is to use the `prompt` function with the object to be documented as argument:

```
> prompt(Ttest)
created file named  Ttest.d  in the current directory
 edit the file and move it to the appropriate .Help directory,
 dropping the .d
```

This creates an outline help document file, here `Ttest.d`, that should be completed using an editor. Although the document is written using a macro package under `nroff` it is easy to provide the required information using only the instructions in the prompt file itself. Further information can be obtained from the `prompt` help document.

By convention the file has the same name as the S object but with a file extension of " `.d` ". To make it readable by the S-PLUS help facility it should be moved or copied into the `.Help` subdirectory of the `.Data` directory and the file extension dropped.

```
> ?prompt      # to review macros and keywords, if need be.
> ! vi Ttest.d  # fill in the outline file using an editor
> ! cp Ttest.d .Data/.Help/Ttest  # place in .Help as 'Ttest'
> ?Ttest                 # check that it reads correctly.
```

Windows

It is possible to create standard **Windows** help facilities (such as those used by our libraries and by the **S-PLUS** system) but this is well beyond the scope of this book. (See the on-line complements.)

It is, however, possible to create simple ASCII help documents that are often all that is needed for many functions, at least as a temporary measure. This is extremely easy. A call to the prompt function such as

```
> prompt(Ttest)
```

will create an outline help file for the object in the directory _Data_Help using the same *mapped* file as that used by the object itself in the _Data directory. The mapped file name is reported in the command window. Using an editor such as notepad the user should then complete the template file.

If a correction is needed to the help file but the mapped file name has been forgotten it may be found by using

```
> true.file.name(Ttest)
```

4.6 Calling the operating system

Sometimes there is an operating-system command that will do exactly what you want. The functions unix, win3 and dos are provided for communication with operating-system commands.

Unix

The syntax of the unix command is

```
unix(command, input=NULL, output.to.S=T)
```

The argument command is a character string which is passed to a Bourne shell to execute (but could invoke another shell itself). If input is specified, it is written to a file used for standard input to the command. By default the standard output is returned line-by-line as a character vector, but if the command needs to interact with the user, supply the argument output.to.S=F when the exit status of the command is returned (as returned by the system call, normally 256 times the usual exit status).

For example, we may test if a file exists and is readable, before attempting to read data from it, using

```
file.exists <- function(name)
    unix(paste("test -r", name), output.to.S=F) == 0
```

Useful functions in conjunction with unix are tempfile which creates a unique name for a temporary file (but not the file itself, dput and dget to write and read objects and unlink to remove files. The functions tempfile and unlink are themselves simple calls to unix:

```
tempfile <- function(pattern = "file")
    paste("/tmp/", pattern, unix("echo $$"), sep = "")
unlink <- function(x)
    invisible(unix(paste("rm", paste("'", x,, "'", sep = "",
                collapse = " ")), output = F))
```

We use these to write a function to compare two objects (a simpler version of the S-PLUS function objdiff):

```
objchk <- function(x, y) {
    old <- tempfile("old"); dput(x, old)
    new <- tempfile("new"); dput(y, new)
    on.exit(unlink(c(old, new)))
    unix(paste("diff", old, new, "| less"), output = F)
    invisible()
}
```

A good use of unix is to manipulate character strings; for example to convert to upper or to lower case we could use the functions

```
to.upper <- function(str) unix('tr "[a-z]" "[A-Z]"',str)
to.lower <- function(str) unix('tr "[A-Z]" "[a-z]"',str)
```

One very useful function based on operating system calls is unix.time, which times its argument and returns a 5-element vector given by the Unix command time, the times (in seconds) of the user, system and elapsed times, plus the user and system times taken by child processes (if any). Further, although this is a function, it is arranged so that assignments within it are done at the level of its call. Note that if the argument returns an invisible result, this results in the return value from unix.time being marked as invisible and not printed. (If this happens unintentionally, use print(.Last.value) .)

The function proc.time returns a 5-element vector giving the times since the beginning of the session.

Windows

The Windows interface function is

```
win3(command, multi=F, trans=F)
```

and the arguments multi determine whether the command is multitasked, and trans whether file paths are converted between Unix–style "/test/file" and MS-DOS–style "\test\file". S-PLUS 4.0 has the command system that differs from win3 only in the way the exit status is returned.

The MS-DOS interface function is

```
dos(command, input, output=T, multi=F, trans=F, redirection=F)
```

By default the input is appended to the command, but if redirection=T, redirection is used (as with the unix command). Using multi=T sets up a MS-DOS box for the command which must be closed explicitly.

For example we could use dos to write a file.exists function:

```
file.exists <- function(name)
    length(dos(paste("attrib", name))) > 0
```

but this operation is already available using the system function access.

The function dos.time is the equivalent of unix.time, but measures elapsed time (in seconds) only. The function proc.time returns the elapsed time of the present S-PLUS session.

4.7   Recursion and handling vectorization

Functions in S are allowed to be recursive, that is they are allowed to call themselves. This idea is both interesting in itself and useful for an understanding of how S organizes its calculations when functions are called, but in practice it should be used cautiously as it can lead to slow and memory-intensive code. In a few cases it can provide a neat and effective solution to an intrinsically recursive problem.

For example consider the problem of generating all possible subsets of size r from a set of elements of size n. A recursive algorithm can be based on the following simple observation. Suppose you single out one of the elements, say the first. Then the subsets of size r consist of those that contain the first and those that do not. The first group may be generated by attaching the first object to each subset of size $r - 1$ selected from the $n - 1$ others, and the second group consists of all subsets of size r from the $n - 1$ others.

If the size of the result is not too large an efficient way to do this is to store the subsets as the rows of an array. The following code will do this:

```
subsets <- function(r, n, v = 1:n)
    if(r <= 0) NULL else
    if(r >= n) v[1:n] else
    rbind(cbind(v[1], subsets(r - 1, n - 1, v[-1])),
                        subsets(r,     n - 1, v[-1]))
```

One potential problem with this is that the name subsets occurs in the body of the function definition. If we re-assign the function body to another name and use the name subsets for something else, or discard it, our function body will cease to work. It would be useful to have a function body that continued to work regardless of the name given it. This is the purpose of the function Recall. The amended function and a check are

```
subsets <- function(r, n, v = 1:n)
    if(r <= 0) NULL else
    if(r >= n) v[1:n] else
    rbind(cbind(v[1], Recall(r - 1, n - 1, v[-1])),
```

```
                            Recall(r,      n - 1, v[-1]))
> subsets(3, 5)
     [,1] [,2] [,3]
[1,]    1    2    3
[2,]    1    2    4
[3,]    1    2    5
[4,]    1    3    4
[5,]    1    3    5
[6,]    1    4    5
[7,]    2    3    4
[8,]    2    3    5
[9,]    2    4    5
[10,]   3    4    5
```

Of course if the size of the result is too large it may have to be generated sequentially or in smaller groups. This is the subject of Exercise 4.5.

Providing vectorized functions

The mathematical functions provided by S-PLUS such as sin, log and dnorm have the useful property that if their first argument is a vector, the result is a vector of the same length. In some functions, such as ifelse or pnorm, the recycling rule applies to all arguments so the length of the result is the length of the longest argument.

```
> pnorm(1, mean = 0, sd = 1:5)
[1] 0.84134 0.69146 0.63056 0.59871 0.57926
```

In designing a general-purpose function some care should be given to making it conform to this convention. Sometimes it is required by other functions; for example the integrate function may be used to find the area under a function of one variable. The integrand is supplied as an S function with the variable of integration, x, as the first argument. The result of integrate is a long list giving various results from the algorithm but only the first component, integral, is of interest to us here.

To make things convenient we will write a wrapper function to give the result a class and provide a print method.

```
> Integrate <- function(...)
  structure(integrate(...), class = "integral")
> print.integral <- function(x, ...) {
    y <- x
    x <- x$integral
    NextMethod("print")
    invisible(y)
  }
```

We can try it out on a known case:

```
> Integrate(sin, 0, pi)
[1] 2
```

It is sometimes claimed that `integrate` cannot integrate constant functions, but the problem is nearly always that the function used does not obey the vector result convention. Compare

```
> Integrate(function(x) 2, 0, 1)
[1] 5.8112e+178
> Integrate(function(x) rep(2, length(x)), 0, 1)
[1] 2
```

done in 3.x. In 4.x this does report the informative error message that the function has not been vectorized.

We can illustrate some of the techniques of writing such vectorized functions by the digamma function. This function is usually written as $\psi(z)$ and is defined by

$$\psi(z) = \frac{d}{dz} \log \Gamma(z)$$

It occurs in statistics in several places, such as estimating the parameters of the gamma distribution or fitting the negative binomial distribution. (It is also used in our function qda.) We will write a function that accepts real or complex arguments, but unless the real part is positive the result will be `NA`.

If the real part, $\mathrm{Re}(z)$, is large enough, say larger than 5, the result may be accurately calculated by an asymptotic expansion derived from Stirling's formula for the gamma function, namely (Abramowitz & Stegun, 1965)

$$\psi(z) = \log z - \frac{1}{2z} - \frac{1}{12z^2} + \frac{1}{120z^4} - \frac{1}{252z^6} + \frac{1}{240z^8} - \frac{1}{132z^{10}} \\ + \frac{691}{32760z^{12}} - \frac{1}{12z^{14}} + \frac{3617}{8160z^{16}} + O(z^{-18})$$

For smaller values of $\mathrm{Re}(z)$ there is a recurrence formula that may be used to relate the calculation to the case of larger z:

$$\psi(z) = \psi(z + k) - \frac{1}{z} - \frac{1}{z+1} - \cdots - \frac{1}{z+k-1}$$

Since we only cater for the case $\mathrm{Re}(z) > 0$ we use this formula with $k = 5$ for small $\mathrm{Re}(z)$.

An S function (in library MASS) for $\psi(z)$ that uses these ideas is

```
digamma <- function(z) {
    if(any(omit <- Re(z) <= 0)) {
        ps <- z; ps[omit] <- NA
        if(any(!omit)) ps[!omit] <- Recall(z[!omit])
        return(ps)
    }
    if(any(small <- Re(z) < 5)) {
```

```
        ps <- z
        x <- z[small]
        ps[small] <- Recall(x + 5) - 1/x - 1/(x + 1) -
                        1/(x + 2) - 1/(x + 3) - 1/(x + 4)
        if(any(!small)) ps[!small] <- Recall(z[!small])
        return(ps)
    }
    x <- (1/z)^2
    tail <- x * (-1/12 + x * (1/120 + x * (-1/252 +
            x * (1/240 + x * (-1/132 + x * (691/32760 +
            x * (-1/12 + 3617 * x/8160)))))))
    log(z) - 1/(2 * z) + tail
}
```

The strategy is

1. Check if any components of z have a non-positive real part and make the result NA in those positions. Fill any other components by a recursive call.

2. Check if any components of z have a small real part. If so fill the corresponding components of the result by a recursive call using the recurrence formula and the other components using a recursive call.

3. Finally implement the truncated asymptotic expansion. Use a nested form to avoid calculating large negative powers and to reduce round-off error through underflow.

Remember that where statements are spread over two lines the first line *must* be syntactically incomplete. For example if in the two lines

```
        ps[small] <- Recall(x + 5) - 1/x - 1/(x + 1) -
                        1/(x + 2) - 1/(x + 3) - 1/(x + 4)
```

the minus sign from the end of the first line were transferred to the beginning of the second, both expressions would be evaluated separately and not combined, and the result of the function would be erroneous where this branch was used. This is a common kind of error, often very difficult to locate. In extreme cases the entire right-hand side might be enclosed within parentheses for safety's sake. When S reproduces an expression such as the nested product term `tail` of the function above in printed code unnecessary bracketing is often generated, presumably to insure against this kind of error.

Consider another problem where the vector result issue is a little more subtle. Suppose we wished to evaluate integrals of the form

$$\int_a^b \prod_{i=1}^k (d_i x + e_i)\, dx$$

An obvious integrand function such as

```
gn <- function(x, d, e) prod(d*x + e)
```

will not work since it always returns a scalar result and employs the recycling rule. The result must have the same length as the first argument, x, and each component of d and e must be used to evaluate the function for every component of x. One way to do this is

```
fn <- function(x, d, e) apply(outer(d, x) + e, 2, prod)
```

To try it out consider a special case. The integrate function may be supplied with a function of several arguments, but the first must be x. Constant values for any remaining arguments can be supplied as (exactly) named arguments to integrate.

```
> Integrate(fn, -4, -2, d = 1:3, e = 4:6)
[1] 4
```

It is an elementary calculus exercise to check that $\int_{-4}^{-2}(x+4)(2x+5)(3x+6)\,dx = 4$. The inappropriate function gn is faster, but again the result is inaccurate (and system-dependent, but reports an error in 4.x).

```
> Integrate(gn, -4, -2, d = 1:3, e = 4:6)
[1] 66.062
```

4.8 Frames

An *evaluation frame* is a list that associates names with values. Their purpose is much the same as dictionaries on the search path. Evaluation frames and object dictionaries are collectively called *databases*.

The concept of a frame has already arisen several times, most explicitly in Section 4.5 where inspect allowed us to walk through, investigate and even manipulate the frames in existence at any one time.

When an S expression is evaluated within a frame, any values needed are obtained by reference to that frame *first*. This first point of reference is called the *local frame* of the expression. To show this in action we can use the function eval which takes an S object of mode expression and (optionally) a local frame:

```
> a <- 1;  b <- 2;  d <- 3
> eval(expression(a + b + d), local=list(a=10, b=20))
[1] 33
> a + b + d
[1] 6
```

Any names (like d in this example) not satisfied by the local frame are next referred to the frame of the top-level expression, frame 1, then the frame of the session, frame 0, and then on to the search path. The sequence of databases consisting of the local frame, frame 1, frame 0 and the search path may be called the *search path* for an object. Notice that the search path for an object at, say, frame 6, does *not* include frames 2 to 5; this can cause unexpected problems if not borne in mind. In particular, objects created in parent functions will *not* be found in this search. This aspect causes puzzles so frequently that we emphasize it:

WARNING!   The S scope rules
Objects not found in the current frame are next referred to the frame
of the top-level expression, frame 1, then the frame of the session,
frame 0 and then on to the search path. Objects created in parent
functions will not *be found in this search.*

An small example where it sometimes seems most puzzling may underline the point. Consider the following elementary function that performs a regression using a simple Box and Cox transformation, defined by y^λ if $\lambda \neq 0$ or $\log y$ if $\lambda = 0$.

```
bcreg <-  function(x, y, lambda)
{                              # WARNING: errors present
    bc <- function(y)
          if (abs(lambda) < 0.001) y^lambda else log(y)
    lsfit(x, bc(y))
}
```

This fails because within the locally defined function `bc` the argument `lambda` of the enclosing function is not visible. There are two ways to correct it. The preferred way is to include `lambda` as a second argument to the internal function `bc` and on the call, but a second way is to get it directly:

```
bcreg <-  function(x, y, lambda)
{
    bc <- function(y) {    # WARNING: artificial
            lambda <- get("lambda", frame = sys.parent())
            if (abs(lambda) < 0.001) y^lambda else log(y)
        }
    lsfit(x, bc(y))
  }
```

This is definitely artificial and not recommended, but it does illustrate that with low-level tools like `get`, discussed in more detail below, variables may be obtained from any frame or position on the search path.

Frame 0 is born with the session and dies with the session. New entries may be added to this frame, but unlike ordinary assignments made during the session they are not permanent. To be permanent an object must reside in an object dictionary on the search path. Frame 0 is used by the system for storing options set by `options` in object `.Options`.

When an expression is evaluated at the interactive level it initiates a new frame called frame 1, usually containing only the unevaluated expression and an auto-print flag specifying whether the result is to be printed or not.

```
> objects(frame=0)
[1] ".Device"              ".Devices"              ".Options"
[4] ".PostScript.Options"
> objects(frame=1)
[1] ".Auto.print" ".Last.expr"
```

If the expression that generated frame 1 contains function calls, these generate frames at higher levels as they are processed. These frames contain the names and values of arguments and local variables within the function. Normal assignments within functions only change the local frame, and so do not affect values for variables outside that frame. This is how recursion is accommodated. Each recursive call establishes its own frame without altering values for the same variables at different recursion levels.

The contents of any database can be inspected and manipulated using the functions in Table 4.3, each of which has an argument `frame` to specify a particular frame number and `where` to specify (alternatively) a position number in the search path. The functions `get` and `exists` by default consult the entire search path. The others consult the local frame (or the working database at the interactive level) unless another focus is specified.

Table 4.3: Functions to access arbitrary databases

`assign`	Creates a new `name=value` pair in a specified database.
`exists`	Tests whether a given object exists in the search path.
`get`	Returns a copy of the object, if it exists; otherwise returns an error.
`objects`	Returns a character vector of the names of objects in the specified database.
`remove`	Deletes specified objects, if they exist, from the specified database.

The function `assign` has an argument `immediate` which defaults to false for databases on disk, but if set to true forces the assignment immediately, circumventing the S backout mechanism. This is useful in error-handling code (as on page 132) and sometimes to save memory.

There is a special assignment operator, `<<-`, which can be used within a function to make an assignment to the working database. If the user of the function is unaware that such an assignment is going to be made it can have disastrous results. We discourage use of the superassignment operator as a general practice, and particularly for functions to be made publicly available, except when the primary purpose of the function is to achieve the assignment side effect, such as with functions like `source` and `fix`.

As another example, the `get` function can be made to 'look at itself':

```
> get(".Last.expr", frame=1)
expression(get(".Last.expr", frame = 1))
> get(".Auto.print",frame=1)
[1] T
```

Notice that the object `.Last.expr` in frame 1 can be used to discover the interactive level expression that enacted the present expression. This can sometimes be useful, for example, if it is necessary to decide if the present top-level expression is an assignment or not.

In frame 1 automatic printing is normally turned on. If an expression or the return value of a function is given as `invisible(value)` it is not automatically printed. The `invisible` function itself is quite short:

```
> invisible
function(x = NULL)
{
    assign(".Auto.print", F, frame = sys.parent(2))
    x
}
```

The function `sys.parent` takes an integer argument n and returns the index of the frame n generations behind the present. The assignment inside `invisible` simply turns off auto-printing in the frame of the parent of the function or expression that called `invisible`. Other functions whose names begin with "`sys.`" are available to determine other aspects of the current memory frames. See the on-line help for `sys.parent` for more details.

Sometimes it is necessary for variables in a function to be made available to functions it calls. The easiest way to do this is to assign the variables in frame 1, since this is part of the globally visible search path throughout the evaluation of the current top-level expression. Care must be taken not to mask other variables further down the search path unintentionally. One way of minimising the chance of this is to use some naming convention for such temporaries so that they are unlikely to conflict. Starting the name with a period is a common convention. The variables `inspect` subliminally inserts into code often have a name both beginning and ending with a period.

Communicating through frame 0 or 1

Consider again the integration example from page 153. Suppose we wished to record the x-values at which the integration routine evaluated the integrand. One way of doing this would be to insert `print` statements in the integrand function, but this would not allow us to manipulate the values afterwards, for example to plot them. We could assign the values to the primary permanent database, but a better plan is to use frame 0 to accumulate the values, at least temporarily. We do via `trace`.

```
> trace(fn, expression(assign(".x",c(.x,x),frame=0)), print=F)
> assign(".x", NULL, frame=0)
NULL
> Integrate(fn, -4, -2, d = 1:3, e = 4:6)
[1] 4
> untrace(fn)
> xyplot(fn(.x, 1:3, 4:6) ~ .x, xlab="x", ylab="fn(x, d, e)")
```

The result is shown in Figure 4.1.

If variables are only needed during the duration of a function call, it is better to assign them to frame 1 (which is on the search path) than to frame 0, as frame 1 exists only during the execution of the current top-level expression.

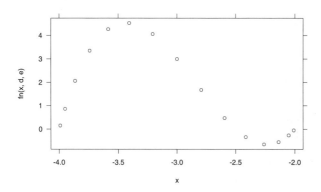

Figure 4.1: The choice of points in a numerical integration.

4.9 Using C and FORTRAN routines

A very important powerful feature of the S environment is that it is not restricted to functions written in the S language but may load and use compiled routines written in either C or FORTRAN.[4] Moreover there are ways in which such externally written and compiled routines may communicate with the S session and make use of S functions and objects.

The details of how to write, compile and load external routines depend very much on the platform on which S-PLUS is running, and may change often, either because of changes to the operating system or in new releases of S-PLUS. Rather than present a static account of the subject in the printed book we have decided treat it in the on-line complements. They contain as complete an account as we currently have and every effort will be made to keep them up to date.

4.10 Working within limited memory

Most S-PLUS programmers find early in their careers that they have written code which exhausts the physical (RAM) memory available to the S-PLUS process, and then proceeds to spend almost all its time in allocating virtual memory. Such code can reduce the fastest workstation to page thrashing, which will reduce severely the size of problem which can be tackled. This section give some hints on using memory more efficiently.

The function `memory.size` is provided to help trace where memory is being allocated. It returns the size of the currently allocated memory in bytes. The function `allocated` gives more detail of the memory usage.

---

[4] There are some S-PLUS student licences where dynamic loading of C and FORTRAN routines is not supported and some combinations of version and platform on which the facility is rather limited.

Loops

A major issue is that S is designed to be able to back out from uncompleted calculations, so that the memory used in intermediate calculations is retained until they are committed. This applies to `for`, `while` and `repeat` loops, for which none of the steps are committed until the whole loop is completed. (In recent versions of S-PLUS attempts are made to release unused memory at the end of each iteration, which has helped considerably.)

It is worth thinking hard if a loop is really necessary. It is rarely necessary to loop over the index of a vector or list: the computation should if at all possible be vectorized (the `ifelse` function is often handy) or a function such as `sapply` or `lapply` can be used. Operations on arrays can often be partially vectorized, and simple matrix operations can be replaced by elementwise or matrix multiplication.

Consider the problem of finding the the distribution of the determinant of a 2×2 matrix where the entries are independent and uniformly distributed digits $0, 1, \ldots, 9$. This amounts to finding all possible values of $ac - bd$ where a, b, c and d are digits. One solution we were shown is

```
val <- NULL
for(a in 0:9)
  for(b in 0:9)
    for(c in 0:9)
      for(d in 0:9)
        val <- c(val, a*b - c*d)
freq <- table(val)
```

This can take a very long time to complete, if ever it does (and will generate many warning messages because of the double use of `c`). A much faster way is to use `outer`:

```
val <- outer(0:9, 0:9, "*")
val <- outer(val, val, "-")
freq <- table(val)
```

(Notice that the third argument to `outer` may be a character string giving the name of a function of two arguments.)

One of us was asked how the following code section, which accumulates 1000 samples (with replacement) from a finite population, could be made to run faster.

```
mat <- matrix(0, 1000, 6)
for(i in 1:1000) mat[i, ] <- sample(x, 6, replace = T)
```

The solution is to choose one sample of size 6000 and place it directly in a matrix:

```
mat <- matrix(sample(x, 6000, replace = T), 1000, 6)
```

but this is only possible for sampling with replacement, of course.

Yet another example comes from the function `Correlogram` in an earlier version of our `spatial` library, where we replaced

```
for(j in 2:length(x)-1) zx <- c(zx, (z[(j+1):n]-zm)*(z[j]-zm))
```

by

```
zscale <- z - zm
zx <- outer(zscale, zscale)
zx <- zx[lower.tri(zx)]
```

Not all calculations can be vectorized, especially those that depend on the result of the previous calculation. The functions cumsum, cumprod, cummax and cummin are sometimes useful to vectorize calculations of this sort.

It was once true that functions such as sapply concealed explicit loops, but this is no longer the case. Using standard constructions is likely to benefit most from future performance improvements in S-PLUS.

If a loop is really necessary

One tip is to package parts of the calculation in functions, as when the function call is completed at least some of the memory used inside the function is released. However, to combat memory build-up, it is essential to return to the session prompt (usually >) from time to time. Thus, if possible, loops should be done at the top level and 'unrolled' as much as possible. This implies

(a) that code should not be run using source, since this is a function to which the delayed commitment rules apply, and

(b) large amounts of memory *may* be saved by replacing

```
for (iter in 1:1000)  myfunction(iter, moreargs)
```

by

```
for (iter in   1:100) {myfunction(iter, moreargs); NULL}
for (iter in 101:200) {myfunction(iter, moreargs); NULL}
       ....
for (iter in 801:900) {myfunction(iter, moreargs); NULL}
for (iter in 901:1000){myfunction(iter, moreargs); NULL}
```

provided this set of commands is typed (or pasted) into the command window, or BATCH is used or (for Unix versions) input is redirected to come from a file or pipe.

Ensuring that the last statement in a loop returns nothing (and NULL does) avoids storing all the last values.

The unrolling of (b) is the rationale of the For function. Running

```
For(iter = 1:1000, myfunction(iter, moreargs),
   grain.size = 100, debug = T)
```

shows us the commands to be run before running them. There will be 1000 calls, grouped (with braces) into groups of 100, so that calculations are committed after every 100. Choosing a suitable grain.size is important for very simple calculations within a loop, but the default value of 1 suffices for calculations taking several seconds or more.

Provided the argument exec=T is given to the For function (the default under Unix), another S-PLUS process is started and used to run the commands of the For loop. Once this has terminated, the databases are synchronized so that the results are visible to the calling S-PLUS process, and of course all the memory used by the child S-PLUS process is released. However, as the results are computed by another process, care may be needed to ensure that the objects needed are visible to the other process. As an example, consider the following simplified version of the function cv.tree.

```
CV.tree <- function(object, rand, FUN = prune.tree)
{
    m <- model.frame(object)
    if(missing(rand))
        rand <- sample(10, length(m[[1]]), replace = T)
    p <- FUN(object)
    expr <- expression({
        tlearn <- tree(model = .m[.rand != i,  , drop = F])
        plearn <- .FUN(tlearn, newdata =
            .m[.rand == i,  , drop = F], k = .pk)
        .cvdev <- .cvdev + plearn$dev
    })[[1]]
    assign(".m", m, w=1); assign(".expr", expr, w=1)
    assign(".FUN", FUN, w=1); assign(".rand", rand, w=1)
    assign(".cvdev", 0, w=1); assign(".pk", p$k, w=1)
    For(i = unique(.rand), eval(.expr), library(MASS, first=T))
    p$dev <- get(".cvdev", w=1)
    remove(c(".m",".FUN",".rand",".cvdev",".pk",".expr"), w=1)
    p
}
```

As the For loop is called from within a function, we have to ensure that variables set within the function and its arguments are stored in the working database; names beginning with a dot are used to help avoid over-writing the user's objects. Note that we must explicitly invoke the libraries we need via the first argument of For.

Because of the way the For function is implemented, nested For functions should be used with care and will always give warnings.

Tips

This subsection is a collection of small points, each of which can help save memory usage.

1. Re-use temporary variables if appropriate. If a variable of length n is used for several different purposes within a function, S-PLUS will usually re-use the allocated storage. Note that this can make code much harder to read, as variables will not have meaningful names—so add some comments.

2. Avoid growing an object. Suppose we want to keep the results in a vector res. Do *not* use

```
res <- NULL
for (iter in 1:1000)
    res <- c(res, myfunction(iter, moreargs))
```

as this forces a copy of res to be made at each iteration; rather use

```
res <- numeric(1000)
for (iter in 1:1000)
    res[iter] <- myfunction(iter, moreargs)
```

This is even more important if the result is a matrix rather than a vector. In some circumstances the final size of res may be unknown, in which case allocate a sufficient size and shrink the vector at the end of the loop. This is part of the reason for the successful speed-up of the Correlogram function given earlier.

3. Remove attributes from frequently used objects. The names attribute of a vector can take as much space as its values.

4. Save useful bits of computations. Good examples are logical and subscripting calculations; for example a classification tree will have

```
ctree <- !is.null(attr(tree, "ylevels"))
```

true, and this assignment can be done within the first if statement of a function which handles both regression and classification trees.

5. Conversely, do not name intermediate results unnecessarily. If an intermediate result is only needed once further down in a calculation, consider weaving it in as an expression at that point rather than as the name of an already evaluated value.

The function remove can be used to remove intermediate results that are no longer needed, but care will be needed to specify the right frame, for example by

```
remove("foo", frame=sys.nframe())
```

6. Using get with an explicit frame or with the argument immediate=T instead of letting the parser find a data object ensures that the object is not retained in memory for possible future use. Using assign with immediate=T (see page 155) or a direct call to dbwrite saves the memory associated with backing-out an assignment.

7. Coding the innermost calculations as C or FORTRAN functions can give rise to dramatic savings in memory usage and computation time. The first edition had both C and S forms of our `spatial` library, but as the C version was 10–15 times faster we have dropped the S version.

4.11 Exercises

4.1.   Write a function to determine which integers (less than 10^{10}) in a given vector are prime.

4.2.   Given two vectors a and b, compute s, where s[i] is the number of a[j] <= b[i], for a much longer than b. Then consider a[j] >= b[i]. (Based on a question of Frank Harrell.)

4.3.   Implement `print` and `plot` method functions for the class `"lda"` of Chapter 13. (Perhaps the `plot` method should give a barchart of the singular values or plot the data on the first few linear discriminants, identifying the groups and marking their means.)

4.4.   Write an S function to increment its argument object. [Not as easy as it looks: based on an example of David Lubinsky, *Statistical Science* **6**, p. 356.]

4.5.   Write a function that generates the all possible subsets of size r from a set of size n successively rather than simultaneously. Given one subset it should produce the next in lexicographic order, or report that no further subsets exist in some convenient way that can be trapped in a program.

4.6.   For each row of a matrix X, we want to find which row of a matrix M is nearest in the sense of Euclidean distance. Use `apply` to find out. (You may assume that there is a unique nearest row.)

4.7.   For designed experiments it is often useful to predict the response at each combination of levels of the treatment factors. Write a function `expand.factors` which takes any number of factor arguments and returns a data frame with the factors as columns and each combination of levels occurring as exactly one row.

Try a direct solution without using `expand.grid`.

Chapter 5

Distributions and Data Summaries

In this chapter we cover a number of topics from classical univariate statistics. Many of the functions used are S-PLUS extensions to S.

5.1 Probability distributions

In this section we confine attention to *univariate* distributions, that is, the distributions of random variables X taking values in the real line $\mathbb{R}$.

The standard distributions used in statistics are defined by probability density functions (for continuous distributions) or probability functions $P(X = n)$ (for discrete distributions). We will refer to both as *densities* since the probability functions can be viewed mathematically as densities (with respect to counting measure). The *cumulative distribution function* or CDF is $F(x) = P(X \leqslant x)$ which is expressed in terms of the density by a sum for discrete distributions and as an integral for continuous distributions. The *quantile function* $Q(u) = F^{-1}(u)$ is the inverse of the CDF where this exists. Thus the quantile function gives the percentage points of the distribution. (For discrete distributions the quantile is the smallest integer m such that $F(m) \geqslant u$.)

S-PLUS has built-in functions to compute the density, cumulative distribution function and quantile function for many standard distributions. In many cases the functions can not be written in terms of standard mathematical functions, and the built-in functions are very accurate approximations. (But this is also true for the numerical calculation of cosines, for example.)

The first letter of the name of the S function indicates the function, so for example dnorm, pnorm, qnorm are respectively, the density, CDF and quantile functions for the normal distribution. The rest of the function name is an abbreviation of the distribution name. The distributions available are listed in Table 5.1.

The first argument of the function is always the observation value q (for quantile) for the densities and CDF functions, and the probability p for quantile functions. Additional arguments specify the parameters, with defaults for 'standard' versions of the distributions where appropriate. Precise descriptions of the parameters are given in the on-line help pages.

Table 5.1: S function names and parameters for standard probability distributions.

Distribution	S name	parameters
beta	beta	shape1, shape2
binomial	binom	size, prob
Cauchy	cauchy	location, scale
chisquared	chisq	df
exponential	exp	rate
F	f	df1, df2
gamma	gamma	shape[1]
geometric	geom	prob
hypergeometric	hyper	m, n, k
log-normal	lnorm	meanlog, sdlog
logistic	logis	location, scale
negative binomial	nbinom	size, prob
normal	norm	mean, sd
normal range	nrange	size
Poisson	pois	lambda
stable	stab	index, skewness
T	t	df
uniform	unif	min, max
Weibull	weibull	shape, scale
Wilcoxon	wilcox	m, n

These functions can be used to replace statistical tables. For example, the 5% critical value for a (two-sided) t test on 11 degrees of freedom is given by qt(0.975, 11), and the p-value associated with a Poisson(25)-distributed count of 32 is given by (by convention) 1 - ppois(31, 25). The functions can be given a vector first argument to calculate several p-values or quantiles.

A function mvrnorm to simulate from a multivariate normal distribution is provided in our library, and **S-PLUS 4.0** has rmvnorm, dmvnorm and pmvnorm.

Q-Q Plots

One of the best ways to compare the distribution of a sample x with a distribution is to use a Q-Q plot. The normal probability plot is the best-known example. For a sample x the quantile function is the inverse of the empirical CDF, that is

$$\text{quantile}\,(p) = \min\,\{z \mid \text{proportion } p \text{ of the data} \leqslant z\,\}$$

The function quantile calculates the quantiles of a single set of data. The function qqplot(x, y, ...) plots the quantile functions of two samples x

---

[1] gamma has an additional parameter rate in S-PLUS 4.0.

and y against each other, and so compares their distributions. The function qqnorm(x) replaces one of the samples by the quantiles of a standard normal distribution. This idea can be applied quite generally. For example, to test a sample against a t_9 distribution we might use

```
plot( qt(ppoints(x),9), sort(x) )
```

where the function ppoints computes the appropriate set of probabilities for the plot. These values are $(i - 1/2)/n$ for $n \geqslant 11$ and $(i - 3/8)/(n + 1/4)$ for $n \leqslant 10$, and are generated in increasing order.

The function qqline helps assess how straight a qqnorm plot is by plotting a straight line through the upper and lower quartiles. To illustrate this we generate 250 points from a t distribution and compare them to a normal distribution (Figure 5.1). This can be done using the base graphics or with Trellis:

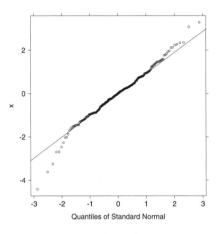

Figure 5.1: Normal probability plot of 250 simulated points from the t_9 distribution, plotted using Trellis.

```
x <- rt(250, 9)
qqnorm(x); qqline(x)
# OR
qqmath(~ x, distribution=qnorm, aspect="xy",
    prepanel = prepanel.qqmathline,
    panel = function(x, y, ...) {
        panel.qqmathline(y, distribution=qnorm, ...)
        panel.qqmath(x, y, ...)
    },
    xlab= "Quantiles of Standard Normal"
)
```

The greater spread of the extreme quantiles for the data is indicative of a long-tailed distribution. The prepanel argument is used to ensure that the y axis covers both the line and the points, and the aspect argument banks the plot to $45°$.

This method is most often applied to residuals from a fitted model. Some people prefer the roles of the axes to be reversed, so that the data goes on the x-axis and the theoretical values on the y-axis; this is achieved for qqnorm by the argument datax=T.

5.2 Generating random data

There are S functions to generate independent random samples from all the probability distributions listed in Table 5.1. These have prefix r and first argument n, the size of the sample required. For most of the functions the parameters can be specified as vectors, allowing the samples to be non-identically distributed. For example, we can generate 100 samples from the contaminated normal distribution in which a sample is from $N(0, 1)$ with probability 0.95 and otherwise from $N(0, 9)$, by

```
contam <- rnorm( 100, 0, (1 + 2*rbinom(100, 1, 0.05)) )
```

The function sample re-samples from a data vector, with or without replacement. It has a number of quite different forms. Here n is an integer, x is a data vector and p is a probability distribution on 1, ..., length(x):

sample(n)	select a random permutation from $1, \ldots, n$.
sample(x)	randomly permute x.
sample(x, replace=T)	a bootstrap sample.
sample(x, n)	sample n items from x without replacement.
sample(x, n, replace=T)	sample n items from x with replacement.
sample(x, n, replace=T, prob=p)	probability sample of n items from x

The last of these provides a way to sample from an arbitrary (finite) discrete distribution; set x to the vector of values and prob to the corresponding probabilities.

The numbers produced by these functions are of course *pseudo*-random rather than genuinely random. The state of the generator is controlled by a set of integers stored in the S object .Random.seed. Whenever sample or an 'r' function is called, .Random.seed is read in, used to initialize the generator, and then its current value is written back out at the end of the function call. If there is no .Random.seed in the current working database, one is created with default values. To re-run a simulation the initial value of .Random.seed needs to be recorded. For some purposes it may be more convenient to use the function set.seed with argument between 1 and 1000, which selects one of 1000 preselected seeds.

Details of the pseudo-random number generator

Some users will want to understand the internals of the pseudo-random number generator; others should skip the rest of this section. Background knowledge on pseudo-random number generators is given by Ripley (1987, Chapter 2). We describe the generator which was current at the time of writing and which had been in use for several years (with a slight change in 1994); it is of course not guaranteed to remain unchanged.

The current generator is based on George Marsaglia's "Super-Duper" package from about 1973. The generator produces a 32-bit integer whose top 31 bits are divided by 2^{31} to produce a real number in $[0, 1)$. The 32-bit integer is produced by a bitwise exclusive-or of two 32-bit integers produced by separate generators. The C code for the S random number generator is (effectively)

```
unsigned long congrval, tausval;
static double xuni()
{
        unsigned long n, lambda = 69069;
        congrval = congrval * lambda;
        tausval ^= tausval >> 15;
        tausval ^= tausval << 17;
        n = tausval ^ congrval;
        return ( (((n>>1) & 017777777777)) / 2147483648.);
}
```

In S-PLUS 3.2 this was been modified to skip zero outcomes, so the sequence of random numbers will differ between versions up to 3.1 and 3.2 and later.

The integer `congrval` follows the congruential generator

$$X_{i+1} = 69069 X_i \bmod 2^{32}$$

as unsigned integer overflows are (silently) reduced modulo 2^{32}, that is, the overflowing bits are discarded. As this is a multiplicative generator (no additive constant), its period is 2^{30} and the bottom bit must always be odd (including for the seed).

The integer `tausval` follows a 32-bit Tausworthe generator (of period 4 292 868 097 $< 2^{32} - 1$):

$$b_i = b_{i-p} \,\text{xor}\, b_{i-(p-q)}, \qquad p = 32, \quad q = 15$$

This follows from a slight modification of Algorithm 2.1 of Ripley (1987, p. 28). (In fact, the period quoted is true only for the vast majority of starting values; for the remaining 2 099 198 non-zero initial values there are shorter cycles.)

For most starting seeds the period of the generator is $2^{30} \times$ 4 292 868 097 $\approx 4.6 \times 10^{18}$, which is quite sufficient for calculations which can be done in a reasonable time in S. The current values of `congrval` and `tausval` are encoded in the vector `.Random.seed`, a vector of 12 integers in the range $0, \ldots, 63$. If x represents `.Random.seed`, we have

$$\text{congrval} = \sum_{i=1}^{6} x_i \, 2^{6(i-1)} \quad \text{and} \quad \text{tausval} = \sum_{i=1}^{6} x_{i+6} \, 2^{6(i-1)}$$

5.3 Data summaries

Standard univariate summaries such as mean, median and var are available. The summary function returns the mean, quartiles and the number of missing values, if non-zero.

The var function will take a data <u>matrix</u> and give the variance-covariance matrix, and cor computes the correlations, either from two vectors or a data matrix. The function cov.wt returns the means and variance matrix and optionally the correlation matrix of a data matrix. As its name implies, the rows of the data matrix can be weighted in these summaries.

The function quantile computes quantiles at specified probabilities, by default $(0, 0.25, 0.5, 0.75, 1)$ giving a "five number summary" of the data vector. This function linearly interpolates, so if $x_{(1)}, \ldots, x_{(n)}$ is the ordered sample,

$$\text{quantile(x, p)} = [1 - (p(n-1) - \lfloor p(n-1) \rfloor)] \, x_{1+\lfloor p(n-1) \rfloor} \\ + [p(n-1) - \lfloor p(n-1) \rfloor]] \, x_{2+\lfloor p(n-1) \rfloor}$$

where $\lfloor \ \rfloor$ denotes the 'floor' or integer part of. (This differs from the definitions of a quantile given earlier.) There are also standard functions max, min and range.

The functions mean and cor will compute trimmed summaries using the argument trim. More sophisticated robust summaries are discussed in Chapter 8.

These functions differ in the way they handle missing values. Functions mean, median, max, min, range and quantile have an argument na.rm which defaults to false, but can be used to remove missing values. The function var and cor fail with missing values prior to S-PLUS 4.0, which allows several options.

Histograms and stem-and-leaf plots

The standard histogram function is hist(x, ...) which plots a conventional histogram. More control is available via the extra arguments; probability=T gives a plot of unit total area rather than of cell counts, and style="old" gives the more conventional representation by outlines of bars rather than grey bars. There is also a Trellis function histogram, but neither is totally satisfactory.

The argument nclass of hist suggests the number of bins, and breaks specifies the breakpoints between bins. The default for nclass is $\log_2 n + 1$. This is known as Sturges' formula, and was based on a histogram of a normal distribution (Scott, 1992, p. 48). Non-normal distributions need more bins; for example Doane's rule is to use $\log_2 n + 1 + \log_2(1 + \hat{\gamma}\sqrt{n/6})$ where $\hat{\gamma}$ is an estimate of the kurtosis. One problem is that nclass is only a suggestion, and it is often exceeded. Another is that the definition of the bins which is of the form $[x_0, x_1], (x_1, x_2], \ldots$, not the convention most people prefer[2].

---

[2] The special treatment of x_0 is governed by the switch include.lowest which defaults to T.

Sturges' formula corresponds to a bin width of $\mathrm{range}(x)/(\log_2 n + 1)$, and this is often too wide. Note that outliers may inflate the range dramatically and so increase the bin width in the centre of the distribution. Two rules based on compromises between the bias and variance of the histogram for a reference normal distribution are to choose bin width as

$$h = 3.5\hat{\sigma}n^{-1/3} \tag{5.1}$$
$$h = 2Rn^{-1/3} \tag{5.2}$$

due to Scott (1979) and Freedman & Diaconis (1981) respectively. Here $\hat{\sigma}$ is the estimated standard deviation and R the inter-quartile range. The Freedman–Diaconis formula is immune to outliers, and chooses rather smaller bins than the Scott formula. We program these to suggest the number of classes and leave `hist` to choose 'pretty' breakpoints.

```
nclass.scott <- function(x)
{
    h <- 3.5 * sqrt(var(x)) * length(x)^(-1/3)
    ceiling(diff(range(x))/h)
}
nclass.FD <- function(x)
{
    r <- quantile(x, c(0.25, 0.75))
    names(r) <- NULL # to avoid the label 75%
    h <- 2 * (r[2] - r[1]) * length(x)^(-1/3)
    ceiling(diff(range(x))/h)
}
hist.scott <- function(x, ...)
    invisible(hist(x, nclass.scott(x),
        xlab = deparse(substitute(x)), ...))
hist.FD <- function(x, ...)
    invisible(hist(x, nclass.FD(x),
        xlab = deparse(substitute(x)), ...))
```

These are not always satisfactory, as Figure 5.2 shows. (The suggested numbers of bins are 6, 5, 25, 5, 42 and 35.) Column `eruptions` of our data frame `faithful` gives the duration (in minutes) of 272 eruptions of the Old Faithful geyser in the Yellowstone National Park (from Härdle, 1991); `chem` is discussed later in this section and `tperm` in Section 5.6. (S-PLUS has a different version of this dataset as `geyser`, from Azzalini & Bowman, 1990.)

Scott (1992, §3.3.1) suggests a lower bound $\sqrt[3]{2n}$ for the number of bins and the upper bound $2.61Rn^{-1/3}$ for the bin width. More sophisticated rules for choosing the bin widths are available which make use of the 'shape' of the histogram; see Scott (1992, §3.3.2). However, we regard it as more important to use a better estimator of the density function and so defer further discussion to Section 5.5.

The Trellis function `histogram` has arguments `breaks`, and `nint` for the suggested number of classes (defaulting to Sturges' rule). It does *not* have a probability argument, plotting the percentage (or the number if `type = "count"`) of

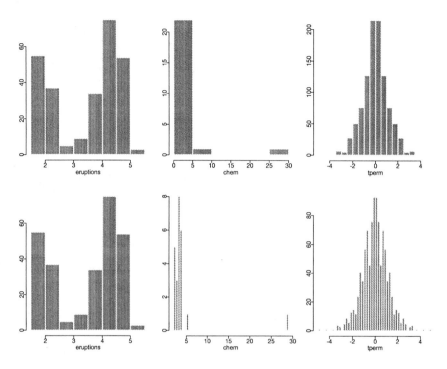

Figure 5.2: Histograms with bin widths chosen by the Scott rule (5.1) (top row) and the Freedman-Diaconis rule (5.2) (bottom row) for datasets faithful$eruptions, chem and tperm.

data items in each bin. Thus *it should not be used* with bins of variable sizes. The bins are chosen to just cover the range of the data and not 'prettified' as for hist.

The beauty of S is that it is easy to write one's own function to plot a histogram. One approach by Härdle (1991) is given in library haerdle as function histogram (predating the Trellis usage). Our function is called truehist (in library MASS). The primary control is by specifying the bin width h, but suggesting the number of bins by argument nbins will give a 'pretty' choice of h. Function truehist is used in Figures 5.5 and later. There is also a Trellis version called histplot for use in conditioning plots.

A *stem-and-leaf* plot is an enhanced histogram. The data are divided into bins, but the 'height' is replaced by the next digits in order. We apply this to swiss.fertility, the standardized fertility measure for each of 47 French-speaking provinces of Switzerland at about 1888.

```
> stem(swiss.fertility)

N = 47    Median = 70.4
Quartiles = 64.4, 79.3

Decimal point is 1 place to the right of the colon
```

```
3 : 5
4 : 35
5 : 46778
6 : 024455555678899
7 : 00222345677899
8 : 0233467
9 : 222
```

Apart from giving a visual picture of the data, this gives more detail. The actual data, in sorted order, are 35, 43, 45, 54, ... and this can be read off the plot. Sometimes the pattern of numbers (all odd? many 0s and 5s?) gives clues. Quantiles can be computed (roughly) from the plot. If there are outliers, they are marked separately:

```
> stem(chem)

N = 24    Median = 3.385
Quartiles = 2.75, 3.7

Decimal point is at the colon

   2 : 22445789
   3 : 00144445677778
   4 :
   5 : 3

High: 28.95
```

(This dataset on measurements of copper in flour, Analytical Methods Committee (1989a), is discussed further in Chapter 8.) Sometimes the result is less successful, and manual override is needed:

```
> stem(abbey)

N = 31    Median = 11
Quartiles = 8, 16

Decimal point is at the colon

   5 : 2
   6 : 59
   7 : 0004
   8 : 00005
   . . . .
  26 :
  27 :
  28 : 0

High:   34   125
```

```
> stem(abbey, scale=-1)

N = 31    Median = 11
Quartiles = 8, 16

Decimal point is 1 place to the right of the colon

    0 : 56777778888899
    1 : 011224444
    1 : 6778
    2 : 4
    2 : 8

High:    34  125
```

Here the `scale` argument sets the backbone to be tens rather than units. The `nl` argument controls the number of rows per backbone unit as 2, 5 or 10. The details of the design of stem-and-leaf plots are discussed by Mosteller & Tukey (1977), Velleman & Hoaglin (1981) and Hoaglin, Mosteller & Tukey (1983).

Boxplots

A *boxplot* is a way to look at the overall shape of a set of data. The central box shows the data between the 'hinges' (roughly quartiles), with the median represented by a line. 'Whiskers' go out to the extremes of the data, and very extreme points are shown by themselves.

```
par(mfrow=c(1,2))
boxplot(chem, sub="chem", range=0.5)
boxplot(abbey, sub="abbey")
par(mfrow=c(1,1))
bwplot(type ~ y | meas, data=fgl.df, scales=list(x="free"),
    strip=function(...) strip.default(..., style=1), xlab="")
```

There is a bewildering variety of optional parameters to `boxplot` documented in the on-line help page. Note how these plots are dominated by the outliers. It is possible to plot boxplots for groups side-by-side; (see Figure 15.19 on page 460) but the Trellis function `bwplot` (see Figures 5.4 and 3.9 on page 93) will probably be preferred.

5.4 Classical univariate statistics

S-PLUS has a section on classical statistics. The same S-PLUS functions are used to perform tests and to calculate confidence intervals.

Table 5.2 shows the amounts of shoe wear in an experiment reported by Box, Hunter & Hunter (1978). There were two materials (A and B) that were

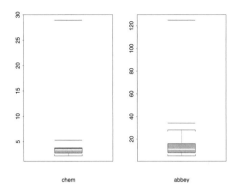

Figure 5.3: Boxplots for the chem and abbey data.

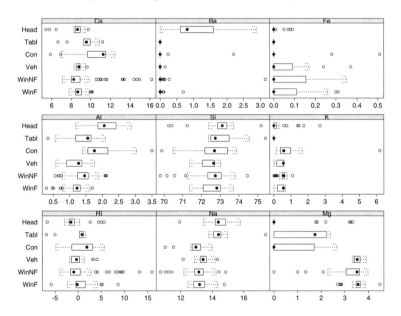

Figure 5.4: Boxplots by type for the fgl dataset.

randomly assigned to the left and right shoes of 10 boys. We will use these data to illustrate one-sample and paired and unpaired two-sample tests. (The rather voluminous output has been edited.)

```
> shoes <- scan(, list(A=0, B=0))
1: 13.2 14.0
    ....
21:
> attach(shoes)
```

This reads in the data and attaches the dataset so we can refer to A and B directly.

Table 5.2: Data on shoe wear from Box, Hunter & Hunter (1978).

boy	A		B	
1	13.2	(L)	14.0	(R)
2	8.2	(L)	8.8	(R)
3	10.9	(R)	11.2	(L)
4	14.3	(L)	14.2	(R)
5	10.7	(R)	11.8	(L)
6	6.6	(L)	6.4	(R)
7	9.5	(L)	9.8	(R)
8	10.8	(L)	11.3	(R)
9	8.8	(R)	9.3	(L)
10	13.3	(L)	13.6	(R)

First we test for a mean of 10 and give a confidence interval for the mean, both for material A.

```
> t.test(A, mu=10)

        One-sample t-Test

data:  A
t = 0.8127, df = 9, p-value = 0.4373
alternative hypothesis: true mean is not equal to 10
95 percent confidence interval:
  8.8764 12.3836
sample estimates:
 mean of x
     10.63

> t.test(A)$conf.int
[1]   8.8764 12.3836
attr(, "conf.level"):
[1] 0.95

> wilcox.test(A, mu = 10)

        Exact Wilcoxon signed-rank test

data:  A
signed-rank statistic V = 34, n = 10, p-value = 0.5566
alternative hypothesis: true mu is not equal to 10
```

Next we consider two-sample paired and unpaired tests, the latter assuming equal variances (the default) or not. Note that we are using this example for illustrative purposes; only the paired analyses are really appropriate.

```
> var.test(A, B)

        F test for variance equality

data:  A and B
F = 0.9474, num df = 9, denom df = 9, p-value = 0.9372
95 percent confidence interval:
 0.23532 3.81420
sample estimates:
 variance of x variance of y
        6.009         6.3427

> t.test(A, B)

        Standard Two-Sample t-Test

data:  A and B
t = -0.3689, df = 18, p-value = 0.7165
95 percent confidence interval:
 -2.7449  1.9249
sample estimates:
 mean of x mean of y
     10.63      11.04

> t.test(A, B, var.equal=F)

        Welch Modified Two-Sample t-Test

data:  A and B
t = -0.3689, df = 17.987, p-value = 0.7165
95 percent confidence interval:
 -2.745  1.925
    ....

> wilcox.test(A, B)

        Wilcoxon rank-sum test

data:  A and B
rank-sum normal statistic with correction Z = -0.5293,
   p-value = 0.5966

> t.test(A, B, paired=T)

        Paired t-Test

data:  A and B
t = -3.3489, df = 9, p-value = 0.0085
95 percent confidence interval:
 -0.68695 -0.13305
```

```
sample estimates:
 mean of x - y
          -0.41

> wilcox.test(A, B, paired=T)

        Wilcoxon signed-rank test

data:  A and B
signed-rank normal statistic with correction Z = -2.4495,
    p-value = 0.0143
```

The sample size is rather small, and one might wonder about the validity of the t–distribution. An alternative for a randomized experiment such as this is to base inference on the permutation distribution of d = B-A. Figure 5.5 shows that the agreement is very good. The computation of the permutations is discussed in Section 5.6.

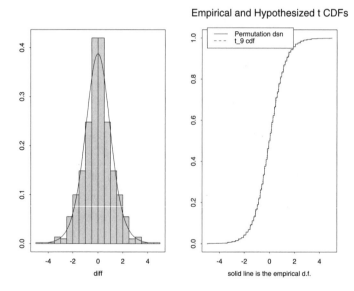

Figure 5.5: Histogram and empirical CDF of the permutation distribution of the paired t–test in the shoes example. The density and CDF of t_9 are shown overlaid.

The full list of classical tests is:

```
binom.test      chisq.test      cor.test          fisher.test
friedman.test   kruskal.test    mantelhaen.test   mcnemar.test
prop.test       t.test          var.test          wilcox.test
```

Many of these have alternative methods—for cor.test there are methods "pearson", "kendall" and "spearman". We have already seen one- and two-sample versions of t.test and wilcox.test, and var.test which compares the variances of two samples. The function cor.test tests for non-zero correlation between two samples, either classically or via ranks.

The Kruskal–Wallis test implemented by kruskal.test is a non-parametric version of a one-way analysis of variance. The Friedman test is a non-parametric version of the analysis of an unreplicated randomized block designed experiment. The remaining tests are for count data.

S-PLUS 3.3 added functions chisq.gof and ks.gof for chi-square and Kolmogorov–Smirnov tests of goodness-of-fit, as well as a function cdf.compare to plot comparisons of cumulative distributions such as the right-hand panel of Figure 5.5.

```
par(mfrow=c(1,2))
truehist(tperm, xlab="diff")
x <- seq(-4,4, 0.1)
lines(x, dt(x,9))
cdf.compare(tperm, distribution="t", df=9)
legend(-5, 1.05, c("Permutation dsn","t_9 cdf"), lty=c(1,3))
```

5.5 Density estimation

The non-parametric estimation of probability density functions is a large topic; several books have been devoted to it, notably Silverman (1986), Härdle (1991), Scott (1992) and Wand & Jones (1995). The methods used are closely related to some of the smoothing methods discussed in Sections 11.1 and 15.1.

The histogram with probability=T is of course an estimator of the density function. We would do better to plot it as a *frequency polygon*, joining the mid-points of the bars by straight lines:

```
frequency.polygon <- function(x, nclass = nclass.freq(x),
    xlab="", ylab="", ...)
{
    hst <- hist(x, nclass, probability=T, plot=F, ...)
    midpoints <- 0.5 * (hst$breaks[-length(hst$breaks)]
                + hst$breaks[-1])
    plot(midpoints, hst$counts, type="l", xlab=xlab, ylab=ylab)
}
nclass.freq <- function(x)
{
    h <- 2.15 * sqrt(var(x)) * length(x)^(-1/5)
    ceiling(diff(range(x))/h)
}
```

As the frequency polygon is smoother than the histogram on which it is based, we can afford to use larger bins. The equivalent of Scott's rule becomes (Scott, 1992, p. 99)

$$h = 2.15\hat{\sigma}n^{-1/5} \tag{5.3}$$

Because of the different power, this formula only gives larger h for $n \geqslant 39$, and the difference is small until n reaches 1000.

The histogram and consequently the frequency polygon depend on the starting point of the grid of bins. The effect can be surprisingly large; see Figure 5.6. The figure also shows that by averaging the histograms and then drawing a frequency polygon we can obtain a much clearer view of the distribution.

```
attach(faithful)
par(mfrow=c(2,3))
truehist(eruptions, h=0.5, x0=0.0, xlim=c(1, 6), ymax=0.6)
truehist(eruptions, h=0.5, x0=0.1, xlim=c(1, 6), ymax=0.6)
truehist(eruptions, h=0.5, x0=0.2, xlim=c(1, 6), ymax=0.6)
truehist(eruptions, h=0.5, x0=0.3, xlim=c(1, 6), ymax=0.6)
truehist(eruptions, h=0.5, x0=0.4, xlim=c(1, 6), ymax=0.6)

breaks <- seq(1, 6.4, 0.1)
counts <- numeric(length(breaks))
for(i in (0:4)) counts[i+(1:50)] <- counts[i+(1:50)] +
    rep(hist(eruptions, , 0.1*i + seq(1, 6, 0.5),
    prob=T, plot=F)$counts, rep(5,10))
plot(breaks+0.05, counts/5, type="l", xlab="eruptions",
    ylab="averaged", bty="n", xlim=c(1, 6), ylim=c(0, 0.6))
```

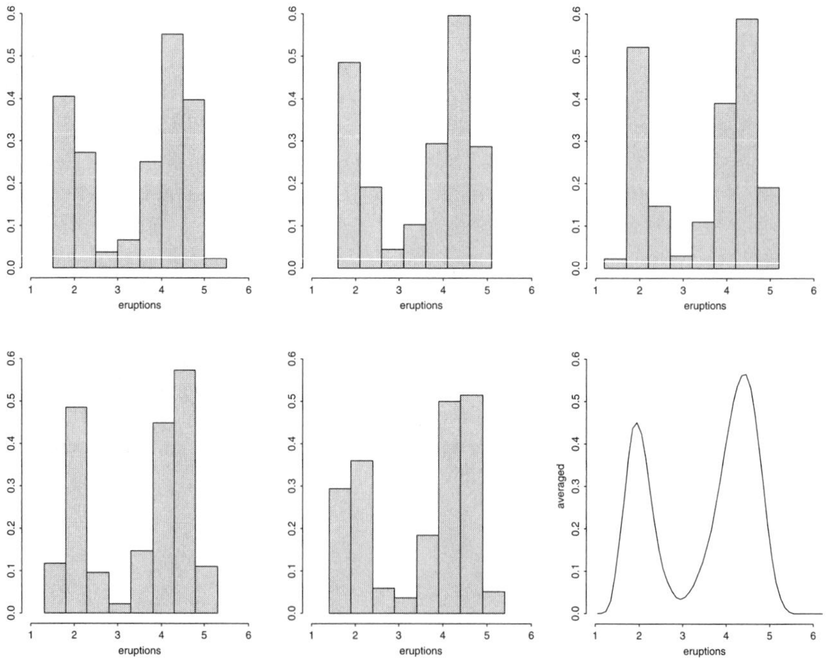

Figure 5.6: Five shifted histograms with bin width 0.5 and the frequency polygon of their average, for the Old Faithful geyser `eruptions` data.

   This idea of an *average shifted histogram* or *ASH* density estimate is a useful
one and is discussed in detail in Scott (1992). Suppose we average over m shifts
by $\delta = h/m$. Another way to look at the result is to take a histogram with bin
width δ, and to compute the histograms of bin width h by aggregating counts
in m adjacent bins. Specifically, if the counts are ν_k, the ASH for a point x in
small bin k is

$$\hat{f}(x) = \frac{1}{nh} \sum_{i=1-m}^{m-1} \left[1 - \frac{|i|}{m}\right] \nu_{k+i} = \frac{1}{nh} \sum_i \left[1 - \frac{|i\delta|}{h}\right]_+ \nu_{k+i}$$

so is a weighted average of the bin counts with a triangular weighting function.
Now suppose we let $m \to \infty$. The counts in bins reduce to zero or one, so the
limit is

$$\hat{f}(x) = \frac{1}{nh} \sum_{j=1}^{n} \left[1 - \frac{|x - x_j|}{h}\right]_+$$

This is a *kernel density* estimate with a triangular kernel. As well as motivating
kernel estimators, ASH and its extension *WARP* discussed in Härdle (1991) provide
computationally attractive approximate algorithms for density estimation.

   The only built-in S function specifically for density estimation is `density`.
This implements a kernel-density smoother of the form

$$\hat{f}(x) = \frac{1}{b} \sum_{j=1}^{n} K\left(\frac{x - x_j}{b}\right) \tag{5.4}$$

for a sample $x_1, \ldots, x_n$, a fixed kernel $K()$ and a bandwidth b. The kernel
is normally chosen to be a probability density function; the default is the normal (argument `window="g"` for Gaussian), with alternatives `"rectangular"`,
`"triangular"` and `"cosine"` (the latter being $(1 + \cos \pi x)/2$ over $[-1, 1]$).[3]
The bandwidth is the length of the non-zero section for the alternatives, and four
times the standard deviation for the normal. (*Note* that these definitions are twice
and four times those most commonly used.)

   The choice of bandwidth is a compromise between smoothing enough to
remove insignificant bumps and not smoothing too much to smear out real peaks.
Mathematically there is a compromise between the bias of $\hat{f}(x)$, which increases
as b is increased, and the variance which decreases. Theory (see (5.6) on page 183)
suggests that the bandwidth should be proportional to $n^{-1/5}$, but the constant of
proportionality depends on the unknown density.

   We can illustrate this effect with the permutation distribution of Figure 5.5,
which is stored in vector `tperm`. Under-smoothing (`width=0.2`) shows spurious
bumps (Figure 5.7); over-smoothing (`width=1.5`) has no bumps in the tails but
reduces the height of the main peak. By default the density is computed at 50
equally-spaced points, which is too few for this example. We use the Trellis
function `densityplot` (which calls `density`) to do the plotting.

---

   [3] established by plotting; Härdle (1991) and Scott (1992) have a different cosine kernel
$(\pi/4) \cos(\pi x/2) I(|x| \leqslant 1)$.

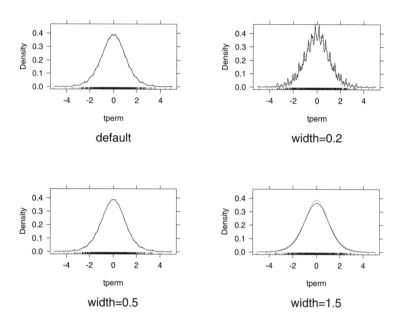

Figure 5.7: Density plots for the permutation distribution for the t statistic for the difference in shoe materials. In each case the reference t_9 density is shown dotted.

```
tperm <- test.t.2(B-A)   # see Section 5.6
dplot <- function(...)
{
    densityplot(~ tperm, ...,
        ylim = c(0, 0.45), from = -5, to = 5, n = 200,
        panel = function(x, y, ...) {
            ourpanel.densityplot(x, y, ...)
            x1 <- seq(-5, 5, 0.1)
            panel.xyplot(x1, dt(x1, 9), lty=2, type="l")
        }
    )
}
ourpanel.densityplot <- function(x, y, type = "l", n = 50,
    window = "gaussian", width = NULL, from = NULL, to = NULL,
    cut = NULL, ...)
{
    denargs <- list(x = x, n = n, window = window)
    if(!is.null(width)) denargs <- c(denargs, list(width = width))
    if(!is.null(from)) denargs <- c(denargs, list(from = from))
    if(!is.null(to)) denargs <- c(denargs, list(to = to))
    if(!is.null(cut)) denargs <- c(denargs, list(cut = cut))
    d <- do.call("density", denargs)
    panel.xyplot(d$x, d$y, type = type, ...)
    rug(jitter(x))
}
```

```
print(dplot(sub=list("default",cex=1.5)),
    split=c(1,2,2,2), more=T)
print(dplot(sub=list("width=0.2",cex=1.5), width=0.2),
    split=c(2,2,2,2), more=T)
print(dplot(sub=list("width=0.5",cex=1.5) ,width=0.5),
    split=c(1,1,2,2), more=T)
print(dplot(sub=list("width=1.5",cex=1.5), width=1.5),
    split=c(2,1,2,2), more=F)
```

By default `densityplot` represents the data by circles, which we have replaced by a call to `rug`.

The default width b is given by a rule of thumb

$$\frac{\text{range}(x)}{2(1 + \log_2 n)}$$

which happens to work quite well in this example (with $b \approx 0.45$). (This seems to be based on Sturges' rule and a rough equivalence with a histogram density estimate of bin width equal to the bandwidth.) There are better-supported rules of thumb such as

$$\hat{b} = 1.06 \min(\hat{\sigma}, R/1.34)n^{-1/5} \tag{5.5}$$

for the IQR R and the Gaussian kernel of bandwidth the standard deviation, to be quadrupled for use in `density` (Silverman, 1986, pp. 45–47).

We can relate the bandwidths of different kernels by the constants given in Table 5.3; kernels with bandwidth $c \times h$ will give estimates of comparable smoothness. This enables rules such as (5.5) to be used for other kernels. Since the rectangular kernel is non-differentiable, its estimates will be subjectively rougher. (Table 5.3 is calculated from results in Härdle (1991) and Scott (1992); it includes other commonly used kernels implemented in S by Härdle (1991).)

Table 5.3: Bandwidth constants for various kernels scaled as used in `density`; kernels with bandwidth $c \times h$ will give comparable smoothness.

Rectangular	1	Triweight	1.711
Triangular	1.393	Gaussian	1.15
Epanechnikov	1.272	Cosine	1.577
Quartic	1.507		

We return to the geyser eruptions data. This is an example in which the default bandwidth, 0.19, is too small, about 12% of the value suggested by (5.5), which is too large (Figure 5.8). Scott (1992, §6.5.1.2) suggests that the value

$$b_{OS} = 1.144\hat{\sigma}n^{-1/5}$$

provides an upper bound on the (Gaussian) bandwidths one would want to consider, again to be quadrupled for use in `density`. The normal reference (5.5) is only slightly below this upper bound, and will be too large for densities with multiple modes.

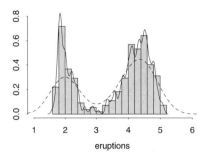 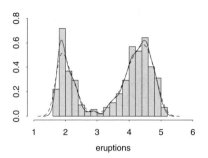

Figure 5.8: Density plots for the Old Faithful eruptions data. Superimposed on a histogram are the default width (left, solid), normal reference bandwidth (left, dashed) and the UCV (right, solid) and BCV (right, dashed) kernel estimates with a Gaussian kernel.

```
attach(faithful)
truehist(eruptions, nbins=15, xlim=c(1,6), ymax=0.8)
lines(density(eruptions, n=200))
bandwidth.nrd <- function(x)
{
    r <- quantile(x, c(0.25, 0.75))
    h <- (r[2] - r[1])/1.34
    4 * 1.06 * min(sqrt(var(x)), h) * length(x)^(-1/5)
}
bandwidth.nrd(eruptions)
[1] 1.5772
lines(density(eruptions, width=1.6, n=200), lty=3)
bandwidth.nrd(tperm)
[1] 1.1536
```

Note that the data in this example are (negatively) serially correlated, so the theory for independent data must be viewed as only a guide.

Bandwidth selection

Ways to find automatically compromise value(s) of b are discussed by Härdle (1991) and Scott (1992). These are based on estimating the *mean integrated square error*

$$MISE = E \int |\hat{f}(x;b) - f(x)|^2 \, \mathrm{d}x = \int E|\hat{f}(x;b) - f(x)|^2 \, dx$$

and choosing the smallest value as a function of b. We can expand $MISE$ as

$$MISE = E \int \hat{f}(x;b)^2 \, \mathrm{d}x - 2E\hat{f}(X;b) + \int f(x)^2 \, \mathrm{d}x$$

where the third term is constant and so can be dropped.

The estimates are based on *cross-validation*, that is using the remaining observations to fit a density at each observation in turn. (See for example Stone,

1974.) Let the fitted value be $\hat{f}(x_i; b)_{-i}$. Then the *unbiased* (or least-squares) *cross-validation* estimates $E\hat{f}(X; b)$ by the mean of $\hat{f}(x_i; b)_{-i}$ to give

$$UCV(b) = \int \hat{f}(x; b)^2 \, \mathrm{d}x - \frac{2}{n} \sum_{i=1}^{n} \hat{f}(x_i; b)_{-i}$$

The minimizer $\hat{b}$ is found to be accurate but rather variable. A more involved approximation gives the *biased cross-validation* criterion $BCV(b)$ which is minimized over $b \leqslant b_{OS}$, and is usually less variable. Both $UCV(b)$ and $BCV(b)$ can have multiple local minima.

Scott (1992, §6.5.13) provides formulae for the Gaussian kernel which we can evaluate via the functions ucv and bcv of our library MASS. (These contain typographic errors which we have corrected.) As the formulae involve calculations using the distances between all pairs of points, they cannot be used for very large datasets (say thousands of observations). The WARP approximation could be used, but these functions take another approach, in which the data are rounded to the nearest point on the grid, and the formulae summed over the much smaller set of distances between pairs which results. (Yet another approach is to use fast Fourier transforms, as described in Appendix D of Wand & Jones, 1995. The versions in the first edition estimate the bandwidth from a sub-sample, then scale the bandwidth proportionally to $n^{-1/5}$.)

```
> c( ucv(eruptions),  bcv(eruptions) )
[1] 0.40766 0.63077
> truehist(eruptions, nbins=15, xlim=c(1,6), ymax=0.8)
> lines(density(eruptions, width=0.41, n=200))
> lines(density(eruptions, width=0.63, n=200), lty=3)
```

which shows both criteria in this example provide visually sensible values, Figure 5.8.

Ideas in bandwidth selection have developed rapidly since 1990, and UCV and BCV are now regarded as 'first generation' rules (Jones *et al.*, 1996). Another approach is to make an asymptotic expansion of $MISE$ of the form

$$MISE = \frac{1}{nb} \int K^2 + \frac{1}{4} b^4 \int (f'')^2 \left\{ \int x^2 K \right\}^2 + O(1/nb + b^4)$$

so if we neglect the remainder, the optimal bandwidth would be

$$b_{AMISE} = \left[\frac{\int K^2}{n \int (f'')^2 \left\{ \int x^2 K \right\}^2} \right]^{1/5} \tag{5.6}$$

The 'direct plug-in' estimators use (5.6), which involves the integral $\int (f'')^2$. This in turn is estimated using the second derivative of a kernel estimator with a different bandwidth, chosen by repeating the process, this time using a reference bandwidth. The 'solve-the-equation' estimators solve (5.6) when the bandwidth for estimating $\int (f'')^2$ is taken as function of b (in practice proportional to $b^{5/7}$). Details are given by Sheather & Jones (1991) and Wand & Jones (1995, §3.6); this is implemented for the Gaussian kernel in our function width.SJ.

```
> c(width.SJ(eruptions, method="dpi"), width.SJ(eruptions))
[1] 0.66109 0.55775
> c( ucv(tperm), bcv(tperm), width.SJ(tperm) )
[1] 1.2958 1.0750 1.1200
Warning messages:
    minimum occurred at one end of the range in: ucv(tperm)
```

There have been a number of comparative studies of bandwidth selection rules (including Park & Turlach, 1992, and Cao *et al.*, 1994) and a review by Jones *et al.* (1996). 'Second generation' rules such as `width.SJ` seem preferred and indeed to be close to optimal.

End effects

Most density estimators will not work well when the density is non-zero at an end of its support, such as the exponential and half-normal densities. (They are designed for continuous densities and this is discontinuity.) One trick is to reflect the density and sample about the endpoint, say a. Thus we compute the density for the sample `c(x, 2a-x)`, and take double its density on $[a, \infty)$ (or $(-\infty, a]$ for an upper endpoint). This will impose a zero derivative on the estimated density at a, but the end effect will be much less severe. For details and further tricks see Silverman (1986, §3.10). The alternative is to modify the kernel near an endpoint (Wand & Jones, 1995, §2.1), but we know of no S implementation of such *boundary kernels*.

Two-dimensional data

It is often useful to look at densities in two dimensions. Visualizing in more dimensions is difficult, although Scott (1992) provides some examples of visualization of three-dimensional densities.

The dataset on the Old Faithful geyser has two components, the `eruptions` duration which we have studied, and `waiting`, the waiting time in minutes until the next eruption. There is also evidence of non-independence of the durations. S provides a function `hist2d` for two-dimensional histograms, but its output is too rough to be useful. We can apply the *ASH* device here too, using a two-dimensional grid of bins; we could use code from Scott's library `ash`[4] available from `statlib`.

We will apply two-dimensional kernel analysis directly; this is most straightforward for the normal kernel aligned with axes, that is with variance $\mathrm{diag}(h_x^2, h_y^2)$. Then the kernel estimate is

$$f(x, y) = \frac{\sum_s \phi((x - x_s)/h_x)\phi((y - y_s)/h_y)}{n\, h_x h_y}$$

which can be evaluated on a grid as XY^T where $X_{is} = \phi((gx_i - x_s)/h_x)$ and (gx_i) are the grid points, and similarly for Y. Our function `kde2d` implements

---

[4] which uses FORTRAN code.

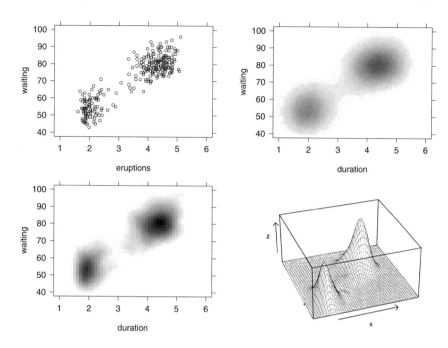

Figure 5.9: Scatter plot and two-dimensional density plots of the bivariate Old Faithful geyser data.

this; the results are shown in Figures 5.9 and 5.10. We reverse the default greylevel postscript colour scheme to make black indicate high (rather than low) levels.

```
attach(faithful)
xyplot(waiting ~ eruptions, xlim=c(1,6), ylim=c(40,100))
f1 <- kde2d(eruptions, waiting, n=50, lims=c(1,6,40,100))
levelplot(z ~ x*y, con2tr(f1),
    xlab="duration", ylab="waiting",
    at = seq(0, 0.045, 0.001), colorkey=F,
    col.regions = rev(trellis.par.get("regions")$col))
f2 <- kde2d(eruptions, waiting, n=50, lims=c(1,6,40,100),
    h = c(width.SJ(eruptions), width.SJ(waiting)) )
levelplot(z ~ x*y, con2tr(f2),
    xlab="duration", ylab="waiting",
    at = seq(0, 0.045, 0.001), colorkey=F,
    col.regions = rev(trellis.par.get("regions")$col))
wireframe(z ~ x*y, con2tr(f2),
    aspect = c(1, 0.5), screen=list(z=20, x=-60), zoom=1.2)

xyplot(eruptions[-1] ~ eruptions[-272], xlim=c(1, 6),
    ylim=c(1, 6),xlab="previous duration", ylab="duration")
f1 <- con2tr(kde2d(eruptions[-272], eruptions[-1],
    h=rep(1.5, 2), n=50, lims=c(1,6,1,6)))
contourplot(z ~ x * y, f1 ,xlab="previous duration",
```

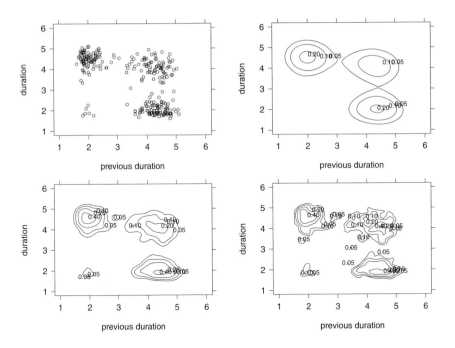

Figure 5.10: Scatter plot and two-dimensional density plots of previous and current duration for the Old Faithful geyser data. The contour plots show the effect of varying the amount of smoothing.

```
        ylab="duration", at = c(0.05, 0.1, 0.2, 0.4) )
f1 <- con2tr(kde2d(eruptions[-272], eruptions[-1],
    h=rep(0.6, 2), n=50, lims=c(1,6,1,6)))
contourplot(z ~ x * y, f1 ,xlab="previous duration",
        ylab="duration", at = c(0.05, 0.1, 0.2, 0.4) )
f1 <- con2tr(kde2d(eruptions[-272], eruptions[-1],
    h=rep(0.4, 2), n=50, lims=c(1,6,1,6)))
contourplot(z ~ x * y, f1 ,xlab="previous duration",
        ylab="duration", at = c(0.05, 0.1, 0.2, 0.4) )
```

An alternative approach using binning and the 2D fast Fourier transform is taken by Wand's library KernSmooth [5] available from statlib.

5.6   Bootstrap and permutation methods

Several modern methods of what is often called *computer-intensive statistics* make use of extensive repeated calculations to explore the sampling distribution of a parameter estimator $\hat{\theta}$. Suppose we have a random sample $x_1, \ldots, x_n$ drawn independently from one member of a parametric family $\{F_\theta \mid \theta \in \Theta\}$

---

[5] which also uses FORTRAN.

of distributions. Suppose further that $\hat{\theta} = T(x)$ is a *symmetric* function of the sample, that is does not depend on the sample order.

The *bootstrap* procedure (Efron, 1982; Efron & Tibshirani, 1993; Davison & Hinkley, 1997) is to take m samples from x *with replacement* and to calculate $\hat{\theta}_j$ for these samples. Note that the new samples consist of an integer number of copies of each of the original data points, and so will normally have ties. Efron's idea was to assess the variability of $\hat{\theta}$ about the unknown true θ by the variability of $\hat{\theta}_i$ about $\hat{\theta}$. In particular, the bias of $\hat{\theta}$ may be estimated by the mean of $\hat{\theta}_i - \hat{\theta}$.

As an example, suppose that we needed to know the median m of the eruptions data. The obvious estimator is the sample median, which is 4. How accurate is this estimator? The large-sample theory says that the median is asymptotically normal with mean m and variance $1/4n\,f(m)^2$. But this depends on the unknown density at the median. We can use our best density estimators to estimate $f(m)$, but as we have seen we can find considerable bias and variability if we are unlucky enough to encounter a peak (as in a unimodal symmetric distribution). Let us try the bootstrap:

```
density(eruptions, n=1, from=4, to=4.01, width=0.41)$y
[1] 0.42081
density(eruptions, n=1, from=4, to=4.01, width=0.63)$y
[1] 0.41028
1/(2*sqrt(length(eruptions))*0.415)
[1] 0.073053
set.seed(101); m <- 1000
res <- numeric(m)
for (i in 1:m) res[i] <- median(sample(eruptions, replace=T))
mean(res - median(eruptions))
[1] -0.014813
sqrt(var(res))
[1] 0.079443
```

which takes about 10 seconds and confirms the adequacy of the large-sample mean and variance for our example (assuming independence).

The bootstrap distribution of $\hat{\theta}_i$ about $\hat{\theta}$ is far from normal (see Figure 5.11). It is a discrete distribution, but its 'shape' is quite different from that of a normal:

```
> truehist(res, nbins=nclass.FD(res))
> lines(density(res, n=200, width=bandwidth.nrd(res)))
> c( ucv(res), bcv(res) )
[1] 0.0095 0.0908
Warning messages:
1: minimum occurred at one end of the range in: ucv(res)
2: minimum occurred at one end of the range in: bcv(res)
> c( width.SJ(res, lower=0.001), width.SJ(res, method="dpi") )
[1] 0.0055 0.00288
> quantile(res, c(0.025, 0.975))
  2.5%  97.5%
3.833 4.1085
```

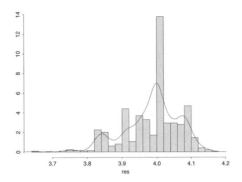

Figure 5.11: Histogram of the bootstrap distribution for the median of the geyser eruptions data, with a smoothed density estimate whose bandwidth comes from the normal reference density.

Note the great discrepancy in the suggested bandwidths: the normal reference gives 0.0727 and b_{OS} is 0.0913. The difficulty arises from the discreteness of the bootstrap distribution, and can also affect data with heavily tied values (for example after rounding).

In fact, in this example the bootstrap resampling can be avoided, for the bootstrap distribution of the median can be found analytically (Efron, 1982, Chapter 10; Staudte & Sheather, 1990, p. 84), at least for odd n.

One approach to a confidence interval for the median is to use the quantiles given here; this is termed the *percentile confidence interval* and sometimes a *Monte Carlo confidence interval* (see Ripley, 1987, p. 176). More complex intervals are discussed in the references, and S code from Efron & Tibshirani (1993) (available from `statlib`; see Appendix C) and Davison & Hinkley (1997) is available to compute these.

S-PLUS 4.0 introduced several functions for working with the bootstrap (and the jackknife). We can repeat our example by

```
> erupt.boot <- bootstrap(eruptions, median, seed=101, B=1000)
> summary(erupt.boot)
    ....
Summary Statistics:
        Observed     Bias   Mean       SE
median         4 -0.01481  3.985  0.07944

Empirical Percentiles:
         2.5%    5%    95%  97.5%
median  3.833 3.833 4.092 4.108

BCa Percentiles:
         2.5%    5%    95%  97.5%
median  3.833 3.833 4.083    4.1
```

The percentiles are given by the functions `limits.emp` and `limits.bca`. There are methods for `plot` and `qqnorm` to examine the distribution of the bootstrapped estimates, and a method for `update` to add further samples. We can use the 'jackknife after bootstrap' technique (Efron & Tibshirani, 1993, §19.4) to assess the accuracy of our bootstrap estimates.

```
> jack.after.bootstrap(erupt.boot, "Bias")
  ....
Functional of Bootstrap Distribution of Parameters:
         Func SE.Func
median -0.01481 0.09499
> jack.after.bootstrap(erupt.boot, "SE")
  ....
Functional of Bootstrap Distribution of Parameters:
         Func SE.Func
median 0.07944 0.05143
```

Permutation tests

Inference for designed experiments is often based on the distribution over the random choices made during the experimental design, on the belief that this randomness alone will give distributions which can be approximated well by those derived from normal-theory methods. There is considerable empirical evidence that this is so, but with modern computing power we can check it for our own experiment, by selecting a large number of re-labellings of our data and computing the test statistics for the re-labelled experiments.

Consider again the shoe-wear data of Section 5.4. The most obvious way to explore the permutation distribution of the t-test of `d = B-A` is to select random permutations. The supplied function `t.test` computes much more than we need. In this case it is simple to write a replacement function to do exactly what we need:

```
attach(shoes)
d <- B - A
ttest <- function(x) mean(x)/sqrt(var(x)/length(x))
n <- 1000
res <- numeric(n)
for(i in 1:n) res[i] <- ttest(x <- d*sign(runif(10)-0.5))
```

which takes around 25 seconds on our PC.

As the permutation distribution has only $2^{10} = 1024$ points we can explore it directly for Figure 5.5:

```
test.t.1 <- function(d) {
    binary <- function(x, digits = if(x > 0) 1 +
                  floor(log(x, base = 2)))     else 1)
    { ans <- 0:(digits - 1); (x %/% 2^ans) %% 2 }
    digits <- length(d)
    n <- 2^digits
```

```
    perm.res <- numeric(n)
    for(i in 1:n) {
        x <- d * 2 * (binary(i, digits = digits) - 0.5)
        perm.res[i] <- ttest(x)
    }
    perm.res
}
tperm <- test.t.1(B-A)
```

This is slower than one would like (30 seconds) despite having been stream-lined considerably.[6] However, as the t-test is the ratio of linear operations, we can do many at once via matrix operations, taking only 0.3 sec! We need a vector form of `binary`:

```
test.t.2 <- function(d) {
# ttest is function(x) mean(x)/sqrt(var(x)/length(x))
    binary.v <- function(x, digits) {
        if(missing(digits)) {
            mx <- max(x)
            digits <- if(mx > 0) 1 + floor(log(mx, base = 2))
                      else 1
        }
        ans <- 0:(digits - 1); lx <- length(x)
        x <- rep(x, rep(digits, lx))
        x <- (x %/% 2^ans) %% 2
        dim(x) <- c(digits, lx)
        x
    }
    digits <- length(d)
    n <- 2^digits; x <- d * 2 * (binary.v(1:n, digits) - 0.5)
    mx <- matrix(1/digits, 1, digits) %*% x
    s <- matrix(1/(digits - 1), 1, digits)
    vx <- s %*% (x - matrix(mx, digits, n, byrow=T))^2
    as.vector(mx/sqrt(vx/digits))
}
```

The speed-up in this example is unusually large, but it is always worth looking for simplifying ideas which allow a large number of calculations to be done by one function.

5.7 Exercises

5.1. Experiment with our dataset `galaxies`. How many modes do you think there are in the underlying density?

---

[6] We are grateful to Bill Dunlap of MathSoft DAPD for his suggestions for this problem.

Chapter 6

Linear Statistical Models

Linear models form the core of classical statistics, and S provides extensive facilities to fit and investigate them. These work with a version of the Wilkinson-Rogers notation (Wilkinson & Rogers, 1973) for specifying models which we discuss in Section 6.2.

The main function for fitting linear models is `lm`, which provides our first example in the book of a style of S functions we shall see repeatedly in later chapters, producing a *fitted model object* which is then analysed using various generic functions.

The analysis of designed experiments by least squares, historically called analysis of variance, can be seen as a special case of linear regression, but there are different conventions for displaying the results. The S function `aov` follows that tradition.

We will consider a simple regression example before exploring the full richness of linear model formulae.

6.1   A linear regression example

Fisher (1947) published a dataset of H. G. O. Holck giving the heart weights, body weights and sex of 47 female and 97 male adult cats. These had been used in bioassay experiments with the muscle-relaxing drug digitalis. The dataset is given in data frame `cats` in our library `MASS` with three variables, `Sex`, `Bwt` and `Hwt`.

It appears that cats with a body weight of at least 2 kilograms were selected. Fisher's interest in the data was to study the relation of heart weight to body weight and to discover if this differs by sex. The truncation to animals of at least 2 kg is a restriction on the scope of inferences possible but not an impediment of principle.

Fisher initially calculated the average ratio of heart weight to body weight for each sex, and the results are very close,

```
> attach(cats)
> tapply(Hwt/Bwt, Sex, mean)
      F       M
  3.9151  3.8945
```

191

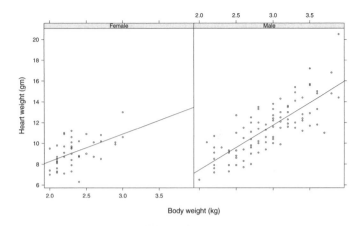

Figure 6.1: Holck's cat weight data with separate regression lines.

but he went on to point out that for separate linear regressions of heart weight on body weight the intercept term is non-significant for males but (borderline) significant for females, so can heart weight be considered proportional to body weight, at least for females over 2 kg?

A plot of the two samples together with the separate regression lines is given in Figure 6.1, generated by

```
Cats <- cats;  levels(Cats$Sex) <- c("Female", "Male")
xyplot(Hwt ~ Bwt | Sex, Cats, aspect = "xy",
  prepanel = prepanel.lmline,
  panel = function(x, y, ...)
    { panel.xyplot(x, y, cex = 0.5);  panel.lmline(x, y) },
  xlab = "Body weight (kg)", ylab = "Heart weight (gm)",
  strip = function(...) strip.default(..., style = 1)
)
```

To explore Fisher's question and others we will need to fit linear regression models, the main function for which is `lm`. The main arguments to `lm` are

```
lm(formula, data, weights, subset, na.action)
```

where

> formula is the model formula (the only required argument),
>
> data is an optional data frame,
>
> weights is a vector of positive weights, if non-uniform weights are needed,
>
> subset is an index vector specifying a subset of the data to be used (by default all items are used),
>
> na.action is a function specifying how missing values are to be handled (by default, missing values are not allowed).

If the argument `data` is specified it gives a data frame from which variables are selected ahead of the search path. Working with data frames and using this argument is strongly recommended.

It should be noted that setting `na.action=na.omit` will allow models to be fitted omitting cases that have missing components on a required variable. If any cases are omitted the fitted values and residual vector will no longer match the original observation vector in length.

Formulae have been discussed in outline in Section 2.5 on page 41. For `lm` the right-hand side specifies the regressors. If operators are used they have a *formula* rather than an arithmetical interpretation, as we shall see.

To fit separate regressions of heart weight on body weight for each sex we may use

```
catsF <- lm(Hwt ~ Bwt, data = cats, subset = Sex=="F")
catsM <- update(catsF, subset = Sex=="M")
```

The first command fits a simple linear regression for the females, as plotted in Figure 6.1. The right-hand side of the formula needs only to specify the variable `Bwt` since an intercept term (corresponding to a column of unities of the model matrix) is always implicitly included. It may be explicitly included using `1 + Bwt`, where the + operator signifies *inclusion* in the model, not addition.

The males could be handled in the same way, but the function `update` is a very convenient way to modify a fitted model. Its first argument is a *fitted model object*, that is, the result of one of the model fitting functions like `lm`. The remaining arguments of `update` specify the desired changes to arguments of the call which generated that object. In this case we simply wish to switch subsets from females to males; the formula and data frame remain the same.

Fitted model objects are lists of the appropriate class, in this case class `lm`. Generic functions are available to perform further operations on the object. These include

> `print`   for a simple display,
> `summary`   for a conventional regression analysis output,
> `coef`   (or `coefficients`) for extracting the regression coefficients,
> `resid`   (or `residuals`) for residuals,
> `fitted`   (or `fitted.values`) for fitted values,
> `deviance`   for the residual sum of squares,
> `anova`   for a sequential analysis of variance table, or a comparison of several hierarchical models,
> `predict`   for predicting means for new data, optionally with standard errors, and
> `plot`   for diagnostic plots.

Many of these method functions are very simple, merely extracting a component of the fitted model object. The only component likely to be accessed for which no extractor function is supplied is `df.residual`, the residual degrees of freedom.

The output from `summary` is self-explanatory. Edited results for the cats fitted models are

```
> summary(catsF)
    ....
Coefficients:
            Value Std. Error t value Pr(>|t|)
(Intercept) 2.981 1.485       2.007   0.051
        Bwt 2.636 0.625       4.215   0.000

Residual standard error: 1.16 on 45 degrees of freedom
> summary(catsM)
    ....
Coefficients:
            Value Std. Error t value Pr(>|t|)
(Intercept) -1.184  0.998     -1.186   0.239
        Bwt  4.313  0.340     12.688   0.000

Residual standard error: 1.56 on 95 degrees of freedom
```

Fisher did not say why he used separate regressions rather than pooling the two samples, but the difference in variance may be one reason. A textbook test for equality of variances does show significance

```
> Fvar <- deviance(catsF)/(Fdf <- catsF$df.resid)
> Mvar <- deviance(catsM)/(Mdf <- catsM$df.resid)
> c(Male=Mvar, Female=Fvar, F=(f <- Fvar/Mvar), "Tail area" =
        2 * if(f < 1) pf(f,Fdf,Mdf) else pf(1/f,Mdf,Fdf))
    Male Female        F Tail area
  2.4238 1.3508 0.55733  0.031043
```

but it is now known this test is highly non-robust to non-normality (see for example Hampel *et al.* (1986, pp. 55, 188)). (Notice that this test cannot be done directly using the function var.test since that does not allow the degrees of freedom to be specified.)

To fit both regression models together in the same call to lm we may use

```
> catsMF <- lm(Hwt ~ Sex/Bwt - 1, data=cats)
> summary(catsMF)
    ....
Coefficients:
          Value Std. Error t value Pr(>|t|)
   SexF   2.981    1.843      1.618   0.108
   SexM  -1.184    0.925     -1.281   0.202
SexFBwt   2.636    0.776      3.398   0.001
SexMBwt   4.313    0.315     13.700   0.000

Residual standard error: 1.44 on 140 degrees of freedom
    ....
```

Notice that the estimates are the same but the standard errors are different because they are now based on the pooled estimate of variance. In this case neither intercept term appears significantly non-zero.

Terms of the form `a/x`, provided `a` is a factor, are best thought of as "separate regression models of type `1 + x` for each level of `a`". In this case an intercept is not needed, since it is replaced by two separate intercepts for the two sexes, and the formula term `- 1` removes it.

We can test for lack of fit by testing our model as a submodel of a larger model less open to challenge, for example a model with a mean parameter for each different body weight that occurs within each sex. As many of the body weights are repeated this still leaves ample degrees of freedom to estimate the error variance from within the body weight classes.

```
> catsMF1 <- lm(Hwt ~ Sex/factor(Bwt), data=cats,
      singular.ok=T)
> anova(catsMF, catsMF1)
Analysis of Variance Table
   . . . .
   Resid. Df    RSS    Test Df Sum of Sq F Value    Pr(F)
1          140 291.05
2          114 228.78 1 vs. 2 26    62.262  1.1932 0.25871
```

When `anova` is used with two or more nested models it gives an analysis of variance table for those models. In this case it fails to show significant evidence of lack of fit. Note that `singular.ok=T` is needed to signal that the model matrix may not be of full column rank; since not all body weights occur for both sexes the corresponding parameters cannot be estimated. The alternative fitting function, `aov`, does not require this argument as it detects and removes redundancies in the model matrix automatically.

In his reappraisal of Fisher's analysis, Aitchison (1986, pp. 227ff) suggested a log transform of both heart and body weights to work in a scale where variance heterogeneity and possibly skewness, were not so marked. A natural hypothesis to investigate on this scale is the simple model implied by Fisher's original calculation, namely $H \propto B$. One way to do this is to fit the model

$$\log(H/B) = \beta_0 + \beta_1 \log B + \epsilon$$

for each sex and test the hypothesis $\beta_1 = 0$:

```
> logcats.lm2 <- lm(log(Hwt/Bwt) ~ Sex/log(Bwt), data=cats)
> summary(logcats.lm2)
   . . . .
Coefficients:
                Value Std. Error t value Pr(>|t|)
   (Intercept)  1.429     0.088    16.248    0.000
           Sex -0.183     0.088    -2.083    0.039
SexFlog(Bwt)   -0.301     0.177    -1.702    0.091
SexMlog(Bwt)    0.099     0.084     1.186    0.237

Residual standard error: 0.134 on 140 degrees of freedom
Multiple R-Squared: 0.0303
F-statistic: 1.46 on 3 and 140 degrees of freedom,
```

the p-value is 0.229

. . . .

The final F-statistic suggests the total contribution to the model of all terms additional to the constant term is negligible, suggesting that in both sexes heart weight is proportional to body weight, *and* that the constant of proportionality is the same for both.[1] This can be checked formally with a two-step sequential analysis of variance. We finish with a few diagnostic plots of the residuals.

```
> logcats.lm1 <- update(logcats.lm2, . ~ Sex)
> logcats.lm0 <- update(logcats.lm1, . ~ 1)
> anova(logcats.lm0, logcats.lm1, logcats.lm2)
     . . . .
           Terms Resid. Df    RSS             Test
1              1      143 2.5818
2            Sex      142 2.5807
3 Sex/log(Bwt)       140 2.5037 +log(Bwt) %in% Sex

   Df Sum of Sq F Value    Pr(F)
1
2   1  0.001119  0.0626  0.80284
3   2  0.077003  2.1529  0.11998

> rs <- resid(logcats.lm0)
> plot(Sex, rs, ylab="log Transform Residuals")
> qqnorm(rs, ylab="log Transform Residuals")
> qqline(rs)
```

When the update function is used to change the formula, the single-dot name, '.', may be used to mean "whatever was in the previous model formula at that point".

The plots shown in Figure 6.2 show little evidence of departures from the assumptions. Similar plots for the untransformed data show some evidence of both skewness and variance heterogeneity.

6.2   Model formulae and model matrices

This section contains some rather technical material and might be skimmed at first reading.

A linear model is specified by the response vector, y, and by the matrix of linear predictors, X. The model formula conveys both pieces of information, the left-hand side providing the response and the right-hand side instructions on how to generate the model matrix according to a particular convention.

---

[1] The Sex coefficient appears to be significant but this is misleading as this term is marginal to log(Bwt) %in% Sex and may be modified by a change of origin.

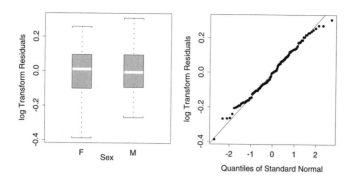

Figure 6.2: Diagnostic plots for the log-transformed cats data.

A multiple regression with three quantitative determining variables might be specified as y ~ x1+x2+x3. This would correspond to a model with a familiar algebraic specification of the form

$$y_i = \beta_0 + \beta_1 x_{i1} + \beta_2 x_{i2} + \beta_3 x_{i3} + \epsilon_i, \qquad i = 1, 2, \ldots, n$$

The model matrix has the partitioned form

$$X = \begin{bmatrix} \mathbf{1} & \boldsymbol{x}_1 & \boldsymbol{x}_2 & \boldsymbol{x}_3 \end{bmatrix}$$

The intercept term (β_0 corresponding to the leading column of ones in X) is implicitly present; its presence may be confirmed by giving a formula such as y ~ 1 + x1 + x2 + x3, but wherever the 1 occurs in the formula the column of ones will always be the first column of the model matrix. It may be omitted and a regression through the origin fitted by giving a - 1 term in the formula, as in y ~ x1 + x2 + x3 - 1.

Factor terms in a model formula are used to specify classifications leading to what are classically known as analysis of variance models. Suppose a is a factor. A one-way analysis of variance model for the classes defined by a might be written in the algebraic form

$$y_{ij} = \mu + \alpha_j + \epsilon_{ij} \qquad i = 1, 2, \ldots, n_j; j = 1, 2, \ldots, k$$

where there are k classes and the n_j is the size of the jth. Let $n = \sum_j n_j$. This specification is over-parametrized, but we could write the model matrix in the form

$$X = \begin{bmatrix} \mathbf{1} & X_{\mathbf{a}} \end{bmatrix}$$

where $X_{\mathbf{a}}$ is an $n \times k$ binary incidence (or 'dummy variable') matrix where each row has a single unity in the column of the class to which it belongs.

The redundancy comes from the fact that the columns of $X_{\mathbf{a}}$ add to $\mathbf{1}$, making X of rank k rather than $k + 1$. One way to resolve the redundancy is to remove the column of ones. This amounts to setting $\mu = 0$, leading to an algebraic specification of the form

$$y_{ij} = \alpha_j + \epsilon_{ij} \qquad i = 1, 2, \ldots, n_j; \; j = 1, 2, \ldots, k$$

so the α_j parameters are the class means. This formulation may be specified by `y ~ a - 1`.

If we do not break the redundancy by removing the intercept term it must be done some other way, since otherwise the parameters are not uniquely identifiable. The way this is done in S is most easily specified in terms of the model matrix. A new model matrix is used of the form

$$X^{\star} = \begin{bmatrix} 1 & X_{\mathsf{a}} C_{\mathsf{a}} \end{bmatrix}$$

where C_{a}, the *contrast matrix* for a, is a $k \times (k-1)$ matrix chosen so that $X^{\star}$ has rank k, the number of columns. A necessary (and usually sufficient) condition for this to be the case is that the square matrix $\begin{bmatrix} 1 & C_{\mathsf{a}} \end{bmatrix}$ be nonsingular.

The reduced model matrix, $X^{\star}$, in turn defines a linear model, but the parameters are often not directly interpretable and an algebraic formulation of the precise model may be difficult to write down. The relationship between the newly defined and original redundant parameters is given by

$$\alpha = C_{\mathsf{a}} \alpha^{\star} \tag{6.1}$$

where α are the original α parameters and $\alpha^{\star}$ are the new.

If c_{a} is a non-zero vector such that $c_{\mathsf{a}}^T C_{\mathsf{a}} = 0$ it can be seen immediately that using $\alpha^{\star}$ as parameters amounts to estimating the original parameters, α subject to the *identification constraint* $c_{\mathsf{a}}^T \alpha = 0$ which is usually sufficient to make them unique. Such a vector (or matrix) c_{a} is called an *annihilator* of C_{a} or a basis for the orthogonal complement of the range of C_{a}.

If we fit the one-way layout model using the formula

```
y ~ a
```

the coefficients we obtain will be estimates of μ and $\alpha^{\star}$. The corresponding constrained estimates of the α may be obtained by multiplying by the contrasts matrix or by using the function `dummy.coef`. As an example consider the Quine data.

```
> quine.Age <- lm(Days ~ Age, quine)
> attach(quine)
> a.star <- coef(quine.Age)
> a.star
  (Intercept)      Age1     Age2     Age3
       16.665   -1.8498   2.6827  0.98035
> a.star <- as.vector(a.star)
> a <- c(mu=a.star[1], alpha=contrasts(Age) %*% a.star[-1])
> a
```

```
       mu  alpha1   alpha2 alpha3 alpha4
    16.665 -1.8132 -5.5128   4.385  2.941
> sum(a[-1])                              # add to zero?
[1] -8.8818e-16
> dummy.coef(quine.Age)
$"(Intercept)":
 (Intercept)
      16.665
$Age:
      F0        F1     F2     F3
  -1.8132 -5.5128  4.385  2.941
```

Notice that the estimates of α sum to zero (up to rounding error). This is because the default contrast matrix used leads to the identification constraint $1^T\alpha = 0$.

Contrast matrices

By default S uses so-called *Helmert* contrast matrices for unordered factors and orthogonal polynomial contrast matrices for ordered factors. The forms of these can be deduced from the following artificial example

```
> N <- factor(Nlevs <- c(0,1,2,4))
> contrasts(N)
  [,1] [,2] [,3]
0   -1   -1   -1
1    1   -1   -1
2    0    2   -1
4    0    0    3
> contrasts(ordered(N))
         .L     .Q        .C
0 -0.67082    0.5 -0.22361
1 -0.22361   -0.5  0.67082
2  0.22361   -0.5 -0.67082
4  0.67082    0.5  0.22361
```

For the `poly` contrasts it can be seen that the corresponding parameters $\alpha^\star$ can be interpreted as the coefficients in an orthogonal polynomial model of degree $r - 1$, *provided* the ordered levels are equally spaced (which is not the case for the example) *and* the class sizes are equal. The $\alpha^\star$ parameters corresponding to the Helmert contrasts also have an easy interpretation, as we see below. Since both the Helmert and polynomial contrast matrices satisfy $1^T C = 0$ the implied constraint on α will be $1^T\alpha = 0$ in both cases.

The default contrast matrices can be changed by resetting the `contrasts` option. This is a character vector of length two giving the names of the functions that generate the contrast matrices for unordered and ordered factors respectively. For example

```
options(contrasts=c("contr.treatment", "contr.poly"))
```

sets the default contrast matrix function for ordinary factors to `contr.treatment` and for ordered factors to `contr.poly` (the original default). Four supplied contrast functions are

`contr.helmert`   for the Helmert contrasts.

`contr.treatment`   for contrasts such that each coefficient represents a comparison of that level with level 1 (omitting level 1 itself). This corresponds to the constraint $\alpha_1 = 0$.

`contr.sum`   where the coefficients are constrained to add to zero, that is in this case the components of $\alpha^\star$ are the same as the first $r - 1$ components of α, with the latter constrained to add to zero.

`contr.poly`   for the equally spaced, equally replicated orthogonal polynomial contrasts.

and others can be written using these as templates (as we do with our function `contr.sdif`). We recommend the use of the treatment contrasts for unbalanced layouts, including generalized linear models and survival models, because the unconstrained coefficients obtained directly from the fit are then easy to interpret.

Notice that the `helmert`, `sum` and `poly` contrasts ensure the rank condition on C is met by choosing C so that the columns of $[\, 1 \; C \,]$ are mutually orthogonal whereas the `treatment` contrasts choose C so that $[\, 1 \; C \,]$ is in echelon form.

Contrast matrices for particular factors may also be set as an attribute of the factor itself. This can be done either by the `contrasts` replacement function or by using the function `C` which takes three arguments, the factor, the matrix from which contrasts are to be taken (or the abbreviated name of a function which will generate such a matrix) and the number of contrasts. On some occasions a p-level factor may be given a contrast matrix with fewer than $p - 1$ columns, in which case it contributes fewer than $p - 1$ degrees of freedom to the model, or the unreduced parameters, α, have additional constraints placed on them apart from the one needed for identification. An alternative method is to use the replacement form with a specific number of contrasts as the second argument. For example suppose we wish to create a factor N2 which would generate orthogonal linear and quadratic polynomial terms, only. Two equivalent ways of doing this would be

```
> N2 <- N
> contrasts(N2, 2) <- poly(Nlevs, 2)
> N2 <- C(N, poly(Nlevs, 2), 2)        # alternative
> contrasts(N2)
            1        2
0 -0.591608   0.56408
1 -0.253546  -0.32233
2  0.084515  -0.64466
4  0.760639   0.40291
```

In this case the constraints imposed on the α parameters are not merely for identification but actually change the model subspace.

Parameter interpretation

The `poly` contrast matrices lead to α^* parameters which are sometimes interpretable as coefficients in an orthogonal polynomial regression. The `treatment` contrasts set $\alpha_1 = 0$ and choose the remaining αs as the α^*s. Other cases are often not so direct, but an interpretation is possible.

To interpret the α^* parameters in general, consider the relationship (6.1). Since the contrast matrix, C, is of full column rank it has a unique left inverse C^+, so we can reverse this relationship to give

$$\alpha^* = C^+\alpha, \quad \text{where} \quad C^+ = (C^TC)^{-1}C^T \tag{6.2}$$

The pattern in the matrix C^+ then provides an interpretation of each unconstrained parameter as a linear function of the (usually) readily appreciated constrained parameters. For example consider the Helmert contrasts for $r = 4$. To exhibit the pattern in C^+ more clearly we use the function `fractions` from the MASS library for rational approximation and display.

```
> fractions(ginv(contr.helmert(n = 4)))
         1      2      3     4
[1,]  -1/2   1/2     0     0
[2,]  -1/6  -1/6   1/3     0
[3,] -1/12 -1/12 -1/12   1/4
```

Hence $\alpha_1^* = \frac{1}{2}(\alpha_2 - \alpha_1)$, $\alpha_2^* = \frac{1}{3}\{\alpha_3 - \frac{1}{2}(\alpha_1 + \alpha_2)\}$ and in general α_j^* is a comparison of α_{j+1} with the average of all preceding αs, divided by $j + 1$. This is a comparison of the (unweighted) mean of class $j + 1$ with that of the preceding classes.

It can sometimes be important to use contrast matrices that give a simple interpretation to the fitted coefficients. This can be done by noting that $(C^+)^+ = C$. For example suppose we wished to choose contrasts so that the $\alpha_j^* = \alpha_{j+1} - \alpha_j$, that is the successive differences of class effects. For $r = 5$, say, the C^+ matrix is then given by

```
> Cp <- diag(-1, 4, 5);  Cp[row(Cp) == col(Cp) - 1] <- 1
> Cp
     [,1] [,2] [,3] [,4] [,5]
[1,]   -1    1    0    0    0
[2,]    0   -1    1    0    0
[3,]    0    0   -1    1    0
[4,]    0    0    0   -1    1
```

Hence the contrast matrix to obtain these linear functions as the estimated coefficients is

```
> fractions(ginv(Cp))
      [,1] [,2] [,3] [,4]
[1,] -4/5 -3/5 -2/5 -1/5
[2,]  1/5 -3/5 -2/5 -1/5
[3,]  1/5  2/5 -2/5 -1/5
[4,]  1/5  2/5  3/5 -1/5
[5,]  1/5  2/5  3/5  4/5
```

Note that again the columns have zero sums, so the implied constraint is that the effects add to zero. (If it were not obvious we could find the induced constraint using our function `Null` to find a basis for the null space of the contrast matrix.)

Since the pattern is obvious from this example, we can write a contrasts matrix function for the general case. To be usable as a component of the `contrasts` option such a function has to conform with a fairly strict convention, but the key computational steps are

```
. . . .
    contr <- col(matrix(nrow = n, ncol = n - 1))
    upper.tri <- !lower.tri(contr)
    contr[upper.tri] <- contr[upper.tri] - n
    contr/n
. . . .
```

The complete function is supplied as `contr.sdif`.

Higher-way layouts

Two- and higher-way layouts may be specified by two or more factors and formula operators. The way the model matrix is generated is then an extension of the conventions for a one-way layout.

If a and b are r- and s-level factors respectively the model formula y ~ a+b specifies an additive model for the two-way layout. Using the redundant specification the algebraic formulation would be

$$y_{ijk} = \mu + \alpha_i + \beta_j + \epsilon_{ijk}$$

and the model matrix would be

$$X = \begin{bmatrix} \mathbf{1} \ X_{\mathrm{a}} \ X_{\mathrm{b}} \end{bmatrix}$$

The reduced model matrix then has the form

$$X^\star = \begin{bmatrix} \mathbf{1} \ X_{\mathrm{a}} C_{\mathrm{a}} \ X_{\mathrm{b}} C_{\mathrm{b}} \end{bmatrix}$$

However if the intercept term is explicitly removed using, say, y ~ a + b - 1, the reduced form is

$$X^\star = \begin{bmatrix} X_{\mathrm{a}} \ X_{\mathrm{b}} C_{\mathrm{b}} \end{bmatrix}$$

Note that this is asymmetric in a and b and order-dependent.

A two-way non-additive model has a redundant specification of the form

$$y_{ijk} = \mu + \alpha_i + \beta_j + \gamma_{ij} + \epsilon_{ijk}$$

The model matrix can be written as

$$X = \begin{bmatrix} \mathbf{1} \ X_{\mathrm{a}} \ X_{\mathrm{b}} \ X_{\mathrm{a}}{:}X_{\mathrm{b}} \end{bmatrix}$$

where we use the notation $A{:}B$ to denote the matrix obtained by taking each column of A and multiplying it elementwise by each column of B. In the example $X_{\mathrm{a}}{:}X_{\mathrm{b}}$ generates an incidence matrix for the sub-classes defined jointly by a and b. Such a model may be specified by the formula

```
    y ~ a + b + a:b
```

or equivalently by y ~ a*b . The reduced form of the model matrix is then

$$X^\star = \begin{bmatrix} 1 & X_a C_a & X_b C_b & (X_a C_a):(X_b C_b) \end{bmatrix}$$

It may be seen that $(X_a C_a):(X_b C_b) = (X_a:X_b)(C_b \otimes C_a)$, where '$\otimes$' denotes the Kronecker product, so the relationship between the γ parameters and the corresponding $\gamma^\star$ s is

$$\gamma = (C_b \otimes C_a)\gamma^\star$$

The identification constraints can be most easily specified by writing γ as an $r \times s$ matrix. If this is done the relationship has the form $\gamma = C_a \gamma^\star C_b^T$ and the constraints have the form $c_a^T \gamma = 0^T$ and $\gamma c_b = 0$, separately.

If the intercept term is removed, however, such as using y ~ a*b -1 the form is different, namely

$$X^\star = \begin{bmatrix} X_a & X_b C_b & X_a^{(-r)}:(X_b C_b) \end{bmatrix}$$

where (somewhat confusingly) $X_a^{(-r)}$ is a matrix obtained by removing the *last* column of X_a. Further, if a model specified as y ~ - 1 + a + a:b the model matrix generated is $\begin{bmatrix} X_a & X_a:(X_b C_b) \end{bmatrix}$. In general addition of a term a:b extends the previously constructed design matrix to a complete non-additive model in some non-redundant way (unless the design is deficient, of course).

Even though a*b expands to a + b + a:b it should be noted that a + a:b is not always the same as a*b - b or even a + b - b + a:b. When used in model fitting functions the last two formulae construct the design matrix for a*b and only then remove any columns corresponding to the b term. (The result is not a statistically meaningful model.) Model matrices are constructed within the fitting functions by arranging the positive terms in order of their complexity, sequentially adding columns to the model matrix according to the redundancy resolution rules and then removing any generated columns corresponding to negative terms. The exception to this rule is the intercept term which is always removed initially. (With update, however, the formula is expanded and all negative terms are removed *before* the model matrix is constructed.)

The model a + a:b generates the same matrix as a/b, which expands to a + b %in% a . There is no compelling reason for the additional operator, %in%, but it does serve to emphasize that the slash operator should be thought of as specifying separate submodels of the form 1 + b for each level of a . The operator behaves like the colon formula operator when the second main effect term is not given, but is conventionally reserved for nested models.

Star products of more than two terms, such as a*b*c, may be thought of as expanding (1 + a):(1 + b):(1 + c) according to ordinary algebraic rules and may be used to define higher way non-additive layouts. There is also a power operator, '^', for generating models up to a specified degree of interaction term. For example (a+b+c)^3 generates the same model as a*b*c but (a+b+c)^2 has the highest order interaction absent.

Combinations of factors and non-factors with formula operators are useful in an obvious way. We have seen already that a/x - 1 generates separate simple linear regressions on x within the levels of a. The same model may be specified as a + a:x - 1, while a*x generates an equivalent model using a different resolution of the redundancy. It should be noted that (x + y + z)^3 does *not* generate a general third-degree polynomial regression in the three variables, as might be expected. This is because terms of the form x:x are regarded as the same as x, not as I(x^2). However, a single term such as x^2 is silently promoted to I(x^2) and interpreted as a power.

6.3 Regression diagnostics

The message in the cats example is relatively easy to discover and we did not have to work hard to find an adequate linear model. There is an extensive literature (e.g. Atkinson, 1985) on examining the fit of linear models to consider whether one or more points are not fitted as well as they should be or have undue influence on the fitting of the model. This can be contrasted with the robust regression methods we discuss in Chapter 8, which automatically take account of anomalous points.

The basic tool for examining the fit is the residuals, and we have already looked for patterns in residuals and assessed the normality of their distribution. The residuals are not independent (they sum to zero if an intercept is present) and they do not have the same variance. Indeed, their variance-covariance matrix is

$$\text{var}\left(e\right) = \sigma^2[I - H] \tag{6.3}$$

where $H = X(X^T X)^{-1} X^T$ is the orthogonal projector matrix onto the model space, or *hat* matrix. If a diagonal entry h_{ii} of H is large, changing y_i will move the fitted surface appreciably towards the altered value. For this reason h_{ii} is said to measure the *leverage* of the observation y_i. The trace of H is p, the dimension of the model space, so 'large' is taken to be greater than two or three times the average, p/n.

Having large leverage has two consequences for the corresponding residual. First, its variance will be lower than average from (6.3). We can compensate for this by re-scaling the residuals to have unit variance. The *standardized residuals* are

$$e_i' = \frac{e_i}{s\sqrt{1 - h_{ii}}}$$

where as usual we have estimated σ^2 by s^2, the residual mean square. Second, if one error is very large, the variance estimate s^2 will be too large, and this deflates all the standardized residuals. Let us consider fitting the model omitting observation i. We then get a prediction for the omitted observation, $\hat{y}_{(i)}$, and an estimate of the error variance, $s_{(i)}^2$, from the reduced sample. The *studentized residuals* are

$$e_i^* = \frac{y_i - \hat{y}_{(i)}}{\sqrt{\text{var}\left(y_i - \hat{y}_{(i)}\right)}}$$

but with σ replaced by $s_{(i)}$. Fortunately, it is not necessary to re-fit the model each time an observation is omitted, since it can be shown that

$$e_i^* = e_i' \Big/ \left[\frac{n - p - e_i'^2}{n - p - 1}\right]^{\frac{1}{2}}$$

Notice that this implies that the standardized residuals, e_i, must be bounded by $\pm\sqrt{n - p}$.

The terminology used here is not universally adopted; in particular studentized residuals are sometimes called *jackknifed* residuals.

It is usually better to compare studentized residuals rather than residuals; in particular we recommend that they are used for normal probability plots.

There are no system functions to compute studentized or standardized residuals, but we have provided functions `studres` and `stdres`. There is a function `hat`, but this expects the model matrix as its argument. (There is a useful function, `lm.influence`, for most of the fundamental calculations. The diagonal of the hat matrix can be obtained by `lm.influence(lmobject)$hat`.)

Scottish hill races

As an example of regression diagnostics, let us return to the data on 35 Scottish hill races in our data frame `hills` considered in Chapter 1. The data come from Atkinson (1986) and are discussed further in Atkinson (1988) and Staudte & Sheather (1990). The columns are the overall race distance, the total height climbed and the record time. In Chapter 1 we considered a regression of `time` on `dist`. We can now include `climb`:

```
> hills.lm <- lm(time ~ dist + climb, hills)
> hills.lm
      ....
Coefficients:
 (Intercept)  dist     climb
      -8.992 6.218 0.011048

Degrees of freedom: 35 total; 32 residual
Residual standard error: 14.676
> plot(fitted(hills.lm), studres(hills.lm))
> identify(fitted(hills.lm), studres(hills.lm),
      row.names(hills))
> qqnorm(studres(hills.lm))
> qqline(studres(hills.lm))
> hills.hat <- lm.influence(hills.lm)$hat
> cbind(hills, lev=hills.hat)[hills.hat > 3/35, ]
                dist climb    time      lev
  Bens of Jura    16  7500 204.617 0.42043
  Lairig Ghru     28  2100 192.667 0.68982
  Ben Nevis       10  4400  85.583 0.12158
Two Breweries     18  5200 170.250 0.17158
  Moffat Chase    20  5000 159.833 0.19099
```

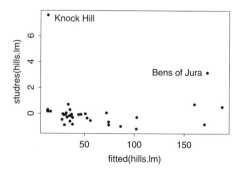
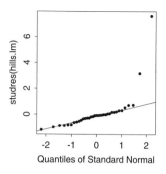

Figure 6.3: Diagnostic plots for Scottish hills data, unweighted model.

so two points have very high leverage, two points have large residuals, and Bens of Jura is in both sets. (See Figure 6.3.)

If we look at Knock Hill we see that the prediction is over an hour less than the reported record:

```
> cbind(hills, pred=predict(hills.lm))["Knock Hill", ]
          dist climb  time    pred
Knock Hill   3   350 78.65  13.529
```

and Atkinson (1988) suggests that the record is one hour out. We drop this observation to be safe:

```
> hills1.lm <- lm(time ~ dist + climb, hills[-18, ])
> hills1.lm
Call:
lm(formula = time ~ dist + climb, data = hills[-18, ])

Coefficients:
 (Intercept)   dist    climb
      -13.53 6.3646 0.011855

Degrees of freedom: 34 total; 31 residual
Residual standard error: 8.8035
```

Since Knock Hill did not have a high leverage, as expected deleting it did not change the fitted model greatly. On the other hand, Bens of Jura had both a high leverage and a large residual and so does affect the fit:

```
> lm(time ~ dist + climb, hills[-c(7,18), ])
 ....
Coefficients:
 (Intercept)   dist     climb
      -10.362 6.6921 0.0080468

Degrees of freedom: 33 total; 30 residual
Residual standard error: 6.0538
```

If we consider this example carefully we find a number of unsatisfactory features. First, the prediction is negative for short races. Extrapolation is often unsafe, but on physical grounds we would expect the model to be a good fit with a zero intercept; indeed hill-walkers use a prediction of this sort (3 miles/hour plus 20 minutes per 1000 feet). We can see from the summary that the intercept is significantly negative:

```
> summary(hills1.lm)
   ....
Coefficients:
               Value Std. Error t value Pr(>|t|)
(Intercept) -13.530    2.649     -5.108   0.000
       dist   6.365    0.361     17.624   0.000
      climb   0.012    0.001      9.600   0.000
   ....
```

Further, we would not expect the predictions of times which range from 15 minutes to over 3 hours to be equally accurate, but rather that the accuracy be roughly proportional to the time. This suggests a log transform, but that would be hard to interpret. Rather we weight the fit using distance as a surrogate for time. We want weights inversely proportional to the variance:

```
> summary(lm(time ~ dist + climb, hills[-18, ],
      weight=1/dist^2))
   ....
Coefficients:
               Value Std. Error t value Pr(>|t|)
(Intercept)  -5.809    2.034     -2.855   0.008
       dist   5.821    0.536     10.858   0.000
      climb   0.009    0.002      5.873   0.000

Residual standard error: 1.16 on 31 degrees of freedom
```

The intercept is still significantly non-zero. If we are prepared to set it to zero on physical grounds, we can achieve the same effect by dividing the prediction equation by distance, and regressing inverse speed (time/distance) on gradient (climb/distance):

```
> lm(time ~ -1 + dist + climb, hills[-18, ], weight=1/dist^2)
Coefficients:
 dist     climb
 4.9 0.0084718

Degrees of freedom: 34 total; 32 residual
Residual standard error (on weighted scale): 1.2786
> hills$ispeed <- hills$time/hills$dist
> hills$grad <- hills$climb/hills$dist
> hills2.lm <- lm(ispeed ~ grad, hills[-18, ])
> hills2.lm
Coefficients:
```

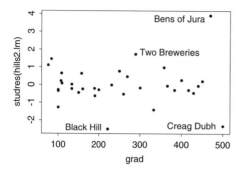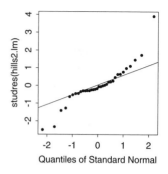

Figure 6.4: Diagnostic plots for Scottish hills data, weighted model.

```
(Intercept)        grad
       4.9  0.0084718

Degrees of freedom: 34 total; 32 residual
Residual standard error: 1.2786
> plot(hills$grad[-18], studres(hills2.lm))
> identify(hills$grad[-18], studres(hills2.lm),
      row.names(hills)[-18])
> qqnorm(studres(hills2.lm))
> qqline(studres(hills2.lm))
> hills2.hat <- lm.influence(hills2.lm)$hat
> cbind(hills[-18,], lev=hills2.hat)[hills2.hat > 1.8*2/34, ]
              dist climb    time  ispeed    grad     lev
Bens of Jura   16  7500 204.617 12.7886  468.75 0.11354
  Creag Dubh    4  2000  26.217  6.5542  500.00 0.13915
```

The two highest-leverage cases are now the steepest two races, and are outliers pulling in opposite directions. We could consider elaborating the model, but this would be to fit only one or two exceptional points; for most of the data we have the formula of 5 minutes/mile plus 8 minutes per 1000 feet. We return to this example in Section 8.4, where robust fits do support a zero intercept.

6.4   Safe prediction

A warning is needed on the use of the `predict` method function when polynomials are used (and also splines, see Chapter 11). We will illustrate this by the dataset `wtloss`, for which a more appropriate analysis is given in Chapter 9. This has a weight loss `Weight` against `Days`. Consider a quadratic polynomial regression model of `Weight` on `Days`. This may be fitted by either of

```
quad1 <- lm(Weight ~ Days + I(Days^2), wtloss)
quad2 <- lm(Weight ~ poly(Days, 2), wtloss)
```

The second uses orthogonal polynomials and is the preferred form on grounds of numerical stability.

Suppose we wished to predict future weight loss. The first step is to create a new data frame with a variable x containing the new values for example

```
new.x <- data.frame(Days = seq(250, 300, 10),
                    row.names = seq(250, 300, 10))
```

The predict method may now be used:

```
> predict(quad1, newdata=new.x)
    250     260     270     280     290     300
 112.51 111.47 110.58 109.83 109.21 108.74
> predict(quad2, newdata=new.x)
    250     260     270     280     290     300
 244.56 192.78 149.14 113.64 86.29 67.081
```

The first form gives correct answers but the second does not!

The reason for this is as follows. The predict method for lm objects works by attaching the estimated coefficients to a new model matrix which it constructs using the formula and the new data. In the first case the procedure will work, but in the second case the columns of the model matrix are for a *different* orthogonal polynomial basis, and so the old coefficients do not apply. The same will hold for any function used to define the model which generates mathematically different bases for old and new data, such as spline bases using bs or ns.

The remedy is to use the method function predict.gam:

```
>  predict.gam(quad2, newdata=new.x)
    250     260     270     280     290     300
 112.51 111.47 110.58 109.83 109.21 108.74
```

This constructs a new model matrix by putting old and new data together, re-estimates the regression using the old data only and predicts using these estimates of regression coefficients. This can involve appreciable extra computation, but the results will be correct for polynomials, but not exactly so for splines since the knot positions will change. As a check, predict.gam compares the predictions with the old fitted values for the original data. If these are seriously different a warning is issued that the process has probably failed.

In our view this is a serious flaw in predict.lm. It would have been better to use the safe method as the default and provide an unsafe argument for the faster method as an option.

6.5   Factorial designs and designed experiments

Factorial designs are powerful tools in the design of experiments. Experimenters often cannot afford to perform all the runs needed for a complete factorial experiment, or they may not all be fitted into one experimental block. To see what can be achieved, consider the following N, P, K (*nitrogen, phosphate, potassium*) factorial experiment on the growth of peas conducted on 6 blocks. The response is yield (in lbs/(1/70)acre-plot).

pk	np	—	nk		n	npk	k	p
49.5	62.8	46.8	57.0		62.0	48.8	45.5	44.2
n	npk	k	p		np	—	nk	pk
59.8	58.5	55.5	56.0		52.0	51.5	49.8	48.8
p	npk	n	k		nk	np	pk	—
62.8	55.8	69.5	55.0		57.2	59.0	53.2	56.0

Half of the design (technically a fractional factorial design) is performed in each of six blocks, so each half occurs three times. (If we consider the variables to take values ± 1, the halves are defined by even or odd parity, equivalently product equal to $+1$ or -1.) Note that the NPK interaction cannot be estimated as it is confounded with block differences, specifically with $(b_2 + b_3 + b_4 - b_1 - b_5 - b_6)$. Suppose the data are entered on file npk.dat. Then an ANOVA table may be computed by

```
> npk1 <- read.table("npk.dat", header=T)
> npk <- data.frame(block=factor(rep(1:6, rep(4,6))),
      N=factor(npk1$N), P=factor(npk1$P), K=factor(npk1$K),
      yield=npk1$yield)
> npk.aov <- aov(yield ~ block + N*P*K, npk)
> npk.aov
    ....
Terms:
                     block      N      P      K    N:P    N:K
  Sum of Squares    343.29 189.28   8.40  95.20  21.28  33.14
  Deg. of Freedom        5      1      1      1      1      1
                      P:K  Residuals
  Sum of Squares     0.48     185.29
  Deg. of Freedom       1         12

Residual standard error: 3.9294
1 out of 13 effects not estimable
Estimated effects are balanced
> summary(npk.aov)
            Df  Sum of Sq  Mean Sq  F Value    Pr(F)
block        5     343.29    68.66    4.447  0.01594
N            1     189.28   189.28   12.259  0.00437
P            1       8.40     8.40    0.544  0.47490
K            1      95.20    95.20    6.166  0.02880
N:P          1      21.28    21.28    1.378  0.26317
N:K          1      33.14    33.14    2.146  0.16865
P:K          1       0.48     0.48    0.031  0.86275
Residuals   12     185.29    15.44
```

```
> alias(npk.aov)
   ....
Complete
       (Intercept) block1 block2 block3 block4 block5 N P K
N:P:K              1      0.33   0.17   -0.3   -0.2
       N:P N:K P:K
N:P:K
> coef(npk.aov)
 (Intercept) block1 block2  block3  block4 block5
      54.875 1.7125 1.6792 -1.8229 -1.0137  0.295
       N        P       K      N:P     N:K     P:K
   2.8083 -0.59167 -1.9917 -0.94167 -1.175 0.14167
```

Note how the `N:P:K` interaction is silently omitted in the summary, although its absence is mentioned in printing `npk.aov`. The `alias` command shows which effect is missing (the particular combinations corresponding to the use of Helmert contrasts of the factor `block`).

Only the `N` and `K` main effects are significant (we ignore blocks whose terms are there precisely because we expect them to be important and so we must allow for them). For two-level factors the Helmert contrast is the same as the sum contrast (up to sign) giving -1 to the first level and $+1$ to the second level. Thus the effects of adding nitrogen and potassium are 5.62 and -3.98 respectively. This interpretation is easier to see with treatment contrasts:

```
> options(contrasts=c("contr.treatment", "contr.poly"))
> npk.aov1 <- aov(yield ~ block + N + K, npk)
> summary.lm(npk.aov1)
   ....
Coefficients:
           Value Std. Error t value Pr(>|t|)
   ....
   N    5.617      1.609     3.490    0.003
   K   -3.983      1.609    -2.475    0.025

Residual standard error: 3.94 on 16 degrees of freedom
```

Note the use of `summary.lm` to give the standard errors. Standard errors of contrasts can also be found from the function `se.contrast`. The full form is quite complex, but a simple use is:

```
> se.contrast(npk.aov1, list(N=="0", N=="1"), data=npk)
Refitting model to allow projection
[1] 1.6092
```

For highly regular designs such as this standard errors may also be found along with estimates of means, effects and other quantities using `model.tables`.

```
> model.tables(npk.aov1, type="means", se=T)
   ....
   N
```

```
     0    1
   52.07 57.68
     . . . .
Standard errors for differences of means
           block      N      K
           2.787  1.609  1.609
   replic. 4.000 12.000 12.000
```

The function `aov.genyates` is very similar to `aov` but uses a generalized Yates' algorithm. It can only be used with equally replicated experimental designs, that is those in which each combination of factor levels occurs equally often, but it can be much faster than `aov` for very large designs. (Equal replication can be checked using the function `replications`.)

Generating designs

The three functions `expand.grid`, `fac.design` and `oa.design` can each be used to construct designs such as our example.

Of these, `expand.grid` is the simplest. It is used in a similar way to `data.frame`: the arguments may be named and the result is a data frame with those names. The columns contain all combinations of values for each argument. If the argument values are numeric the column is numeric; if they are anything else, for example character, the column is a factor. Consider an example:

```
> mp <- c("-","+")
> NPK <- expand.grid(N=mp, P=mp, K=mp)
> NPK
  N P K
1 - - -
2 + - -
3 - + -
4 + + -
5 - - +
6 + - +
7 - + +
8 + + +
```

Note that the first column changes fastest and the last slowest. This is a single complete replicate.

Our example used three replicates, each split into two blocks so that the block comparison is confounded with the highest order interaction. We can construct such a design in stages. First we find a half replicate to be repeated three times and form the contents of three of the blocks. The simplest way to do this is to use `fac.design`:

```
blocks13 <- fac.design(levels=c(2,2,2),
    factor=list(N=mp, P=mp, K=mp), rep=3, fraction=1/2)
```

The first two arguments give the numbers of levels and the factor names and level labels. The third argument gives the number of replications (default 1). The `fraction` argument may only be used for 2^p factorials. It may be given either as a small negative power of 2, as here, or as a *defining contrast formula*. When `fraction` is numerical the function chooses a defining contrast which becomes the `fraction` attribute of the result. For half replicates the highest order interaction is chosen to be aliased with the mean. To find the complementary fraction for the remaining three blocks we need to use the defining contrast formula form for `fraction`:

```
blocks46 <- fac.design(levels=c(2,2,2),
    factor=list(N=mp, P=mp, K=mp), rep=3, fraction=~ -N:P:K)
```

(This will be explained below.) To complete our design we put the blocks together, add in the `block` factor and randomize:

```
NPK <- design(block = factor(rep(1:6, rep(4,6))),
    rbind(blocks13, blocks46))
i <- order(runif(6)[NPK$block], runif(24))
NPK <- NPK[i,]   # Randomized
```

Using `design` instead of `data.frame` creates an object of class `design` that inherits from `data.frame`. For most purposes designs and data frames are equivalent, but some generic functions such as `plot`, `formula` and `alias` have useful `design` methods.

Defining contrast formulae resemble model formulae in syntax only; the meaning is quite distinct. There is no left-hand side. The right-hand side consists of colon products of factors only, separated by + or − signs. A plus (or leading blank) specifies that the treatments with *positive* signs for that contrast are to be selected and a minus those with *negative* signs. A formula such as ~A:B:C−A:D:E specifies a quarter replicate consisting of the treatments which have a positive sign in the ABC interaction and a negative sign in ADE.

Box, Hunter & Hunter (1978, §12.5) consider a 2^{7-4} design used for an experiment in riding up a hill on a bicycle. The seven factors are Seat (up or down), Dynamo (off or on), Handlebars (up or down), Gears (low or medium), Raincoat (on or off), Breakfast (yes or no) and Tyre pressure (hard or soft). A resolution III design was used, so the main effects are not aliased with each other. Such a design cannot be constructed using a numerical fraction in `fac.design` so the defining contrasts have to be known. Box *et al.* use the design relations:

$$D = AB, \quad E = AC, \quad F = BC, \quad G = ABC$$

which mean that ABD, ACE, BCF and $ABCG$ are all aliased with the mean, and form the defining contrasts of the fraction. Whether we choose the positive or negative halves is immaterial here.

```
> lev <- rep(2,7)
> factors <- list(S=mp, D=mp, H=mp, G=mp, R=mp, B=mp, P=mp)
```

```
> Bike <- fac.design(lev, factors, fraction =
  ~ S:D:G + S:H:R + D:H:B + S:D:H:P)
> Bike
  S D H G R B P
1 - - - - - - -
2 - + + + + - -
3 + - + + - + -
4 + + - - + + -
5 + + + - - - +
6 + - - + + - +
7 - + - + - + +
8 - - + - + + +

Fraction:  ~ S:D:G + S:H:R + D:H:B + S:D:H:P
```

(We chose P for pressure rather than T for tyres since T and F are reserved identifiers.)

We may check the symmetry of the design using `replications`:

```
> replications(~.^2, data = Bike)
  S D H G R B P S:D S:H S:G S:R S:B S:P D:H D:G D:R D:B D:P H:G
  4 4 4 4 4 4 4   2   2   2   2   2   2   2   2   2   2   2   2
H:R H:B H:P G:R G:B G:P R:B R:P B:P
  2   2   2   2   2   2   2   2   2
```

Fractions may be specified either in a call to `fac.design` or subsequently using the `fractionate` function.

The third function, `oa.design`, provides some resolution III designs (also known as *main effect plans* or *orthogonal arrays*) for factors at 2 or 3 levels. Only low-order cases are provided, but these are the most useful in practice.

6.6   An unbalanced four-way layout

Aitkin (1978) discussed an observational study of S. Quine. The response is the number of days absent from school in a year by children from a large town in rural New South Wales, Australia. The children were classified by four factors, namely

Age	4 levels: primary, first, second or third form.
Eth	2 levels: aboriginal or non-aboriginal.
Lrn	2 levels: slow or average learner.
Sex	2 levels: male or female.

The dataset is included in the paper of Aitkin (1978) and is available as data frame `quine` in our library MASS. This has been explored several times already, but we now consider a more formal statistical analysis.

There were 146 children in the study. The frequencies of the combinations of factors are

```
> attach(quine)
> table(Lrn, Age, Sex, Eth)
, , F, A                          , , F, N
    FO F1 F2 F3                       FO F1 F2 F3
AL   4  5  1  9               AL   4  6  1 10
SL   1 10  8  0               SL   1 11  9  0

, , M, A                          , , M, N
    FO F1 F2 F3                       FO F1 F2 F3
AL   5  2  7  7               AL   6  2  7  7
SL   3  3  4  0               SL   3  7  3  0
```

(The output has been slightly rearranged to save space.) The classification is unavoidably very unbalanced. There are no slow learners in the fourth form, but all 28 other cells are non-empty. In his paper Aitkin considers a normal analysis on the untransformed response, but in the reply to the discussion he chooses a transformed response, $\log(\text{Days} + 1)$.

A casual inspection of the data shows that homoscedasticity is likely to be an unrealistic assumption on the original scale, so our first step is to plot the cell variances and standard deviations against the cell means.

```
Means <- tapply(Days, list(Eth, Sex, Age, Lrn), mean)
Vars  <- tapply(Days, list(Eth, Sex, Age, Lrn), var)
SD <- sqrt(Vars)
par(mfrow=c(1,2))
plot(Means, Vars, xlab="Cell Means", ylab="Cell Variances")
plot(Means, SD, xlab="Cell Means", ylab="Cell Std Devn.")
```

Missing values are silently omitted from the plot. Interpretation of the result in Figure 6.5 requires some caution because of the small and widely different degrees of freedom on which each variance is based. Nevertheless the approximate linearity of the standard deviations against the cell means suggests a logarithmic transformation or something similar is appropriate. (See for example Rao, 1973, §6g).

Some further insight on the transformation needed is provided by considering a model for the transformed observations

$$y^{(\lambda)} = \begin{cases} (y^\lambda - 1)/\lambda & \lambda \neq 0, \\ \log y & \lambda = 0. \end{cases}$$

where here $y = \text{Days} + \alpha$. (The positive constant α is added to avoid problems with zero entries.) Rather than include α as a second parameter we first consider Aitkin's choice of $\alpha = 1$. Box & Cox (1964) show that the profile likelihood function for λ is

$$\widehat{L}(\lambda) = \text{const} - \tfrac{n}{2} \log \text{RSS}(z^{(\lambda)})$$

where $z^{(\lambda)} = y^{(\lambda)}/\dot{y}^{\lambda-1}$, $\dot{y}$ is the geometric mean of the observations and $\text{RSS}(z^{(\lambda)})$ is the residual sum of squares for the regression of $z^{(\lambda)}$.

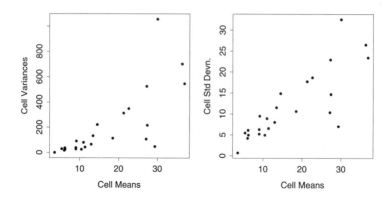

Figure 6.5: Two diagnostic plots for the Quine data.

Box & Cox suggest using the profile likelihood function for the largest linear model to be considered as a guide in choosing a value for λ, which will then remain fixed for any remaining analyses. Ideally other considerations from the context will provide further guidance in the choice of λ, and in any case it is desirable to choose easily interpretable values such as square-root, log or inverse.

Our MASS library function boxcox calculates and (optionally) displays the Box–Cox profile likelihood function, together with a horizontal line showing what would be an approximate 95% likelihood ratio confidence interval for λ. The function is generic and several calling protocols are allowed but a convenient one to use here is with the same arguments as lm together with an additional (named) argument, lambda, to provide the sequence at which the marginal likelihood is to be evaluated. (By default the result is extended using a spline interpolation.)

Since the dataset has four empty cells the full model Eth*Sex*Age*Lrn has a rank-deficient model matrix. Hence we must use either aov or lm with singular.ok = T to fit the model.

```
boxcox(Days+1 ~ Eth*Sex*Age*Lrn, data = quine, singular.ok = T,
    lambda = seq(-0.05, 0.45, len = 20))
```

Alternatively the first argument may be a fitted model object which supplies all needed information apart from lambda. The result is shown on the left-hand side of Figure 6.6 which strongly suggests that a log transformation is not optimal when $\alpha = 1$ is chosen. An alternative one-parameter family of transformations that could be considered in this case is

$$t(y, \alpha) = \log(y + \alpha)$$

Using the same analysis as presented in Box & Cox (1964) the profile log likelihood for α is easily seen to be

$$\widehat{L}(\alpha) = \text{const} - \tfrac{n}{2} \log \text{RSS}\{\log(y + \alpha)\} - \sum \log(y + \alpha)$$

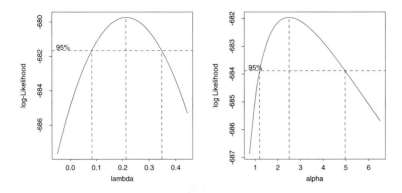

Figure 6.6: Profile likelihood for a Box–Cox transformation model with displacement $\alpha = 1$, left, and a displaced log transformation model, right.

It is interesting to see how this may be calculated directly using low-level tools, in particular the functions `qr` for the QR-decomposition and `qr.resid` for orthogonal projection onto the residual space. Readers are invited to look at our functions `logtrans.default` and `boxcox.default`.

```
logtrans(Days ~ Age*Sex*Eth*Lrn, data = quine,
      alpha = seq(0.75, 6.5, len=20), singular.ok = T)
```

The result is shown in the right-hand panel of Figure 6.6. If a displaced log transformation is chosen a value $\alpha = 2.5$ is suggested, and we adopt this in our analysis. Note that $\alpha = 1$ is outside the notional 95% confidence interval. It can also be checked that with $\alpha = 2.5$ the log transform is well within the range of reasonable Box–Cox transformations to choose.

Model selection

The complete model, `Eth*Sex*Age*Lrn`, has a different parameter for each identified group and hence contains all possible simpler models for the mean as special cases, but has little predictive or explanatory power. For a better insight into the mean structure we need to find more parsimonious models. Before considering tools to prune or extend regression models it is useful to make a few general points on the process itself.

Marginality restrictions

In regression models it is usually the case that not all terms are on an equal footing as far as inclusion or removal is concerned. For example in a quadratic regression on a single variable x one would normally consider removing only the highest degree term, x^2, first. Removing the first degree term while the second degree one is still present amounts to forcing the the fitted curve to be flat at $x = 0$, and unless there were some good reason from the context to do this it would be an arbitrary imposition on the model. Another way to view this is to note that if we write a polynomial regression in terms of a new variable $x^\star = x - \alpha$ the model remains in predictive terms the same, but only the highest-order coefficient remains invariant. If, as is usually the case, we would like our model selection procedure not to depend on the arbitrary choice of origin we must work only with the highest-degree terms at each stage.

The linear term in x is said to be *marginal* to the quadratic term, and the intercept term is marginal to both. In a similar way if a second-degree term in two variables, $x_1 x_2$, is present, any linear terms in either variable or an intercept term are marginal to it.

There are circumstances where a regression through the origin does make sense, but in cases where the origin is arbitrary one would normally only consider regression models where for each term present all terms marginal to it are also present.

In the case of factors the situation is even more clear-cut. A two-factor interaction, a:b, is marginal to any higher order interaction that contains a and b. Fitting a model such as a + a:b leads to a model matrix where the columns corresponding to a:b are extended to compensate for the absent marginal term, b, and the fitted values are the same as if it were present. Fitting models with marginal terms removed such as with a*b - b generates a model with no readily understood statistical meaning[2] but updating models specified in this way using update changes the model matrix so that the absent marginal term again has no effect on the fitted model. In other words removing marginal factor terms from a fitted model is either statistically meaningless or futile in the sense that the model simply changes its parametrization to something equivalent.

Variable selection for the Quine data

The anova function when given a single fitted model object argument constructs a *sequential* analysis of variance table. That is, a sequence of models is fitted by expanding the formula, arranging the terms in increasing order of marginality and including one additional term for each row of the table. The process is order-dependent for non-orthogonal designs and several different orders may be needed to appreciate the analysis fully if the non-orthogonality is severe. For an orthogonal design the process is not order dependent provided marginality restrictions are obeyed.

To explore the effect of adding or dropping terms from a model the two functions add1 and drop1 are usually more convenient. These allow the effect of respectively adding or removing individual terms from a model to be assessed, where the model is defined by a fitted model object given as the first argument. For add1 a second argument is required to specify the scope of the terms considered for inclusion. This may be a formula or an object defining a formula for a larger model. Terms are included or removed in such a way that the marginality principle

---

[2] Marginal terms are sometimes removed in this way in order to calculate what are known as "Type III sums of squares" but we have yet to see a situation where this makes compelling statistical sense.

for factor terms is obeyed; for purely quantitative regressor variables this has to be managed by the user.

Both functions are generic and the method for `lm` objects by default displays the change to the Mallows' C_p statistic. This is defined as

$$C_p = \text{RSS} + 2\sigma^2 p$$

where p is the dimension of the model space. It may be viewed as the residual sum of squares penalised for the model complexity. On page 221 we show it may also be viewed as an approximation to the variable part of the Akaike Information Criterion. If σ^2 is not known, as is usually the case, some estimate of it must be used. This may be supplied as a `scale` argument; by default the estimate of error variance from the starting model is taken, which may highly inflated if the model is grossly inadequate or unrealistically low if the model is overfitted. In the case of the Quine data we will use the estimate from the complete factorial model, but in general this requires a careful judgement by the user.

For an example consider removing the four-way interaction from the complete model and assessing which thee-way terms might be dropped next. We will fit the model using `aov` since we expect benign rank deficiencies, but `lm` method functions are more convenient otherwise.

```
> quine.hi <- aov(log(Days + 2.5) ~ .^4, quine)
> s2 <- summary.lm(quine.hi)$sigma^2 ; s2
> [1] 0.53653
> quine.nxt <- update(quine.hi, . ~ . - Eth:Sex:Age:Lrn)
> drop1.lm(quine.nxt, scale = s2)
Single term deletions
. . . .
```

	Df	Sum of Sq	RSS	Cp
<none>			64.099	91.998
Eth:Sex:Age	3	0.9739	65.073	89.753
Eth:Sex:Lrn	1	1.5788	65.678	92.504
Eth:Age:Lrn	2	2.1284	66.227	91.981
Sex:Age:Lrn	2	1.4662	65.565	91.319

Clearly dropping `Eth:Sex:Age` most reduces C_p but dropping `Eth:Sex:Lrn` would increase it. Note that only non-marginal terms are included.

Alternatively we could start from the simplest model and consider adding terms to reduce C_p; in this case the choice of scale parameter is important, since the simple-minded choice is inflated and may over-penalise complex models.

```
> quine.lo <- aov(log(Days+2.5) ~ 1, quine)
> add1(quine.lo, quine.hi, scale = s2)
Single term additions
. . . .
```

	Df	Sum of Sq	RSS	Cp
<none>			106.79	107.86
Eth	1	10.682	96.11	98.25
Sex	1	0.597	106.19	108.34

```
Age  3      4.747 102.04 106.33
Lrn  1      0.004 106.78 108.93
```

It appears that only Eth and Age might be useful, although in fact all factors are needed since some higher-way interactions lead to large decreases in the residual sum of squares.

Automated model selection

Rather than step manually through a sequence of models using add1, drop1 and update the function step may be used to automate the process. It requires a fitted model to define the starting process, (one somewhere near the final model is probably advantageous), a list of two formulae defining the upper (most complex) and lower (most simple) models for the process to consider and a scale estimate. If a large model is selected as the starting point, the scope and scale arguments have generally reasonable defaults, but for a small model where the process is probably to be one of adding terms, they will usually need both to be supplied. (A further argument, direction, may be used to specify whether the process should only add terms, only remove terms, or do either as needed.)

By default the function produces a verbose account of the steps it takes which we turn off here for reasons of space, but which the user will often care to note. The anova component of the result shows the sequence of steps taken and the reduction in C_p achieved.

```
> quine.stp <- step(quine.nxt,
     scope = list(upper = ~Eth*Sex*Age*Lrn, lower = ~1),
     scale = s2, trace = F)
> quine.stp$anova
....
                Step Df Deviance Resid. Df Resid. Dev    AIC
1                                     120     64.099 91.998
2 - Eth:Sex:Age   3   0.9739        123     65.073 89.753
3 - Sex:Age:Lrn   2   1.5268        125     66.600 89.134
4 - Eth:Age:Lrn   2   2.0960        127     68.696 89.084
5      - Eth:Age  3   3.0312        130     71.727 88.896
```

At this stage we might want to look further at the final model from a significance point of view. The result of step has the same class as its starting point argument, so in this case drop1 may be used, with the default scale, to check each remaining non-marginal term for significance.

```
> drop1(quine.stp)
Single term deletions
Model:
log(Days + 2.5) ~ Eth + Sex + Age + Lrn + Eth:Sex + Eth:Lrn +
        Sex:Age + Sex:Lrn + Age:Lrn + Eth:Sex:Lrn
             Df Sum of Sq    RSS F Value    Pr(F)
     <none>               71.727
    Sex:Age  3   11.566 83.292   6.987 0.000215
    Age:Lrn  2    2.912 74.639   2.639 0.075279
Eth:Sex:Lrn  1    6.818 78.545  12.357 0.000605
```

The term `Age:Lrn` is not significant at the conventional 5% significance level. This suggests, correctly, that selecting terms on the basis of C_p can be somewhat permissive in its choice of terms, being roughly equivalent to choosing an F-cutoff of 2. If we removed this term we would obtain a model equivalent to `Sex/(Age + Eth*Lrn)` which is the same as that found by Aitkin (1978), apart from his choice of $\alpha = 1$ for the displacement constant. (However when we consider a negative binomial model for the same data in Section 7.4 a more extensive model seems to be needed.)

Standard diagnostic checks on the residuals from our final fitted model show no strong evidence of any failure of the assumptions, as the reader may wish to verify.

It can also be verified that had we started from a very simple model and worked forwards we would have stopped much sooner with a much simpler model, even using the same scale estimate. This is because the major reductions in the residual sum of squares only occur when the third-order interaction `Eth:Sex:Lrn` is included.

There are other tools for model selection called `stepwise` and `leaps`, but these only apply for quantitative regressors. There is also no possibility of ensuring marginality restrictions are obeyed.

Akaike Information Criterion (AIC)

It is helpful to relate this definition of AIC to that most commonly used and given by Akaike (1974), namely

$$\text{AIC} = -2 \text{ maximized log likelihood} + 2 \,\#\,\text{parameters}$$

Since the log-likelihood is defined only up to a constant depending on the data, this is also true of AIC.

For a regression model with n observations, p parameters and normally-distributed errors the log-likelihood is

$$L(\beta, \sigma^2; y) = \text{const} - \tfrac{n}{2} \log \sigma^2 - \tfrac{1}{2\sigma^2} \|y - X\beta\|^2$$

and on maximizing over β we have

$$L(\widehat{\beta}, \sigma^2; y) = \text{const} - \tfrac{n}{2} \log \sigma^2 - \tfrac{1}{2\sigma^2} \text{RSS}$$

Thus if σ^2 is *known*, we can take

$$\text{AIC} = \frac{\text{RSS}}{\sigma^2} + 2p$$

but if σ^2 is *unknown*,

$$L(\widehat{\beta}, \widehat{\sigma}^2; y) = \text{const} - \tfrac{n}{2} \log \widehat{\sigma}^2 - \tfrac{n}{2}, \qquad \widehat{\sigma}^2 = \text{RSS}/n$$

and so

$$\text{AIC} = n \log(\text{RSS}/n) + 2p$$

If we now expand this about the residual sum of squares, s^2, for an initial model, we find

$$\text{AIC} = \text{AIC}_0 + n \log\left(\frac{\text{RSS}}{ns^2}\right) + 2(p - p_0) \approx \text{AIC}_0 + \left(\frac{\text{RSS}}{s^2} - n\right) + 2(p - p_0)$$

so we can regard $\text{RSS}/s^2 + 2p$ as an approximation to the variable part of AIC. The value used by `step` is s^2 times this.

Our function `stepAIC` computes the Akaike definition of AIC, but for linear models there will be little difference, except that `stepAIC` will be slower.

Chapter 7

Generalized Linear Models

Generalized linear models (GLMs) extend linear models to accommodate both non-normal response distributions and transformations to linearity. (We will assume that Chapter 6 has been read before this chapter.) The essay by Firth (1991) gives a good introduction to GLMs; the comprehensive reference is McCullagh & Nelder (1989).

A generalized linear model may be described by the following assumptions:

- There is a response, y, observed independently at fixed values of stimulus variables $x_1, \ldots, x_p$.
- The stimulus variables may only influence the distribution of y through a single linear function called the *linear predictor* $\eta = \beta_1 x_1 + \cdots + \beta_p x_p$.
- The distribution of y has density of the form

$$f(y_i; \theta_i, \varphi) = \exp\left[A_i \left\{y_i \theta_i - \gamma(\theta_i)\right\} / \varphi + \tau(y_i, \varphi/A_i)\right] \qquad (7.1)$$

where φ is a *scale parameter* (possibly known), A_i is a *known* prior weight and parameter θ_i depends upon the linear predictor.

- The mean, μ, is a smooth invertible function of the linear predictor:

$$\mu = m(\eta), \qquad \eta = m^{-1}(\mu) = \ell(\mu)$$

The inverse function, $\ell(.)$, is called the *link function*.

Note that θ is also an invertible function of μ, in fact $\theta = (\gamma')^{-1}(\mu)$ as we show below. If φ were known the distribution of y would be a one-parameter canonical exponential family. An unknown φ is handled as a nuisance parameter by moment methods.

GLMs allow a unified treatment of statistical methodology for several important classes of models. We will consider a few examples.

Gaussian For a normal distribution $\varphi = \sigma^2$ and we can write

$$\log f(y) = \frac{1}{\varphi} \left\{y\mu - \tfrac{1}{2}\mu^2 - \tfrac{1}{2}y^2\right\} - \tfrac{1}{2}\log(2\pi\varphi)$$

so $\theta = \mu$ and $\gamma(\theta) = \theta^2/2$.

Poisson For a Poisson distribution with mean μ we have

$$\log f(y) = y \log \mu - \mu - \log(y!)$$

so $\theta = \log \mu$, $\varphi = 1$ and $\gamma(\theta) = \mu = e^{\theta}$.

Binomial For a binomial distribution with fixed number of trials a and parameter p we take the response to be $y = s/a$ where s is the number of "successes". The density is

$$\log f(y) = a \left[y \log \frac{p}{1-p} + \log(1-p) \right] + \log \binom{a}{ay}$$

so we take $A_i = a_i$, $\varphi = 1$, θ to be the logit transform of p and $\gamma(\theta) = -\log(1-p) = \log(1 + e^{\theta})$.

The generalized linear model families handled by the functions supplied in S include `gaussian`, `binomial`, `poisson`, `inverse.gaussian` and `Gamma` response distributions.

Each response distribution allows a variety of link functions to connect the mean with the linear predictor. Those automatically available are given in Table 7.1. The combination of response distribution and link function is called the *family* of the generalized linear model.

Table 7.1: Families and link functions. The default link is denoted by D.

	Family Name				
Link	binomial	Gamma	gaussian	inverse.-gaussian	poisson
logit	D				
probit	●				
cloglog	●				
identity		●	D		●
inverse		D			
log		●			D
1/mu^2				D	
sqrt					●

For n observations from a GLM the log-likelihood is

$$l(\theta, \varphi; Y) = \sum_i \left[A_i \left\{ y_i \theta_i - \gamma(\theta_i) \right\} / \varphi + \tau(y_i, \varphi/A_i) \right] \tag{7.2}$$

and this has score function for θ of

$$U(\theta) = A_i \left\{ y_i - \gamma'(\theta_i) \right\} / \varphi \tag{7.3}$$

From this it is easy to show that

$$E(y_i) = \mu_i = \gamma'(\theta_i) \qquad \text{var}(y_i) = \frac{\varphi}{A_i} \gamma''(\theta_i)$$

(See, for example, McCullagh & Nelder, 1989, §2.2.) It follows immediately that

$$E\left(\frac{\partial^2 l(\theta, \varphi; y)}{\partial\theta\,\partial\varphi}\right) = 0$$

Hence θ and φ, or more generally β and φ, are *orthogonal parameters*.

The function defined by $V(\mu) = \gamma''(\theta(\mu))$ is called the *variance function*.

For each response distribution the link function $\ell = (\gamma')^{-1}$ for which $\theta \equiv \eta$ is called the *canonical link*. If X is the model matrix, so $\eta = X\beta$, it is easy to see that with φ known, $A_i \equiv 1$ and the canonical link that $X^T y$ is a minimal sufficient statistic for β. Also using (7.3) the score equations for the regression parameters β reduce to

$$X^T y = X^T \widehat{\mu} \tag{7.4}$$

This relation is sometimes described by saying that the "observed and fitted values have the same marginal totals". Equation (7.4) is the basis for certain simple fitting procedures, for example Stevens' algorithm for fitting an additive model to a non-orthogonal two-way layout by alternate row and column sweeps (Stevens, 1948), and the Deming & Stephan (1940) or *iterative proportional scaling* algorithm for fitting log-linear models to frequency tables (see Darroch & Ratcliff, 1972).

Table 7.2 shows the canonical links and variance functions. The canonical link function is the default link for the families catered for by the S software (except for the inverse Gaussian, where the factor of 2 is dropped).

Table 7.2: Canonical (default) links and variance functions.

Family	Canonical link	Name	Variance	Name
binomial	$\log(\mu/(1-\mu))$	logit	$\mu(1-\mu)$	mu(1-mu)
Gamma	$-1/\mu$	inverse	μ^2	mu^2
gaussian	μ	identity	1	constant
inverse.gaussian	$-2/\mu^2$	1/mu^2	μ^3	mu^3
poisson	$\log\mu$	log	μ	mu

Iterative estimation procedures

Since explicit expressions for the maximum likelihood estimators are not usually available estimates must be calculated iteratively. It is convenient to give an outline of the iterative procedure here; for a more complete description the reader is referred to McCullagh & Nelder (1989, §2.5, pp. 40ff) or Firth (1991, §3.4). The scheme is sometimes known by the acronym IWLS, for *iterative weighted least squares*.

An initial estimate to the linear predictor is found by some guarded version of $\widehat{\eta}_0 = \ell(y)$. (Guarding is necessary to prevent problems such as taking logarithms of zero.) Define *working weights* W and *working values* z by

$$W = \frac{A}{V}\left(\frac{d\mu}{d\eta}\right)^2, \qquad z = \eta + \frac{y - \mu}{d\mu/d\eta}$$

Initial values for W_0 and z_0 can be calculated from the initial linear predictor.

At iteration k a new approximation to the estimate of β is found by a weighted regression of the working values z_k on X with weights W_k. This provides a new linear predictor and hence new working weights and values which allow the iterative process to continue. The difference between this process and a Newton-Raphson scheme is that the Hessian matrix is replaced by its expectation. This statistical simplification was apparently first used in Fisher (1925), and is often called *Fisher scoring*. The estimate of the large sample variance of $\widehat{\beta}$ is $\widehat{\varphi}(X^T\widehat{W}X)^{-1}$, which is available as a by-product of the iterative process. It is easy to see that the iteration scheme for $\widehat{\beta}$ does not depend on the scale parameter φ.

This iterative scheme depends on the response distribution only through its mean and variance functions. This has led to ideas of *quasi-likelihood*, implemented in the family `quasi`.

The analysis of deviance

A *saturated model* is one in which the parameters θ_i, or almost equivalently the linear predictors η_i, are free parameters. It is then clear from (7.3) that the maximum likelihood estimator of θ_i is obtained by $y_i = \gamma'(\widehat{\theta}_i) = \widehat{\mu}_i$ which is also a special case of (7.4). Denote the saturated model by S.

Assume temporarily that the scale parameter φ is known and has value $\varphi = 1$. Let M be a model involving $p < n$ regression parameters, and let $\hat{\theta}_i$ be the maximum likelihood estimate of θ_i under M. Twice the log-likelihood ratio statistic for testing M within S is given by

$$D_M = 2 \sum_{i=1}^{n} A_i \left[\left\{ y_i \theta(y_i) - \gamma\left(\theta(y_i)\right) \right\} - \left\{ y_i \widehat{\theta}_i - \gamma(\widehat{\theta}_i) \right\} \right] \qquad (7.5)$$

The quantity D_M is called the *deviance* of model M, even when the scale parameter is unknown or is known to have a value other than one. In the latter case D_M/φ, the difference in twice the log-likelihood, is known as the *scaled deviance*. (Confusingly, sometimes either is called the *residual deviance*, for example McCullagh & Nelder, p. 119.)

For a Gaussian family with identity link the scale parameter φ is the variance and D_M is the residual sum of squares. Hence in this case the scaled deviance has distribution

$$D_M/\varphi \sim \chi^2_{n-p} \qquad (7.6)$$

leading to the customary unbiased estimator

$$\widehat{\varphi} = \frac{D_M}{n-p} \qquad (7.7)$$

In other cases the distribution (7.6) for the deviance under M may be approximately correct suggesting $\widehat{\varphi}$ as an approximately unbiased estimator of φ. It should be noted that sufficient (if not always necessary) conditions under which (7.6) becomes approximately true are that the individual distributions for the components y_i should become closer to normal form and the link effectively closer to an identity link. The approximation will often *not* improve as the sample size n increases since the number of parameters under S also increases and the usual likelihood ratio approximation argument does not apply. Nevertheless, (7.6) may sometimes be a good approximation, for example in a binomial GLM with large values of a_i. Firth (1991, p. 69) discusses this approximation, including the extreme case of a binomial GLM with only one trial per case, that is with $a_i = 1$.

Let $M_0 \subset M$ be a submodel with $q < p$ regression parameters and consider testing M_0 within M. If φ is known, by the usual likelihood ratio argument under M_0 we have a test given by

$$\frac{D_{M_0} - D_M}{\varphi} \overset{\cdot}{\sim} \chi^2_{p-q} \tag{7.8}$$

where $\overset{\cdot}{\sim}$ denotes 'is approximately distributed as'. The distribution is exact only in the Gaussian family with identity link. If φ is not known, by analogy with the Gaussian case it is customary to use the approximate result

$$\frac{(D_{M_0} - D_M)}{\widehat{\varphi}\,(p-q)} \overset{\cdot}{\sim} F_{p-q,n-p} \tag{7.9}$$

although this must be used with some caution in non-Gaussian cases.

Some of the fitting functions use a quantity AIC defined by

$$D_M + 2p\,\widehat{\varphi} \tag{7.10}$$

(Chambers & Hastie, 1992, p. 234). If the scale parameter is one, this corresponds up to an additive constant to the commonly used definition of AIC (Akaike, 1974) of

$$AIC = -2 \text{ maximized log likelihood} + 2 \,\#\, \text{parameters}$$

and if φ is known but not one it also differs by a factor. In both cases choosing models to minimize AIC will give the same result. However, as we saw in Section 6.6, the concepts differ when φ is estimated. It is important when using (7.10) to hold $\widehat{\varphi}$ constant when comparing models.

7.1 Functions for generalized linear modelling

The linear predictor part of a generalized linear model may be specified by a model formula using the same notation and conventions as linear models. Generalized linear models also require the family to be specified, that is the response distribution, the link function and perhaps the variance function for quasi models.

The fitting function is glm for which main arguments are

```
glm(formula, family, data, weights, control)
```

The `family` argument is usually given as the name of one of the standard family functions listed under "Family Name" in Table 7.1. Where there is a choice of links, the name of the link may also be supplied in parentheses as a parameter, for example `binomial(link=probit)`. (The variance function for the `quasi` family may also be specified in this way.) For user-defined families (such as our `neg.bin` discussed in Section 7.4) other arguments to the family function may be allowed or even required.

Prior weights A_i may be specified using the `weights` argument. For binomial models these are implied and should not be specified separately.

The iterative process can be controlled by many parameters. The only ones which are at all likely to need altering are `maxit`, which controls the maximum number of iterations and the default value of 10 is occasionally too small, `trace` which will often be set as `trace=T` to trace the iterative process, and `epsilon` which controls the stopping rule for convergence. The convergence criterion is to stop if

$$|\text{deviance}^{(i)} - \text{deviance}^{(i-1)}| < \epsilon(\text{deviance}^{(i-1)} + \epsilon)$$

(This comes from reading the S code and is not as described in Chambers & Hastie, 1992, p. 243.) It is quite often necessary to reduce the tolerance `epsilon` whose default value is 10^{-4}. Under some circumstances the convergence of the IWLS algorithm can be extremely slow, when the change in deviance at each step can be small enough for premature stopping to occur with the default ϵ.

Generic functions with methods for `glm` objects include `coef`, `resid`, `print`, `summary` and `deviance`. It is useful to have a `glm` method function to extract the variance-covariance matrix of the estimates. This can be done using part of the result of `summary`:

```
vcov.glm <- function(obj) {
    so <- summary(obj, corr = F)
    so$dispersion * so$cov.unscaled
}
```

Our library `MASS` contains the generic function, `vcov`, and methods for objects of classes `lm` and `nls` as well as `glm`.

For `glm` fitted model objects the `anova` function allows an additional argument `test` to specify which test is to be used. Two possible choices are `test="Chisq"` for chi-squared tests using (7.8) and `test="F"` for F-tests using (7.9). The default is `test="Chisq"` for the binomial and Poisson families, otherwise `test="F"`.

Prediction and residuals

The `predict` method function for `glm` has a `type` argument to specify what is to be predicted. The default is `type="link"` which produces predictions of the linear predictor η. Predictions on the scale of the mean μ (for example, the fitted values $\widehat{\mu}_i$) are specified by `type="response"`.

For `glm` models there are four types of residual that may be requested, known as *deviance, working, Pearson* and *response* residuals. The response residuals are simply $y_i - \widehat{\mu}_i$. The Pearson residuals are a standardized version of the response residuals, $(y_i - \widehat{\mu}_i)/\sqrt{\widehat{V}_i}$. The working residuals come from the last stage of the iterative process, $(y_i - \widehat{\mu}_i) \; / \; \mathrm{d}\mu_i/\mathrm{d}\eta_i$. The deviance residuals d_i are defined as the signed square roots of the summands of the deviance (7.5) taking the same sign as $y_i - \widehat{\mu}_i$.

For Gaussian families all four types of residual are identical. For binomial and Poisson GLMs the sum of the squared Pearson residuals is the Pearson chi-squared statistic, which often approximates the deviance, and the deviance and Pearson residuals are usually then very similar.

Method functions for the `resid` function have an argument `type` which defaults to `type="deviance"` for objects of class `glm`. Other values are `"response"`, `"pearson"` or `"working"`; these may be abbreviated to the initial letter. Deviance residuals are the most useful for diagnostic purposes.

Concepts of leverage and its effect on the fit are as important for GLMs as they are in linear regression, and are discussed in some detail by Davison & Snell (1991) and extensively for binomial GLMs by Collett (1991). On the other hand, they seem less often used, as GLMs are most often used either for simple regressions or for contingency tables where, as in designed experiments, high-leverage can not occur.

Robust versions of GLM models are discussed in Chapter 8 on page 259.

Model formulae

The model formula language for GLMs is slightly more general than that described in Section 6.2 for linear models in that the function `offset` which may be used. Its effect is to to evaluate its argument and to add it to the linear predictor, equivalent to enclosing the argument in `I( )` and forcing the coefficient to be one.

Note that `offset` can be used with linear models, but it is completely ignored, without any warning. Of course, with linear models the same effect can be achieved by subtracting the offset from the dependent variable, but if the more intuitive formula with offset is desired, it can be used with `glm` and the `gaussian` family. For example the model for the log-transformed cat data on page 195 could have been fitted as

```
glm(log(Hwt) ~ Sex*log(Bwt) + offset(log(Bwt)), data = cats)
```

where the `gaussian` family is used by default. Sometimes as here a variable is included both as a free regressor and as an offset to allow a test of the hypothesis that the regression coefficient is one or, by extension, any specific value. (In this example the interaction term would need to be removed first, of course.) Another reason to use an offset rather than an adjusted dependent variable is to allow direct predictions from the fitted model object.

The default Gaussian family

A call to glm with the default gaussian family achieves the same purpose as a call to lm but less efficiently. The gaussian family is not provided with a choice of links, so no argument is allowed. If a problem requires a Gaussian family with a non-identity link, this can usually be handled using the quasi family. (Indeed yet another, even more inefficient way to emulate lm with glm is to use the family quasi(link = identity, variance = constant).) Although the gaussian family is the default, it is virtually never used in practice other than when an offset is needed.

7.2   Binomial data

Consider first a small example. Collett (1991, p. 75) reports an experiment on the toxicity to the tobacco budworm *Heliothis virescens* of doses of the pyrethroid *trans*-cypermethrin to which the moths were beginning to show resistance. Batches of twenty moths of each sex were exposed for 3 days to the pyrethroid and the number in each batch which were dead or knocked down was recorded. The results were

			dose			
sex	1	2	4	8	16	32
Male	1	4	9	13	18	20
Female	0	2	6	10	12	16

The doses were in μg. We fit a logistic regression model using $\log_2$ (dose) since the doses are powers of two. To do so we must specify the numbers of trials of a_i. This is done using glm with the binomial family in one of three ways:

1. If the response is a numeric vector it is assumed to hold the data in ratio form, $y_i = s_i/a_i$, in which case the a_is must be given as a vector of weights using the weights argument. (If the a_i are all one the default weights suffices.)

2. If the response is a logical vector or a two-level factor it is treated as a 0/1 numeric vector and handled as above.

3. If the response is a two-column matrix it is assumed that the first column holds the number of successes, s_i, and the second holds the number of failures, $a_i - s_i$, for each trial. In this case no weights argument is required.

The non-intuitive third form seems to allow the fitting function to select a better starting value, and for this reason we tend to favour it.

   In all cases the response is the relative frequency, $y_i = s_i/a_i$, so the means μ_i are the probabilities p_i. Hence fitted yields probabilities, not binomial means.

   Since we have binomial data we use the second possibility:

```
> options(contrasts=c("contr.treatment", "contr.poly"))
> ldose <- rep(0:5, 2)
> numdead <- c(1, 4, 9, 13, 18, 20, 0, 2, 6, 10, 12, 16)
> sex <- factor(rep(c("M", "F"), c(6, 6)))
> SF <- cbind(numdead, numalive=20-numdead)
> budworm.lg <- glm(SF ~ sex*ldose, family=binomial)
> summary(budworm.lg)
    ....
Coefficients:
                Value Std. Error   t value
(Intercept) -2.99354    0.55253  -5.41789
        sex  0.17499    0.77816   0.22487
      ldose  0.90604    0.16706   5.42349
  sex:ldose  0.35291    0.26994   1.30735
    ....
    Null Deviance: 124.88 on 11 degrees of freedom
Residual Deviance: 4.9937 on 8 degrees of freedom
    ....
```

This shows slight evidence of a difference in slope between the sexes. Note that we use treatment contrasts to make interpretation easier. Since female is the first level of `sex` (they are in alphabetical order) the parameter for `sex:ldose` represents the increase in slope for males just as the parameter for `sex` measures the increase in intercept. We can plot the data and the fitted curves by

```
plot(c(1,32), c(0,1), type="n", xlab="dose",
    ylab="prob", log="x")
text(2^ldose, numdead/20,as.character(sex))
ld <- seq(0, 5, 0.1)
lines(2^ld, predict(budworm.lg, data.frame(ldose=ld,
    sex=factor(rep("M", length(ld)), levels=levels(sex))),
    type="response"))
lines(2^ld, predict(budworm.lg, data.frame(ldose=ld,
    sex=factor(rep("F", length(ld)), levels=levels(sex))),
    type="response"))
```

see Figure 7.1. Note that when we set up a factor for the new data we must specify all the levels or both lines would refer to level one of `sex`. (Of course, here we could have predicted for both sexes and plotted separately, but it helps to understand the general mechanism needed.)

The apparently non-significant sex effect in this analysis has to be interpreted carefully. Since we are fitting separate lines for each sex, it tests the (uninteresting) hypothesis that the lines do not differ at zero log-dose. If we re-parametrize to locate the intercepts at dose 8 we find

```
> budworm.lgA <- update(budworm.lg, . ~ sex*I(ldose-3))
> summary(budworm.lgA)$coefficients
                Value Std. Error t value
(Intercept) -0.27543    0.23049 -1.1950
        sex  1.23373    0.37694  3.2730
```

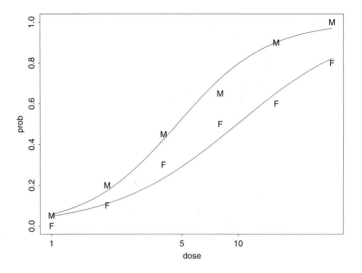

Figure 7.1: Tobacco budworm destruction *versus* dosage of *trans*-cypermethrin by sex.

```
I(ldose - 3)    0.90604    0.16706  5.4235
sex:I(ldose - 3)    0.35291    0.26994  1.3074
```

which shows a significant difference between the sexes at dose 8. The model fits very well as judged by the residual deviance, (4.9937 is a small value for a χ^2_8 variate, and the estimated probabilities are based on a reasonable number of trials), so there is no suspicion of curvature. We can confirm this by an analysis of deviance:

```
> anova(update(budworm.lg, . ~ . + sex*ldose^2), test="Chisq")
....
Terms added sequentially (first to last)
                Df Deviance Resid. Df Resid. Dev Pr(Chi)
        NULL                       11     124.88
         sex  1     6.08          10     118.80 0.01370
       ldose  1   112.04           9       6.76 0.00000
 I(ldose^2)  1     0.91           8       5.85 0.34104
  sex:ldose  1     1.24           7       4.61 0.26552
sex:I(ldose^2)  1     1.44         6       3.17 0.23028
```

This isolates a further two degrees of freedom which if curvature were appreciable would most likely be significant, but are not. Note how `anova` when given a single fitted model object produces a sequential analysis of deviance, which will nearly always be order-dependent for `glm` objects. The additional argument `test="Chisq"` to the `anova` method may be used to specify tests using equation (7.8). The default corresponds to `test="none"`. The other possible choices are `"F"` and `"Cp"`, neither of which is appropriate here.

Our analysis so far suggests a model with parallel lines (on logit scale) for each sex. We will estimate doses corresponding to a given probability of death.

The first step is to re-parametrize the model to give separate intercepts for each sex and a common slope.

```
> budworm.lg0 <- glm(SF ~ sex + ldose - 1, family=binomial)
> summary(budworm.lg0)$coefficients
          Value Std. Error t value
  sexF -3.4732    0.46829 -7.4166
  sexM -2.3724    0.38539 -6.1559
  ldose 1.0642    0.13101  8.1230
```

Let ξ_p be the log-dose for which the probability of response is p. (Historically $2^{\xi_{0.5}}$ was called the "50% lethal dose" or LD50.) Clearly

$$\xi_p = \frac{\ell(p) - \beta_0}{\beta_1}, \qquad \frac{\partial \xi_p}{\partial \beta_0} = -\frac{1}{\beta_1}, \qquad \frac{\partial \xi_p}{\partial \beta_1} = -\frac{\ell(p) - \beta_0}{\beta_1^2} = -\frac{\xi_p}{\beta_1}$$

where β_0 and β_1 are the slope and intercept respectively. Our library MASS contains functions to calculate and print $\widehat{\xi_p}$ and its asymptotic standard error, namely

```
dose.p <- function(obj, cf = 1:2, p = 0.5) {
  eta <- family(obj)$link(p)
  b <- coef(obj)[cf]
  x.p <- (eta - b[1])/b[2]
  names(x.p) <- paste("p = ", format(p), ":", sep = "")
  pd <-  - cbind(1, x.p)/b[2]
  SE <- sqrt(((pd %*% vcov(obj)[cf, cf]) * pd) %*% c(1, 1))
  structure(x.p, SE = SE, p = p, class = "glm.dose")
}
print.glm.dose <- function(x, ...) {
  M <- cbind(x, attr(x, "SE"))
  dimnames(M) <- list(names(x), c("Dose", "SE"))
  x <- M
  NextMethod("print")
}
```

Notice how the `family` function can used to extract the link function, which can then be used to obtain values of the linear predictor. For females the values of ξ_p at the quartiles may be calculated as

```
> dose.p(budworm.lg0, cf = c(1,3), p = 1:3/4)
             Dose      SE
p = 0.25: 2.2313 0.24983
p = 0.50: 3.2636 0.22971
p = 0.75: 4.2959 0.27462
```

For males the corresponding log-doses are lower.

In biological assays the *probit* link used to be more conventional and the technique was called *probit analysis*. Unless the estimated probabilities are concentrated in the tails, probit and logit links tend to give similar results with the values of ξ_p near the centre almost the same. We can demonstrate this for the budworm assay by fitting a probit model and comparing the estimates of ξ_p for females.

```
> dose.p(update(budworm.lg0, family=binomial(link=probit)),
          cf = c(1,3), p = 1:3/4)
              Dose      SE
p = 0.25: 2.1912 0.23841
p = 0.50: 3.2577 0.22405
p = 0.75: 4.3242 0.26685
```

The differences are insignificant. This occurs because the logistic and standard normal distributions can approximate each other very well, at least between the 10th and 90th percentiles, by a simple scale change in the abscissa. (See for example Cox & Snell, 1989, p. 21.) The menarche data frame in our MASS library has data with a substantial proportion of the responses at very high probabilities and provides an example where probit and logit models appear noticeably different.

A binary data example: Low birth weight in infants

Hosmer & Lemeshow (1989) give a dataset on 189 births at a US hospital, with the main interest being in low birth weight. The following variables are available in our data frame birthwt :

low	birth weight less than 2.5 kg (0/1)
age	age of mother in years
lwt	weight of mother (lbs) at last menstrual period
race	white / black / other
smoke	smoking status during pregnancy (0/1)
ptl	number of previous premature labours
ht	history of hypertension (0/1)
ui	has uterine irritability (0/1)
ftv	number of physician visits in the first trimester
bwt	actual birth weight (grams)

Although the actual birth weights are available, we shall concentrate on predicting if the birth weight is low from the remaining variables. The dataset contains a small number of pairs of rows which are identical apart from the ID; it is possible that these refer to twins but identical birth weights seem unlikely.

   We will use a logistic regression with a binomial (in fact 0/1) response. It is worth considering carefully how to use the variables. It is unreasonable to expect a linear response with ptl. Since the numbers with values greater than one are so small we reduce it to a indicator of past history. Similarly, ftv can be reduced to three levels. With non-Gaussian GLMs it is usual to use treatment contrasts.

```
> options(contrasts=c("contr.treatment", "contr.poly"))
> attach(birthwt)
> race <- factor(race, labels=c("white", "black", "other"))
> table(ptl)
   0  1 2 3
 159 24 5 1
> ptd <- factor(ptl > 0)
```

```
> table(ftv)
   0  1  2 3 4 6
 100 47 30 7 4 1
> ftv <- factor(ftv)
> levels(ftv)[-(1:2)] <- "2+"
> table(ftv)  # as a check
   0  1 2+
 100 47 42
> bwt <- data.frame(low=factor(low), age, lwt, race,
     smoke=(smoke>0), ptd, ht=(ht>0), ui=(ui>0), ftv)
> detach("birthwt"); rm(race, ptd, ftv)
```

We can then fit a full logistic regression, and omit the rather large correlation matrix from the summary.

```
> birthwt.glm <- glm(low ~ ., family=binomial, data=bwt)
> summary(birthwt.glm, correlation=F)
    ....
Coefficients:
                 Value Std. Error   t value
(Intercept)   0.823013  1.2440732   0.66155
        age  -0.037234  0.0386777  -0.96267
        lwt  -0.015653  0.0070759  -2.21214
  raceblack   1.192409  0.5357458   2.22570
  raceother   0.740681  0.4614609   1.60508
      smoke   0.755525  0.4247645   1.77869
        ptd   1.343761  0.4804445   2.79691
         ht   1.913162  0.7204344   2.65557
         ui   0.680195  0.4642156   1.46526
       ftv1  -0.436379  0.4791611  -0.91071
      ftv2+   0.179007  0.4562090   0.39238

    Null Deviance: 234.67 on 188 degrees of freedom
Residual Deviance: 195.48 on 178 degrees of freedom
```

Since the responses are binary, even if the model is correct there is no guarantee that the deviance will have even an approximately chi-squared distribution, but since the value is about in line with its degrees of freedom there seems no serious reason to question the fit. Rather than select a series of sub-models by hand, we make use of the step function. This uses an approximation to AIC to select models. We first allow this to drop variables, then to choose pairwise interactions. By default argument trace is true and produces voluminous output.

```
> birthwt.step <- step(birthwt.glm, trace=F)
> birthwt.step$anova
Initial Model:
low ~ age + lwt + race + smoke + ptd + ht + ui + ftv
Final Model:
low ~ lwt + race + smoke + ptd + ht + ui
```

```
      Step Df Deviance Resid. Df Resid. Dev    AIC
1                             178     195.48 217.48
2 - ftv  2   1.3582          180     196.83 214.83
3 - age  1   1.0179          181     197.85 213.85
> birthwt.step2 <- step(birthwt.glm, ~ .^2 + I(scale(age)^2)
    + I(scale(lwt)^2), trace = F)
> birthwt.step2$anova
Initial Model:
low ~ age + lwt + race + smoke + ptd + ht + ui + ftv
Final Model:
low ~ age + lwt + smoke + ptd + ht + ui + ftv + age:ftv
    + smoke:ui

        Step Df Deviance Resid. Df Resid. Dev    AIC
1                               178     195.48 217.48
2  + age:ftv -2  -12.475        176     183.00 209.00
3 + smoke:ui -1   -3.057        175     179.94 207.94
4     - race  2    3.130        177     183.07 207.07

> summary(birthwt.step2, corr=F)$coef
                Value Std. Error  t value
(Intercept) -0.582520   1.418729 -0.41059
        age  0.075535   0.053880  1.40190
        lwt -0.020370   0.007465 -2.72878
      smoke  0.780057   0.419249  1.86061
        ptd  1.560205   0.495741  3.14722
         ht  2.065549   0.747204  2.76437
         ui  1.818252   0.664906  2.73460
       ftv1  2.920800   2.278843  1.28170
      ftv2+  9.241693   2.631548  3.51188
    ageftv1 -0.161809   0.096472 -1.67726
   ageftv2+ -0.410873   0.117548 -3.49537
   smoke:ui -1.916401   0.970786 -1.97407
> table(bwt$low, predict(birthwt.step2) > 0)
   FALSE TRUE
0    116   14
1     28   31
```

Note that although both `age` and `ftv` were previously dropped, their interaction is now included, the slopes on `age` differing considerably within the three `ftv` groups. The AIC criterion penalizes terms less severely than a likelihood ratio or Wald's test would, and so although adding the term `smoke:ui` reduces the AIC its t-statistic is only just significant at the 5% level. We also considered three-way interactions, but none were chosen.

Residuals are not always very informative with binary responses but at least none are particularly large here.

We can examine the linearity in age and mother's weight more flexibly using generalized additive models with smooth terms. This will be considered in Section 11.1 after generalized additive models have been discussed, but no serious evidence of non-linearity on the logistic scale arises, suggesting the model chosen here is acceptable.

An alternative approach is to predict the actual live birth weight and later threshold at 2.5 kilograms. This is left as an exercise for the reader; surprisingly it produces somewhat worse predictions with around 52 errors. We also consider a tree-based prediction rule in Section 14.3, with a smaller AIC but similar error rate.

Problems with binomial GLMs

We warned that `step` uses an approximation to AIC to choose models. There are examples in which that approximation is extremely poor and leads to misleading results.

There is a little-known phenomenon for binomial GLMs that was pointed out by Hauck & Donner (1977). The standard errors and t values derive from the Wald approximation to the log-likelihood, obtained by expanding the log-likelihood in a second-order Taylor expansion at the maximum likelihood estimates. If there are some $\hat{\beta}_i$ which are large, the curvature of the log-likelihood at $\hat{\beta}$ can be much less than near $\beta_i = 0$, and so the Wald approximation underestimates the change in log-likelihood on setting $\beta_i = 0$. This happens in such a way that as $|\hat{\beta}_i| \to \infty$, the t statistic tends to zero. Thus highly significant coefficients according to the likelihood ratio test may have non-significant t ratios. (Curvature of the log-likelihood surface may to some extent be explored *post hoc* using the `glm` method for the `profile` generic function and the associated `plot` and `pairs` methods supplied with our `MASS` library. The `profile` generic function is discussed in Chapter 9 on page 276.)

The method `step.glm` for `step` is based on a local linearization about $\hat{\beta}$, turning a GLM into a linear model. Thus the significance of one-term additions is judged via the score statistic approximation to the log-likelihood and hence AIC, but the significance of one-term deletions is judged via the Wald approximation. This can lead `step.glm` to choose to delete extremely significant terms. Fortunately `step.glm` checks that the next model has a lower AIC, finds that it does not and stops the stepwise process; unfortunately there may well be other terms that could be deleted or added that would reduce AIC. The only solution is to calculate the exact AIC, which is the method used by the function `stepAIC` in our library. Optionally this uses the linear approximation for screening additions, which can be useful when many possible additions are being tested at an already satisfactory model.

There is one fairly common circumstance in which both convergence problems and the Hauck–Donner phenomenon (and trouble with `step`) can occur. This is when the fitted probabilities are extremely close to zero or one. Consider a medical diagnosis problem with thousands of cases and around fifty binary explanatory variables (which may arise from coding fewer categorical factors); one of these indicators is rarely true but always indicates that the disease is present. Then the fitted probabilities of cases with that indicator should be one, which can only be achieved by taking $\widehat{\beta}_i = \infty$. The result from glm will be warnings and an estimated coefficient of around ± 10. (Such cases are not hard to spot, but might occur during a series of stepwise fits.) There has been fairly extensive discussion of this in the statistical literature, usually claiming the non-existence of maximum likelihood estimates; see Santer & Duffy (1989, p. 234). However, the phenomenon was discussed much earlier (as a desirable outcome) in the pattern recognition literature (Duda & Hart, 1973; Ripley, 1996).

It is straightforward to fit binomial GLMs by direct maximization (see page 293) and this provides a route to study several extensions of logistic regression.

7.3 Poisson models

The canonical link for the Poisson family is log, and the major use of this family is to fit surrogate Poisson log-linear models to what is actually multinomial frequency data. Such log-linear models have a large literature. (For example, Plackett, 1974; Bishop, Fienberg & Holland, 1975; Haberman, 1978, 1979; Goodman, 1978; Whittaker, 1990.) The surrogate Poisson models approach we adopt here is in line with that of McCullagh & Nelder (1989).

It is convenient to divide the factors classifying a multi-way frequency table into *response* and *stimulus* factors. Stimulus factors have their marginal totals fixed in advance (or for the purposes of inference). The main interest lies in the conditional probabilities of the response factor given the stimulus factors.

It is well known that the conditional distribution of a set of independent Poisson random variables given their sum is multinomial with probabilities given by the ratios of the Poisson means to their total. This result is applied to the counts for the multi-way response within each combination of stimulus factor levels. This allows models for multinomial data with a multiplicative probability specification to be fitted and tested using Poisson log-linear models.

Identifying the multinomial model corresponding to any surrogate Poisson model is straightforward. Suppose A, B, ... are factors classifying a frequency table. The *minimum model* is the interaction of all stimulus factors, and must be included for the analysis to respect the fixed totals over response factors. This model is usually of no interest, corresponding to a uniform distribution over response factors independent of the stimulus factors. Interactions between response and stimulus factors indicate interesting structure. For large models is often helpful to use a graphical structure to represent the conditional independencies (Whittaker, 1990; Lauritzen, 1996).

A four-way contingency table

As an example of a surrogate Poisson model analysis consider the data in Table 7.3. This shows the result of a detergent brand preference study where the respondents are also classified according to the temperature and softness of their washing water and whether or not they were previous users of brand M. The dataset was reported by Ries & Smith (1963) and analysed by Cox & Snell (1989).

Table 7.3: Numbers preferring brand X of a detergent to brand M.

M user?	No				Yes			
Temperature	Low		High		Low		High	
Preference	X	M	X	M	X	M	X	M
Water softness								
Hard	68	42	42	30	37	52	24	43
Medium	66	50	33	23	47	55	23	47
Soft	63	53	29	27	57	49	19	29

We can enter this dataset by

```
detg <- cbind(expand.grid(Brand=c("X","M"),
  Temp=c("Low","High"), M.user=c("N","Y"),
  Soft=c("Hard","Medium","Soft")),
  Fr= c(68,42,42,30,37,52,24,43, 66,50,33,23,47,55,23,47,
        63,53,29,27,57,49,19,29))
detg$Soft <- ordered(detg$Soft,
  levels=c("Soft","Medium","Hard"))
```

The levels of Soft are specified to ensure their ordering is correct.

We study how the proportion of users preferring one or other brand varies with the other factors. In our terminology Brand is the only response factor and M.user, Temp and Soft are stimulus factors. The minimum model is M.user*Temp*Soft. The aim is to arrive at the simplest model not contradicted by the data. The best way to achieve this is to be guided by any insights the research worker can provide and to test a sequence of models suggested by the context. In the absence of such detailed information, and partly for illustration, we note here what the automatic stepwise procedure suggests as a final model. Note that we include the minimum model within the lower bound on the scope of step, and start with the simplest interesting model, which has probability of choice independent of the stimulus factors.

```
> detg.m0 <- glm(Fr ~ M.user*Temp*Soft + Brand,
                 family=poisson, data=detg)
> detg.m0
    ....
Degrees of Freedom: 24 Total; 11 Residual
Residual Deviance: 32.826
> detg.step <- step(detg.m0, list(lower=formula(detg.m0),
    upper= ~ .^3), scale=1, trace=F)
> detg.step$anova
Initial Model:
```

```
Fr ~ M.user * Temp * Soft + Brand

Final Model:
Fr ~ M.user + Temp + Soft + Brand + M.user:Temp + M.user:Soft +
     Temp:Soft + M.user:Brand + Temp:Brand + M.user:Temp:Soft +
     Brand:M.user:Temp

                  Step Df Deviance Resid. Df Resid. Dev    AIC
1                                         11    32.826 58.826
2         + M.user:Brand -1  -20.581      10    12.244 40.244
3           + Temp:Brand -1   -3.800       9     8.444 38.444
4 + Brand:M.user:Temp -1     -2.788        8     5.656 37.656
```

The effect of M.user is overwhelmingly significant, but there is a suggestion of
an effect of Temp and a M.user × Temp interaction.

It is helpful to re-fit the model keeping together terms associated with the
minimum model, which we do using terms with keep.order = T and omit
these in our display:

```
> detg.mod <- glm(terms(Fr ~ M.user*Temp*Soft +
                    Brand*M.user*Temp, keep.order=T),
                    family=poisson, data=detg)
> summary(detg.mod, correlation=F)
Coefficients:
                      Value Std. Error    t value
    ....
          Brand -0.306470   0.109420   -2.80087
   Brand:M.user  0.407566   0.159608    2.55354
    Brand:Temp   0.044106   0.184629    0.23889
Brand:M.user:Temp  0.444267  0.266730    1.66560

    Null Deviance: 118.63 on 23 degrees of freedom
Residual Deviance: 5.656 on 8 degrees of freedom
```

From the sign of the M.user term previous users of brand M are less likely to
prefer brand X. The interaction term, though non-significant, suggests that for
M users this proportion differs for those who wash at low and high temperatures.

Fitting by iterative proportional scaling

The function loglin fits log-linear models by iterative proportional scaling. This
starts with an array of fitted values that has the correct multiplicative structure,
(for example with all values equal to 1) and makes multiplicative adjustments
so that the observed and fitted values come to have the same marginal totals, in
accordance with equation (7.4). (See Darroch & Ratcliff, 1972, .) This is usually
very much faster than GLM fitting but is less flexible.

To use loglin we need to form the frequencies into an array. A simple way
to do this is to construct a matrix subscript from the factors:

```
attach(detg)
detg.tab <- table(M.user, Temp, Soft, Brand)
names(dimnames(detg.tab)) <- c("M.user","Temp","Soft","Brand")
detg.tab[cbind(M.user, Temp, Soft, Brand)] <- Fr
detg.ips <- loglin(detg.tab,  margin=list(c(1,2,3), c(1,2,4)) )
c(detg.ips$df, detg.ips$lrt, detg.ips$pearson)
[1] 8.000 5.656 5.650
```

The model is specified by giving the margins to be fitted, which correspond to the
two third-order interactions in the final `glm` model. The fitted values are returned
as a table in component `fit` if argument `fit = T`. The three components of the
result which we have printed are the degrees of freedom, the residual deviance,
and the (Pearson) chi-squared statistic.

This function may be made much easier to use by enclosing it within a wrapper
function to give the result a class, such as

```
LogLin <- function(...) structure(loglin(...), class="loglin")
```

and writing method functions for `print`, `summary`, `anova` and so on to manipulate it. With a little more work it may be configured to accept its input as a
formula and data frame rather than as array and margin list. We discuss aspects
of programming with formulae and data frames in the on-line complements. The
details are left as an advanced exercise.

Equivalence to logistic analysis

When there is just one response factor with two levels, as here, it is also possible
to treat it as binomial data. It is important to note that if we do so using the logit
link the results are exactly equivalent to the log-linear analysis:

```
> attach(detg)
> deterg <- cbind(detg[Brand=="X", -1],
      M = detg[Brand=="M","Fr"])
> detach()
> names(deterg)[4] <- "X"
> detg.lg <- glm(cbind(M, X) ~ M.user*Temp,
      family=binomial, data=deterg)
> summary(detg.lg, correlation=F)
Coefficients:
                Value Std. Error t value
(Intercept) -0.306470    0.10941 -2.8011
     M.user  0.407566    0.15960  2.5537
       Temp  0.044106    0.18462  0.2389
 M.user:Temp 0.444267    0.26670  1.6658

    Null Deviance: 32.826 on 11 degrees of freedom
Residual Deviance: 5.656 on 8 degrees of freedom
```

The null deviance is for a model with just an intercept, which corresponds to
the surrogate Poisson model `detg.m0`. For simple models such as this the best
summary is often the table of fitted probabilities:

```
> cbind(deterg, p=predict(detg.lg, type="response"))
    Temp M.user    Soft  X  M       p
1   Low      N    Hard 68 42 0.42398
3  High      N    Hard 42 30 0.43478
5   Low      Y    Hard 37 52 0.52525
7  High      Y    Hard 24 43 0.64324
    ....
```

The use of surrogate Poisson models can demand large amounts of CPU time and memory to fit parameters in the minimum model that are not used in the interpretation as a multinomial model. It may be preferable to use direct maximization of the log likelihood, as in the `multinom` function in our library `nnet`.

7.4 A negative binomial family

Once the link function and its derivative, the variance function, the deviance function and some method for obtaining starting values are known the fitting procedure is the same for all generalized linear models. This particular information is all taken from the *family object*, which in turn makes it fairly easy to handle a new GLM family by writing a *family object generator* function. We illustrate the procedure with a family for negative binomial models with known shape parameter, a restriction which will be relaxed later.

Using the negative binomial distribution in modelling is important in its own right and has a long history. An excellent modern reference is Lawless (1987). The variance is greater than the mean, suggesting that it might be useful for describing frequency data where this is a prominent feature, often loosely called "overdispersed Poisson data".

The negative binomial can arise from a two-stage model for the distribution of a discrete variable Y. We suppose there is an unobserved random variable E having a gamma distribution gamma$(\theta)/\theta$, that is with mean 1 and variance $1/\theta$. Then the model postulates that conditionally on E, Y is Poisson with mean μE. Thus:

$$Y \mid E \sim \text{Poisson}(\mu E), \qquad \theta E \sim \text{gamma}(\theta)$$

The marginal distribution of Y is then negative binomial with probability function, mean and variance given by

$$\mathrm{E}(Y) = \mu, \quad \text{var}(Y) = \mu + \mu^2/\theta, \quad f_Y(y; \theta, \mu) = \frac{\Gamma(\theta + y)}{\Gamma(\theta)\,y!} \frac{\mu^y\,\theta^\theta}{(\mu + \theta)^{\theta + y}}$$

If θ is known, as for the present we assume it is, this distribution has the general form (7.1).

A function `make.family` is available to piece together the information required for the `family` argument of the `glm` fitting function. It requires three main arguments `name`   a character string giving the name of the family,

`link`   a list supplying information about the link function, its inverse and derivative and an initialization expression,

`variance`   a list specifying the variance and deviance functions.

The two datasets `glm.links` and `glm.variances` are matrices of lists with examples that can be used as templates. We have in mind a negative binomial model for the Quine data introduced in Section 6.6, so we consider only a `log` link. The following function provides the family, passing the parameter θ as a required argument.

```
neg.bin <- function(theta = stop("theta must be given"))
{
    nb.lnk <- list(names = "Log: log(mu)",
        link = function(mu) log(mu),
        inverse = function(eta) exp(eta),
        deriv = function(mu) 1/mu,
        initialize = expression(mu <- y + (y == 0)/6) )
    nb.var <- list(
        names = "mu + mu^2/theta",
        variance = substitute(function(mu, th = .Theta)
            mu * (1 + mu/th), list(.Theta = theta)),
        deviance = substitute(
            function(mu, y, A, residuals = F, th = .Theta)
            {
                devi <- 2 * A * (y * log(pmax(1, y)/mu) -
                    (y + th) * log((y + th)/(mu + th)))
                if(residuals) sign(y - mu) * sqrt(abs(devi))
                else sum(devi)
            }, list(.Theta = theta) )
    )
    make.family("Negative Binomial", link = nb.lnk,
        variance = nb.var)
}
```

The `names` components are cosmetic and only used by the `print` method.

A negative binomial model for the Quine data

A Poisson model for the Quine data has an excessively large deviance:

```
> glm(Days ~ .^4, family=poisson, data=quine)
    ....
Degrees of Freedom: 146 Total; 118 Residual
Residual Deviance: 1173.9
```

Inspection of the mean–variance relationship in Figure 6.5 suggests a negative binomial model with $\theta \approx 2$ might be appropriate. We will assume at first that $\theta = 2$ is known. A negative binomial model may be fitted by

```
quine.nb <- glm(Days ~ .^4, family=neg.bin(2), data=quine)
```

The standard generic functions may now be used to fit sub-models, produce analysis of variance tables, and so on. For example let us check the final model found in Chapter 6. (The output has been edited.)

```
> quine.nb0 <- update(quine.nb, . ~ Sex/(Age + Eth*Lrn))
> anova(quine.nb0, quine.nb, test="Chi")
  Resid. Df Resid. Dev  Test  Df Deviance  Pr(Chi)
1      132    198.51
2      118    171.98 1 vs. 2  14   26.527 0.022166
```

which suggests that model to be an over-simplification in the fixed-θ negative binomial setting.

Consider now what happens when θ is estimated rather than held fixed. The function `negative.binomial` supplied with our library MASS is similar to `neg.bin` defined above, but allows more links. We have also included a function `glm.nb`, a modification of `glm` which incorporates maximum likelihood estimation of θ. This has summary and anova methods; the latter produces likelihood ratio tests for the sequence of fitted models. (For deviance tests to be applicable the θ parameter has to be held constant for all fitted models.)

The following models summarize the results of a manual selection (`step` does not work with `glm.nb`, although `stepAIC` does), and indicate model `quine.nb2`.

```
> quine.nb1 <- glm.nb(Days ~ Sex/(Age + Eth*Lrn), data=quine)
> quine.nb2 <- update(quine.nb1, . ~ . + Sex:Age:Lrn)
> quine.nb3 <- update(quine.nb2, Days ~ .^4)
> anova(quine.nb1, quine.nb2, quine.nb3)
Likelihood ratio tests of Negative Binomial Models
   ....
  theta Resid. df  2 x log-lik.  Test  df LR stat. Pr(Chi)
1 1.5980      132        10254
2 1.6869      128        10262 1 vs 2   4    7.627 0.10623
3 1.9284      118        10278 2 vs 3  10   16.074 0.09754
```

Testing the first model within the third is still borderline significant, so model 2, which is equivalent to `Sex*Lrn/(Age + Eth)`, should be retained. (Starting from a complete model `stepAIC` arrives at `Lrn/(Age + Eth + Sex)^2` but the extra three-way interaction is not significant.)

The estimate of θ and its standard error are available from the summary or as components of the fitted model object:

```
> c(theta=quine.nb2$theta, SE=quine.nb2$SE)
  theta      SE
 1.6869 0.22677
```

We can perform some diagnostic checks by examining the deviance residuals:

```
rs <- resid(quine.nb2, type="deviance")
plot(predict(quine.nb2), rs, xlab="Linear predictors",
    ylab="Deviance residuals")
abline(h=0, lty=2)
qqnorm(rs, ylab="Deviance residuals")
qqline(rs)
```

The result is shown in Figure 7.2. There is perhaps a little skewness, but no indication of serious violations of the model assumptions.

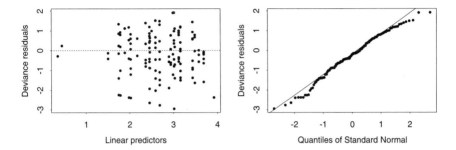

Figure 7.2: Two residual plots for the deviance residuals of a negative binomial model for the Quine data.

Chapter 8

Robust Statistics

Outliers are sample values which cause surprise in relation to the majority of the sample. This is not a pejorative term; outliers may be correct, but they should always be checked for transcription errors. They can play havoc with standard statistical methods, and many *robust* and *resistant* methods have been developed since 1960 to be less sensitive to outliers.

The sample mean $\bar{y}$ can be upset completely by a single outlier; if any data value $y_i \to \pm\infty$, then $\bar{y} \to \pm\infty$. This contrasts with the sample median, which is little affected by moving any single value to $\pm\infty$. We say that the median is *resistant* to *gross errors* whereas the mean is not. In fact the median will tolerate up to 50% gross errors before it can be made arbitrarily large; we say its *breakdown point* is 50% whereas that for the mean is 0%. Although the mean is the optimal estimator of the location of the normal distribution, it can be substantially sub-optimal for distributions close to the normal. Robust methods aim to have high efficiency in a neighbourhood of the assumed statistical model.

There are now a number of books on robust statistics. Huber (1981) is rather theoretical, Hampel *et al.* (1986) and Staudte & Sheather (1990) less so. Rousseeuw & Leroy (1987) is principally concerned with regression, but is very practical. A series of books edited by Hoaglin, Mosteller and Tukey (Hoaglin *et al.*, 1983, 1985, 1991) provide a wealth of angles on modern developments in resistant statistics. The chapters by Goodall (1983) and Iglewicz (1983) provide concise introductions to univariate robust statistics. Marazzi (1993) documents a set of Fortran routines for robust statistics that have a (rather complex) interface to S-PLUS.[1]

The earlier techniques of outlier rejection and their relation to modern methods are discussed by Barnett & Lewis (1985).

Why will it not suffice to screen data and remove outliers? There are several aspects to consider:

1. Users, even expert statisticians, do not always screen the data. One of us met this 15 years ago with an example of multiple regression. The data had originally been entered on punched cards, and for one of the regressors about

---

[1] These routines are available from `statlib` (see Appendix C) and include code for the resistant methods discussed in Section 8.4.

half the observations were two columns out and so were read as 100 times their true value. Several cohorts of students and some eminent teachers had used the example without reporting the problem.

2. The sharp decision to keep or reject an observation is wasteful. We can do better by down-weighting dubious observations than by rejecting them, although we may wish to reject completely wrong observations.

3. It can be difficult or even impossible to spot outliers in multivariate or highly structured data.

4. Rejecting outliers affects the distribution theory, which ought to be adjusted. In particular, variances will be under-estimated from the 'cleaned' data.

Robust and resistant methods have been provided by S and S-PLUS for some years, and a bewildering variety of functions are available. We consider first univariate problems, then regression and finally robust covariance estimation for use in multivariate techniques. Robust time-series methods are considered briefly in Chapter 15.

8.1   Univariate samples

For a fixed underlying distribution, we define the *relative efficiency* of an estimator $\tilde{\theta}$ relative to another estimator $\hat{\theta}$ by

$$RE(\tilde{\theta}; \hat{\theta}) = \frac{\text{variance of } \hat{\theta}}{\text{variance of } \tilde{\theta}}$$

since $\hat{\theta}$ needs only RE times as many observations as $\tilde{\theta}$ for the same precision, approximately. The asymptotic relative efficiency (ARE) is the limit of the RE as the sample size $n \to \infty$. (It may be defined more widely via asymptotic variances.) If $\hat{\theta}$ is not mentioned, it is assumed to be the optimal estimator. There is a difficulty with biased estimators whose variance can be small or zero. One solution is to use the mean square-error, another to rescale by $\theta/E(\hat{\theta})$. Iglewicz (1983) suggests using $\text{var}(\log \hat{\theta})$ (which is scale-free) for estimators of scale.

We can apply the concept of ARE to the mean and median. At the normal distribution $ARE(\text{median; mean}) = 2/\pi \approx 64\%$. For longer-tailed distributions the median does better; for the t distribution with 5 degrees of freedom (which is often a better model of error distributions than the normal) $ARE(\text{median; mean}) \approx 96\%$.

The following example from Tukey (1960) is more dramatic. Suppose we have n observations $Y_i \sim N(\mu, \sigma^2), i = 1, \ldots, n$ and we want to estimate σ^2. Consider $\hat{\sigma}^2 = s^2$ and $\tilde{\sigma}^2 = d^2\pi/2$ where

$$d = \frac{1}{n}\sum_i |Y_i - \overline{Y}|$$

and the constant is chosen since for the normal $d \to \sqrt{2/\pi}\,\sigma$. The $ARE(\tilde{\sigma}^2; s^2)$ = 0.876. Now suppose that each Y_i is from $N(\mu, \sigma^2)$ with probability $1 - \epsilon$ and from $N(\mu, 9\sigma^2)$ with probability ϵ. (Note that both the overall variance and the variance of the uncontaminated observations are proportional to σ^2.) We have

ϵ (%)	$ARE(\tilde{\sigma}^2; s^2)$
0	0.876
0.1	0.948
0.2	1.016
1	1.44
5	2.04

Since the mixture distribution with $\epsilon = 1\%$ is indistinguishable from normality for all practical purposes, the optimality of s^2 is very fragile. We say it lacks *robustness of efficiency.*

There are better estimators of σ than $d\sqrt{\pi/2}$ (which has breakdown point 0%). Two alternatives are proportional to

$$IQR = X_{(3n/4)} - X_{(n/4)}$$
$$MAD = \underset{i}{\text{median}}\,\{|Y_i - \underset{j}{\text{median}}\,(Y_j)|\}$$

(Order statistics are linearly interpolated where necessary.) At the normal,

$$MAD \to \text{median}\,\{|Y - \mu|\} \approx 0.6745\sigma$$
$$IQR \to \sigma\big[\Phi^{-1}(0.75) - \Phi^{-1}(0.25)\big] \approx 1.35\sigma$$

(We will now refer to $MAD/0.6745$ as the MAD estimator.) Both are not very efficient but are very resistant to outliers in the data. The MAD estimator has ARE 37% at the normal (Staudte & Sheather, 1990, p. 123).

The function mad calculates the rescaled MAD. A centre other than the median can be specified. There is a choice of median if n is even. By default mad uses the central median (the linear interpolant) but by specifying the parameter low=T it will use the low median, the smaller of the two order statistics which the span the 50% point. (By analogy the high median would be the larger of the two.)

Consider n independent observations Y_i from a location family with pdf $f(y - \mu)$ for a function f symmetric about zero, so it is clear that μ is the centre (median, mean if it exists) of the distribution of Y_i. We also think of the distribution as being not too far from the normal. There are a number of obvious estimators of μ, including the sample mean, the sample median, and the MLE.

The *trimmed mean* is the mean of the central $1 - 2\alpha$ part of the distribution, so αn observations are removed from each end. This is implemented by the function mean with the argument trim specifying α. Obviously, trim=0 gives the mean and trim=0.5 gives the median (although it is easier to use the function median). (If αn is not an integer, the integer part is used.)

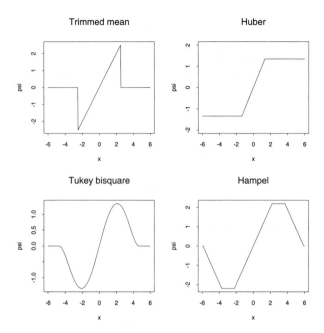

Figure 8.1: The ψ-functions for four common M-estimators.

Most of the location estimators we consider are *M-estimators*. The name derives from 'MLE-like' estimators. If we have density f, we can define $\rho = -\log f$. Then the MLE would solve

$$\min_{\mu} \sum_i -\log f(y_i - \mu) = \min_{\mu} \sum_i \rho(y_i - \mu)$$

and this makes sense for functions ρ not corresponding to pdfs. Let $\psi = \rho'$ if this exists. Then we will have $\sum_i \psi(y_i - \hat{\mu}) = 0$ or $\sum_i w_i (y_i - \hat{\mu}) = 0$ where $w_i = \psi(y_i - \hat{\mu})/(y_i - \hat{\mu})$. This suggests an iterative method of solution similar to that for GLMs.

Examples of M-estimators

The mean corresponds to $\rho(x) = x^2$, and the median to $\rho(x) = |x|$. (For even n any median will solve the problem.) The function

$$\psi(x) = \begin{cases} x & |x| < c \\ 0 & \text{otherwise} \end{cases}$$

corresponds to *metric trimming* and large outliers have no influence at all. The function

$$\psi(x) = \begin{cases} -c & x < -c \\ x & |x| < c \\ c & x > c \end{cases}$$

is known as *metric Winsorizing*[2] and brings in extreme observations to $\mu \pm c$. The corresponding $-\log f$ is

$$\rho(x) = \begin{cases} x^2 & \text{if } |x| < c \\ c(2|x| - c) & \text{otherwise} \end{cases}$$

and corresponds to a density with a Gaussian centre and double-exponential tails. This estimator is due to Huber. Note that its limit as $c \to 0$ is the median, and as $c \to \infty$ the limit is the mean. The value $c = 1.345$ gives 95% efficiency at the normal.

Tukey's *biweight* has

$$\psi(t) = t \left[1 - \left(\frac{t}{R} \right)^2 \right]_+^2$$

where $[\,]_+$ denotes the positive part of. This implements 'soft' trimming. The value $R = 4.685$ gives 95% efficiency at the normal.

Hampel's ψ has several linear pieces,

$$\psi(x) = \text{sgn}(x) \begin{cases} |x| & 0 < |x| < a \\ a & a < |x| < b \\ a(c - |x|)/(c - b)a & b < |x| < c \\ 0 & c < |x| \end{cases}$$

for example with $a = 2.2s, b = 3.7s, c = 5.9s$. Figure 8.1 illustrates these functions.

There is a scaling problem with the last four choices, since they depend on a scale factor (c, R, or s). We can apply the estimator to rescaled results, that is

$$\min_\mu \sum_i \rho \left(\frac{y_i - \mu}{s} \right)$$

for a scale factor s, for example the MAD estimator.

Alternatively, we can estimate s in a similar way. The MLE for density $s^{-1} f((x - \mu)/s)$ gives rise to the equation

$$\sum_i \psi \left(\frac{y_i - \mu}{s} \right) \left(\frac{y_i - \mu}{s} \right) = n$$

which is not resistant (and is biased at the normal). We modify this to

$$\sum_i \chi \left(\frac{y_i - \mu}{s} \right) = (n - 1)\gamma$$

for bounded χ, where γ is chosen for consistency at the normal distribution, so $\gamma = E\,\chi(N)$. The main example is "Huber's proposal 2" with

$$\chi(x) = \psi(x)^2 = \min(|x|, c)^2 \tag{8.1}$$

----

[2] A term attributed by Dixon (1960) to Charles P. Winsor.

In very small samples we need to take account of the variability of $\hat{\mu}$ in performing the Winsorizing. As $Y_i - \hat{\mu}$ has approximate scale factor $s\sqrt{(1 - 1/n)}$, we might replace c by $c\sqrt{(1 - 1/n)}$. (This is not done internally in the software provided.)

If the location μ is known we can apply these estimators with $n - 1$ replaced by n to estimate the scale s alone.

Another approach to scale estimation is the A-estimators described by Iglewicz (1983). The bi-weight A-estimator is

$$s_{bi} = \frac{\sqrt{n}\left[\sum(x_i - \hat{\mu})^2(1 - u_i^2)_+^4\right]^{1/2}}{\left[\sum(1 - u_i^2)_+(1 - 5u_i^2)\right]}$$

for $u_i = (x_i - \hat{\mu})/(c \times MAD)$, which will be re-scaled to be correct at the normal. For $c = 6$ this has efficiency over 80% for a wide range of distributions.

Implementation

The current S-PLUS function for location M-estimation is `location.m` which has parameters including

```
location.m(x, location, scale, weights, na.rm=F,
    psi.fun="bisquare", parameters)
```

The initial values specified by `location` defaults to the low median, and `scale` is by default fixed as MAD (with the high median). The `psi.fun` can be Tukey's bi-square, `"huber"` or a user-supplied function. The `parameters` are passed to the `psi.fun` and default to 5 for bi-square and 1.45 for Huber, for efficiency at the normal of about 96%. Hampel's ψ function can be implemented by the example (using C) given on the help page.

For scale estimation there are the functions `mad` described above, `scale.a` and `scale.tau`. All are centred at the median by default. The function `scale.tau` implements Huber's scale based on (8.1) with parameter 1.95 chosen for 80% efficiency at the normal. The function `scale.a` uses Tukey's bisquare A-estimate of scale with parameter 3.85, again chosen for 80% efficiency at the normal. (Both have an optional argument `tuning` to set the parameter.) The author of the `scale.tau` function seems unaware that for the Huber function integration by parts gives

$$E(\chi(N)) = 2c^2 + 2(1 - c^2)\Phi(c) - 2c\,\phi(c) - 1$$

and so the function uses numerical integration. (See, for example, Marazzi, 1993, p. 105.)

Examples

We give two datasets taken from analytical chemistry (Abbey, 1988; Analytical Methods Committee, 1989a,b). The dataset abbey contains 31 determinations of nickel content ($\mu g\ g^{-1}$) in SY-3, a Canadian syenite rock, and chem contains 24 determinations of copper ($\mu g\ g^{-1}$) in wholemeal flour. These data are part of a larger study which suggests $\mu = 3.68$.

We also supply S functions huber and hubers for the Huber M-estimator with MAD and "proposal 2" scale respectively, with default $c = 1.5$.

```
> sort(chem)
 [1]  2.20  2.20  2.40  2.40  2.50  2.70  2.80  2.90  3.03
[10]  3.03  3.10  3.37  3.40  3.40  3.40  3.50  3.60  3.70
[19]  3.70  3.70  3.70  3.77  5.28 28.95
> mean(chem)
[1] 4.2804
> median(chem)
[1] 3.385
> location.m(chem)
[1] 3.1452
    ....
> location.m(chem, psi.fun="huber")
[1] 3.2132
    ....
> mad(chem)
[1] 0.52632
> scale.tau(chem)
[1] 0.639
> scale.a(chem)
[1] 0.64411
> scale.tau(chem, 3.68)
[1] 0.91578
> scale.a(chem, 3.68)
[1] 0.85875
> unlist(huber(chem))
      mu       s
  3.2067 0.52632
> unlist(hubers(chem))
      mu       s
  3.2055 0.67365
```

The sample is clearly highly asymmetric with one value than appears to be out by a factor of 10. It was checked and reported as correct by the laboratory.

```
> sort(abbey)
 [1]    5.2    6.5    6.9    7.0    7.0    7.0    7.4    8.0    8.0
[10]    8.0    8.0    8.5    9.0    9.0   10.0   11.0   11.0   12.0
[19]   12.0   13.7   14.0   14.0   14.0   16.0   17.0   17.0   18.0
[28]   24.0   28.0   34.0  125.0
> mean(abbey)
```

```
[1] 16.006
> median(abbey)
[1] 11
> location.m(abbey)
[1] 10.804
> location.m(abbey, psi.fun="huber")
[1] 11.517
> unlist(hubers(abbey))
      mu       s
 11.732 5.2585
> unlist(hubers(abbey, 2))
      mu       s
 12.351 6.1052
> unlist(hubers(abbey, 1))
      mu       s
 11.365 5.5673
```

Note how reducing the constant reduces the estimate of location, as this sample (like many in analytical chemistry) has a long right tail.

8.2 Median polish

Consider a two-way layout. The additive model is

$$\hat{y}_{ij} = \mu + \alpha_i + \beta_j, \qquad \alpha. = \beta. = 0$$

The least squares fit corresponds to choosing the parameters μ, α_i and β_j so that the row and column sums of the residuals are zero.

Means are not resistant. Suppose we use medians instead. That is, we seek a fit of the same form, but with median (α_i) = median (β_j) = 0 and median$_i$ (e_{ij}) = median$_j$ (e_{ij}) = 0. This is no longer a set of linear restrictions, so there may be many solutions. The median polish algorithm (Mosteller & Tukey, 1977; Emerson & Hoaglin, 1983) is to augment the table with row and column effects as

$$
\begin{array}{cccc}
e_{11} & \cdots & e_{1c} & a_1 \\
\vdots & \ddots & \vdots & \vdots \\
e_{r1} & \cdots & e_{rc} & a_r \\
b_1 & \cdots & b_r & m
\end{array}
$$

where initially $e_{ij} = y_{ij}$, $a_i = b_j = m = 0$. At all times we maintain

$$y_{ij} = m + a_i + b_j + e_{ij}$$

In a *row sweep* for each row we subtract the median of columns $1, \ldots, c$ from those columns and add it to the last column. For a *column sweep* for each column we subtract the median of rows $1, \ldots, r$ from those rows and add it to the bottom row.

Median polish operates by alternating row and column sweeps until the changes made become small or zero (or the human computer gets tired!). (Often just two pairs of sweeps are recommended.) The answer may depend on whether rows or columns are tried first and is very resistant to outliers. Using means rather medians will give the least-squares decomposition without iteration.

An example

The table below gives specific volume (*cc/gm*) of rubber at four temperatures ($°C$) and six pressures (*kg/cm^2 above atmo*). These data were published by Wood & Martin (1964, p. 260), and used by Mandel (1969) and Emerson & Wong (1985).

			Pressure			
Temperature	500	400	300	200	100	0
0	1.0637	1.0678	1.0719	1.0763	1.0807	1.0857
10	1.0697	1.0739	1.0782	1.0828	1.0876	1.0927
20	1.0756	1.0801	1.0846	1.0894	1.0944	1.0998
25	1.0786	1.0830	1.0877	1.0926	1.0977	1.1032

The default `trim=0.5` option of `twoway` performs median polish. We have, after multiplying by 10^4,

			Pressure				
Temperature	500	400	300	200	100	0	a_i
0	7.0	4.5	1.5	-1.5	-6.5	-9.0	-96.5
10	3.0	1.5	0.5	-0.5	-1.5	-3.0	-32.5
20	-3.0	-1.5	-0.5	0.5	1.5	3.0	32.5
25	-4.5	-4.0	-1.0	1.0	3.0	5.5	64.0
b_j	-111.0	-67.5	-23.5	23.5	72.5	125.0	$m = 10837.5$

This is interpreted as
$$y_{ij} = m + a_i + b_j + e_{ij}$$
and the body of the table contains the residuals e_{ij}. These have both row medians and column medians zero. Originally the value for temperature 0, pressure 400 was entered as 1.0768; the only change was to increase the residual to 94.5×10^{-4} which was easily spotted.

Note the pattern of residuals in the table; this suggests a need for transformation. Note also how linear the row and column effects are in the factor levels. Emerson & Wong (1985) fit Tukey's 'one degree of freedom for non-additivity' model
$$y_{ij} = m + a_i + b_j + e_{ij} + ka_ib_j \tag{8.2}$$
by plotting the residuals against a_ib_j/m and estimating a power transformation y^λ with $\lambda = 1 - mk$ estimated as -6.81. As this is such an awkward power, they thought it better to retain the model (8.2).

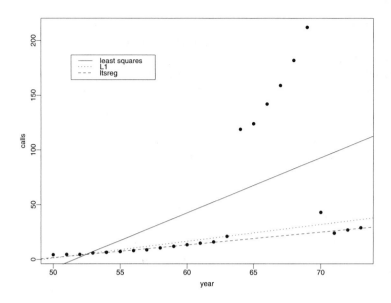

Figure 8.2: Millions of phone calls in Belgium, 1950-73, from Rousseeuw & Leroy (1987), with three fitted lines.

8.3 Robust regression

There are a number of ways to perform robust regression in S and S-PLUS, but all have drawbacks. First consider an example. Rousseeuw & Leroy (1987) give data on annual numbers of Belgian 'phone calls, given in our dataset phones.

```
attach(phones)
phones.lm <- lm(calls ~ year, phones)
plot(year, calls)
abline(phones.lm$coef)
abline(l1fit(year, calls), lty=2)
abline(ltsreg(year, calls), lty=3)
legend(locator(1), legend=c("least squares", "L1", "ltsreg"),
    lty=1:3)
```

Figure 8.2 shows the least squares line, the L_1 regression and the least trimmed squares regression (Section 8.4). The ltsreg line is $-56.45 + 1.16\,\text{year}$. Rousseeuw & Leroy's investigations showed that for 1964–9 the total length of calls (in minutes) had been recorded rather than the number, with each system being used during 1963 and 1970.

Next some theory. In a regression problem there are two possible sources of errors, the observations y_i and the corresponding row vector of p regressors, x_i. Most robust methods in regression only consider the first, and in some cases (designed experiments?) errors in the regressors can be ignored. This is the case for M-estimators, the only ones we shall consider in this section.

Consider a regression problem with n cases $(y_i, \boldsymbol{x}_i)$ from the model

$$y = \boldsymbol{x}\beta + \epsilon$$

for a p-variate row vector $\boldsymbol{x}$.

L_1 regression

The median is the L_1 estimator in the location problem. A natural generalization of the median to regression is:

$$\min_b \sum_{i=1}^{n} |y_i - \boldsymbol{x}_i b|$$

The solution is not necessarily unique (as with the median). There is a solution which fits exactly at p observations.

To compute the solutions, we may reduce the problem to one of linear programming. Define non-negative variables e_i^+ and e_i^- and set

$$e_i^+ - e_i^- = e_i = y_i - \boldsymbol{x}_i b$$

Then we minimize $\sum_i (e_i^+ + e_i^-)$ subject to $e_i^+, e_i^- \geqslant 0$ and the equation above. Clearly either e_i^+ or e_i^- will be zero, and so $(e_i^+ + e_i^-) = |e_i|$.

Like the median, L_1 regression is not very efficient. In fact, asymptotically we have

$$\text{var}(b) \approx \frac{1}{4f(0)^2}(X^T X)^{-1}$$

(Bassett & Koenker, 1978) so the ARE is the same as that of the median.

Bloomfield & Steiger (1983) is a good general reference for L_1 regression; Narula & Wellington (1982) provide a concise survey for (then) known results.

The S function `l1fit` implements L_1 regression. Its call is

```
l1fit(x, y, intercept=T, print=T)
```

where `x` is the design matrix, `y` the vector of dependent observations and an intercept is added to `x` by default. The parameter `print` refers to warnings and diagnostics.

M-estimators

If we assume a scaled pdf $f(e/s)/s$ for ϵ and set $\rho = -\log f$, the maximum likelihood estimator minimizes

$$\sum_{i=1}^{n} \rho\left(\frac{y_i - \boldsymbol{x}_i b}{s}\right) + n \log s \tag{8.3}$$

Suppose for now that s is known. Let $\psi = \rho'$. Then the MLE b of β solves

$$\sum_{i=1}^{n} x_i \psi \left(\frac{y_i - x_i b}{s} \right) = 0 \qquad (8.4)$$

Let $r_i = y_i - x_i b$ denote the residuals.

The solution to equation (8.4) or to minimising over (8.3) can be used to define an M-estimator of β. Note that L_1 regression corresponds to the Laplace or double-exponential distribution.

There are a number of versions of asymptotic theory for M-estimators. Several of the most precise forms are given by Huber (1981, §7.6). Let X as usual be the design matrix and let

$$\kappa = 1 + \frac{p}{n} \frac{\text{var}(\psi')}{(E\psi')^2}$$

evaluated at the distribution of ϵ (and in practice estimated from the distribution of residuals). Then Huber gives three forms of estimators of the variance-covariance matrix of b, of which we use

$$\kappa^2 \frac{[1/(n-p)] \sum \psi(r_i)^2}{[(1/n) \sum \psi'(r_i)]^2} (X^T X)^{-1} \qquad (8.5)$$

A simplification occurs with Huber's ψ function, for which ψ' picks out the un-Winsorized observations. Let m denote the proportion of these, and S the variance (with divisor $n - p$) of the Winsorized residuals. Then the simplified forms are

$$\kappa = 1 + \frac{p(1-m)}{nm}, \qquad \kappa^2 \frac{S}{m^2} (X^T X)^{-1}$$

A common way to solve (8.4) is by iterated re-weighted least squares (IWLS or IRLS), with weights

$$w_i = \psi \left(\frac{y_i - x_i b}{s} \right) \Bigg/ \left(\frac{y_i - x_i b}{s} \right) \qquad (8.6)$$

The iteration is only guaranteed to converge for *convex* ρ functions, and for redescending functions (such as those of Tukey and Hampel), equation (8.4) may have multiple roots. In such cases it is usual to choose a good starting point (such as a fully-iterated Huber estimator) and perform a small and fixed number of iterations.

Unknown scale

Of course, in practice the scale s is not known. A simple and very resistant scale estimator is the MAD about some centre. This is applied to the residuals about zero, either to the current residuals within the loop or to the residuals from a prior very resistant fit (e.g. L_1 or those of Section 8.4).

Alternatively, we can estimate s in an MLE-like way. Finding a stationary point of (8.3) with respect to s gives

$$\sum_i \psi\left(\frac{y_i - \boldsymbol{x}_i b}{s}\right)\left(\frac{y_i - \boldsymbol{x}_i b}{s}\right) = n$$

which is not resistant (and is biased at the normal). As in the univariate case we modify this to

$$\sum_i \chi\left(\frac{y_i - \boldsymbol{x}_i b}{s}\right) = (n - p)\gamma \tag{8.7}$$

The function `rreg`

The oldest robust regression function is `rreg` which was re-designed for the 1990 release of S (Heiberger & Becker, 1992). Its usage is, for example,

```
rreg(stack.x, stack.loss, iter=20,
    method=function(x)wt.huber(x, c=1.5))
```

This function does not use model formulae, and no functions are provided to analyse its output. In particular there is no way to calculate standard errors for the parameter estimates. It uses the iterated MAD scale estimate, and a wide range of ψ functions are provided via S functions. The default is to use a converged Huber estimator followed by the Tukey bisquare, both with constants (which can not be changed) chosen to give 95% efficiency at the normal.

There is a simplified version, `rbiwt`, for use with a single regressor.

`robust` family generator for `glm`

This uses the iterative re-weighted least squares mechanism of `glm` to fit Huber M-regressions with the weights given by (8.6) and default tuning parameter $c = 1.345$. Occasionally `glm` fails to converge in the default 10 iterations, and the iteration control mechanism is not explained well. The form needed is

```
glm(calls ~ year, robust(maxit=50), data=phones)
```

The `maxit` and `control` parameters to `glm` do not change the iteration limit. The tuning parameter is set by the argument `k=` to `robust`.

The standard errors in the summary functions are given by

$$\text{var}\left(b\right) = s^2 (X^T W X)^{-1}$$

where W is a diagonal matrix of the current weights. This is often a reasonable estimate, if less accurate than (8.5), but can be poor as our example below and Street, Carroll & Ruppert (1988) show.

The scale estimation is by $MAD/0.67$ (*sic*) of the residuals from the least squares fit. Because it is not iterated, it can seriously over-estimate the scale. The `summary` and `print` functions for `glm` do not know much about `robust` and so are inappropriate. In particular, they base the scale estimate on the usual GLM method of

$$\sum_i \rho\left(\frac{y_i - x_i b}{s}\right) = (n - p) \tag{8.8}$$

Note that this is not resistant (has breakdown point zero) as ρ is unbounded for the Huber function, and can be a serious over-estimate.

Our class "rlm"

An alternative method is provided in our main library. This introduces a new class rlm and model-fitting function rlm, building on lm. The syntax in general follows lm, and the current version of the S code is based on that for the lm class and of rreg. Huber's M-estimator is used, and the summary function computes standard errors from (8.5). The default tuning parameter is $c = 1.345$. By default the scale s is estimated by iterated MAD, but at the iteration number of the sw parameter it switches to solving (8.7). (Typically sw = 3 works well if this scale estimator is desired.) The default iteration limit is 20, but this may be increased by the maxit parameter.

To show that these differences do matter, consider again the Belgian 'phones data.

```
> summary(lm(calls ~ year, data=phones))
              Value Std. Error  t value Pr(>|t|)
(Intercept) -260.059  102.607    -2.535   0.019
       year    5.041    1.658     3.041   0.006
Residual standard error: 56.2 on 22 degrees of freedom
> summary(rlm(calls ~ year, maxit=50, data=phones))
              Value Std. Error  t value
(Intercept) -102.812   26.680    -3.853
       year    2.045    0.431     4.743
Residual standard error: 9.08 on 22 degrees of freedom
> summary(rlm(calls ~ year, sw=3, data=phones))
              Value Std. Error  t value
(Intercept) -227.828  101.783    -2.238
       year    4.451    1.645     2.707
Residual standard error: 57.2 on 22 degrees of freedom
> summary(glm(calls ~ year, robust, data=phones))
              Value Std. Error t value
(Intercept) -216.0149   92.7864 -2.3281
       year    4.2308    1.5077  2.8061
(Dispersion Parameter for Robust ... 2487.1)
Residual Deviance: 64349 on 22 degrees of freedom
> summary(rlm(calls ~ year, k=0.25, maxit=50, data=phones))
              Value Std. Error t value
(Intercept) -74.057   15.277    -4.848
       year    1.499    0.247     6.071
Residual standard error: 3.96 on 22 degrees of freedom
> summary(glm(calls ~ year, robust(k=0.25, maxit=20,
```

```
                 data=phones)))
              Value Std. Error t value
(Intercept) -105.5421    45.90532 -2.2991
       year    2.0959     0.76789  2.7294
(Dispersion Parameter for Robust ... 478.73)
Residual Deviance: 19211 on 22 degrees of freedom
> attach(phones)
> rreg(year, calls, method=wt.hampel)
$coef:
  (Intercept)       x
     -248.77 4.8362
> rreg(year, calls, init=l1fit(year,calls)$coef,
      method=wt.hampel)
$coef:
  (Intercept)       x
     -52.382 1.1006
```

Note how the robust GLM method inflates the quoted standard errors for severe Winsorizing. The dispersion parameter is also over-estimated because the scale is not iterated.

As Figure 8.2 shows, in this example there is a batch of outliers from a different population in the late 1960s, and these should probably be rejected, which the Huber M-estimators do not. The final two fits show how crucially the iterated re-descending Hampel estimator depends on the initial scale.

8.4 Resistant regression

As an attempt to get a very resistant regression fit, Rousseeuw suggested

$$\min_{b} \operatorname{median}_{i} |y_i - x_i b|^2$$

called the *least median of squares* (LMS) estimator. The square is necessary if n is even, when the central median is taken.

This fit is very resistant, and needs no scale estimate. Note that unlike the robust regression it can reject values that fit badly because their x_i are outliers. It is however very inefficient, converging at rate $1/\sqrt[3]{n}$. Further, it displays marked sensitivity to central data values: see Hettmansperger & Sheather (1992) and Davies (1993, §2.3).

Rousseeuw later suggested least trimmed squares (LTS) regression:

$$\min_{b} \sum_{i=1}^{q} |y_i - x_i b|^2_{(i)}$$

as this is more efficient, but shares the same extreme resistance. The sum is over the smallest $q = \lfloor n/2 \rfloor + \lfloor (p+1)/2 \rfloor$ squared residuals. (Other authors differ; for example Marazzi (1993, p. 200) has $q = \lfloor n/2 \rfloor + 1$ and Rousseeuw & Leroy (1987, p.132) mention both choices.)

LMS and LTS are very suitable methods for finding a starting point for rejecting outliers or for a few steps of a M-estimator; LTS is preferable to LMS.

S-PLUS implementation

Both methods are implemented in S-PLUS. The essential parts of their calls are

```
lmsreg(x, y, intercept=T)
ltsreg(x, y, intercept=T)
```

where x is the design matrix, y the dependent variable and by default a column of 1's is added to x. Both algorithms involve an approximate random search, so it would be more accurate to say that they approximate LMS and LTS. As a consequence, the results given here will not be reproducible. In S-PLUS 4.0 there are formula-based methods for lmsreg and ltsreg.

Brownlee's stack loss data

We consider Brownlee's (1965) much-studied stack loss data, given in the S datasets stack.x and stack.loss. The data are from the operation of a plant for the oxidation of ammonia to nitric acid, measured on 21 consecutive days. There are 3 explanatory variables (air flow to the plant, cooling water inlet temperature, and acid concentration) and the response, 10 times the percentage of ammonia lost.

```
> summary(lm(stack.loss ~ stack.x))
Residuals:
   Min    1Q Median    3Q Max
 -7.24 -1.71 -0.455  2.36 5.7

Coefficients:
                     Value Std. Error t value Pr(>|t|)
      (Intercept) -39.920     11.896  -3.356    0.004
   stack.xAir Flow   0.716      0.135   5.307    0.000
  stack.xWater Temp  1.295      0.368   3.520    0.003
  stack.xAcid Conc. -0.152      0.156  -0.973    0.344

Residual standard error: 3.24 on 17 degrees of freedom

> lmsreg(stack.x, stack.loss)
$coef:
 Intercept Air Flow Water Temp  Acid Conc.
     -34.5  0.71429    0.35714 -3.3255e-17

   Min.    1st Qu. Median  Mean 3rd Qu.  Max.
 -7.643 -1.776e-15 0.3571 1.327   1.643 9.714

$wt:
 [1] 0 0 0 0 1 1 1 1 1 1 1 1 1 1 1 1 1 1 1 1 0
> ltsreg(stack.x, stack.loss)
$coefficients:
 (Intercept) Air Flow Water Temp Acid Conc.
     -35.435  0.76316    0.32619  -0.011591
```

```
 Min. 1st Qu. Median   Mean 3rd Qu.  Max.
-8.269  -0.361 0.2133 0.9869   1.283 9.313
```

The weights returned by `lmsreg` indicate that five points have large residuals and should be considered as possible outliers. For each of `lmsreg` and `ltsreg` we give the results of `summary` applied to their residuals. Now consider M-estimators:

```
> l1fit(stack.x, stack.loss)
$coefficients:
 Intercept Air Flow Water Temp Acid Conc.
    -39.69  0.83188     0.57391    -0.06087


   Min. 1st Qu. Median    Mean 3rd Qu.  Max.
 -9.481  -1.217      0 0.08944  0.5275 7.635
> stack.rl <- rlm(stack.loss ~ stack.x)
> summary(stack.rl)
Residuals:
   Min    1Q Median   3Q Max
 -8.92 -1.73 0.0617 1.54 6.5

Coefficients:
                    Value Std. Error t value
      (Intercept) -41.027     9.793    -4.189
   stack.xAir Flow   0.829     0.111     7.470
 stack.xWater Temp   0.926     0.303     3.057
 stack.xAcid Conc.  -0.128     0.129    -0.994

Residual standard error: 2.44 on 17 degrees of freedom
> stack.rl$w:
 [1] 1.000 1.000 0.786 0.505 1.000 1.000 1.000
 [8] 1.000 1.000 1.000 1.000 1.000 1.000 1.000
[15] 1.000 1.000 1.000 1.000 1.000 1.000 0.368
```

The L_1 method fits observations 2, 8, 16 and 18 exactly. The component `w` returned by `rlm` contains the final weights in (8.6). Although all methods seem to agree about observation 21, they differ in their view of the early observations. Atkinson (1985, pp. 129–136, 267–8) discusses this example in some detail, as well as the analyses performed by Daniel & Wood (1980). They argue for a logarithmic transformation, dropping acid concentration and fitting interactions or products of the remaining two regressors. However, the question of outliers and change of model are linked, since most of the evidence for changing the model comes from the possible outliers.

Rather than fit a parametric model we examine the points in the air flow – water temp space, using the robust fitting option of `loess` (discussed briefly in Chapter 11); see Figure 8.3.

```
x1 <- stack.x[,1]; x2 <- stack.x[,2]
stack.loess <- loess(log(stack.loss) ~ x1*x2, span=0.5,
```

```
    family="symmetric")
stack.plt <- expand.grid(x1=seq(50,80,0.5), x2=seq(17,27,0.2))
stack.plt$z <- as.vector(predict(stack.loess, stack.plt))
dupls <- c(2,7,8,11)
contourplot(z ~ x1*x2, stack.plt, aspect=1,
   xlab="Air flow", ylab="Water temp",
   panel = function(x, y, subscripts, ...){
       panel.contourplot(x, y, subscripts, ...)
       panel.xyplot(x1, x2)
       text(x1[-dupls] + par("cxy")[1] ,
           x2[-dupls] + 0.5* par("cxy")[2],
           as.character(seq(x1)[-dupls]), cex=0.7)
   })
```

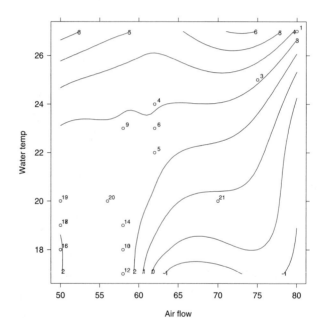

Figure 8.3: Fitted surface for Brownlee's stack loss data on log scale using `loess`.

This shows clearly that the 'outliers' are also outlying in this space. (There are duplicate points; in particular points 1 and 2 are coincident.)

Scottish hill races revisited

We return to the data on Scottish hill races studied in the introduction and Section 6.3. There we saw one gross outlier and a number of other extreme observations. Many of the robust and resistant methods do not allow weighted fits.

```
> hills.lm
Coefficients:
```

```
(Intercept)   dist     climb
     -8.992 6.218 0.011048

Degrees of freedom: 35 total; 32 residual
Residual standard error: 14.676
> hills1.lm # omitting Knock Hill
Coefficients:
 (Intercept)    dist     climb
     -13.53 6.3646 0.011855

Degrees of freedom: 34 total; 31 residual
Residual standard error: 8.8035
> rlm(time ~ dist + climb, hills)
Coefficients:
 (Intercept)    dist      climb
    -9.6067 6.5507 0.0082959

Degrees of freedom: 35 total; 32 residual
Scale estimate: 5.21
> summary(rlm(time ~ dist + climb, hills, weights=1/dist^2))
Coefficients:
               Value Std. Error t value
(Intercept)  -1.270    1.843     -0.689
       dist   5.389    0.257     20.929
      climb   0.007    0.001      7.922

Residual standard error: 4.24 on 32 degrees of freedom

> attach(hills)
> rreg(cbind(dist, climb), time)$coef
 (Intercept)    dist     climb
    -8.2671 6.6358 0.006631
> rreg(cbind(dist, climb), time, wx=1/dist^2)$coef
 (Intercept)    dist     climb
    -1.3783 5.1361 0.007578
> ltsreg(cbind(dist, climb), time)$coef
 (Intercept)    dist      climb
    -1.2426 4.8894 0.0083503
```

Notice that the intercept is no longer significant in the robust weighted fit.

If we move to the model for inverse speed:

```
> summary(hills2.lm) # omitting Knock Hill
Coefficients:
               Value Std. Error t value Pr(>|t|)
(Intercept)    4.900    0.474    10.344   0.000
       grad    0.008    0.002     5.022   0.000

Residual standard error: 1.28 on 32 degrees of freedom
> ltsreg(grad, ispeed)$coef
```

```
(Intercept)        x
    4.7488 0.0080721
> rreg(grad, ispeed)$coef
(Intercept)        x
    5.1522 0.0072142
> summary(rlm(ispeed ~ grad, hills))
Coefficients:
              Value Std. Error t value
(Intercept)   5.176  0.381      13.593
       grad   0.007  0.001       5.431

Residual standard error: 0.869 on 33 degrees of freedom
```

The results are in close agreement with the least-squares results removing Knock Hill.

8.5 Multivariate location and scale

Somewhat counter-intuitively, it does not suffice to apply a robust location estimator to each component of a multivariate mean (Rousseeuw & Leroy, 1987, p. 250), and it is easier to consider the estimation of mean and variance simultaneously.

Multivariate variances are very sensitive to outliers. One method for robust covariance estimation is available via the function cov.mve. Let there be n observations of p variables. The method seeks an ellipsoid containing $\lfloor (n + p + 1)/2 \rfloor$ points of minimum volume. Having found such an ellipsoid by a random search, it returns a (product-moment) covariance estimate of those points whose Mahalanobis distance from the mean computed via the ellipsoid covariance is not too large (specifically within the 97.5% point). This estimator was proposed by Rousseeuw & van Zomeren (1990). Note that the centre of the ellipsoid and the mean of the 'cleaned' data provide reasonably robust estimates of the population mean. Inspecting the code shows that cov.mve returns the mean of the 'cleaned' data as its component center.

An alternative approach is to extend the idea of M-estimation to this setting, fitting a multivariate t_ν distribution for a small number ν of degrees of freedom. This is implemented in our function cov.trob; the theory behind the algorithm used is given in Kent et al. (1994); Ripley (1996). Normally cov.trob is much faster than cov.mve, but it lacks the latter's extreme resistance.

Chapter 9

Non-linear Models

In linear regression the mean surface in sample space is a plane; in non-linear regression it may be an arbitrary curved surface but in all other respects the models are same. Fortunately in practice the mean surface in most non-linear regression models will be approximately planar in the region of highest likelihood, allowing some good approximations based on linear regression techniques to be used, but non-linear regression models can still present tricky computational and inferential problems.

A thorough treatment of non-linear regression is given in Bates & Watts (1988). Another encyclopaedic reference is Seber & Wild (1989), and the books of Ratkowsky (1983, 1990) and Ross (1990) also offer some practical statistical advice. The S software is described by Bates & Chambers (1992) who state that its methods are based on those described in Bates & Watts (1988). An alternative approach based on their nls2 library is described by Huet *et al.* (1996).

In this chapter we consider first non-linear regression and then more general minimization and maximum likelihood estimation. Non-linear mixed models are considered in Section 10.4.

We start with a simple example. Obese patients on a weight reduction programme tend to lose adipose tissue at a diminishing rate as the treatment progresses. Our dataset wtloss has kindly been supplied by Dr T. Davies (personal communication). The two variables are Days, the time (in days) since start of the programme, and Weight, weight in kilograms measured under standard conditions. The dataset pertains to a male patient, aged 48, height 193 cm ($6'4''$) with a large body frame. The results are illustrated in Figure 9.1, produced by

```
attach(wtloss)
# alter margin 4; others are default
oldpar <- par(mar=c(5.1, 4.1, 4.1, 4.1))
plot(Days, Weight, type="p",ylab="Weight (kg)")
Wt.lbs <- pretty(range(Weight*2.205))
axis(side=4, at=Wt.lbs/2.205, lab=Wt.lbs, srt=90)
mtext("Weight (lb)", side=4, line=3)
par(oldpar) # restore settings
```

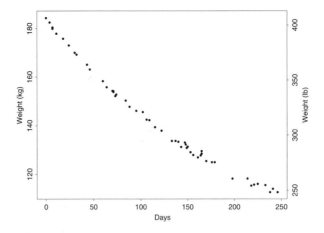

Figure 9.1: Weight loss from an obese patient.

Although polynomial regression models may describe such data very well within the time range, they can fail spectacularly outside this range (see Exercise 9.1 on page 295). A more useful model with some theoretical and empirical support is exponential of the form

$$y = \beta_0 + \beta_1 2^{-t/\theta} + \epsilon \tag{9.1}$$

Notice that all three parameters have a ready interpretation, namely

β_0 is the ultimate lean weight, or asymptote,
β_1 is the total amount to be lost and
θ is the time taken to lose half the amount remaining to be lost,

which allows us to find rough estimates directly from the plot of the data.

The parameters β_0 and β_1 are called *linear parameters* since the second partial derivative of the model function with respect to them is identically zero. Other parameters for which this is not the case, such as θ in this example, are called *non-linear parameters*.

9.1   Fitting non-linear regression models

The general form of a non-linear regression model is

$$y = \eta(x, \beta) + \epsilon \tag{9.2}$$

where x is a vector of covariates, β is a p–component vector of unknown parameters and ϵ is a $N(0, \sigma^2)$ error term. In the weight loss example the parameter vector is $\beta = (\beta_0, \beta_1, \theta)^T$. (As x plays little part in the discussion that follows we will often omit it from the notation.)

Suppose y is a sample vector of size n and $\eta(\beta)$ is its mean vector. It is easy to show that the maximum likelihood estimate of β is a least-squares estimate, that is a minimizer of $\|y - \eta(\beta)\|^2$. The variance parameter, σ^2, is then estimated by the residual mean square as in linear regression.

For varying β the vector $\eta(\beta)$ traces out a p–dimensional surface in $\mathbb{R}^n$ that we refer to as the *solution locus*. The parameters β define a coordinate system within the solution locus. From this point of view a linear regression model is one for which the solution locus is a plane through the origin and the coordinate system within it defined by the regression coefficients is affine, that is it has no curvature. The computational problem in both cases is then to find the coordinates of the point on the solution locus closest to the sample vector y in the sense of Euclidean distance.

The process of fitting non-linear regression models in S is similar to that for fitting linear models, with two important differences:

1. There is no explicit formula for the estimates, so iterative procedures are required, for which initial values are needed.

2. Linear regression model formulae are not adequate to specify non-linear regression models. A more flexible protocol is needed.

The primary S function for fitting a non-linear regression model is `nls`. We can fit the weight loss model by

```
> wtloss.st <- c(b0=90, b1=95, th=120)
> wtloss.fm <- nls(Weight ~ b0 + b1*2^(-Days/th),
    data = wtloss, start = wtloss.st, trace = T)
67.5435 : 90 95 120
40.1808 : 82.7263 101.305 138.714
39.2449 : 81.3987 102.658 141.859
39.2447 : 81.3737 102.684 141.911
> wtloss.fm
Residual sum of squares : 39.245
parameters:
     b0      b1      th
 81.374 102.68 141.91
formula: Weight ~ b0 + b1 * 2^( - Days/th)
52 observations
```

The arguments to `nls` are:

`formula` A non-linear model formula. The form is `response ~ mean`, as usual, but the right-hand side can have either of two forms. The standard one is an ordinary algebraic expression containing both parameters and determining variables, where the operators have their usual arithmetical meaning. (The second form is used with the `plinear` fitting algorithm, discussed in Section 9.3 on page 275.)

`data` An optional data frame for the variables (and sometimes parameters).

start A list or numeric vector specifying the starting values for the parameters
in the model.

The names for the components of start are also used to specify which
of the names occurring on the right-hand side of the model formula are of
parameters. All others are then taken to be determining variables.[1]

control An optional argument allowing some features of the default iterative
procedure to be changed.

algorithm An optional character string argument allowing a particular fitting
algorithm to be specified. The default procedure is specified by using
"default".

trace An argument allowing tracing information from the iterative procedure to
be printed. By default none is printed but we often prefer to have it shown.

In our example the names of the parameters were specified as b0, b1 and th.
The initial values of 90, 95 and 120 were found by inspection of Figure 9.1.
From the trace output the procedure is seen to converge in three iterations.

The nls function has no weights argument, but non-linear regressions with
known weights may be handled by writing the formula as ~ sqrt(W)*(y - M)
rather than y ~ M. (The algorithm minimises the sum of squared differences
between left- and right-hand sides and an empty left-hand side counts as zero.) If
the weights are now known but W contains unknown parameters as well as eta,
the log-likelihood function has an extra term and the problem must be handled by
the more general optimisation methods such as those discussed in Section 9.7 on
pages 286ff. See Problem 9.5.

Using function derivative information

Most non-linear regression fitting algorithms operate in outline as follows. The
first-order Taylor-series approximation to η_k at an initial value $\beta^{(0)}$ is

$$\eta_k(\beta) \approx \eta_k(\beta^{(0)}) + \sum_{j=1}^{p} (\beta_j - \beta_j^{(0)}) \left.\frac{\partial \eta_k}{\partial \beta_j}\right|_{\beta=\beta^{(0)}}$$

In vector terms these may be written

$$\eta(\beta) \approx \omega^{(0)} + Z^{(0)}\beta \tag{9.3}$$

where

$$Z_{kj}^{(0)} = \left.\frac{\partial \eta_k}{\partial \beta_j}\right|_{\beta=\beta^{(0)}} \quad \text{and} \quad \omega_k^{(0)} = \eta_k(\beta^{(0)}) - \sum_{j=1}^{p} \beta_j^{(0)} Z_{kj}^{(0)}$$

Equation (9.3) defines the tangent plane to the surface at the coordinate point
$\beta = \beta^{(0)}$. The process consists of regressing the observation vector, y, onto the

---

[1] In S-PLUS 3.4 and earlier (and possibly later!) there is a bug which may be avoided if the order
in which the parameters appear in the start vector is the same as the order in which they first appear
in the model. It is as if the names attribute were ignored.

tangent plane defined by $Z^{(0)}$ with *offset* vector $\omega^{(0)}$ to give a new approximation, $\beta = \beta^{(1)}$, and iterating to convergence. For a linear regression the offset vector is 0 and the matrix $Z^{(0)}$ is the model matrix X, a constant matrix, so the process converges in one step. In the non-linear case the next approximation is

$$\beta^{(1)} = \left(Z^{(0)\,T} Z^{(0)}\right)^{-1} Z^{(0)\,T} \left(y - \omega^{(0)}\right)$$

With the `default` algorithm the Z matrix is computed approximately by numerical methods unless formulae for the first derivatives are supplied. Providing derivatives often (but not always) improves convergence.

Derivatives can only be provided as an attribute of the model. The standard way to do this is to write an **S** function to calculate the mean vector, η, and the Z matrix. The result of the function is η with the Z matrix included as a `gradient` attribute.

For our simple example the three derivatives are

$$\frac{\partial \eta}{\partial \beta_0} = 1, \qquad \frac{\partial \eta}{\partial \beta_1} = 2^{-x/\theta}, \qquad \frac{\partial \eta}{\partial \theta} = \frac{\log(2)\,\beta_1 x 2^{-x/\theta}}{\theta^2}$$

so an **S** function to specify the model including derivatives is

```
expn <- function(b0, b1, th, x) {
    temp <- 2^(-x/th)
    model.func <- b0 + b1 * temp
    Z <- cbind(1, temp, (b1 * x * temp * log(2))/th^2)
    dimnames(Z) <- list(NULL, c("b0","b1","th"))
    attr(model.func, "gradient") <- Z
    model.func
}
```

Note that the gradient matrix must have column names matching those of the corresponding parameters.

We can fit our model again using first derivative information:

```
> wtloss.gr <- nls(Weight ~ expn(b0, b1, th, Days),
    data = wtloss, start = wtloss.st, trace = T)
67.5435 : 90 95 120
40.1808 : 82.7263 101.305 138.714
39.2449 : 81.3987 102.658 141.859
39.2447 : 81.3738 102.684 141.911
```

This appears to make no difference to the speed of convergence, but tracing the function `expn` shows that 6 function evaluations are required when derivatives are supplied compared with 21 if they are not supplied.

Functions such as `expn` can often be generated automatically using the symbolic differentiation function `deriv`. It is called with three arguments:

(a) the model formula, with the left-hand side optionally left blank,

(b) a character vector giving the names of the parameters, and

(c) an empty function with an argument specification as required for the result.

An example makes the process clearer. For the weight loss data with the exponential model, we can use:

```
expn1 <- deriv(y ~ b0 + b1 * 2^(-x/th), c("b0", "b1", "th"),
               function(b0, b1, th, x) {})
```

The result is the function

```
expn1 <- function(b0, b1, th, x)
{
    .expr3 <- 2^(( - x)/th)
    .value <- b0 + (b1 * .expr3)
    .grad <- array(0, c(length(.value), 3),
                list(NULL, c("b0", "b1", "th")))
    .grad[, "b0"] <- 1
    .grad[, "b1"] <- .expr3
    .grad[, "th"] <- b1 *
        (.expr3 * (0.693147180559945 * (x/(th^2))))
    attr(.value, "gradient") <- .grad
    .value
}
```

Self-starting non-linear regressions

Very often reasonable starting values for a non-linear regression can be calculated
by some simple and fairly automatic procedure. The nlme software[2] contains
an extension of the nls function that allows a procedure to be specified for
calculating initial values in lieu of providing the values themselves.

Setting up such a *self-starting* non-linear model is somewhat technical and
requires some attention to detail.

(a) Prior to S-PLUS 3.4 the nlme library must be attached with first = T as
it contains a modified nls function needed to recognize self-starting models.

(b) The response function for the non-linear model must be specified as an explicit
S function.

(c) This model *function* is given a class attribute, "selfStart", and an
initial attribute which is another S function defining the initial value
procedure. The initial value function must have an identical argument sequence to the model function, and return a named list of initial values with
the same names as the parameters used on the call.

(d) The initial value procedure will usually require the response variable as well.
This may be obtained as the object .nls.initial.response, guaranteed
to exist in frame 1.

---

[2] Included in S-PLUS 3.4 and later, and also available from the statlib archive as a library.

The function `selfStart` can be used like `deriv` to generate a self-starting model function, including first derivative information.

Consider again a negative exponential decay model such as that used in the weight loss example but this time written in the more usual exponential form:

$$y = \beta_0 + \beta_1 \exp(-x/\theta) + \epsilon$$

One effective initial value procedure follows.

(i) Fit an initial quadratic regression in x.

(ii) Find the fitted values, say y_0, y_1 and y_2 at three equally spaced points x_0, $x_1 = x_0 + \delta$, and $x_2 = x_0 + 2\delta$,

(iii) Equate the three fitted values to their expectation under the non-linear model to give an initial value for θ as

$$\theta_0 = \delta \Big/ \log \left(\frac{y_0 - y_1}{y_1 - y_2} \right)$$

(iv) Initial values for β_0 and β_1 can then be obtained by linear regression of y on $\exp(-x/\theta_0)$.

An S function to implement this procedure (with a few extra checks) called `negexp.ival` is supplied in the MASS library. We will not reproduce the function here, but interested readers should study it carefully, particularly for the way it finds the names used for its parameters using `sys.call` and for how it retrieves the response vector from frame 1.

We can make a self-starting model with both first derivative information and this initial value routine by

```
negexp <- selfStart(model = ~ b0 + b1*exp(-x/th),
       initial = negexp.ival, parameters = c("b0", "b1", "th"),
       template = function(x, b0, b1, th) {})
```

where the first, third and fourth arguments are the same as for `deriv`. We may now fit the model without explicit initial values.

```
> wtloss.ss <- nls(Weight ~ negexp(Days, B0, B1, theta),
                 data = wtloss, trace = T)
      B0      B1   theta
  82.713 101.49 200.16
  39.5453 : 82.7131 101.495 200.160
  39.2450 : 81.3982 102.659 204.652
  39.2447 : 81.3737 102.684 204.734
```

(The first two lines of output come from the initial value procedure and the last three from the `nls` trace.)

The `nlme` collection of software contains four examples of self-starting models, a two-term exponential model, a first-order compartmental model and three- and four-parameter logistic models.[3]

---

[3] If `nlme` is being used as a library under Unix, these will be in library `nlmedata`.

9.2   Non-linear fitted model objects and method functions

The result of a call to `nls` is an object of class `nls`. The standard method functions are available, the most important of which are

`print` to print an abbreviated synopsis (usually called implicitly),

`summary` to print a summary of the fitting process results,

`coef` to extract the estimated mean parameters,

`fitted` to extract the fitted value,

`residuals` to extract residuals,

`profile` to explore the the least- squares surface in the vicinity of the estimate.

For our example above the summary function gives:

```
> summary(wtloss.gr)
Formula: Weight ~ expn1(b0, b1, th, Days)
Parameters:
      Value Std. Error t value
b0  81.374     2.2690  35.863
b1 102.684     2.0828  49.302
th 141.911     5.2945  26.803
Residual standard error: 0.894937 on 49 degrees of freedom
Correlation of Parameter Estimates:
        b0      b1
b1 -0.989
th -0.986   0.956
```

Surprisingly, no working deviance method function exists but such a method function is easy to write and is included in the `MASS` library. It merely requires

```
> deviance.nls <- function(object) sum(object$residuals^2)
> deviance(wtloss.gr)
[1] 39.245
```

Library `MASS` also has a generic function `vcov` that will extract the estimated variance matrix of the mean parameters:

```
> vcov(wtloss.gr)
        b0       b1      th
b0   5.1484  -4.6745 -11.841
b1  -4.6745   4.3379  10.543
th -11.8414  10.5432  28.032
```

Another surprising omission prior to S-PLUS 4.0 is a `predict` method.[4]

---

[4] Professor Douglas Bates has kindly permitted us to include his `predict.nls` and `anova.nls` methods in our `MASS` library.

9.3 Taking advantage of linear parameters

If all non-linear parameters were known the model would be linear and standard linear regression methods could be used. This simple idea lies behind the `plinear` algorithm. It requires a different form of model specification that combines linear and non-linear model protocols. In this case the right-hand-side expression specifies a *matrix* whose columns are functions of the non-linear parameters. The linear parameters are then implied as the regression coefficients for the columns of the matrix. Initial values are only needed for the non-linear parameters.

In our weight loss example there are two linear parameters, β_0 and β_1. To fit the model using the `plinear` algorithm we use:

```
> wtloss.pl <- nls(Weight ~ cbind(1, 2^(-Days/th)), wtloss,
                algorithm="plinear", start=c(th=120), trace=T)
58.0817 : 120
39.5380 : 138.778
39.2448 : 141.855
39.2447 : 141.911
> summary(wtloss.pl)$parameters
      Value Std. Error t value
th 141.911    5.2945   26.803
    81.374    2.2690   35.863
   102.684    2.0828   49.302
```

Notice that the linear parameters are unnamed.

There are several advantages in using the partially linear algorithm. It can be much more stable than methods that do not take advantage of linear parameters, it requires fewer initial values and it can often converge from poor starting positions where other procedures fail.

In some cases it will be convenient to use a separately calculated numeric matrix. For example, consider fitting a negative exponential model with separate linear parameters for each level of a factor, say `f` (in data frame `dat`), but common θ. Setting

```
L <- model.matrix(~ f - 1, data = dat)
```

the model function by be specified as `cbind(L, L*exp(-x/th))` irrespective of the number of levels in `f`.

It is also possible to specify this model in the default algorithm form using *indexed* parameters as `b0[f] + b1[f] * exp(-x/th)`. In this case the starting values must be given as a *list* of length three, and the `b0` and `b1` components must be numeric vectors of the appropriate length. Using the `plinear` algorithm requires only one initial value for `th`, and if `f` has many levels it may be much faster and more stable as well.

9.4   Confidence intervals for parameters

For some non-linear regression problems the standard error may be an inadequate summary of the uncertainty of a parameter estimate and an asymmetric confidence interval may be more appropriate. In non-linear regression it is customary to invert the "extra sum of squares" to provide such an interval, although the result is usually almost identical to the likelihood ratio interval.

Suppose the parameter vector is $\boldsymbol{\theta} = (\theta_1, \boldsymbol{\theta}_2)^T$ and without loss of generality we wish to test an hypothesis $H_0 : \theta_1 = \theta_{10}$. Let $\widehat{\boldsymbol{\theta}}$ be the overall least-squares estimate, $\widehat{\boldsymbol{\theta}}_{2|1}$ be the conditional least-squares estimate of $\boldsymbol{\theta}_2$ with $\theta_1 = \theta_{10}$ and put $\widehat{\widehat{\boldsymbol{\theta}}}(\theta_{10}) = (\theta_{10}, \widehat{\boldsymbol{\theta}}_{2|1})^T$. The extra sum of squares principle then leads to the test statistic:

$$F(\theta_{10}) = \frac{\mathrm{RSS}\{\widehat{\widehat{\boldsymbol{\theta}}}(\theta_{10})\} - \mathrm{RSS}(\widehat{\boldsymbol{\theta}})}{s^2}$$

which under the null hypothesis has an approximately $F_{1,\,n-p}$–distribution. (Here RSS denotes the residual sum of squares.) Similarly the *signed square root*

$$\tau(\theta_{10}) = \mathrm{sign}(\theta_{10} - \widehat{\theta}_1)\sqrt{F(\theta_{10})} \qquad (9.4)$$

has an approximate t_{n-p}–distribution under H_0. The confidence interval for θ_1 is then the set $\{\theta_1 \mid -t < \tau(\theta_1) < t\}$, where t is the appropriate percentage point from the t_{n-p}–distribution.

The `profile` function is generic. The method for `nls` objects explores the residual sum of squares surface by taking each parameter, θ_i, in turn and fixing it at a series of values above and below $\widehat{\theta}_i$ so that (if possible) $\tau(\theta_i)$ varies from 0 by at least some pre-set value t in both directions. (If the sum of squares function "flattens out" in either direction, or if the fitting process fails in some way this may not be achievable, in which case the profile is incomplete.) The value returned is a list of data frames, one for each parameter, and named by the parameter names. Each data frame component consists of

(a) a vector, `tau`, of values of $\tau(\theta_i)$, and

(b) a matrix, `par.vals`, giving the values of the corresponding parameter estimates $\widehat{\widehat{\boldsymbol{\theta}}}(\theta_i)$, sometimes called the 'parameter traces'.

The argument `which` of `profile` may be used to specify only a subset of the parameters to be so profiled. (Note that profiling is not available for fitted model objects in which the `plinear` algorithm is used.)

Finding a confidence interval then requires the `tau` values for that parameter to be interpolated and the set of θ_i values to be found (here always of the form $\underline{\theta}_i < \theta_i < \overline{\theta}_i$). The following function is a simple implementation included in the `MASS` library. As a similar operation may be needed for other forms of fitted model we define a generic function.

```
confint <- function(object, ...) UseMethod("confint")
confint.nls <- function(object, parm, level = 0.95) {
  if(is.character(parm))
    parm <- match(parm, names(coef(object)), nomatch = 0)
  pro <- profile(object, which = parm,
                 alphamax = (1 - level)/4)[[parm]]
  x <- pro[, "par.vals"][, parm]
  y <- pro$tau
  n <- length(object$fitted.values) - length(object$parameters)
  cutoff <- qt(1 - (1 - level)/2, n)
  if(max(y) < cutoff || min(y) > - cutoff)
    stop("the profiling did not extend far enough")
  sp <- spline(x, y)
  approx(sp$y, sp$x, xout = c( - cutoff, cutoff))$y
}
```

The weight loss data, continued

For the person on the weight loss programme an important question is how long it might be before he achieves some goal weight. If δ_0 is the time to achieve a predicted weight of w_0, then solving the equation $w_0 = \beta_0 + \beta_1 2^{-\delta_0/\theta}$ gives

$$\delta_0 = -\theta \log_2\{(w_0 - \beta_0)/\beta_1\}$$

We will now find a confidence interval for δ_0 for various values of w_0 by the method outlined above.

The first step is to re-write the model function using δ_0 as one of the parameters; the most convenient parameter for it to replace is θ and the new expression of the model becomes

$$y = \beta_0 + \beta_1 \left(\frac{w_0 - \beta_0}{\beta_1}\right)^{x/\delta_0} + \epsilon \tag{9.5}$$

In order to find a confidence interval for δ_0 we need to fit the model with this parametrization. First we build a model function:

```
expn2 <- deriv(~b0 + b1*((w0 - b0)/b1)^(x/d0),
               c("b0","b1","d0"), function(b0, b1, d0, x, w0) {})
```

It is also convenient to have a function that builds an initial value vector from the estimates of the present fitted model:

```
wtloss.init <- function(obj, w0) {
  p <- coef(obj)
  d0 <- - log((w0 - p["b0"])/p["b1"], 2) * p["th"]
  c(p[c("b0", "b1")], d0 = as.vector(d0))
}
```

(Note the use of `as.vector` to remove unwanted names.) The main calculations are done in a short loop:

```
> out <- NULL
> w0s <- c(110, 100, 90)
> for(w0 in w0s) {
     fm <- nls(Weight ~ expn2(b0, b1, d0, Days, w0),
               wtloss, start = wtloss.init(wtloss.gr, w0))
     out <- rbind(out, c(coef(fm)["d0"], confint(fm, "d0")))
  }
> dimnames(out) <- list(paste(w0s,"kg:"), c("d0","low","high"))
> out
         d0     low    high
110 kg: 261.51 256.23 267.50
100 kg: 349.50 334.74 368.02
 90 kg: 507.09 457.56 594.97
```

As the weight decreases the confidence intervals become much wider and more asymmetric. For weights closer to the estimated asymptotic weight the profiling procedure can fail.

A bivariate region: the Stormer viscometer data

The following example comes from Williams (1959). The Stormer viscometer measures the viscosity of a fluid by measuring the time taken for an inner cylinder in the mechanism to perform a fixed number of revolutions in response to an actuating weight. The viscometer is calibrated by measuring the time taken with varying weights while the mechanism is suspended in fluids of accurately known viscosity. The dataset comes from such a calibration, and theoretical considerations suggest a non-linear relationship between time T, weight w and viscosity V of the form

$$T = \frac{\beta_1 v}{w - \beta_2} + \epsilon$$

where β_1 and β_2 are unknown parameters to be estimated. Note that β_1 is a linear parameter and β_2 is non-linear. The dataset is given in Table 9.1.

Table 9.1: The Stormer viscometer calibration data. The body of the table shows the times in seconds.

Weight									
					Viscosity (poise)				
(grams)	14.7	27.5	42.0	75.7	89.7	146.6	158.3	161.1	298.3
20	35.6	54.3	75.6	121.2	150.8	229.0	270.0		
50	17.6	24.3	31.4	47.2	58.3	85.6	101.1	92.2	187.2
100				24.6	30.0	41.7	50.3	45.1	89.0,86.5

Williams suggested that a suitable initial value may be obtained by writing the regression model in the form

$$wT = \beta_1 v + \beta_2 T + (w - \beta_2)\epsilon$$

and regressing wT on v and T using ordinary linear regression. With the data available in a data frame `stormer` with variables `Viscosity`, `Wt` and `Time`, we may proceed as follows:

```
> fm0 <- lm(Wt*Time ~ Viscosity + Time - 1, data=stormer)
> b0 <- coef(fm0);  names(b0) <- c("b1", "b2")
> b0
     b1      b2
28.876 2.8437
> storm.fm <- nls(Time ~ b1*Viscosity/(Wt-b2), data=stormer,
          start=b0, trace=T)
885.365 : 28.8755 2.84373
825.110 : 29.3935 2.23328
825.051 : 29.4013 2.21823
```

Since there are only two parameters we can display a confidence region for the regression parameters as a contour map. To this end put:

```
bc <- coef(storm.fm)
se <- sqrt(diag(vcov(storm.fm)))
dv <- deviance(storm.fm)
```

Define $d(\beta_1, \beta_2)$ as the sum of squares function:

$$d(\beta_1, \beta_2) = \sum_{i=1}^{23} \left(T_i - \frac{\beta_1 v_i}{w_i - \beta_2} \right)^2$$

Then `dv` contains the minimum value, $d_0 = d(\widehat{\beta}_1, \widehat{\beta}_2)$, the residual sum of squares or model deviance.

If β_1 and β_2 are the true parameter values the "extra sum of squares" statistic

$$F(\beta_1, \beta_2) = \frac{(d(\beta_1, \beta_2) - d_0)/2}{d_0/21}$$

is approximately distributed as $F_{2,21}$. An approximate confidence set contains those values in parameter space for which $F(\beta_1, \beta_2)$ is less than the 95% point of the $F_{2,21}$ distribution. We will construct a contour plot of the $F(\beta_1, \beta_2)$ function and mark off the confidence region. The result is shown in Figure 9.2.

A suitable region for the contour plot is three standard errors either side of the least-squares estimates in each parameter. Since these ranges are equal in their respective standard error units it is useful to make the plotting region square.

```
par(pty = "s")
b1 <- bc[1] + seq(-3*se[1], 3*se[1], length = 51)
b2 <- bc[2] + seq(-3*se[2], 3*se[2], length = 51)
bv <- expand.grid(b1, b2)
```

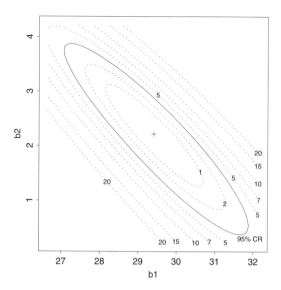

Figure 9.2: The Stormer data. The F–statistic surface and a confidence region for the regression parameters.

The simplest way to calculate the sum of squares function is to use `apply` function:

```
ssq <- function(b)
        sum((Time - b[1] * Viscosity/(Wt-b[2]))^2)
dbetas <- apply(bv, 1, ssq)
```

However using a function such as `outer` makes better use of the vectorizing facilities of S, and in this case a direct calculation is the most efficient:

```
attach(stormer)
cc <- matrix(Time - rep(bv[,1],rep(23, 2601)) *
        Viscosity/(Wt - rep(bv[,2], rep(23, 2601))), 23)
dbetas <- matrix(drop(rep(1, 23) %*% cc^2), 51)
```

The F–statistic array is then:

```
fstat <- matrix( ((dbetas - dv)/2) / (dv/21), 51, 51)
```

We can now produce a contour map of the F–statistic, taking care that the contours occur at relatively interesting levels of the surface. Note that the confidence region contour is at about 3.5:

```
> qf(0.95, 2, 21)
[1] 3.4668
```

Our intention is to produce a slightly non-standard contour plot and for this the old style plotting functions are more flexible than Trellis graphics functions. Rather than use `contour` to set up the plot directly, we begin with a call to `plot`:

```
plot(b1, b2, type="n")
lev <- c(1,2,5,7,10,15,20)
contour(b1, b2, fstat, levels=lev, labex=0.75, lty=2, add=T)
contour(b1, b2, fstat, levels=qf(0.95,2,21), add=T, labex=0)
text(31.6,0.3,"95% CR", adj=0, cex=0.75)
points(bc[1], bc[2], pch=3, mkh=0.1)
```

Since the likelihood function has the same contours as the F–statistic, the near elliptical shape of the contours is an indication that the approximate theory based on normal linear regression is probably accurate, although more than this is needed to be confident. (See the next section.) Given the way the axis scales have been chosen, the elongated shape of the contours shows that the estimates $\widehat{\beta}_1$ and $\widehat{\beta}_2$ are highly (negatively) correlated.

Note that finding a bivariate confidence region for two regression parameters where there are several others present can be a difficult calculation, at least in principle, since each point of the F–statistic surface being contoured must be calculated by optimizing with respect to the other parameters.

Bootstrapping

An alternative way to explore the distribution of the parameter estimates is to use the bootstrap. With regressions we have two alternatives: to treat the regressors as random and resample the pairs (x, y) or to treat the regressors as fixed and resample from the residuals (Efron & Tibshirani, 1993, pp. 111–5). This was a designed experiment, so we illustrate residual bootstrapping using the functions from S-PLUS 4.0.

```
> storm.fm <- nls(Time ~ b*Viscosity/(Wt - c), stormer,
        start = c(b=29.401, c=2.2183))
> storm.bf <- function(rs) {
        assign("Tim", fitted(storm.fm) + rs, frame = 1)
        nls(Tim ~ (b * Viscosity)/(Wt - c), stormer,
                start = coef(storm.fm))$parameters
    }
> rs <- scale(resid(storm.fm), scale = F)
> storm.boot <- bootstrap(rs, storm.bf, B = 1000)
> summary(storm.fm)$parameters
      Value Std. Error t value
b 29.4010    0.91553 32.1135
c  2.2183    0.66553  3.3332
> storm.boot
    ....
Summary Statistics:
    Observed    Bias   Mean     SE
b     28.72  0.7172 29.433 0.8191
c      2.48 -0.2806  2.199 0.6065

> limits.emp(storm.boot)
Empirical Percentiles:
```

```
        2.5%      5%      95%   97.5%
 b  27.8073  28.109  30.816  31.006
 c   0.9743   1.187   3.229   3.375
```

As this is a non-linear model the residuals may have a non-zero mean and we follow Huet *et al.* (1996, p.36) in centring them.

9.5   Assessing the linear approximation

Measures of local curvature

It is convenient to separate two sources of curvature, that of the solution locus itself, the *intrinsic curvature*, and that of the coordinate system within the solution locus, the *parameter-effects curvature*. The intrinsic curvature is fixed by the data and solution locus, but the parameter-effects curvature additionally depends upon the parametrization.

Summary measures for both kinds of *relative* curvature were proposed by Beale (1960) and elegantly interpreted by Bates & Watts (1980, 1988, §7.3). (The measures are relative to the estimated standard error of y and hence scale free.) The two measures are denoted by c^θ and c^ι for the parameter-effects and intrinsic root-mean-square curvatures respectively. If F is the $F_{p,n-p}$ critical value, Bates & Watts suggest that a value of $c\sqrt{F} > 0.3$ should be regarded as indicating unacceptably high curvature of either kind. Readers are referred to Bates & Watts (1988) or Seber & Wild (1989, §4.3) for further details.

Calculating curvature measures requires both first and second derivatives of the solution locus with respect to the parameters at each observation. The second derivatives must be supplied as an $n \times p \times p$ array where the ith $p \times p$ "face" provides the symmetric matrix of second partial derivatives $\partial^2 \eta_i(\beta)/\partial\beta_j\partial\beta_k$. This may be supplied as a `hessian` attribute of the value of the model function along with the `gradient`. (Unfortunately the `nls` fitting function can make no use of any `hessian` information.) The function `deriv3` in the MASS library is an extension of `deriv` that generates a model function that produces a value together with `gradient` and `hessian` attributes. (This function was written by David Smith and may also be found on `statlib`.)

The function `rms.curv` supplied with our library can be used to calculate and display $c^\theta\sqrt{F}$ and $c^\iota\sqrt{F}$. The only required argument is an `nls` fitted model object, provided the model function has both `gradient` and `hessian` attributes. Consider our weight loss example.

```
> expn3 <- deriv3(~ b0 + b1*2^(-x/th), c("b0","b1","th"),
      function(x, b0, b1, th) {})
> wtloss.he <- nls(Weight ~ expn3(Days, b0, b1, th),
      wtloss, start = coef(wtloss.gr))
> rms.curv(wtloss.he)
Parameter effects: c^theta x sqrt(F) = 0.1679
        Intrinsic: c^iota  x sqrt(F) = 0.0101
```

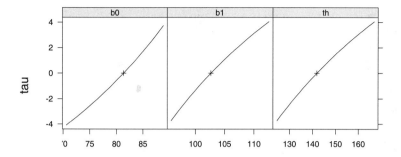

Figure 9.3: Profile plots for the negative exponential weight loss model.

Although this result is acceptable, a lower parameter-effects curvature would be preferable (see Exercise 9.4).

Profiles

A more global way of assessing departures from the planar assumption (both of the surface itself and of the coordinate system within the surface) is to look at the low-dimensional profiles of the sum of squares function, as discussed in Section 9.4. Indeed this is the original intended use for the `profile` function.

For coordinate directions along which the approximate linear methods are accurate, a plot of the non-linear t–statistics, $\tau(\beta_i)$ against β_i over several standard deviations on either side of the maximum likelihood estimate should be straight. (The $\tau(\beta_i)$ were defined in equation (9.4) on page 276.) Any deviations from straightness serve as a warning that the linear approximation may be misleading in that direction.

Appropriate plot methods for objects produced by the `profile` function are available (and our library `MASS` contains an enhanced version using Trellis). For the weight loss data we can use (Figure 9.3)

```
plot(profile(wtloss.gr))
```

A limitation of these plots is that they only examine each coordinate direction separately. Profiles can also be defined for two or more parameters simultaneously. These present much greater difficulties of computation and to some extent of visual assessment. An example is given in the previous section with the two parameters of the `stormer` data model. In Figure 9.2 an assessment of the linear approximation requires checking both that the contours are approximately elliptical and that each one-dimensional cross-section through the minimum is approximately quadratic. In one dimension it is much easier visually to check the linearity of the signed square root than the quadratic nature of the sum of squares itself.

The profiling process itself actually accumulates much more information than that shown in the plot. As well as recording the value of the optimized likelihood for a fixed value of a parameter, β_j, it also finds and records the values of all other parameters for which the conditional optimum is achieved. The `pairs` method function for `profile` objects supplied in our MASS library can display this so-called *profile trace* information in graphical form, which can shed considerably more light on the local features of the log-likelihood function near the optimum. This is discussed in more detail in the on-line complements.

9.6   Constrained non-linear regression

All the functions for regression we have seen so far assume that the parameters can take any real value. Constraints on the parameters can be often incorporated by reparametrization (such as $\beta_i = e^\theta$, possibly thereby making a linear model non-linear), but this can be undesirable if, for example, the constraint is $\beta_i \geqslant 0$ and 0 is expected to occur. Two functions for constrained regression are provided in recent versions of S-PLUS.

The function `nnls.fit` performs linear least squares subject to the constraint that all the coefficients are non-negative. If some of the coefficients are unconstrained, we can use the usual 'trick' of including both x and -x. For the cats data of Section 6.1 we can use

```
> attach(cats); Bwt <- Bwt[Sex=="F"]; Hwt <- Hwt[Sex=="F"]
> nnls.fit(cbind(1, -1, Bwt), Hwt)
$coefficients:
              Bwt
 2.9813 0 2.6364
     . . . .
```

This constrains the slope to be non-negative, but does not constrain the intercept. The fitted object includes residuals and some diagnostics of the algorithm.

The function `nlregb` performs non-linear regression with range bounds on the parameters, so the parameters are constrained to lie within a hypercube. The user must specify a function giving the *residuals* whose sum of squares is to be minimized. First derivatives of the residuals (called the *Jacobian* in the function documentation) may be supplied, otherwise the derivatives are approximated by finite differences. We can use `nlregb` for the weight loss example by

```
attach(wtloss)
nlregb(nrow(wtloss), c(90,95,120), function(x)
  Weight-x[1]-x[2]*2^(-Days/x[3]), lower=rep(0,3))$parameters
[1]   81.374 102.684 141.910
detach()
```

Of course, in this example the positivity constraints on the parameters are unlikely to be needed. It is also possible to pass the data via extra arguments, which we illustrate whilst supplying derivatives.

```
wtloss.r <- function(x, Weight, Days)
    Weight - x[1] - x[2] * 2^(-Days/x[3])
wtloss.rg <- function(x, Weight, Days) {
```

```
      temp <- 2^(-Days/x[3])
      -cbind(1, temp, x[2]*Days*temp*log(2)/x[3]^2)
}
wtloss.nl <- nlregb(nrow(wtloss), c(90, 95, 120),
      wtloss.r,  wtloss.rg, lower = rep(0,3),
      Weight = wtloss$Weight, Days = wtloss$Days)
```

No facility is provided for finding the estimated variance matrix of the parameters in a `nlregb` fit. The non-constant part of the negative log-likelihood is $E = \sum_i r_i(\theta)^2/2\sigma^2$ for residuals $r_i(\theta)$, when the error variance σ^2 is assumed known. Then

$$\frac{\partial^2 E}{\partial\theta_s\partial\theta_t} = \frac{1}{\sigma^2}\left[\sum_i \frac{\partial r_i}{\partial\theta_s}\frac{\partial r_i}{\partial\theta_t} + \sum_i r_i\frac{\partial^2 r_i}{\partial\theta_s\partial\theta_t}\right]$$

The first term is the Gauss-Newton approximation, and under the usual modelling assumptions is the Fisher information (as the residuals have mean zero). We can base an estimated variance matrix on this term (as used by `summary.nls` and hence `vcov.nls`) by

```
vcov1 <- function(object) {
    gr <- object$jacobian
    df <- length(object$resid) - length(object$param)
    sum(object$resid^2)/df * solve(t(gr) %*% gr)
}
```

As usual, we substitute $\hat\theta$ for θ and s^2 for σ^2.

We can use a finite-difference approximation to the second derivative to find the observed information more precisely, using the function `vcov.nlreg` in our library. (Note: there is no class `nlreg`, so this must be called directly.)

```
> sqrt(diag(vcov1(wtloss.nl)))
[1] 2.2690 2.0828 5.2945
> sqrt(diag(vcov.nlregb(wtloss.nl, method="Fisher")))
[1] 2.2690 2.0828 5.2945
> sqrt(diag(vcov.nlregb(wtloss.nl, method="observed")))
[1] 2.2653 2.0795 5.2856
```

In our example (and many others) the correction is negligible. Where there is a difference, the approach via the observed information is thought to be more accurate (see Efron & Hinkley, 1978).

The rationale for using either the observed or Fisher information is based on asymptotic theory which assume that the data were in fact generated by the non-linear regression for a particular set of parameters. If we drop this (dubious) assumption we can regard the estimate of the 'least false' parameter (that which fits best the true data generating mechanism), and use the appropriate asymptotic theory developed by Huber (1967) and White (1982). (See also Ripley, 1996, pp. 31–36.) We can use

```
> sqrt(diag(vcov.nlregb(wtloss.nl, method="Huber")))
[1] 2.2616 2.0763 5.2766
```

Note that the theory used here is unreliable if the solution $\hat{\theta}$ is close to or on the boundary; the asymptotic theory is delicate in that case and `vcov.nlreg` issues a warning.

The function `vcov.nlreg` is quite long and employs some intricate S programming ideas, so may be of interest to S programmers. If the `nlregb` fit did not supply the Jacobian, a local linear approximation is used, and only the Fisher information method is available.

```
> wtloss.nl0 <- nlregb(nrow(wtloss), c(90,95,120),
      wtloss.r, lower = rep(0,3),
      Weight = wtloss$Weight, Days = wtloss$Days)
> sqrt(diag(vcov.nlregb(wtloss.nl0)))
[1] 2.2690 2.0827 5.2945
```

9.7 General optimization and maximum likelihood estimation

Each of `nls` and `nlregb` has an analogue for general minimization, and there are also functions for univariate searches. Most are documented in the section `Non-linear Regression` of the help system, rather than `Optimization`.

Univariate functions

The functions `optimize` and `uniroot` work with continuous functions of a single variable: `optimize` can find a (local) minimum (the default) or a (local) maximum whereas `uniroot` finds a zero. Both search within an interval specified either by the argument `integer` or arguments `lower` and `upper`. Both are based on safeguarded polynomial interpolation (Brent, 1973; Nash, 1990).

Using ms

The `ms` function is a general function for minimizing quantities that can be written as sums of one or more terms. The call to `ms` is very similar to that for `nls`, but the interpretation of the formula is slightly different. It has no response, and the quantity on the right-hand side specifies the entire sum to be minimized. Hence the parameter estimates from the call

```
fm <- nls(y ~ fn(a,b,c,x), start=p0, data=dat)
```

should in principle be the same as those from

```
gm <- ms( ~ (y - fn(a,b,c,x))^2, start=p0, data=dat)
```

The resulting object is of class `ms`, for which few method functions are available. The four main ones are `print` for basic printing (mostly used implicitly),

`coef` for the parameter estimates,

`summary` for a succinct report of the fitted results,

`profile` for an exploration of the region near the minimum.

The fitted object has additional components, two of independent interest being

`gm$value` the minimum value of the sum, and

`gm$pieces` the vector of summands at the minimum.

Note that `fitted` and `resid` are not available since in this general context they have no unique meaning.

The most common statistical use of `ms` is to minimize negative log-likelihood functions, thus finding maximum likelihood estimates. The fitting algorithm can make use of first- and second-derivative information, the first- and second-derivative arrays of the summands rather than of a model function. In the case of a negative log-likelihood function, these are arrays whose "row sums" are the negative score vector and observed information matrix respectively. The first-derivative information is supplied as a `gradient` attribute of the formula, as described for `nls` in Section 9.1. Second-derivative information may be supplied via a `hessian` attribute.

The convergence criterion for `ms` is that *one of* the relative changes in the parameter estimates is small, the relative change in the achieved value is small *or* that the optimized function value is close to zero. The latter is only appropriate for deviance-like quantities, and can be eliminated by setting the argument

```
control = ms.control(f.tolerance=-1)
```

Fitting a mixture model

The waiting times between eruptions in the data `faithful` are strongly bimodal. Figure 9.4 shows a histogram with a density estimate superimposed (with bandwidth chosen by the Sheather-Jones method described in Chapter 5).

```
attach(faithful)
truehist(waiting, xlim=c(35,105), ymax=0.045)
width.SJ(waiting)
[1] 10.017
wait.dns <- density(waiting, 200, width=10)
lines(wait.dns, lty=2)
```

From inspection of this figure a mixture of two normal distributions would seem to be a reasonable descriptive model for the marginal distribution of waiting times. We now consider how to fit this by maximum likelihood. The observations are not independent since successive waiting times are strongly negatively correlated. In this section we propose to ignore this, both for simplicity and because ignoring this aspect is not likely seriously to misrepresent the information in the sample on the marginal distribution in question.

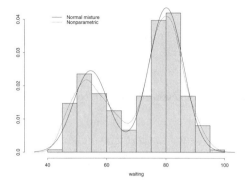

Figure 9.4: Histogram for the waiting times between successive eruptions for the "Old Faithful" geyser, with non-parametric and parametric estimated densities superimposed.

Useful references for mixture models include Everitt & Hand (1981), Titterington, Smith & Makov (1985) and McLachlan & Basford (1988). Everitt & Hand describe the EM algorithm for fitting a mixture model, which is simple (and provides an interesting S programming exercise), but we shall consider here a more direct function minimization method which can be faster and more reliable (Redner & Walker, 1984; Ingrassia, 1992).

If y_i, $i = 1, 2, \ldots, n$ is a sample waiting time the log-likelihood function for a mixture of two normal components is

$$L(\pi, \mu_1, \sigma_1, \mu_2, \sigma_2) = \sum_{i=1}^{n} \log \left[\frac{\pi}{\sigma_1} \phi \left(\frac{y_i - \mu_1}{\sigma_1} \right) + \frac{1 - \pi}{\sigma_2} \phi \left(\frac{y_i - \mu_2}{\sigma_2} \right) \right]$$

We will estimate the parameters by minimizing $-L$.

It will be helpful in this example to use both first- and second-derivative information and we would like to use `deriv3` to produce a model function for us. Both `deriv` and `deriv3` use a system function, `D`, to generate symbolic derivatives but as it stands it does not differentiate functions involving the normal distribution or density functions. To overcome this we need to make minor changes to the functions `D`, which does the actual differentiation, and the ancillary function, `make.call`. Extended versions are supplied in our library `MASS`, which must be attached with `first=T` to make them effective. Comparing these with the system versions will show how to make similar extensions if needed in the future. At most two or three statements are involved. Note that these extensions only allow calls to `pnorm` and `dnorm` with one argument. Thus calls such as `pnorm(x,u,s)` must be written as `pnorm((x-u)/s)`.

We can now generate a function which calculates the summands of $-L$ and also returns first- and second-derivative information for each summand by:

```
lmix2 <- deriv3(
    ~ -log(p*dnorm((x-u1)/s1)/s1 + (1-p)*dnorm((x-u2)/s2)/s2),
    c("p", "u1", "s1", "u2", "s2"),
    function(x, p, u1, s1, u2, s2) NULL)
```

which took 40 seconds on the PC.

Initial values for the parameters could be obtained by the method of moments described in Everitt & Hand (1981, pp. 31ff) but for well-separated components as we have here, simply choosing initial values by reference to the plot suffices. For the initial value of π we take the proportion of the sample below the density low point at 70.

```
> p0 <- c(p=mean(waiting < 70), u1=50, s1=5, u2=80, s2=5)
> p0
        p u1 s1 u2 s2
0.37868 50  5 80  5
```

We will trace the iterative process, using a trace function that produces just one line of output per iteration:

```
tr.ms <- function(info, theta, grad, scale, flags, fit.pars) {
    cat(round(info[3], 3), ":", signif(theta), "\n")
    invisible()
}
```

Note that the trace function must have the same arguments as the standard trace function, `trace.ms`.

We can now fit the mixture model:

```
> wait.mix2 <- ms(~ lmix2(waiting, p, u1, s1, u2, s2),
    start=p0, data = faithful, trace = tr.ms)
1077.343 : 0.378676 50 5 80 5
1073.094 : 0.342766 56.2689 3.6252 79.8122 5.7471
1043.704 : 0.340921 54.6143 4.15463 79.6473 6.24173
1036.185 : 0.349106 54.5222 4.92496 79.9016 6.08841
1034.165 : 0.359554 54.6176 5.59894 80.0729 5.87407
1034.004 : 0.360617 54.6104 5.83976 80.0865 5.87119
1034.002 : 0.360882 54.6148 5.87079 80.091 5.86778
1034.002 : 0.360886 54.6149 5.87122 80.0911 5.86773
```

For future reference, this took 2.5 seconds. Without any derivative information `ms` converges from this starting point in 32 iterations and 4.9 seconds whereas with first derivatives only it requires 36 iterations and 4.2 seconds.

We can now add the parametric density estimate to our original histogram plot.

```
dmix2 <- function(x, p, u1, s1, u2, s2)
                p * dnorm(x, u1, s1) + (1-p) * dnorm(x, u2, s2)
cf <- coef(wait.mix2)
attach(structure(as.list(cf), names = names(cf)))
wait.fdns <- list(x = wait.dns$x,
                  y = dmix2(wait.dns$x, p, u1, s1, u2, s2))
lines(wait.fdns)
par(usr = c(0,1,0,1))
legend(0.1, 0.9, c("Normal mixture", "Nonparametric"),
    lty = c(1,2), bty = "n")
```

The computations for Figure 9.4 are now complete.

The parametric and nonparametric density estimates are in fair agreement on the right component, but there is a suggestion of disparity on the left. We can informally check the adequacy of the parametric model by a Q-Q plot. First we solve for the quantiles using a Newton method:

```
> pmix2 <- deriv(~ p*pnorm((x-u1)/s1) + (1-p)*pnorm((x-u2)/s2),
                 "x", function(x, p, u1, s1, u2, s2) {})
> pr0 <- (seq(along = waiting) - 0.5)/length(waiting)
> x0 <- x1 <- as.vector(sort(waiting)) ; del <- 1; i <- 0
> while((i <- 1 + 1) < 10 && abs(del) > 0.0005) {
    pr <- pmix2(x0, p, u1, s1, u2, s2)
    del <- (pr - pr0)/attr(pr, "gradient")
    x0 <- x0 - del
    cat(format(del <- max(abs(del))), "\n")
}
1.9953
1.1259
0.35148
0.029407
0.00019061
> detach(2)
```

The plot is shown in Figure 9.5 and confirms the adequacy of the mixture model.

```
par(pty = "s")
plot(x0, x1, xlim = range(x0, x1), ylim = range(x0, x1),
     xlab = "Model quantiles", ylab = "Waiting time")
abline(0,1)
```

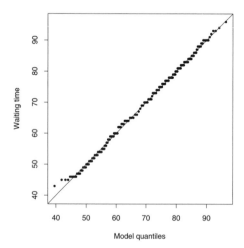

Figure 9.5: Sorted waiting times against normal mixture model quantiles for the "Old Faithful" eruptions data.

Another advantage of using second-derivative information is that it provides an estimate of the (large sample) variance matrix of the parameters. The fitted model object has a component, `hessian`, giving the sum of the second derivative array over the observations and since the function we minimized is the negative of the log-likelihood, this component is the observed information matrix. Its inverse is calculated (but incorrectly labelled[5]) by `summary.ms`

```
vmat <- summary(wait.mix2)$Information
cbind(coef(wait.mix2), sqrt(diag(vmat)))
        [,1]       [,2]
 p   0.36089 0.031165
u1  54.61486 0.699675
s1   5.87122 0.537322
u2  80.09107 0.504595
s2   5.86773 0.400962
```

It appears that a simpler model of two normal components with equal variances would have been adequate. We can easily check this via a score test or a likelihood ratio test:

```
> as.vector((cf["s1"] - cf["s2"]) /
      sqrt(c(0,0,1,0,1) %*% vmat %*% c(0,0,1,0,1)))
[1] 0.0061413
> wait.mix2e <- ms(~ -log(p * dnorm(waiting, u1, s1) +
    (1-p) * dnorm(waiting, u2, s1)), start = cf[-5],
    data = faithful)
> coef(wait.mix2e)
       p      u1     s1    u2
 0.36085 54.614 5.8691 80.09
> 2*(wait.mix2e$value - wait.mix2$value)
> [1] 2.107e-05
```

The S-PLUS dataset `geyser` has data of the same type from the same source. A normal mixture model also fits this dataset well, but the variances of the two components are significantly different, which suggests the agreement here is fortuitous.

General facilities for minimization

The function `nlmin` finds an unconstrained local minimum of a function. For the mixture problem we can use

```
mix.f <- function(p) {
    e <- p[1]*dnorm((waiting-p[2])/p[3])/p[3] +
         (1-p[1])*dnorm((waiting-p[4])/p[5])/p[5]
    -sum(log(e)) }
waiting.init <- c(mean(waiting < 70), 50, 5, 80, 5)
nlmin(mix.f, waiting.init, print.level=1)
```

---

[5] and differently from the help page which has `information`.

The argument `print.level=1` provides some tracing of the iterations (here 14, one less than the default limit). This fit took 3.7 seconds. There is no way to supply derivatives.

The most general minimization routine is `nlminb`, which can find a local minimum of a twice-differentiable function within a hypercube in parameter space. Either the gradient or gradient plus Hessian can be supplied; if no gradient is supplied it is approximated by finite differences. The underlying algorithm is a quasi-Newton optimizer, or a Newton optimizer if the Hessian is supplied. We can fit our mixture density (using zero, one and two derivatives) and giving the constraints that we have ignored hitherto.

```
attach(faithful)
mix.obj <- function(p, x) {
    e <- p[1]*dnorm((x-p[2])/p[3])/p[3] +
        (1-p[1])*dnorm((x-p[4])/p[5])/p[5]
    -sum(log(e)) }
mix.nl0 <- nlminb(waiting.init, mix.obj,
    scale = c(10, rep(1,4)), lower = c(0, -Inf, 0, -Inf, 0),
    upper = c(1, rep(Inf, 4)), x = waiting)

lmix2a <- deriv(
    ~ -log(p*dnorm((x-u1)/s1)/s1 + (1-p)*dnorm((x-u2)/s2)/s2),
    c("p", "u1", "s1", "u2", "s2"),
    function(x, p, u1, s1, u2, s2) NULL)
mix.gr <- function(p, x) {
    u1 <- p[2]; s1 <- p[3]; u2 <- p[4]; s2 <- p[5]; p <- p[1]
    e <- lmix2a(x, p, u1, s1, u2, s2)
    rep(1, length(x)) %*% attr(e, "gradient") }
mix.nl1 <- nlminb(waiting.init, mix.obj, mix.gr,
    scale = c(10, rep(1,4)), lower = c(0, -Inf, 0, -Inf, 0),
    upper = c(1, rep(Inf, 4)), x = waiting)
mix.grh <- function(p, x) {
    e <- lmix2(x, p[1], p[2], p[3], p[4], p[5])
    g <- attr(e, "gradient")
    g <- rep(1, length(x)) %*% g
    H <- apply(attr(e, "hessian"), c(2,3), sum)
    list(gradient=g, hessian=H[row(H) <= col(H)]) }
mix.nl2 <- nlminb(waiting.init, mix.obj, mix.grh, T,
    scale = c(10, rep(1,4)), lower = c(0, -Inf, 0, -Inf, 0),
    upper = c(1, rep(Inf, 4)), x = waiting)
```

These fits took 4.8, 2.4 and 3.2 seconds. (It is not unusual for `nlminb` fits to be comparatively slow, but here we have calculated the derivatives in an inelegant way.) We use a `scale` parameter to set the step length on the first parameter much smaller than the others, which can speed convergence. Generally it is helpful to have the scale set so that the range of uncertainty in (scale × parameter) is about one.

It is also possible to supply a separate function to calculate the Hessian; however it is supplied, it is a vector giving the lower triangle of the Hessian in row-first order (unlike S matrices).

nlminb only computes the Hessian at the solution if a means to compute Hessians is supplied (and even then it returns a scaled version of the Hessian). Thus it provides no help in using the observed information to provide approximate standard errors for the parameter estimates in a maximum likelihood estimation problem. For nlminb we use a finite-difference approximation to the Hessian if it is not available, via the function vcov.nlminb in our library.

```
> sqrt(diag(vcov.nlminb(mix.nl0)))
[1] 0.03154 0.70429 0.56639 0.49766 0.41288
> sqrt(diag(vcov.nlminb(mix.nl1)))
[1] 0.031165 0.699675 0.537323 0.504595 0.400962
> sqrt(diag(vcov.nlminb(mix.nl2)))
[1] 0.031165 0.699675 0.537322 0.504595 0.400962
```

The differences reflect a difference in philosophy: when no derivatives are available we seek a quadratic approximation to the negative log-likelihood over the scale of random variation in the parameter estimate rather than at the MLE.

Note that the theory used here is unreliable if the parameter estimate is close to or on the boundary; vcov.nlminb issues a warning.

Fitting GLM models

The Fisher scoring algorithm is not the only possible way to fit GLMs. Indeed, the function glm allows other methods to be used via its method argument, although we know of no such methods and the return values required of a method are geared to the IWLS algorithm.

The purpose of this section is to point out how easy it is to fit specific GLMs by direct maximization of the likelihood, and the code can easily be modified for other fit criteria.

Logistic regression

We consider maximum likelihood fitting of a binomial logistic regression. For other link functions just replace plogis and dlogis, for example by pnorm and dnorm for a probit regression.

```
logitreg <- function(x, y, wt=rep(1, length(y)),
               intercept = T, start, trace = T, ...)
{
    fmin <- function(x, y, w, beta) {
        eta <- x %*% beta
        p <- plogis(eta)
        g <- -2 * dlogis(eta) * ifelse(y, 1/p, -1/(1-p))
        value <- -2 * w * ifelse(y, log(p), log(1-p))
        attr(value, "gradient") <- x*g*w
        value
    }
    if(is.null(dim(x))) dim(x) <- c(length(x), 1)
    dn <- dimnames(x)[[2]]
```

```
        if(!length(dn)) dn <- paste("Var", 1:ncol(x), sep="")
        p <- ncol(x) + intercept
        if(intercept) {
            x <- cbind(1, x)
            dn <- c("(Intercept)", dn)
        }
        if(missing(start)) start <- rep(0, p)
        fit <- ms(~ fmin(x, y, wt, beta), start=list(beta=start),
                    trace=trace, ...)
        names(fit$par) <- dn
        cat("\nCoefficients:\n")
        print(fit$parameters)
        cat("Residual Deviance:", format(fit$value), "\n")
        invisible(fit)
    }
```

This can easily be modified to allow constraints on the parameters (by using
nlminb rather than ms; see below), the handling of missing values by multiple
imputation (Ripley, 1996), the possibility of errors in the recorded x or y values
(Copas, 1988) and shrinkage and approximate Bayesian estimators, none of which
can readily be handled by the glm algorithm.

Constrained Poisson regressions

Workers on the backcalculation of AIDS have used a Poisson regression model
with identity link and non-negative parameters (for example Rosenberg & Gail,
1991). The history of AIDS is divided into J time intervals $[T_{j-1}, T_j)$, and the
data are counts Y_j of cases in each time interval. Then for a Poisson incidence
process of rate ν, the counts are independent Poisson random variables with mean

$$EY_j = \mu_j = \int_0^{T_j} \left[F(T_j - s) - F(T_{j-1} - s) \right] \nu(s) \, ds$$

where F is the cumulative distribution of the incubation time. If we model ν by
a step function

$$\nu(s) = \sum_i I(t_{i-1} \leqslant s < t_i)\beta_i$$

then $EY_j = \sum x_{ji}\beta_i$ for constants x_{ji} which are considered known.

This is a Poisson regression with identity link, no intercept and constraint
$\beta_i \geqslant 0$. The deviance is $2 \sum_j \mu_j - y_j - y_j \log(\mu_j/y_j)$.

```
    AIDSfit <- function(y, z, start=rep(mean(y), ncol(z)), ...)
    {
        deviance <- function(beta, y, z) {
            mu <- z %*% beta
            2 * sum(mu - y - y*log(mu/y)) }
        grad <- function(beta, y, z) {
            mu <- z %*% beta
            2 * t(1 - y/mu) %*% z }
        nlminb(start, deviance, grad, lower = 0, y = y, z = z, ...)
    }
```

As an example, we consider the history of AIDS in Belgium using the numbers of cases for 1981 to 1993 (as reported by WHO in March 1994). We use 2-year intervals for the step function and do the integration crudely using a Weibull model for the incubation times.

```
Y <- scan()
12 14 33 50 67 74 123 141 165 204 253 246 240

library(nnet) # for class.ind
s <- seq(0, 13.999, 0.01); tint <- 1:14
X <- expand.grid(s, tint)
Z <- matrix(pweibull(pmax(X[,2] - X[,1],0), 2.5, 10),length(s))
Z <- Z[,2:14] - Z[,1:13]
Z <- t(Z) %*% class.ind(factor(floor(s/2))) * 0.01
round(AIDSfit(Y, Z)$param)
515   0 140 882   0   0   0
rm(s, X, Y, Z)
```

This has infections only in 1981–2 and 1985–8. We will not pursue the medical implications!

9.8 Exercises

This section suggests ways to explore further the examples used in this chapter.

9.1. For the weight loss example compare the negative exponential model with quadratic and cubic polynomial regression alternative models, in particular check the behaviour of each model under extrapolation into the future.

9.2. Fit the negative exponential weight loss model in the 'goal weight' form, equation (9.5), for the three goal weights, $w_0 = 110$, 100 and 90 kg. Plot the profiles and compare the local curvature measures.

9.3. The model used in connection with the Stormer data may also be expressed as a generalized linear model. To do this we write

$$\frac{\beta_1 v}{w - \beta_2} = \frac{1}{\gamma_1 z_1 + \gamma_2 z_2}$$

where $\gamma_1 = 1/\beta_1$, $\gamma_2 = \beta_2/\beta_1$, $z_1 = w/v$ and $z_2 = -1/v$. This has the form of a generalized linear model with inverse link. Fit the model in this form using a quasi family with inverse link and constant variance function.

Back-transform the estimated coefficients and show that they agree with the values obtained using the non-linear regression approach.

Also compute the estimated standard errors and verify that they also agree with the values obtained directly by the non-linear regression approach.

Finding the standard errors is more challenging. We need first to find the variance matrix from the generalized linear model. The large sample variance matrix for $\widehat{\beta}$ is related to that for $\widehat{\gamma}$ by $\mathrm{Var}[\widehat{\beta}] = J\mathrm{Var}[\widehat{\gamma}]J^T$ where J is the Jacobian matrix of the inverse of the parameter transformation:

$$J = \begin{bmatrix} \partial\beta_1/\partial\gamma_1 & \partial\beta_1/\partial\gamma_2 \\ \partial\beta_2/\partial\gamma_1 & \partial\beta_2/\partial\gamma_2 \end{bmatrix} = \begin{bmatrix} -1/\gamma_1^2 & 0 \\ -\gamma_2/\gamma_1^2 & 1/\gamma_1 \end{bmatrix}$$

(To achieve close agreement you may need to tighten the convergence criteria for the glm fit, for example by setting eps=1.0e-10.)

9.4. *Stable parameters*

Ross (1970) has suggested using *stable parameters* for non-linear regression, mainly to achieve estimates which are as near to uncorrelated as possible. It turns out that in many cases stable parameters also define a coordinate system within the solution locus with a small curvature.

The idea is to use the means at p well-separated points in sample space as the parameters. Writing the regression function in terms of the stable parameters is often intractable, but in the case of a negative exponential decay model of the type we considered for the weight loss data it is possible if the points are chosen equally spaced.

If the three mean parameters, μ_i are chosen at x–points $x_0 + i\delta_x$, $i = 0, 1, 2$, show that the model may be written explicitly as:

$$\eta = \frac{\mu_0\mu_2 - \mu_1^2}{\mu_0 - 2\mu_1 + \mu_2} + \frac{(\mu_0 - \mu_1)^2}{\mu_0 - 2\mu_1 + \mu_2} \left(\frac{\mu_1 - \mu_2}{\mu_0 - \mu_1}\right)^{(x-x_0)/\delta_x}$$

Fit the negative exponential decay model to the weight loss data using this parametrization and choosing, say, $x_0 = 40$ days and $\delta_x = 80$ days. Look at the characteristics of the fit, including the correlations between the parameter estimates. Explain in heuristic terms why they are relatively low.

Examine the profiles of the fit and check for straightness. Also look at the local relative curvature measures[6]. Verify numerically that the intrinsic curvature is unchanged (as it must be), but that the parametric curvature is much reduced.

9.5. *Heteroscedastic regression models*

A common heteroscedastic regression model specifies that the observations have constant coefficient of variation, that is $Y \sim N(\mu, \theta\mu^2)$ where $\theta > 0$ and μ depends on regressor variables according to some linear model perhaps with a link function such as $\mu = \exp\eta$. Write a function to fit such models and try it out using the Quine data. Compare with the negative binomial models fitted in Section 7.4, page 243ff.

---

[6] For this rather complicated-looking model specification, using deriv3 to produce a model function with gradient and hessian attributes may cause memory overflow problems on some machines.

Chapter 10

Random and Mixed Effects

We collect together several ways to handle linear and non-linear models with random effects, possibly as well as fixed effects.

10.1 Random effects and variance components

The function `raov` may be used for balanced designs with only random effects, and gives a conventional analysis including the estimation of variance components. The function `varcomp` is more general, and may be used to estimate variance components for balanced or unbalanced mixed models.

To illustrate these functions we take a portion of the data from the data frame `coop` of our library `MASS`. This is a co-operative trial in analytical chemistry taken from Analytical Methods Committee (1987). Seven specimens were sent to 6 laboratories, each 3 times a month apart for duplicate analysis. The response is the concentration of (unnamed) analyte in g/kg. We use the data from specimen 1 shown in Table 10.1.

Table 10.1: Analyte concentrations (g/kg) from 6 laboratories and 3 batches.

			Laboratory			
Batch	1	2	3	4	5	6
1	0.29	0.40	0.40	0.9	0.44	0.38
	0.33	0.40	0.35	1.3	0.44	0.39
2	0.33	0.43	0.38	0.9	0.45	0.40
	0.32	0.36	0.32	1.1	0.45	0.46
3	0.34	0.42	0.38	0.9	0.42	0.72
	0.31	0.40	0.33	0.9	0.46	0.79

The purpose of the study was to assess components of variation in co-operative trials. For this purpose laboratories and batches are regarded as random, so a model for the response for laboratory i, batch j and duplicate k is

$$y_{ijk} = \mu + \xi_i + \beta_{ij} + \epsilon_{ijk}$$

where ξ, β and ϵ are independent random variables with zero means and variances σ_{L}^2, σ_{B}^2 and σ_e^2 respectively. For l laboratories, b batches and $r = 2$ duplicates a nested analysis of variance gives:

Source of variation	Degrees of freedom	Sum of squares	Mean square	E(MS)
Between laboratories	$l-1$	$br\sum_i(\bar{y}_i - \bar{y})^2$	MS_{L}	$br\sigma_{\mathrm{L}}^2 + r\sigma_{\mathrm{B}}^2 + \sigma_e^2$
Batches within laboratories	$l(b-1)$	$r\sum_{ij}(\bar{y}_{ij} - \bar{y}_i)^2$	MS_{B}	$r\sigma_{\mathrm{B}}^2 + \sigma_e^2$
Replicates within batches	$lb(r-1)$	$\sum_{ijk}(y_{ij} - \bar{y}_{ij})^2$	MS_e	σ_e^2

So the unbiased estimators of the variance components are

$$\hat{\sigma}_{\mathrm{L}}^2 = (br)^{-1}(\mathrm{MS}_{\mathrm{L}} - \mathrm{MS}_{\mathrm{B}}), \quad \hat{\sigma}_{\mathrm{B}}^2 = r^{-1}(\mathrm{MS}_{\mathrm{B}} - \mathrm{MS}_e), \quad \hat{\sigma}_e^2 = \mathrm{MS}_e$$

The model is fitted in the same way as an analysis of variance model with `raov` replacing `aov`:

```
> summary(raov(Conc ~ Lab/Bat, data = coop, subset = Spc=="S1"))
                 Df Sum of Sq Mean Sq Est. Var.
Lab              5    1.8902  0.37804  0.060168
Bat %in% Lab 12       0.2044  0.01703  0.005368
Residuals    18       0.1134  0.00630  0.006297
```

The same variance component estimates can be found using `varcomp`, but as this allows mixed models we need first to declare which factors are random effects using the `is.random` function. All factors in a data frame are declared to be random effects by

```
is.random(coop) <- T
```

If we use `Spc` as a factor it may need to be treated as fixed. Individual factors can also have their status declared in the same way. When used as an unassigned expression `is.random` reports the status of (all the factors in) its argument:

```
> is.random(coop$Spc) <- F
> is.random(coop)
 Lab Spc Bat
  T   F   T
```

We can now estimate the variance components:

```
> varcomp(Conc ~ Lab/Bat, data = coop, subset = Spc=="S1")
Variances:
       Lab Bat %in% Lab Residuals
  0.060168     0.0053681 0.0062972
```

The fitted values for such a model are technically the grand mean, but the `fitted` method function calculates them as if it were a fixed effects model, so giving best linear unbiased predictors (BLUPs; see the review paper by Robinson, 1991). Residuals are also calculated as differences of observations from the BLUPs. An examination of the residuals and fitted values points up laboratory 4 batches 1 & 2 as possibly suspect, and these will inflate the estimate of σ_e^2. (However laboratory 4 overall and laboratory 6 batch 3 are outliers for the other variance components, but a simple residual analysis will not expose this. See Analytical Methods Committee, 1987.)

Variance component estimates are known to be very sensitive to aberrant observations; we can get some check on this by repeating the analysis with a robust estimating method. The result is very different:

```
> varcomp(Conc ~ Lab/Bat, data = coop, subset = Spc=="S1",
    method = c("winsor", "minque0"))
Variances:
      Lab Bat %in% Lab Residuals
 0.0040043   -0.00065681 0.0062979
```

(A related robust analysis is given in Analytical Methods Committee, 1989b.)

The possible estimation methods are `"minque0"` for minimum norm quadratic estimators (the default), `"ml"` for maximum likelihood and `"reml"` for residual (or reduced or restricted) maximum likelihood. (See, for example, Rao, 1971a,b; Rao & Kleffe, 1988.) Method `"winsor"` specifies that the data are 'cleaned' before further analysis. As in our example, a second method may be prescribed as the one to use for estimation of the variance components on the 'cleaned' data. The method used, Winsorizing, is discussed in Chapter 8; unfortunately the tuning constant is not adjustable, and is set for rather mild data cleaning.

10.2 Multistratum models

Multistratum models occur where there is more than one source of random variation in an experiment, as in a split-plot experiment, and have been most used in agricultural field trials. The sample information and the model for the means of the observations may be partitioned into so-called *strata*. In orthogonal experiments such as split-plot designs each parameter may be estimated in one and only one stratum, but in non-orthogonal experiments such as a balanced incomplete block design with "random blocks", treatment contrasts are estimable both from the between-block and the within-block stratum. Combining the estimates from two strata is known as "the recovery of interblock information", for which the interested reader should consult a comprehensive reference such as Scheffé (1959, pp. 170ff). In this section we consider the separate stratum analyses only.

The details of the computational scheme employed for multistratum experiments are given by Heiberger (1989).

A split-plot experiment

Our example was first used by Yates (1935) and is discussed further in Yates (1937). It has also been used by many authors since; see, for example, John (1971, §5.7). The experiment involved varieties of oats and manure (nitrogen), conducted in 6 blocks of 3 whole plots. Each whole plot was divided into 4 subplots. Three varieties of oats were used in the experiment with one variety being sown in each whole plot (this being a limitation of the seed drill used), while four levels of manure (0, 0.2, 0.4 and 0.6 cwt per acre) were used, one level in each of the four subplots of each whole plot.

The data are shown in Table 10.2, where the blocks are labelled I-VI, V_i denotes the ith variety and N_j denotes the jth level of nitrogen. The dataset is available in our library as the data frame `oats`, with variables `B`, `V`, `N` and `Y`. Since the levels of `N` are ordered it is appropriate to create an ordered factor:

```
oats$Nf <- ordered(oats$N, levels=sort(levels(oats$N)))
```

Table 10.2: A split-plot field trial of oat varieties.

		N_1	N_2	N_3	N_4			N_1	N_2	N_3	N_4
	V_1	111	130	157	174		V_1	74	89	81	122
I	V_2	117	114	161	141	IV	V_2	64	103	132	133
	V_3	105	140	118	156		V_3	70	89	104	117
	V_1	61	91	97	100		V_1	62	90	100	116
II	V_2	70	108	126	149	V	V_2	80	82	94	126
	V_3	96	124	121	144		V_3	63	70	109	99
	V_1	68	64	112	86		V_1	53	74	118	113
III	V_2	60	102	89	96	VI	V_2	89	82	86	104
	V_3	89	129	132	124		V_3	97	99	119	121

The strata for the model we will use here are:

1. A 1-dimensional stratum corresponding to the total of all observations,
2. A 5-dimensional stratum corresponding to contrasts between block totals,
3. A 12-dimensional stratum corresponding to contrasts between variety (or equivalently whole plot) totals within the same block, and
4. A 54-dimensional stratum corresponding to contrasts within whole plots.

(We use the term 'contrast' here to mean a linear function with coefficients adding to zero, and thus representing a comparison.)

Only the overall mean is estimable within stratum 1 and since no degrees of freedom are left for error, this stratum is suppressed in the printed summaries (although a component corresponding to it is present in the fitted model object). Stratum 2 has no information on any treatment effect. Information on the V main effect is only available from stratum 3, and the N main effect and N × V interaction are only estimable within stratum 4.

Multistratum models may be fitted using aov (only), and are specified by a model formula of the form

$$response ~\sim~ mean.formula ~+~ \text{Error}\,(strata.formula)$$

In our example the *strata.formula* is B/V, specifying strata 2 and 3; the fourth stratum is included automatically as the "within" stratum, the residual stratum from the strata formula.

The fitted model object is of class aovlist, which is a list of fitted model objects corresponding to the individual strata. Each component is an object of class aov (although in some respects incomplete). There is no anova method function for class aovlist. The appropriate display function is summary which displays separately the anova tables for each stratum (except the first).

```
> oats.aov <- aov(Y ~ Nf*V + Error(B/V), data=oats)
> summary(oats.aov)

Error: B
          Df Sum of Sq Mean Sq F Value Pr(F)
Residuals  5     15875  3175.1

Error: V %in% B
          Df Sum of Sq Mean Sq F Value    Pr(F)
V          2    1786.4  893.18  1.4853 0.27239
Residuals 10    6013.3  601.33

Error: Within
          Df Sum of Sq Mean Sq F Value  Pr(F)
Nf         3     20020  6673.5  37.686 0.0000
Nf:V       6       322    53.6   0.303 0.9322
Residuals 45      7969   177.1
```

Polynomial components

There is a clear nitrogen effect, but no evidence of variety differences nor interaction. Since the levels of Nf are quantitative, it is natural to consider partitioning the sums of squares for nitrogen and its interactions into polynomial components. Here the details on factor coding become important.

Since the levels of nitrogen are equally spaced, and we chose an ordered factor, the default contrast matrix is appropriate. We can obtain an analysis of variance table with the degrees of freedom for nitrogen partitioned into components by giving an additional argument to the summary function:

```
> summary(oats.aov, split=list(Nf=list(L=1, Dev=2:3)))
    ....
```

```
Error: Within
                Df Sum of Sq Mean Sq F Value    Pr(F)
Nf               3     20020    6673   37.69  0.00000
    Nf: L        1     19536   19536  110.32  0.00000
    Nf: Dev      2       484     242    1.37  0.26528
Nf:V             6       322      54    0.30  0.93220
    Nf:V: L      2       168      84    0.48  0.62476
    Nf:V: Dev    4       153      38    0.22  0.92786
Residuals       45      7969     177
```

The `split` argument is a list with the name of each component that of some factor appearing in the model. Each component is also a list with components of the form name=int.seq where name is the name for the table entry and int.seq is an integer sequence giving the contrasts to be grouped under that name in the ANOVA table.

Residuals in multistratum analyses: Projections

Residuals and fitted values from the individual strata are available in the usual way by accessing each component as a fitted model object. Thus `fitted(oats.aov[[4]])` and `resid(oats.aov[[4]])` are vectors of length 54 representing fitted values and residuals from the last stratum, based on 54 orthonormal linear functions of the original data vector. It is not possible to associate them uniquely with the plots of the original experiment, which for field trials would sometimes be very useful.

The function `proj` takes a fitted model object and finds the projections of the original data vector onto the subspaces defined by each line in the analysis of variance tables (including, for multistratum objects, the suppressed table with the grand mean only). The result is a list of matrices, one for each stratum, where the column names for each are the component names from the analysis of variance tables. If the argument `qr=T` has been set when the model was initially fitted this calculation is considerably faster. Two diagnostic plots are shown in Figure 10.1, computed by

```
oats.fm <- update(oats.aov, qr=T)   # not strictly necessary
plot(fitted(oats.fm[[4]]), studres(oats.fm[[4]]))
oats.pr <- proj(oats.fm)
qqnorm(oats.pr[[4]][,"Residuals"], ylab="Stratum 4 residuals")
qqline(oats.pr[[4]][,"Residuals"])
```

Tables of means and components of variance

A better appreciation of the results of the experiment comes from looking at the estimated marginal means. The function `model.tables` calculates tables of the effects, means or residuals and optionally their standard errors. Tables of means are only available for balanced designs, and the standard errors calculated are for differences of means. We now refit the model omitting the V:N interaction, but

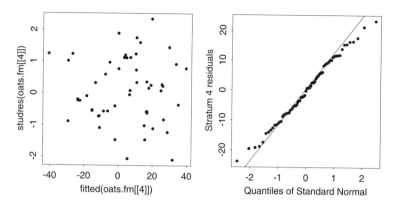

Figure 10.1: Two diagnostic plots for the oats data. (The exact plot obtained varies between releases of S-PLUS.)

retaining the V main effect (since that might be considered a blocking factor). Since the model.tables calculation also requires the projections we should fit the model with either qr=T or proj=T set to avoid later refitting.

```
> oats.aov <- aov(Y ~ N + V + Error(B/V), data = oats, qr = T)
> model.tables(oats.aov, type = "means", se = T)
Tables of means
    ....
 N
 0.0cwt 0.2cwt 0.4cwt 0.6cwt
  79.39  98.89  114.2  123.4
 V
 Golden.rain Marvellous Victory
       104.5      109.8   97.63

 Standard errors for differences of means
             N       V
           4.25   7.079
replic.   18.00  24.000
```

When interactions are present the table is a little more complicated.

Finally we use the **S-PLUS** function varcomp to estimate the variance components associated with the lowest three strata. To do this we need to declare B to be a random factor:

```
> is.random(oats$B) <- T
> varcomp(Y ~ N + V + B/V, data = oats)
Variances:
      B V %in% B Residuals
 214.48  109.69    162.56
    ....
```

In this simple balanced design the estimates are the same as those obtained by equating the residual mean squares to their expectations. The result suggests that the main blocking scheme, at least, has been reasonably effective.

10.3   Linear mixed effects models

S-PLUS 3.4 and later contain the `nlme` software of Pinheiro, Bates and Lindstrom for fitting linear and nonlinear mixed effects models. A version of the software is available from the `statlib` archive (see Appendix C) for both Unix and Windows.

The simplest case where a mixed effects model might be considered is when there are two stages of sampling. At the first stage units are selected at random from a population and at the second stage a number of measurements is made on each unit sampled in the first stage. For example, in the simple split-plot analysis of variance model the blocks are regarded as being chosen at random from a population of blocks and the plots within the blocks give the repeated measurements within the primary sampled unit. Another common situation giving rise to a mixed effects model is when measurements are made sequentially in time or space on the primary sampling units. Common names for such datasets analysed by mixed effects models include *repeated measures* data and *longitudinal* data. A good reference for both linear and nonlinear mixed effects models is Davidian & Giltinan (1995); Pinheiro, Bates and Lindstrom give Laird & Ware (1982) as their primary reference.

The *gasoline* data (Prater, 1956) was originally used to build an estimation equation for yield of the refining process. The possible predictor variables were the specific gravity, vapour pressure and ASTM 10% point made on the crude oil itself and the volatility of the desired product measured as the ASTM end point. The dataset is included in our library as `petrol`. Several authors (including Daniel & Wood, 1980; Hand *et al.*, 1993) have noted that the three measurements made on the crude oil occur in only ten discrete sets and have inferred that only ten crude oil samples were involved and several yields with varying ASTM end points were measured on each crude oil sample. We adopt this view here.

A natural way to inspect the data is to plot the yield against the ASTM end point for each sample, for example, using Trellis graphics by

```
xyplot(Y ~ EP | No, data = petrol,
    xlab = "ASTM end point (deg. F)",
    ylab = "Yield as a percent of crude",
    panel = function(x, y) {
        m <- sort.list(x)
        panel.grid()
        panel.xyplot(x[m], y[m], type = "b", cex = 0.5)
    })
```

The result shown in Figure 10.2 shows a fairly consistent and generally linear rise in yield with ASTM end point but with some variation in intercept that may correspond to differences in the covariates made on the crude oil itself. In fact the

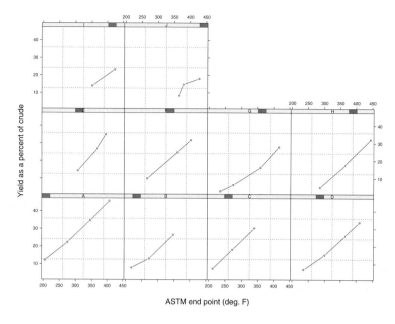

Figure 10.2: Prater's gasoline data: yield versus ASTM end point within samples.

intercepts appear to be steadily decreasing and since the samples are arranged so that the value of V10 is increasing this suggests a significant regression on V10 with negative coefficient.

If we regard the ten crude oil samples as fixed the most general model we might consider would have separate simple linear regressions of Y on EP for each crude oil sample. First we centre the determining variables so that the origin is in the centre of the design space, thus making the intercepts simpler to interpret.

```
> Petrol <- petrol
> names(Petrol)
[1] "No"  "SG"  "VP"  "V10" "EP"  "Y"
> Petrol[, 2:5] <- scale(as.matrix(Petrol[, 2:5]), scale = F)
> pet1.lm <- lm(Y ~ No/EP - 1, Petrol)
> matrix(round(coef(pet1.lm),2), 2, 10, byrow = T, dimnames =
    list(c("b0","b1"),levels(Petrol$No)))
```

	A	B	C	D	E	F	G	H	I	J
b0	32.75	23.62	28.99	21.13	19.82	20.42	14.78	12.68	11.62	6.18
b1	0.17	0.15	0.18	0.15	0.23	0.16	0.14	0.17	0.13	0.13

There is a large variation in intercepts, as expected, and some variation in slopes. We test if the slopes can be considered constant; as they can we adopt the simpler parallel-line regression model.

```
> pet2.lm <- lm(Y ~ No - 1 + EP, Petrol)
> anova(pet2.lm, pet1.lm)
```

```
        Terms RDf    RSS  Df Sum of Sq F Value    Pr(F)
No - 1 + EP   21 74.132
  No/EP - 1   12 30.329   9   43.803 1.9257 0.1439
```

We still need to explain the variation in intercepts; we try a regression on SG, VP and V10.

```
> pet3.lm <- lm(Y ~ SG + VP + V10 + EP, Petrol)
> anova(pet3.lm, pet2.lm)
                  Terms RDf    RSS  Df Sum of Sq F Value      Pr(F)
SG + VP + V10 + EP   27 134.80
       No - 1 + EP   21  74.13   6   60.672 2.8645 0.033681
```

(Notice that SG, VP and V10 are constant within the levels of No so these two models are genuinely nested.) The result suggests that differences between intercepts are *not* adequately explained by such a regression.

A promising way of generalizing the model is to assume that the ten crude oil samples form a random sample from a population where the intercepts after regression on the determining variables depend on the sample. The model we will investigate has the form

$$y_{ij} = \mu + e_i + \beta_1\, \text{SG}_i + \beta_2\, \text{VP}_i + \beta_3\, \text{V10}_i + \beta_4\, \text{EP}_{ij} + \epsilon_{ij}$$

where i denotes the sample and j the observation on that sample, and $e_i \sim N(0, \sigma_1^2)$ and $\epsilon_{ij} \sim N(0, \sigma^2)$, independently. This allows for the possibility that predictions from the fitted equation will need to cope with two sources of error, one associated with the sampling process for the crude oil sample and the other with the measurement process within the sample itself.

The fitting function for linear mixed effects models is lme. Its arguments are numerous but the main ones are the first four that specify the fixed effects model, the random effects, the factor defining the 'clusters' over which the random effects vary, all as formulae, and the usual data frame. (The model assumes that the observational units are collected into 'clusters' such as the ten samples of petrol, that some of the coefficients in a linear model may vary between clusters, and the clusters should be regarded as a random sample. Only one level of clusters is allowed unlike multistratum models, although nested random effects can be simulated, as we will see below.)

For our example we can use

```
pet3.lme <- lme(Y ~ SG + VP + V10 + EP, random = ~ 1,
    cluster = ~ No, data = Petrol)
```

(Fits by lme are very much slower than those by lm.) The output from the summary method for the result is voluminous: a subset is

```
> summary(pet3.lme)
Estimation Method: RML
Convergence at iteration: 6
Restricted Loglikelihood: -76.191
```

```
Restricted AIC: 166.38
Restricted BIC: 176.64
    ....
  Standard Deviation(s) of Random Effect(s)
  (Intercept)
      1.4447
  Cluster Residual Variance: 3.5052

Fixed Effects Estimate(s):
                Value Approx. Std.Error z ratio(C)
(Intercept)  19.70679            0.568273    34.6784
        SG    0.21940            0.146938     1.4931
        VP    0.54586            0.520527     1.0487
       V10   -0.15424            0.039962    -3.8597
        EP    0.15718            0.005588    28.1276
    ....
Random Effects (Conditional Modes):
  (Intercept)
A   -0.059438
    ....
```

The estimate of the residual variance within clusters is $\hat{\sigma}^2 = 3.5052$ and the variance of the random effects between clusters is $\hat{\sigma}_1^2 = (1.4447)^2 = 2.0873$. The 'conditional modes' are BLUPs of the random effects.

The (default) estimation method used is REML. The parameters in the variance structure (such as the variance components) are estimated by maximizing the marginal likelihood of the residuals from a least-squares fit of the linear model, and then the fixed effects are estimated by maximum likelihood assuming that the variance structure is known, which amounts to fitting by generalized least squares. (This is a reasonable procedure since taking the residuals from *any* generalized least-squares fit will lead to the same marginal likelihood: McCullagh & Nelder (1989, pp. 247, 282–3).)

With such a small sample and so few clusters the difference between REML and maximum likelihood estimation is appreciable.

```
> pet3.lme <- update(pet3.lme, est.method = "ML")
> summary(pet3.lme)
Estimation Method: ML
Convergence at iteration: 5
Loglikelihood: -67.692
AIC: 149.38
BIC: 159.64
    ....
  Standard Deviation(s) of Random Effect(s)
  (Intercept)
      0.92889

  Cluster Residual Variance: 3.339
```

```
Fixed Effects Estimate(s):
                Value Approx. Std.Error z ratio(C)
(Intercept)  19.69430            0.4392097    44.8403
         SG   0.22133            0.1128129     1.9619
         VP   0.54940            0.4048627     1.3570
        V10  -0.15269            0.0313863    -4.8649
         EP   0.15634            0.0053946    28.9806
    ....
```

Both tables of fixed effects suggests that only V10 and EP may be useful predictors (although the z ratio columns assume that the asymptotic theory is appropriate, in particular that the estimates of variance components are accurate). We can check this by fitting it as a submodel and performing a likelihood ratio test. However for testing hypotheses on the fixed effects the estimation method *must* be set to ML to ensure the test compares likelihoods based on the same data.

```
> pet4.lme <- update(pet3.lme, fixed = Y ~ V10 + EP)
> anova(pet4.lme, pet3.lme)
    ....
          Model Df    AIC    BIC  Loglik     Test Lik.Ratio
pet4.lme      1  5 149.61 156.94 -69.806
pet3.lme      2  7 149.38 159.64 -67.692 1 vs. 2     4.2285
          P value
pet4.lme
pet3.lme 0.12072
> coef(pet4.lme)
    (Intercept)      V10       EP
A        21.054 -0.21081 0.15759
    ....
```

Note how the coef method combines the slopes (the fixed effects) and the (random effect) intercept for each sample. The coefficients of the fixed effects may be extracted by the function fixed.effects.

The AIC column gives (as in Section 6.6) minus twice the log-likelihood plus twice the number of parameters, and BIC refers to the criterion of Schwarz (1978) with penalty $\log n$ times the number of parameters.

Finally we check if we need both random regression intercepts and slopes on EP, so we fit the model

$$y_{ij} = \mu + e_i + \beta_3 \, V10_i + (\beta_4 + \eta_i) \, EP_{ij} + \epsilon_{ij}$$

where (e_i, η_i) and ϵ_{ij} are independent, but e_i and η_i can be correlated.

```
> pet5.lme <- update(pet4.lme, random = ~ 1 + EP)
> anova(pet4.lme, pet5.lme)
    ....
          Model Df    AIC    BIC  Loglik     Test Lik.Ratio
pet4.lme      1  5 149.61 156.94 -69.806
pet5.lme      2  7 153.61 163.87 -69.805 1 vs. 2 0.0025239
          P value
pet4.lme
pet5.lme 0.99874
```

The test is non-significant, leading us to retain the simpler model. Notice that the degrees of freedom for testing this hypothesis is $7 - 5 = 2$. The additional two parameters are the variance of the random slopes and the covariance between the random slopes and random intercepts.

Prediction and fitted values

A little thought must be given to fitted and predicted values for a mixed effects model. The result of applying `fitted` to a `lme` object is a matrix of *two* columns labelled `population` and `cluster`. The 'population' fitted values are the prediction from the fixed effects, obtained by setting all the random terms to zero. The 'cluster' fitted values set the random terms associated with the cluster (in our examples e_i and η_i) to the BLUP values, whilst setting the residual random term to zero. Thus for our final model `pet4.lme` the fitted values are

$$\hat{\mu} + \hat{\beta}_3 \, \text{V10}_i + \hat{\beta}_4 \, \text{EP}_{ij} \qquad \text{and} \qquad \hat{\mu} + \hat{e}_i + \hat{\beta}_3 \, \text{V10}_i + \hat{\beta}_4 \, \text{EP}_{ij}$$

The `predict` method `predict.lme` is rather non-standard. It requires the argument `data` (not `newdata`) for the new data, and has an `na.action` argument (which can cause difficulty in matching which cases have been predicted). Most importantly, it has a `cluster` argument like that of `lme`, which if present assigns the new data to the clusters used in the fit. If a `cluster` argument is present the prediction has a component `fit` with columns `cluster` and `population` and a component `cluster` naming the cluster used. Otherwise just the `population` prediction is given.

Relation to multistratum models

The multistratum models of Section 10.2 are also linear mixed effects models, and showing how they might be fitted with `lme` allows us to demonstrate some further features of that function.

The final model we used for the `oats` dataset corresponds to the linear model

$$y_{bnv} = \mu + \alpha_n + \beta_v + \eta_b + \zeta_{bv} + \epsilon_{bnv}$$

with three random terms corresponding to plots, subplots and observations. We can code this in `lme` by

```
> oats$sp <- model.matrix(~ V - 1, oats)
> options(contrasts = c("contr.treatment", "contr.poly"))
> oats.lme <- lme(Y ~ N + V, random = ~ sp, cluster = ~ B,
      data = oats, re.block = list(1, 2:4),
      re.structure = c("unrestricted", "identity"),
      control = lme.control(tol=1e-10, ms.tol=1e-10))
> summary(oats.lme)
    ....
Variance/Covariance Components Estimate(s):
 Block: 1
```

```
Structure: unrestricted
Standard Deviation(s) of Random Effect(s)
(Intercept)
   14.645

Block: 2
Structure: identity
Standard Deviation(s) of Random Effect(s)
spGolden.rain spMarvellous spVictory
      10.473         10.473     10.473

Cluster Residual Variance: 162.56

Fixed Effects Estimate(s):
              Value Approx. Std.Error z ratio(C)
(Intercept) 79.9167           8.2204    9.72176
   NO.2cwt  19.5000           4.2500    4.58829
   NO.4cwt  34.8333           4.2500    8.19617
   NO.6cwt  44.0000           4.2500   10.35306
VMarvellous  5.2917           7.0789    0.74753
   VVictory -6.8750           7.0789   -0.97119
```

We first set up an indicator matrix for the subplot. The `random` formula expands to 1 + sp so gives rise to four columns for η_b, ζ_{b1}, ζ_{b2} and ζ_{b3} and hence there are four variances for the random effects. We want the last three to be the same but to be estimated independently of the first. To do so we first separate the random components into two blocks (which also forces independence between the blocks) and then ask that in the second block the covariance matrix is proportional to the identity, so all the ζ_{bv} are independent and have the same variance.

The variance components are estimated as $14.6450^2 = 214.48$, $10.4735^2 = 109.69$ and 162.56. (The convergence criteria were tightened to get this close an agreement with the result of `varcomp`.) The standard errors for treatment differences also agree.

Covariance structures for random effects and residuals

It is time to specify precisely what models `lme` considers. Let i index the clusters and j the measurements on each cluster. Then the linear mixed-effects model is

$$y_{ij} = \mu_{ij} + z_{ij}\eta_i + \epsilon_{ij}, \qquad \mu_{ij} = x_{ij}\beta \tag{10.1}$$

Here x and z are row vectors of explanatory variables associated with each measurement. The ϵ_{ij} are independent for each cluster but possibly correlated within clusters, so

$$\text{var}(\epsilon_{ij}) = \sigma^2 g(\mu_{ij}, z_{ij}, \theta), \qquad \text{corr}(\epsilon_{ij}) = \Gamma(\alpha) \tag{10.2}$$

Thus the variances are allowed to depend on the means and other covariates. The random effects η_i are independent of the ϵ_{ij} with a variance matrix

$$\text{var}(\eta_i) = D(\alpha_\eta) \tag{10.3}$$

Almost all examples will be much simpler than this, and by default $g \equiv 1$, $\Gamma = I$ and D is unrestricted; in the previous subsection we imposed restrictions on D.

Full details of how to specify the components of a linear mixed-effects model can be obtained by `help("lme.formula")`. It is possible for users to supply their own functions to compute many of these components. The free parameters $(\sigma^2, \theta, \alpha, \alpha_\eta)$ in the specification of the error distribution are estimated, by REML by default.

We start by considering the specification of $D(\alpha_\eta)$. The effect of `re.block` is to split the random effect variables into independent blocks given as a list of vectors, each vector giving the numbers or names of the random effect variables. (The variables are numbered in the order they will occur when `random` is expanded, normally starting with the implied intercept.) This splits D into a (possibly permuted) block-diagonal form. The parameter `re.structure` is a vector giving for each block the type of covariance structure, as one of

`"unstructured"`	general (the default)
`"diagonal"`	independent components
`"identity"`	iid components
`"compsymm"`	equal diagonal and equal off-diagonal elements of the variance matrix; $\kappa[(1 - \rho)I + \rho\mathbf{11}^T]$
`"ar1"`	the covariance matrix of an AR(1) process, with $c_{ij} = \kappa\rho^{\vert i-j \vert}$

The last option is only appropriate if the random effects have been specified in a time (or linear spatial) ordering.

The `compsymm` structure is also known as a *uniform correlation* model (Diggle *et al.*, 1994, p. 56) and occurs when there is a common underlying random effect plus independent individual effects. This suggests working with random effects $\eta_b + \zeta_{bv}$ as an alternative way to fit the `oats` model. We used

```
> options(contrasts=c("contr.helmert", "contr.poly"))
> oats$Nc <- C(oats$N, contr.treatment)
> oats$Vc <- C(oats$V, contr.treatment)
> oats1.lme <- lme(Y ~ Nc + Vc, random = ~ V - 1,
     cluster = ~ B, data = oats, re.structure = "compsymm",
     control = lme.control(tol=1e-10, ms.tol=1e-10))
> summary(oats1.lme)
    ....
  VGolden.rain VMarvellous VVictory
       18.005      18.005   18.005
  Correlation of Random Effects
            VGolden.rain VMarvellous
VMarvellous 0.66162
   VVictory 0.66162                  0.66162
```

(The contrasts have to be set up in this way to work round a bug in `lme`.) Note that $\mathrm{var}(\eta_b + \zeta_{bv}) = 214.48 + 109.69 = 18.005^2$ and $\mathrm{corr}(\eta_b + \zeta_{bu}, \eta_b + \zeta_{bv}) = 214.48/(214.48 + 109.69) = 0.6616$.

The variance function $g(\mu, z, \theta)$ can be specified as an parametrized function of the covariates via lhe arguments `var.function`, `var.covariate` and `var.estimate`.

For repeated measures studies the cluster is usually one individual and the measurements on each cluster form a time series, so we would also like to specify a covariance structure for the 'residual' terms, that is to have $\Gamma \neq I$. The times of the measurements can be specified by the argument `serial.covariate` possibly transformed by `serial.covariate.transformation`, but by default the measurements are taken to be equally spaced in the order in which they occur in the data frame. The argument `serial.structure` defines the time-series model as one of

`"identity"`	iid components (the default)
`"compsymm"`	equal diagonal and equal off-diagonal elements of the variance matrix
`"ar1"`	the covariance matrix of an $AR(1)$ process
`"ar1.continuous"`	the covariance matrix of an $AR(1)$ process with $0 \leqslant \rho < 1$ (also known as an exponential correlation model)
`"ar2"`	the covariance matrix of an $AR(2)$ process
`"ma1"`	the covariance matrix of an $MA(1)$ process
`"ma2"`	the covariance matrix of an $MA(2)$ process
`"arma11"`	the covariance matrix of an $ARMA(1, 1)$ process

With all the time-series specifications except `ar1.continuous` the time measurement must be integer-valued.

Diggle, Liang & Zeger (1994) given an example of repeated measurements on the log-size[1] of 79 Sitka spruce trees, 54 of which were grown in ozone-enriched chambers and 25 of which were controls. The size was measured five times in 1988, at roughly monthly intervals. (The time is given in days since 1 January 1988. There were a further eight measurements in 1989 given in data frame `Sitka89`.) We consider a random effect of the tree and serial correlation of the residual terms for each tree.

We first consider a general curve on the five times, for each of the two groups, then a linear difference between the two groups. Taking `ordered(Time)` parametrizes the curve by polynomial components.

```
> options(contrasts = c("contr.treatment", "contr.poly"))
> sitka.lme <- lme(size ~ treat*ordered(Time), random = ~1,
     cluster = ~tree,   data = Sitka,
     serial.structure = "ar1.continuous",
     serial.covariate = ~ Time)

> summary(sitka.lme)
    ....
Variance/Covariance Components Estimate(s):
```

---

[1] by convention this is log height plus twice log diameter.

```
Structure: unstructured
Parametrization: matrixlog
Standard Deviation(s) of Random Effect(s)
(Intercept)
  0.0065704

Cluster Residual Variance: 0.41815

Serial Correlation Structure: ar1.continuous
Serial Correlation Parameter(s): 0.97274

Fixed Effects Estimate(s):
                        Value Approx. Std.Error z ratio(C)
          (Intercept)  4.985120           0.126543   39.39459
                treat -0.211157           0.153058   -1.37959
      ordered(Time).L  1.197112           0.047749   25.07104
      ordered(Time).Q -0.134058           0.025682   -5.22000
      ordered(Time).C -0.040857           0.019227   -2.12499
    ordered(Time) ^ 4 -0.027299           0.016251   -1.67986
 treatordered(Time).L -0.178566           0.057754   -3.09185
 treatordered(Time).Q -0.026447           0.031063   -0.85141
 treatordered(Time).C -0.014249           0.023255   -0.61272
treatordered(Time) ^ 4  0.012403          0.019656    0.63101
```

```
> attach(Sitka)
> Sitka$treatslope <- Time * (treat=="ozone")
> detach()
> sitka.lme2 <- update(sitka.lme,
      fixed = size ~ ordered(Time) + treat + treatslope)
> summary(sitka.lme2)$fixed.table
                        Value Approx. Std.Error z ratio(C)
       (Intercept)  4.9895188          0.12614292   39.5545
   ordered(Time).L  1.2021921          0.04673324   25.7246
   ordered(Time).Q -0.1452428          0.01455217   -9.9808
   ordered(Time).C -0.0505964          0.01078304   -4.6922
 ordered(Time) ^ 4 -0.0166451          0.00913756   -1.8216
             treat  0.2316414          0.20483261    1.1309
        treatslope -0.0022195          0.00066857   -3.3198
```

```
# fitted curve for ozone-enriched
> fitted(sitka.lme2)[1:5, 1]
[1] 4.0598 4.4696 4.8407 5.1763 5.3132
# and fitted curve for controls
> fitted(sitka.lme2)[301:305, 1]
[1] 4.1656 4.6241 5.0552 5.4485 5.6542
```

This shows a significant difference between the two groups, which varies linearly with time. (This can be confirmed by maximum-likelihood-based tests.) The correlation parameter seems high, but estimates the correlation at observations one *day* apart; at the average spacing between observations of 26.5 days the estimated correlation is $0.97274^{26.5} \approx 0.48$.

10.4   Non-linear mixed effects models

Non-linear mixed effects models are fitted by the function `nlme`. Most of its many
arguments are specified in the same way as for `lme`, and the class of models that
can be fitted is based on those described by equations (10.1 to 10.3) on page 310.
The difference is that

$$y_{ij} = \mu_{ij} + \epsilon_{ij}, \qquad \mu_{ij} = f(\boldsymbol{x}_{ij}, \boldsymbol{\beta}, \boldsymbol{\eta}_i) \tag{10.4}$$

so the conditional mean is a non-linear function specified by giving both fixed and
random parameters. Which parameters are fixed and which are random is specified
by the arguments `fixed` and `random`, each of which is a list of formulae with
left-hand side the parameter name. Parameters can be in both lists, and random
effects will have mean zero unless they are. For example, we can specify a simple
exponential growth curve for each tree in the `Sitka` data with random intercept
and asymptote by

```
options(contrasts=c("contr.treatment", "contr.poly"))
sitka.nlme <- nlme(size ~ A + B*(1 - exp(-(Time-100)/C)),
    fixed = list(A ~ treat, B ~ treat, C ~ .),
    random = list(A ~ ., B ~ .),
    cluster = ~ tree,   data = Sitka,
    start   = list(fixed = c(2, 0, 4, 0, 100)),
    serial.structure = "ar1.continuous",
    serial.covariate = ~ Time,
    verbose = T)
```

Here the 'shape' of the curve is taken as common to all trees. It is necessary to
specify starting values for all the fixed parameters (in the order they occur) and
starting values can be specified for other parameters.

   It is not usually possible to find the exact likelihood of such a model, as the
non-linearity prevents integration over the random effects. Various linearization
schemes have been proposed; `nlme` uses the strategy of Lindstrom & Bates (1990),
linearization about the conditional modes of the random effects $\boldsymbol{\eta}_i$. As `nlme`
calls `lme` iteratively and `lme` fits can be slow, `nlme` fits can be very slow (and
also memory-intensive). Adding the argument `verbose = T` helps to monitor
progress. Finding good starting values can help a great deal. Our starting values
were chosen by inspection of the growth curves.

```
> summary(sitka.nlme)
Estimation Method: ML
Convergence at iteration: 4
Approximate Loglikelihood: 63.927
AIC: -107.85
   ....
 Serial Correlation Parameter(s): 0.48774

Fixed Effects Estimate(s):
A.(Intercept)    2.28945          0.20723    11.04811
```

```
          A.treat   0.17512        0.21381    0.81907
    B.(Intercept)   3.89149        0.18037   21.57523
          B.treat  -0.56702        0.21363   -2.65419
                C  79.44503        5.25692   15.11248
```

Note that unlike `lme`, the default estimation method is maximum likelihood, not REML. We can test for the difference in the treatment groups by

```
> summary(update(sitka.nlme,
      fixed = list(A ~ ., B ~ ., C ~ .),
      start = list(fixed=c(2.3, 3.9, 79))))
  ....
Approximate Loglikelihood: 59.433
AIC: -102.86
  ....
```

Blood pressure in rabbits

As a more extensive example, consider the data in Table 10.3 described in Ludbrook (1994)[2]. To quote from the paper:

> Five rabbits were studied on two occasions, after treatment with saline (control) and after treatment with the 5-HT$_3$ antagonist MDL 72222. After each treatment ascending doses of phenylbiguanide (PBG) were injected intravenously at 10 minute intervals and the responses of mean blood pressure measured. The goal was to test whether the cardiogenic chemoreflex elicited by PBG depends on the activation of 5-HT$_3$ receptors.

The response is the *change* in blood pressure relative to the start of the experiment. The data set is a data frame `Rabbit` in our library.

Treatment	Rabbit	6.25	12.5	25	50	100	200
Placebo	1	0.50	4.50	10.00	26.00	37.00	32.00
	2	1.00	1.25	4.00	12.00	27.00	29.00
	3	0.75	3.00	3.00	14.00	22.00	24.00
	4	1.25	1.50	6.00	19.00	33.00	33.00
	5	1.50	1.50	5.00	16.00	20.00	18.00
MDL 72222	1	1.25	0.75	4.00	9.00	25.00	37.00
	2	1.40	1.70	1.00	2.00	15.00	28.00
	3	0.75	2.30	3.00	5.00	26.00	25.00
	4	2.60	1.20	2.00	3.00	11.00	22.00
	5	2.40	2.50	1.50	2.00	9.00	19.00

Dose of Phenylbiguanide (μg) spans the last six columns.

Table 10.3: Data from a blood pressure experiment with five rabbits.

There are logically three strata of variation

[2] We are grateful to Professor Ludbrook for supplying us with the numerical data.

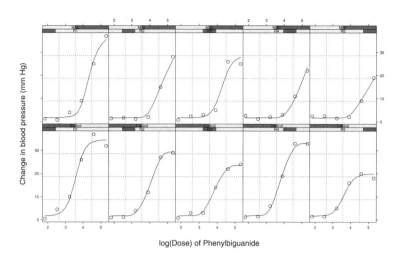

log(Dose) of Phenylbiguanide

Figure 10.3: Data from a cardiovascular experiment using 5 rabbits on two occasions: on control and with treatment. On each occasion the animals are given increasing doses of Phenylbiguanide at about 10 minutes apart. The response is the change in blood pressure. The fitted curve is derived from a model on page 321.

1. between animals,
2. within animals between occasions, and
3. within animals within occasions.

To simplify matters without severely distorting the inference we will take a cluster to be a set of measurements made on one animal on one occasion, leading to a factor `Run`, with ten levels. We will then take `Animal` and `Treatment` to be factors leading to fixed effects in the model. The consequences of this slight simplification are not important for the questions of interest, namely the effect, if any, of the treatment on the animal response profiles.

A plot of the data shown in Figure 10.3 shows the blood pressure rising with the dose of PBG on each occasion generally according to a sigmoid curve. Evidently the main effect of the treatment is to suppress the change in blood pressure at lower doses of PBG, hence translating the response curve a distance to the right. There is a great deal of variation, though, so there may well be other effects as well.

Ludbrook suggests a four-parameter logistic response function in log(Dose) (as is commonly used in this context) and we adopt his suggestion. This function has the form

$$f(\alpha, \beta, \lambda, \theta, x) = \alpha + \frac{\beta - \alpha}{1 + \exp[(x - \lambda)/\theta]}$$

Notice that $f(\alpha, \beta, \lambda, \theta, x)$ and $f(\beta, \alpha, \lambda, -\theta, x)$ are identically equal in x; to resolve this lack of identification we will require that θ be positive. For our example this makes α the right asymptote, β the left asymptote (or 'baseline'), λ

the $\log(\text{Dose})$ (LD50) leading to a response exactly halfway between asymptotes and θ is an abscissa scale parameter determining the rapidity of ascent.

A preliminary analysis using `nlsList`

As a first step we might fit separate non-linear regressions in each of the 10 clusters and compare the parameter estimates. The `nlme` function `nlsList` does precisely this, and works best if the model function is a self-starting `nls` model. The `nlme` software contains a self-starting model function[3] for the four-parameter logistic response to the logarithm of a dose with name and arguments as in `fpl(Dose, A, B, ld50, scal)`. It requires a small modification to ensure that the value of of the scale parameter is always positive. The extra statement is

```
if(val[[4]] < 0) val <- c(val[c(2, 1, 3)], list( -val[[4]]))
```

and comes third to last in the initialization routine.

The function `nlsList` requires a model, a definition of clusters and a data frame for the variables. If the model is self-starting these are sufficient.[4] (Prior to S-PLUS 3.4, library `nlme` needs to be attached with `first=T`.)

```
> R.nlsList <- nlsList(BPchange ~ fpl(Dose, A, B, ld50, scal),
      cluster = ~ Run, data = Rabbit)
Error in nls(resp ~ cbind(1, 1/(1 + exp((1..: singular gradient
      matrix
Dumped
Error in nls(resp ~ cbind(1, 1/(1 + exp((1..: singular gradient
      matrix
Dumped
> M1 <- coef(R.nlsList)
> M1
        A       B    ld50     scal
C1 34.789 1.8091 3.5611 0.30926
C2 29.683 1.4841 4.0382 0.27792
C3 23.759 1.5994 3.8581 0.26935
C4 34.198 1.4077 3.8426 0.30502
C5 19.024 1.4144 3.5374 0.22894
M1 41.817 1.1296 4.4688 0.41050
M2 28.612 1.3676 4.6049 0.18381
M3     NA     NA     NA      NA
M4 24.149 1.9063 4.7033 0.26620
M5     NA     NA     NA      NA
```

Note that in two of the clusters the model fitting process fails.

The coefficients for the separate fitted models show a large variation in the right asymptote and small variations in baseline and scale. The `ld50` parameter varies across clusters but appears to be consistently higher for the MDL group than for the controls.

---

[3] in library `nlmedata` on Unix and library `nlme` under Windows in the `statlib` versions.

[4] For a later use with `nlme` it is essential that the parameter names given to `fpl` are exactly those of its definition.

The function `fixed.effects` averages coefficients over clusters (for which parameter estimates are available) to give a crude overall model estimate:

```
> fixed.effects(R.nlsList)
      A       B     ld50     scal
 29.504 1.5148 4.0768 0.28138
```

It is interesting to compare these results with a fixed effects non-linear regression model with separate A and ld50 parameters across runs but common B and scal:

```
> R.nls <- nls(BPchange ~ A[Run] + (B - A[Run])/
      (1 + exp((log(Dose) - ld50[Run])/scal)), data = Rabbit,
      start = list(A=rep(29.5, 10), B=1.5, ld50=rep(4.1, 10),
                   scal=0.28))
> b <- as.vector(coef(R.nls))
> M2 <- cbind(b[1:10], b[11], b[12:21], b[22])
> dimnames(M2) <- dimnames(M1)
> M2
         A      B     ld50     scal
C1 34.351 1.6515 3.5481 0.27383
C2 29.646 1.6515 4.0417 0.27383
C3 23.804 1.6515 3.8613 0.27383
C4 33.876 1.6515 3.8468 0.27383
C5 19.335 1.6515 3.5630 0.27383
M1 37.592 1.6515 4.3883 0.27383
M2 30.682 1.6515 4.6632 0.27383
M3 27.672 1.6515 4.2249 0.27383
M4 24.276 1.6515 4.6994 0.27383
M5 21.402 1.6515 4.7547 0.27383
```

(Notice how parameters in a `nls` fit may be given an indexing factor, but the start vector must then be given as a named list rather than a single named vector.)

The A and ld50 parameter estimates are much the same as for the separate regressions, so constraining B and scal to be fixed appears not to make very much difference.

Non-linear mixed effects models

The `nlme` function has a method to specify the model via a `nlsList` object in place of model specification and starting values. In our example this gives

```
> nlme(R.nlsList, verbose=T)
    ....
Variance/Covariance Components Estimate(s):
  Standard Deviation(s) of Random Effect(s)
        A       B     ld50     scal
 6.5809 0.22784 0.39839 0.094246
    ....
Cluster Residual Variance: 1.3479
```

```
Fixed Effects Estimate(s):
      A      B   1d50    scal
27.532 1.7728 4.1332 0.20991
```

(This fit is slow: 40 seconds on our PC.) This gives all the parameters both fixed and random effects, and allows the variance matrix of the random effects to be unrestricted. It does not take account of the matching between animals and the treatment groups. Rather than update this fit, we will use the standard call specifying the model function, and also supplying first derivative information. We start by considering the two treatment groups separately.

```
Fpl <- deriv(~ A + (B-A)/(1 + exp((log(d) - 1d50)/th)),
    c("A","B","1d50","th"), function(d, A, B, 1d50, th) {})
c1 <- fixed.effects(R.nlsList)
Rc.nlme <- nlme(BPchange ~ Fpl(Dose, A, B, 1d50, th),
    fixed = list(A ~ ., B ~ ., 1d50 ~ ., th ~ .),
    random = list(A ~ ., 1d50 ~ .),
    cluster = ~ Animal, data = Rabbit,
    subset = Rabbit$Treatment=="Control",
    start = list(fixed=c1), verbose = T)
Rm.nlme <- update(Rc.nlme, subset = Rabbit$Treatment=="MDL")
```

Unlike other statistical model functions, in the current software the `subset` is not evaluated with reference to the `data` argument. Unlike `nls`, the parameter names given in a `nlme` call to a function such as `Fpl` must exactly match those in the function definition. We defer consideration of serial correlation to an exercise.

We may now look at the results of the separate analyses.

```
> Rc.nlme
Variance/Covariance Components Estimate(s):
  Standard Deviation(s) of Random Effect(s)
       A     1d50
  5.7708 0.17964
     ....
Fixed Effects Estimate(s):
      A      B   1d50      th
28.332 1.5133 3.7743 0.28956

> Rm.nlme
Variance/Covariance Components Estimate(s):
  Standard Deviation(s) of Random Effect(s)
       A     1d50
  5.3655 0.19004
     ....
Fixed Effects Estimate(s):
      A      B   1d50      th
27.521 1.7839 4.5257 0.24237
```

This suggests a combined model in which the distribution of the random effects do not depend on the treatment. We use Run as the cluster which corresponds to independent random effects for the two treatments on the same animal, an assumption we will relax later. Initially we allow all the parameter means to differ by group.

```
> options(contrasts=c("contr.treatment", "contr.poly"))
> c1 <- c(fixed.effects(R.nlsList), 0)
> R.nlme1 <- nlme(BPchange ~ Fpl(Dose, A, B, ld50, th),
      fixed = list(A ~ Treatment, B ~ Treatment,
                   ld50 ~ Treatment, th ~ Treatment),
      random = list(A ~ ., ld50 ~ .),
      cluster = ~ Run, data = Rabbit,
      start = list(fixed=c1[c(1,5,2,5,3,5,4,5)]),
      verbose = T)
> summary(R.nlme1)
Fixed Effects Estimate(s):
```

	Value	Approx. Std.Error	z ratio(C)
A.(Intercept)	28.329502	2.583745	10.96451
A.Treatment	-0.804005	3.729161	-0.21560
B.(Intercept)	1.525682	0.479335	3.18291
B.Treatment	0.271823	0.600600	0.45259
ld50.(Intercept)	3.778247	0.088848	42.52469
ld50.Treatment	0.746732	0.125979	5.92742
th.(Intercept)	0.289326	0.030206	9.57843
th.Treatment	-0.048937	0.042637	-1.14776

This suggests that only `ld50` depends on the treatment, confirmed by

```
> R.nlme2 <- update(R.nlme1,
      fixed = list(A ~ ., B ~ ., ld50 ~ Treatment, th ~ .),
      start = list(fixed=c1[c(1:3,5,4)]))
> anova(R.nlme2, R.nlme1)
        Model Df    AIC     BIC  Loglik     Test Lik.Ratio
R.nlme2     1  9 284.01 302.86 -133.01
R.nlme1     2 12 289.23 314.36 -132.62 1 vs. 2   0.78355
> summary(R.nlme2)$fixed
```

	Value	Approx. Std.Error	z ratio(C)
A	28.12091	1.868489	15.0501
B	1.67604	0.293042	5.7194
ld50.(Intercept)	3.77960	0.088530	42.6929
ld50.Treatment	0.75711	0.125649	6.0256
th	0.26909	0.021717	12.3909

We next test if we should allow the distribution of the random effects to depend on the treatment, by switching to `Animal` as the cluster and allowing an unrestricted covariance structure of the random effects for both treatments.

```
> Rabbit$tr <- model.matrix(~ Treatment - 1, Rabbit)
> R.nlme3 <- update(R.nlme2, cluster = ~ Animal,
               random = list(A ~ tr-1, ld50 ~ tr-1))
> anova(R.nlme2, R.nlme3)
```

```
           Model Df    AIC    BIC  Loglik    Test Lik.Ratio
R.nlme2       1  9 284.01 302.86 -133.01
R.nlme3       2 16 292.54 326.05 -130.27 1 vs. 2    5.4709
> summary(R.nlme3)
Variance/Covariance Components Estimate(s):

  Standard Deviation(s) of Random Effect(s)
  A.trControl A.trMDL 1d50.trControl 1d50.trMDL
     5.6969  5.3571          0.17363      0.20317
  Correlation of Random Effects
                  A.trControl    A.trMDL 1d50.trControl
       A.trMDL   0.759589
  1d50.trControl -0.093337    -0.119991
     1d50.trMDL  0.064579     -0.539194  0.022326

Fixed Effects Estimate(s):
                      Value Approx. Std.Error z ratio(C)
      A.(Intercept) 28.24089           2.394750   11.7928
                  B  1.65561           0.296067    5.5920
   1d50.(Intercept)  3.77880           0.084365   44.7910
      1d50.Treatment  0.76774          0.118025    6.5049
                 th  0.27394           0.021589   12.6891
```

This makes very little difference. We conclude that the treatment produces a systematic shift in the 1d50 parameter to the right, but may not affect the asymptote. The baseline and scale parameters may be considered fixed. (Refinement of the covariance model is left as an exercise, as it is peripheral to the primary interest in studying the effects of Treatment.)

Finally we may display the results and a spline approximation to the (BLUP) fitted curve.

```
R2 <- fitted(R.nlme2)$cluster
xyplot(BPchange ~ log(Dose) | Animal * Treatment, Rabbit,
    xlab = "log(Dose) of Phenylbiguanide",
    ylab = "Change in blood pressure (mm Hg)",
    subscripts = T, aspect = "xy", panel =
      function(x, y, subscripts) {
        panel.grid()
        panel.xyplot(x, y)
        sp <- spline(x, R2[subscripts])
        panel.xyplot(sp$x, sp$y, type="l")
      })
```

The result, shown in Figure 10.3 on page 316, seems to fit the data very well.

10.5 Exercises

10.1. Add the fitted lines for the final model for the petrol data to Figure 10.2.

10.2. Find a way to plot the `Sitka` data which facilitates comparison of the growth curves for the two treatment groups.

Add to your plot the fitted mean growth curve and some 95% confidence intervals.

10.3. Consider how to explore the assumptions made for the `lme` model for the `Sitka` data. Are the `plot` methods for `lme` objects helpful in this? In particular, how would you check the adequacy of the $AR(1)$ model for ϵ_{ij}? How much is this likely to matter?

10.4. The object `Sitka89` contains the 1989 data on the same 79 Sitka trees measured on 8 days in 1989. Analyse the 1989 data separately, and then in conjunction with the 1988 data.

10.5. Investigate restricting the covariance matrix of the random effects in model `R.nlme3`.

10.6. For the rabbit data, the PGB was administered in increasing doses at 10 minute intervals. Consider how to allow for serial correlation between measurements in an `nlme` fit and investigate if this might have any effect on the estimates and conclusions. (It may help to start by considering each treatment separately.)

10.7. Consider models that use a multiplicative effect of treatment on the asymptote `A` in the `Rabbit` example, and explore the assumption of a constant residual variance.

10.8. The default estimation method for `lme` is REML but the default method for `nlme` is ML. Why do you suppose this is the case? Investigate the difference changing from ML to REML makes in both the small but complex `Rabbit` data set and the larger but simpler `Sitka` set.

Note that the estimation method can be changed using `update`:

```
fmRML <- update(fmML, est.method="RML",
          start=list(fixed=fixed.effects(fmML),
                     random=random.effects(fmML)))
```

but even starting from this initial value set the calculations can still be quite lengthy. For serially correlated structures such as `sitka.nlme` you may also want to set an initial value for `alpha`.

Chapter 11

Modern Regression

S-PLUS has a 'Modern Regression Module' which contains functions for a number of regression methods. These are not necessarily non-linear in the sense of Chapter 9, which refers to a non-linear parametrization, but they do allow non-linear functions of the independent variables to be chosen by the procedures. The methods are all fairly computer-intensive, and so are only feasible in the era of plentiful computing power (and hence are 'modern'). Some of these methods are part of the S modelling language, and others have been added by S-PLUS. As the latter predate the modelling language and have not been updated, the functions of this chapter do not have a consistent style and user interface.

There are few texts covering this material. Although it does not cover all our topics in equal detail, for what it does cover Hastie & Tibshirani (1990) is an excellent reference. Chambers & Hastie (1992) cover gam and loess in considerably greater detail than this chapter. Thisted (1988) gives a brief computationally-oriented introduction to many of the methods of this chapter.

11.1   Additive models and scatterplot smoothers

For linear regression we have a dependent variable Y and a set of predictor variables $X_1, \ldots, X_p$, and model

$$Y = \alpha + \sum_{j=1}^{p} \beta_j X_j + \epsilon$$

Additive models replace the linear function $\beta_j X_j$ by a non-linear function, to get

$$Y = \alpha + \sum_{j=1}^{p} f_j(X_j) + \epsilon \tag{11.1}$$

Since the functions f_j are rather general, they can subsume the α. Of course, it will not be useful to allow an arbitrary function f_j, and it will help to think of it as a *smooth* function.

The classical way to introduce non-linear functions of dependent variables is to add a limited range of transformed variables to the model, for example

323

quadratic and cubic terms, or to split the range of the variable and use a piecewise constant function. (In S these can be achieved by the functions poly and cut.) A 'modern' alternative is to use *spline* functions. (Green & Silverman, 1994, provide a gentle introduction to splines.) We will only need cubic splines. Divide the real line by an ordered set of points $\{z_i\}$ known as *knots*. On the interval $[z_i, z_{i+1}]$ the spline is a cubic polynomial, and it is continuous and has continuous first and second derivatives, imposing 3 conditions at each knot. With n knots, $n + 4$ parameters are needed to represent the cubic spline (from $4(n + 1)$ for the cubic polynomials minus $3n$ continuity conditions). Of the many possible parametrizations, that of B-splines has desirable properties. The S function bs generates a matrix of B-splines, and so can be included in a linear-regression fit.

A restricted form of B-splines known as *natural splines* and implemented by the S function ns, is linear on $(-\infty, z_1]$ and $[z_n, \infty)$ and thus would have n parameters. However, ns adds an extra knot at each of the maximum and minimum of the data points, and so has $n+2$ parameters, dropping the requirement for the derivative to be continuous at z_1 and z_n. The functions bs and ns may have the knots specified or be allowed to choose the knots as quantiles of the empirical distribution of the variable to be transformed, by specifying the number df of parameters.

Prediction from models including splines (and indeed the orthogonal polynomials generated by poly) needs care, as the basis for the functions depends on the observed values of the independent variable. If predict.lm is used, it will form a new set of basis functions and then erroneously apply the fitted coefficients. The function predict.gam will work more nearly correctly. (It uses both the old and new data to choose the knots.)

Splines are by no means the only way to fit smooth functions in (11.1), and indeed as we shall see, fitting as part of the regression is not the only way to use splines to find a smooth term f_j. To consider these approaches we need to digress to consider the simplest case of just one independent variable, so the relationship between x and y can be shown on a scatterplot. The single-variable methods are used as building-blocks in fitting multi-variable additive models.

Scatterplot smoothing

These methods depend on being able to estimate 'smooth' functions. They make use of scatterplot smoothers, which, given a scatterplot of (x, y) values, draw a smooth curve against x. It is not necessary to understand the exact details, and there are several alternative methods, including splines, running means and running lines. However, some users will want to understand the smoother being used.

Our running example is a set of data on 133 observations of acceleration against time for a simulated motorcycle accident, taken from Silverman (1985). A series of smoothers is shown in Figure 11.1, obtained by the following code.

```
attach(mcycle)
par(mfrow = c(3,2))
```

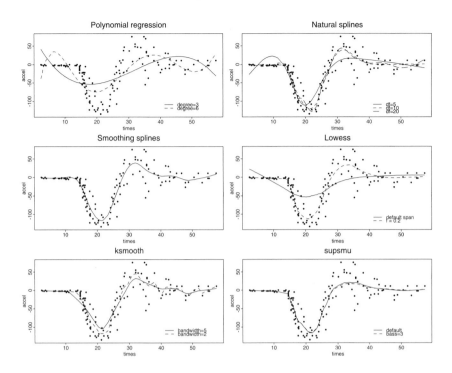

Figure 11.1: Scatterplot smoothers for the simulated motorcycle data given by Silverman (1985).

```
plot(times, accel, main="Polynomial regression")
lines(times, fitted(lm(accel ~ poly(times, 3))))
lines(times, fitted(lm(accel ~ poly(times, 6))), lty=3)
legend(40, -100, c("degree=3", "degree=6"), lty=c(1,3),bty="n")
plot(times, accel, main="Natural splines")
lines(times, fitted(lm(accel ~ ns(times, df=5))))
lines(times, fitted(lm(accel ~ ns(times, df=10))), lty=3)
lines(times, fitted(lm(accel ~ ns(times, df=20))), lty=4)



legend(40, -100, c("df=5", "df=10", "df=20"), lty=c(1,3,4),
    bty="n")
plot(times, accel, main="Smoothing splines")
lines(smooth.spline(times, accel))
plot(times, accel, main="Lowess")
lines(lowess(times, accel))
lines(lowess(times, accel, 0.2), lty=3)
legend(40, -100, c("default span", "f = 0.2"), lty=c(1,3),
    bty="n")
plot(times, accel, main ="ksmooth")
lines(ksmooth(times, accel,"normal", bandwidth=5))
lines(ksmooth(times, accel,"normal", bandwidth=2), lty=3)
```

```
legend(40, -100, c("bandwidth=5", "bandwidth=2"), lty=c(1,3),
    bty="n")
plot(times, accel, main ="supsmu")

lines(supsmu(times, accel))
lines(supsmu(times, accel, bass=3), lty=3)
legend(40, -100, c("default", "bass=3"), lty=c(1,3), bty="n")
```

We have already discussed polynomials and regression splines, and Figure 11.1 shows how much better splines are than polynomials at adapting to general smooth curves. (The df parameter controls the number of terms in the regression spline.)

Suppose we have n pairs (x_i, y_i). A *smoothing spline* minimizes a compromise between the fit and the degree of smoothness of the form

$$\sum w_i [y_i - f(x_i)]^2 + \lambda \int (f''(x))^2 \, dx$$

over all (measurably twice-differentiable) functions f. It is a cubic spline with knots at the x_i, but does not interpolate the data points for $\lambda > 0$ and the degree of fit is controlled by λ. The S function smooth.spline allows λ[1] or the (equivalent) degrees of freedom to be specified, otherwise it will choose the degree of smoothness automatically by cross-validation (see page 182). There are various definitions of equivalent degrees of freedom; see Green & Silverman (1994, pp. 27–8) and Hastie & Tibshirani (1990, Appendix B). That used in smooth.spline is the trace of the smoother matrix S; as fitting a smoothing spline is a linear operation, there is an $n \times n$ matrix S such $\hat{y} = Sy$. (In a regression fit S is the 'hat matrix' (page 205; this has trace equal to the number of free parameters.)

For $\lambda = 0$ the smoothing spline will interpolate the data points if the x_i are distinct. There are simpler methods to fit interpolating cubic splines, implemented in the function spline. This can be useful to draw smooth curves thorough the result of some expensive smoothing algorithm: we could also linear interpolation implemented by approx.

The algorithm used by lowess is quite complex; it uses robust locally linear fits. A window is placed about x; data points that lie inside the window are weighted so that nearby points get the most weight and a robust weighted regression is used to predict the value at x. The parameter f controls the window size and is the proportion of the data which is included. The default, f=2/3, is often too large for scatterplots with appreciable structure. The S function loess is an extension of the ideas of lowess which will work in one, two or more dimensions in a similar way. The function scatter.smooth plots a loess line on a scatter plot, using loess.smooth.

A *kernel smoother* is of the form

$$\hat{y}_i = \sum_{j=1}^{n} y_i K \left(\frac{x_i - x_j}{b} \right) \Big/ \sum_{j=1}^{n} K \left(\frac{x_i - x_j}{b} \right) \tag{11.2}$$

---

[1] more precisely λ with the (x_i) scaled to $[0, 1]$ and the weights scaled to average 1.

where b is a bandwidth parameter, and K a kernel function, as in density estimation. In our example we use the function `ksmooth` and take K to be a standard normal density. The critical parameter is the bandwidth b. Other S code (calling C code), including methods for bandwidth choice, is given by Härdle (1991) and in the library `haerdle` available from `statlib`. The function `ksmooth` seems rather slow, and the faster alternatives in the libraries `haerdle` and `KernSmooth` are recommended for large problems.

The smoother `supsmu` is the one used by the S-PLUS functions for modern regression. It is based on a symmetric k-nearest neighbour linear least squares procedure. (That is, $k/2$ data points on each side of x are used in a linear regression to predict the value at x.) This is run for three values of k, $n/2$, $n/5$ and $n/20$, and cross-validation is used to choose a value of k for each x which is approximated by interpolation between these three. Larger values of the parameter `bass` (up to 10) encourage smooth functions.

Kernel regression can be seen as local fitting of a constant. Theoretical work (Wand & Jones, 1995) has shown advantages in local fitting of polynomials, especially those of odd order. (Of course, the advantages of local fitting of lines have been demonstrated by `lowess` and `supsmu` for many years and have been well-known in the time-series literature.) Library `KernSmooth` contains code (using FORTRAN) for local polynomial fitting and bandwidth-selection.

There are problems in which the main interest is in estimating the first or second derivative of a smoother. One idea is to fit locally a polynomial of high enough order and report its derivative (implemented in library `KernSmooth`). We can differentiate a spline fit, although it may be desirable to fit splines of higher order than cubic: library `pspline` provides a suitable generalization of `smooth.spline`.

Fitting additive models

As we have seen, smooth terms parametrized by regression splines can be fitted by `lm`. For smoothing splines it would be possible to set up a penalized least-squares problem and minimize that, but there would be computational difficulties in choosing the smoothing parameters simultaneously (Wahba, 1990; Wahba *et al.*, 1995). Instead an iterative approach is often used.

The *backfitting* algorithm fits the smooth functions f_j in (11.1) one at a time by taking the residuals

$$Y - \sum_{k \neq j} f_k(X_k)$$

and smoothing them against X_j using one of the scatterplot smoothers of the previous subsection. The process is repeated until it converges. Linear terms in the model (including any linear terms in the smoother) are fitted by least squares.

This procedure is implemented in S by the function `gam`. The model formulae are extended to allow the terms `s(x)` and `lo(x)` which respectively specify a smoothing spline and a loess smoother. These have similar parameters to the scatterplot smoothers; for `s()` the default degrees of freedom is 4, and for `lo()`

the window width is controlled by span with default 0.5. There is a plot method, plot.gam, which shows the smooth function fitted for each term in the additive model.

Our dataset rock contains measurements on four cross-sections of each of 12 oil-bearing rocks; the aim is to predict permeability y (a property of fluid flow) from the other three measurements. As permeabilities vary greatly (6.3–1300), we use a log scale. The measurements are the end product of a complex image-analysis procedure and represent the total area, total perimeter and a measure of 'roundness' of the pores in the rock cross-section.

We first fit a linear model, then a full additive model. In this example convergence of the backfitting algorithm is unusually slow, so the control limits must be raised. The plots are shown in Figure 11.2.

```
rock.lm <- lm(log(perm) ~ area + peri + shape, data=rock)
summary(rock.lm)
rock.gam <- gam(log(perm) ~ s(area) + s(peri) + s(shape),
    control=gam.control(maxit=50, bf.maxit=50), data=rock)
summary(rock.gam)
anova(rock.lm, rock.gam)
par(mfrow=c(2,2))
plot(rock.gam, se=T)
rock.gam1 <- gam(log(perm) ~ area + peri + s(shape), data=rock)
par(mfrow=c(2,2))
plot(rock.gam1, se=T)
anova(rock.lm, rock.gam1, rock.gam)
```

It is worth showing the output from summary.gam and the analysis of variance table from anova(rock.lm, rock.gam):

```
Deviance Residuals:
     Min        1Q  Median       3Q     Max
 -1.6855 -0.46962 0.12531 0.54248 1.2963


(Dispersion Parameter ... 0.7446 )

     Null Deviance: 126.93 on 47 degrees of freedom
Residual Deviance: 26.065 on 35.006 degrees of freedom

Number of Local Scoring Iterations: 1

DF for Terms and F-values for Nonparametric Effects

            Df Npar Df Npar F   Pr(F)
(Intercept)  1
   s(area)   1       3 0.3417 0.79523
   s(peri)   1       3 0.9313 0.43583
  s(shape)   1       3 1.4331 0.24966

Analysis of Variance Table
```

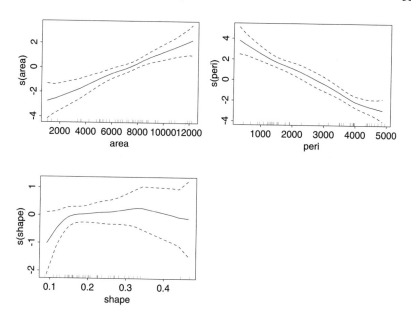

Figure 11.2: The results of fitting an additive model to the dataset `rock` with smooth functions of all three predictors. The dashed lines are approximate 95% pointwise confidence intervals. The tick marks show the locations of the observations on that variable.

```
                          Terms Resid. Df    RSS    Test
1              area + peri + shape    44.000 31.949
2 s(area) + s(peri) + s(shape)       35.006 26.065 1 vs. 2
        Df Sum of Sq F Value    Pr(F)
1
2 8.9943     5.8835   0.8785 0.55311
```

This shows that each smooth term has one linear and three non-linear degrees of freedom. The reduction of RSS from 31.95 (for the linear fit) to 26.07 is not significant with an extra 9 degrees of freedom, but Figure 11.2 shows that only one of the functions appears non-linear, and even that is not totally convincing. Although suggestive, the non-linear term for `peri` is not significant. With just that non-linear term (Figure 11.3) the RSS is 29.00.

We can also fit a smooth term for `shape` by regression splines (see Figure 11.3):

```
rock.ns <- lm(log(perm) ~ area + peri + ns(shape, df=4), rock)
summary(rock.ns)
plot.gam(rock.ns, se=T)
```

The fit is very similar to that using smoothing splines. Note how `plot.gam` can still be used for regression splines fitted by `lm`.

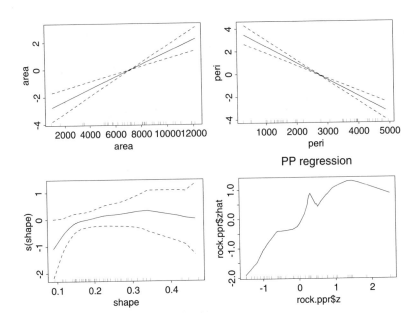

Figure 11.3: Additive model functions fitted to `rock` with linear `area`, linear `peri` and a smooth function of `shape`. The bottom right plot is of the smooth function fitted by a one-term projection pursuit regression.

Generalized additive models

These stand in the same relationship to additive models as generalized linear models do to regression models. Consider, for example, a logistic regression model. There are n observational units, each of which records a random variable Y_i with a Binomial (n_i, p_i) distribution, and (p_i) is determined by

$$\text{logit}\,(p) = \alpha + \sum_{j=1}^{p} \beta_j X_j \tag{11.3}$$

This is readily generalized to the logistic additive model:

$$\text{logit}\,(p) = \alpha + \sum_{j=1}^{p} f_j(X_j) \tag{11.4}$$

and the full form of a GAM is then obvious; replace the linear predictor in a GLM by an additive predictor. (Direct minimization with smoothing splines has been used here too: Wahba *et al.*, 1995.)

Low birth weights

We return to the data on low birth weights in data frame `birthwt` studied in Section 7.2. We will examine the linearity in age and mother's weight, using generalized additive models and including smooth terms.

```
> attach(bwt)
> age1 <- age*(ftv=="1"); age2 <- age*(ftv=="2+")
> birthwt.gam <- gam(low ~ s(age) + s(lwt) + smoke + ptd +
      ht + ui + ftv + s(age1) + s(age2) + smoke:ui, binomial,
      bwt, bf.maxit=25)
> summary(birthwt.gam)

Residual Deviance: 170.35 on 165.18 degrees of freedom

DF for Terms and Chi-squares for Nonparametric Effects

          Df Npar Df Npar Chisq  P(Chi)
 s(age)    1    3.0      3.1089 0.37230
 s(lwt)    1    2.9      2.3392 0.48532
 s(age1)   1    3.0      3.2504 0.34655
 s(age2)   1    3.0      3.1472 0.36829

> table(low, predict(birthwt.gam) > 0)
    FALSE TRUE
0    115   15
1     28   31
> plot(birthwt.gam, ask=T, se=T)
```

Creating the variables age1 and age2 allows us to fit smooth terms for the *difference* in having one or more visits in the first trimester. Both the summary and the plots show no evidence of non-linearity. Note that the convergence of the fitting algorithm is slow in this example, so we increased the control parameter bf.maxit from 10 to 25. The parameter ask=T allows us to choose plots from a menu. Our choice of plots is shown in Figure 11.4.

11.2 Projection-pursuit regression

Now suppose that the explanatory vector $X = (X_1, \ldots, X_p)$ is of high dimension. The additive model (11.1) may be too flexible as it allows a few degrees of freedom per X_j, yet it does not cover the effect of interactions between the independent variables. Projection pursuit regression (Friedman & Stuetzle, 1981) applies an additive model to projected variables. That is, it is of the form:

$$Y = \alpha_0 + \sum_{j=1}^{M} f_j(\alpha_j^T X) + \epsilon \tag{11.5}$$

for vectors α_j, and a dimension M to be chosen by the user. Thus it uses an additive model on predictor variables which are formed by projecting X in M carefully chosen directions. For large enough M such models can approximate

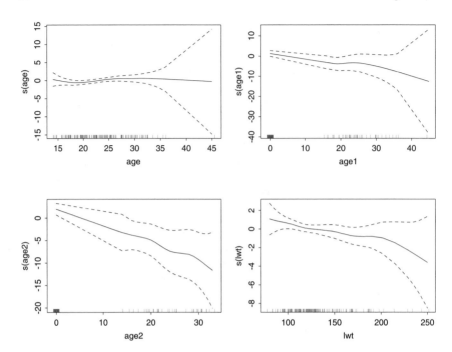

Figure 11.4: Plots of smooth terms in a generalized additive model for the data on low birth weight. The dashed lines indicate plus and minus two pointwise standard deviations.

(uniformly on compact sets and in many other senses) arbitrary continuous functions of X (e.g. Diaconis & Shahshahani, 1984). The terms of (11.5) are called ridge functions, since they are constant in all but one direction.

The S-PLUS function `ppreg` fits (11.5) by least squares, and constrains the vectors α_k to be of unit length. It first fits M_{max} terms sequentially, then prunes back to M by at each stage dropping the least effective term and re-fitting. The function returns the proportion of the variance explained by all the fits from $M, \ldots, M_{max}$. Rather than a model formula, the arguments are an X matrix and a y vector (or matrix, since multiple dependent variables can be included). In many cases a suitable X matrix can be obtained by calling `model.matrix`, but care is needed not to duplicate intercept terms.

For the `rock` example we have:

```
> attach(rock)
> x <- cbind(area, peri, shape)
# or  model.matrix(~ -1 + area + peri + shape)
> rock.ppr <- ppreg(x, log(perm), 1, 5)
> SStot <- var(log(perm))*(length(perm)-1)
> SStot*rock.ppr$esq
[1] 17.4248  11.3637  7.1111  5.9324  3.1139
> rock.ppr$allalpha[1,,1] # first alpha for first fit
[1]  0.00041412 -0.00116210  0.99999928
```

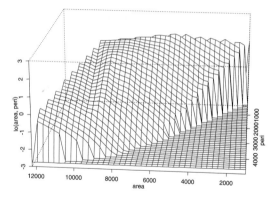

Figure 11.5: Perspective plot of `lo` term fitted to `rock` dataset. Internally `plot.gam`
used `persp` to make this plot.

```
> rock.ppr$allalpha[1,,1]*sqrt(diag(var(x)))
[1]   1.111428 -1.663737  0.083496
> plot(rock.ppr$z, rock.ppr$zhat, type="l")
```

This shows that even a single smooth projection term fits significantly better than
any of the additive models. The combination appears at first sight to be dominated
by shape, but the variables are on very different scales, have standard errors (2680,
1430, 0.084), and the variable appears to be a contrast between `area` and `peri`.
The non-linear function is shown in Figure 11.3. (This fit is very sensitive to
rounding error, and will vary from system to system.)

The fit improves steadily with extra terms, but there are only 48 observations,
and the number of parameters is approximately $1 + \sum_{i=1}^{M}[2 + \mathrm{df}_i - 1]$ where
df_i is the equivalent degrees of freedom for the ith smooth fit. (The projection
can be rescaled, so accounts for 2 not 3 degrees of freedom, and each smooth fit
can fit a constant term.) Unfortunately df_i is not specified, but the plot suggests
$\mathrm{df}_i \approx 6\text{–}8$, so using more than two or three terms would be over-fitting.

It is also possible to use `gam` with `lo` to include smooth functions of two or
more variables. (This again fits robustly a locally weighted linear surface.) We
test out the linear term in `area-peri` space. We deliberately under-smooth (by
choosing a small value of `span`) and obtain a RSS of 10.22 on roughly 24 degrees
of freedom.

```
rock.gam2 <- gam(log(perm) ~ shape+lo(area, peri, span=0.2))
summary(rock.gam2)
anova(rock.lm, rock.gam2)
plot(rock.gam2, eye=c(-10000, 50000, 20), pty="s")
```

The perspective plot for the `lo` term is shown in Figure 11.5. The plot confirms
a broadly linear fitted surface, but with some suspicion of a curvature in the
orthogonal direction, which is confirmed by a PP regression fit on just the first two
variables.

```
> rock.ppr2 <- ppreg(x[, 1:2], log(perm), 1, 3)
> SStot*rock.ppr2$esq
[1] 20.064 15.309 14.145
> rock.ppr2$allalpha[1,,1]*sqrt(diag(var(x[, 1:2])))
[1]    932.37 -1342.49
> rock.ppr2$allalpha[1:2,,2]*sqrt(diag(var(x[, 1:2])))
          [,1]     [,2]
[1,]    960.55 -2506.1
[2,] -1012.56  1012.1
```

Note that dropping the third variable changes drastically the scaling of a unit-length combination α.

We can even include all pairwise combinations of variables. The very different scaling causes problems with the plots, so we re-scale.

```
area1 <- area/10000; peri1 <- peri/10000
rock.gam3 <- gam(log(perm) ~ lo(area1, peri1) +
  lo(area1, shape) + lo(peri1, shape),
  control=gam.control(maxit=50, bf.maxit=50))
summary(rock.gam3)
anova(rock.lm, rock.gam3)
par(mfrow=c(1,3))
plot(rock.gam3, pty="s")
```

Figure 11.6 looks like the sum of ridge functions, so it is not surprising that PP regression can fit very well. The RSS is 13.83 on about 25 degrees of freedom.

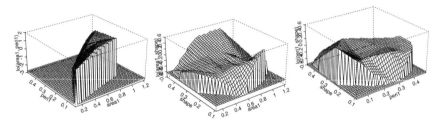

Figure 11.6: Perspective plots of the pairwise terms `lo` terms in the model for `rock`.

11.3 Response transformation models

If we want to predict Y, it may be better to transform Y as well, so we have

$$\theta(Y) = \alpha + \sum_{j=1}^{p} f_j(X_j) + \epsilon \qquad (11.6)$$

for an invertible smooth function $\theta()$, for example the log function we used for the `rock` dataset.

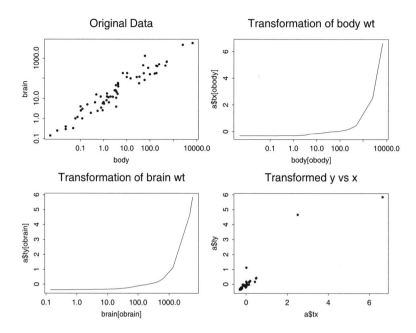

Figure 11.7: Plots of ACE fit of the `mammals` dataset.

The ACE (Alternating Conditional Expectation) algorithm of Breiman & Friedman (1985) chooses the functions θ and $f_1, \ldots, f_j$ to maximize the correlation between the predictor $\alpha + \sum_{j=1}^{p} f_j(X_j)$ and $\theta(Y)$. Tibshirani's (1988) procedure AVAS (additivity and variance stabilising transformation) fits the same model (11.6), but with the aim of achieving constant variance of the residuals for monotone θ.

The S-PLUS functions `ace` and `avas` fit (11.6). Both allow the functions f_j and θ to be constrained to be monotone or linear, and the functions f_j to be chosen for circular or categorical data. Thus these functions provide another way to fit additive models (with linear θ) but they do not provide measures of fit nor standard errors. They do, however, automatically choose the degree of smoothness.

Our first example is data on body weights (*kg*) and brain weights (*g*) of 62 mammals, from Weisberg (1985, pp. 144–5), supplied in our data frame `mammals`. For these variables we would expect log transformations will be appropriate. In this example AVAS succeeds but ACE does not, as Figures 11.7 and 11.8 show:

```
attach(mammals)
a <- ace(body, brain)
o1<- order(body); o2 <- order(brain)
par(mfrow=c(2,2))
plot(body, brain, main="Original Data", log="xy")
plot(body[o1], a$tx[o1], main="Transformation of body wt",
    type="l", log="x")
```

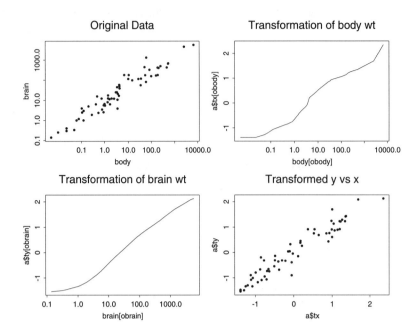

Figure 11.8: Plots of AVAS fit of the mammals dataset.

```
plot(brain[o2], a$ty[o2], main="Transformation of brain wt",
    type="l", log="x")
plot(a$tx, a$ty, main="Transformed y vs  x")

a <- avas(body, brain)
par(mfrow=c(2,2))
plot(body, brain, main="Original Data", log="xy")
plot(body[o1], a$tx[o1], main="Transformation of body wt",
    type="l", log="x")
plot(brain[o2], a$ty[o2], main="Transformation of brain wt",
    type="l", log="x")
plot(a$tx, a$ty, main="Transformed y vs x")
```

For the `rock` dataset we can use the following code. The S function `rug` produces the ticks to indicate the locations of data points, as used by `gam.plot`.

```
attach(rock)
x <- cbind(area, peri, shape)
o1 <- order(area); o2 <- order(peri); o3 <- order(shape)
a <- avas(x, perm)
par(mfrow=c(2,2))
plot(area[o1], a$tx[o1,1], type="l")          see Figure 11.9
rug(area)
plot(peri[o2], a$tx[o2,2], type="l")
rug(peri)
```

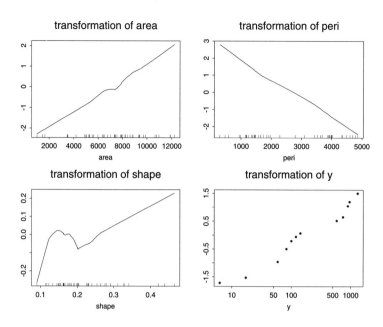

Figure 11.9: AVAS on permeabilities

```
plot(shape[o3], a$tx[o3,3], type="l")
rug(shape)
plot(perm, a$ty, log="x")                        note log scale
a <- avas(x, log(perm))                          looks like log(y)
a <- avas(x, log(perm), linear=0)                so force θ(y) = log(y)
# repeat plots
```

Here AVAS indicates a log transformation of permeabilities (expected on physical grounds) but little transformation of area or perimeter.

11.4   Neural networks

Feed-forward neural networks provide a flexible way to generalize linear regression functions. General references are Bishop (1995); Hertz, Krogh & Palmer (1991) and Ripley (1993, 1996).

We start with the simplest but most common form with one hidden layer as shown in Figure 11.10. The input units just provide a 'fan-out' and distribute the inputs to the 'hidden' units in the second layer. These units sum their inputs, add a constant (the 'bias') and take a fixed function ϕ_h of the result. The output units are of the same form, but with output function ϕ_o. Thus

$$y_k = \phi_o \left(\alpha_k + \sum_h w_{hk}\, \phi_h \left(\alpha_h + \sum_i w_{ih}\, x_i \right) \right) \qquad (11.7)$$

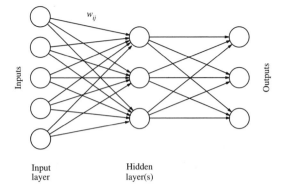

Input layer

Hidden layer(s)

Inputs

Outputs

Figure 11.10: A generic feed-forward neural network.

The 'activation function' ϕ_h of the hidden layer units is almost always taken to be the logistic function

$$\ell(z) = \frac{\exp(z)}{1 + \exp(z)}$$

and the output units are linear, logistic or threshold units. (The latter have $\phi_o(x) = I(x > 0)$.) Note the similarity to projection pursuit regression (*cf* (11.5)), which has linear output units but general smooth hidden units. (However, arbitrary smooth functions can be approximated by sums of re-scaled logistics.)

The general definition allows more than one hidden layer, and also allows 'skip-layer' connections from input to output when we have

$$y_k = \phi_o \left(\alpha_k + \sum_{i \to k} w_{ik} x_i + \sum_{j \to k} w_{jk} \phi_h \left(\alpha_j + \sum_{i \to j} w_{ij} x_i \right) \right) \qquad (11.8)$$

which allows the non-linear units to perturb a linear functional form.

We can eliminate the biases α_i by introducing an input unit 0 which is permanently at $+1$ and feeds every other unit. The regression function f is then parametrized by the set of weights w_{ij}, one for every link in the network (or zero for links which are absent).

The original biological motivation for such networks stems from McCulloch & Pitts (1943) who published a seminal model of a neuron as a binary thresholding device in discrete time, specifically that

$$n_i(t) = H \left(\sum_{j \to i} w_{ji} n_j(t-1) - \theta_i \right)$$

the sum being over neurons j connected to neuron i. Here H denotes the Heaviside or threshold function $H(x) = I(x > 0)$, $n_i(t)$ is the output of neuron i at time t, and $0 < w_{ij} < 1$ are attenuation weights. Thus the effect is to threshold a weighted sum of the inputs at value θ_i. Real neurons are now known to be more complicated; they have a graded response rather than the simple thresholding of the McCulloch–Pitts model, work in continuous time, and can perform more general non-linear functions of their inputs, for example logical functions. Nevertheless, the McCulloch–Pitts model has been extremely influential in the development of artificial neural networks.

Feed-forward neural networks can equally be seen as a way to parametrize a fairly general non-linear function. Such networks *are* rather general: Cybenko (1989), Funahashi (1989), Hornik, Stinchcombe & White (1989) and later authors have shown that neural networks with linear output units can approximate any continuous function f uniformly on compact sets, by increasing the size of the hidden layer.

The approximation results are non-constructive, and in practice the weights have to be chosen to minimize some fitting criterion, for example least squares

$$E = \sum_p \|t^p - y^p\|^2$$

where t^p is the target and y^p the output for the pth example pattern. Other measures have been proposed, including for $y \in [0, 1]$ 'maximum likelihood' (in fact minus the logarithm of a conditional likelihood) or equivalently the Kullback–Leibler distance, which amount to minimizing

$$E = \sum_p \sum_k \left[t_k^p \log \frac{t_k^p}{y_k^p} + (1 - t_k^p) \log \frac{1 - t_k^p}{1 - y_k^p} \right] \tag{11.9}$$

This is half the deviance for a logistic model with linear predictor given by (11.7) or (11.8).

One way to ensure that f is smooth is to restrict the class of estimates, for example by using a limited number of spline knots. Another way is *regularization* in which the fit criterion is altered to

$$E + \lambda C(f)$$

for example, with a penalty C on the second derivatives of f such as

$$\int \sum_{i,o} \frac{\partial^2 y_o}{\partial x_i^2} \, dx \approx \frac{1}{P} \sum_{i,o,p} \frac{\partial^2 y_o}{\partial x_i^2}$$

for P patterns, as is used in the derivation of smoothing splines. *Weight decay*, specific to neural networks, uses as penalty the sum of squares of the weights w_{ij}. (This only makes sense if the inputs are rescaled to range about $[0, 1]$ to be comparable with the outputs of internal units.) The use of weight decay seems both to help the optimization process and to avoid over-fitting. Arguments in Ripley (1993, 1994) based on a Bayesian interpretation suggest $\lambda \approx 10^{-4} - 10^{-2}$ depending on the degree of fit expected, for least-squares fitting to variables of range one and $\lambda \approx 0.01 - 0.1$ for the entropy fit.

Software to fit feed-forward neural networks with a single hidden layer but allowing skip-layer connections (as in (11.8)) is provided in our library `nnet`. The format of the call is to the fitting function `nnet` is

```
nnet(x, y, weights, size, Wts, linout=F, entropy=F, softmax=F,
     skip=F, rang=0.7, decay=0, maxit=100, trace=T)
```

The parameters are

x	matrix of dependent variables.
y	vector or matrix of independent variables.
weights	weights for training cases in E; default 1.
size	number of units in the hidden layer.
Wts	optional initial vector for w_{ij}.
linout	logical for linear output units.
entropy	logical for entropy rather than least-squares fit.
softmax	logical for log-probability models.
skip	logical for links from inputs to outputs.
rang	if `Wts` is missing, use random weights from `runif(n,-rang, rang)`.
decay	parameter λ.
maxit	maximum of iterations for the optimizer.
trace	logical for output from the optimizer. Very reassuring!

There are `predict`, `print` and `summary` methods for neural networks, and a function `nnet.Hess` to compute the Hessian with respect to the weight parameters and so check if a secure local minimum has been found. For our `rock` example we have

```
> attach(rock)
> area1 <- area/10000; peri1 <- peri/10000
> rock.x <- cbind(area1, peri1, shape)
> set.seed(5555)
> rock.nn <- nnet(rock.x, log(perm), size=3, decay=1e-3,
      linout=T,skip=T,  maxit=1000)
# weights:  19
initial  value 1820.342503
    ....
final  value 12.757497
> summary(rock.nn)
a 3-3-1 network with 19 weights
options were - skip-layer connections  linear output units
  decay=0.001
  b->h1 i1->h1 i2->h1 i3->h1
   4.47 -11.16  15.31   -8.78
  b->h2 i1->h2 i2->h2 i3->h2
   9.15 -14.68  18.45  -22.93
  b->h3 i1->h3 i2->h3 i3->h3
   1.22  -9.80   7.10   -3.77
```

```
    b->o   h1->o   h2->o   h3->o   i1->o   i2->o   i3->o
   8.78  -16.06    8.63    9.66   -1.99   -4.15    1.65
> sum((log(perm) - predict(rock.nn, rock.x))^2)
[1] 10.447
> eigen(nnet.Hess(rock.nn, rock.x, log(perm)), T)$values
 [1] 1.3527e+03 7.7746e+01 4.8631e+01 1.8276e+01 1.0255e+01
 [6] 4.0966e+00 1.2148e+00 8.5386e-01 5.3286e-01 3.5416e-01
[11] 1.3670e-01 7.7035e-02 2.8582e-02 1.3385e-02 1.0144e-02
[16] 7.0201e-03 5.9288e-03 4.0461e-03 3.3190e-03
```

The quoted values include the weight decay term. The eigenvalues of the Hessian suggest that a secure local minimum has been achieved. In the summary the b refers to the 'bias' unit (input unit 0), and i, h and o to input, hidden and bias units.

The input and outputs can also be specified by a formula: we can use

```
rock1 <- data.frame(perm, area=area1, peri=peri1, shape)
rock.nn1 <- nnet(log(perm) ~ area + peri + shape, data=rock1,
                 size=3, decay=1e-3, linout=T, skip=T, maxit=1000)
```

More extensive examples of the use of neural networks are given in Chapter 17 and in the on-line complements.

11.5 Conclusions

We have considered a large, perhaps bewildering, variety of extensions to linear regression. These can be thought of as belonging to two classes, the 'black-box' fully automatic and maximally flexible routines represented by projection pursuit regression and neural networks, and the small steps under full user control of additive models. Although the latter gain in interpretation, as we saw in the rock example they may not be general enough.

There are at least two other, intermediate approaches. Tree-based models (Chapter 14) use piecewise-constant functions and can build interactions by successive partitions. Friedman's (1991) MARS explicitly has products of simple functions of each variable. (S software for MARS is available in the library mda from statlib; it is described in the on-line complements to this chapter.)

What is best for any particular problem depends on its aim, in particular whether prediction or interpretation is paramount. The methods of this chapter are powerful tools with very little distribution theory to support them so it is very easy to over-fit and over-explain features of the data. Be warned!

Chapter 12

Survival Analysis

S-PLUS contains extensive survival analysis facilities written by Terry Therneau (Mayo Foundation). The functions in S-PLUS 3.3 and 3.4 are modified versions of code available from `statlib` as `survival4` (see Appendix C for further information); S-PLUS 3.2 and earlier versions used the rather different `survival2`. As the code for `survival4` is available[1], we strongly recommend its use.

Survival analysis is concerned with the distribution of lifetimes, often of humans but also of components and machines. There are two distinct levels of mathematical treatment in the literature. Cox & Oakes (1984), Kalbfleisch & Prentice (1980), Lawless (1982), Miller (1981) and Collett (1994) take a traditional and mathematically non-rigorous approach. The modern mathematical approach based on continuous-parameter martingales is given by Fleming & Harrington (1991) and Andersen *et al.* (1993). (The latter is needed here only to justify some of the distribution theory and for the concept of martingale residuals.)

Let T denote a lifetime random variable. It will take values in $(0, \infty)$, and its continuous distribution may be specified by a cumulative distribution function F with a density f. (Mixed distributions can be considered, but many of the formulae used by the software need modification.) For lifetimes it is more usual to work with the *survivor function* $S(t) = 1 - F(t) = P(T > t)$, the *hazard function* $h(t) = \lim_{\Delta t \to 0} P(t \leqslant T < t + \Delta t \mid T \geqslant t)/\Delta t$ and the *cumulative hazard function* $H(t) = \int_0^t h(s)\, ds$. These are all related; we have

$$h(t) = \frac{f(t)}{S(t)}, \qquad H(t) = -\log S(t)$$

Common parametric distributions for lifetimes are (Kalbfleisch & Prentice, 1980) the exponential, with $S(t) = \exp -\lambda t$ and hazard λ, the Weibull with

$$S(t) = \exp -(\lambda t)^\alpha, \qquad h(t) = \lambda \alpha (\lambda t)^{\alpha - 1}$$

the log-normal, the gamma and the log-logistic with

$$S(t) = \frac{1}{1 + (\lambda t)^\tau}, \qquad h(t) = \frac{\lambda \tau (\lambda t)^{\tau - 1}}{1 + (\lambda t)^\tau}$$

---

[1] a compiled version for S-PLUS 3.2 for Windows is available *via* our WWW sites.

The major distinguishing feature of survival analysis is *censoring*. An individual case may not be observed on the whole of its lifetime, so that, for example, we may only know that it survived to the end of the trial. More general patterns of censoring are possible, but all lead to data for each case of the form either of a precise lifetime or the information that the lifetime fell in some interval (possibly extending to infinity).

Clearly we must place some restrictions on the censoring mechanism, for if cases were removed from the trial just before death we would be misled. Consider right censoring, in which the case leaves the trial at time C_i, and we know either T_i if $T_i \leqslant C_i$ or that $T_i > C_i$. *Random censoring* assumes that T_i and C_i are independent random variables, and therefore in a strong sense that censoring is uninformative. This includes the special case of *type I* censoring, in which the censoring time is fixed in advance, as well as trials in which the patients enter at random times but the trial is reviewed at a fixed time. It excludes *type II* censoring in which the trial is concluded after a fixed number of failures. Most analyses (including all those based solely on likelihoods) are valid under a weaker assumption which Kalbfleisch & Prentice (1980, §5.2) call *independent* censoring in which the hazard at time t conditional on the whole history of the process only depends on the survival of that individual to time t. (Independent censoring does cover type II censoring.) Conventionally the time recorded is $\min(T_i, C_i)$ together with the indicator variable for observed death $\delta_i = I(T_i \leqslant C_i)$. Then under independent right censoring the likelihood for parameters in the lifetime distribution is

$$L = \prod_{\delta_i=1} f(t_i) \prod_{\delta_i=0} S(t_i) = \prod_{\delta_i=1} h(t_i)S(t_i) \prod_{\delta_i=0} S(t_i) = \prod_{i=1}^{n} h(t_i)^{\delta_i} S(t_i)$$

$$(12.1)$$

Usually we are not primarily interested in the lifetime distribution *per se*, but how it varies between groups (usually called *strata* in the survival context) or on measurements on the cases, called *covariates*. In the more complicated problems the hazard will depend on covariates which vary with time, such as blood pressure measurements, or changes of treatments.

The function `Surv(times, status)` is used to describe the censored survival data to the routines in the `survival4` library, and always appears on the left side of a model formula. In the simplest case of right censoring the variables are $\min(T_i, C_i)$ and δ_i (logical or 0/1 or 1/2). Further forms allow left and interval censoring. The results of printing the object returned by `Surv` are the vector of the information available, either the lifetime or an interval.

We consider three small running examples. Uncensored data on survival times for leukaemia (Feigl & Zelen, 1965; Cox & Oakes, 1984, p. 9) are in data frame `leuk`. This has two covariates, the white blood count `wbc`, and `ag` a test result which returns 'present' or 'absent'. Two-sample data (Gehan, 1965; Cox & Oakes, 1984, p. 7) on remission times for leukaemia are given in data frame `gehan`. This trial has 42 individuals in matched pairs, and no covariates (other than the treatment group).[2] Data frame `motors` contains the results of an accelerated life test experiment with ten replicates at each of four temperatures reported by Nelson & Hahn (1972) and Kalbfleisch & Prentice (1980, pp. 4–5). The times are given in hours, but all but one is a multiple of 12, and only 14 values occur

   17 21 22 56 60 70 73.5 115.5 143.5 147.5833 157.5 202.5 216.5 227

in days, which suggests that observation was not continuous. Thus this is a good example to test the handling of ties.

12.1 Estimators of survivor curves

The estimate of the survivor curve for uncensored data is easy; just take one minus the empirical distribution function. For the leukaemia data we have

```
attach(leuk)
plot(survfit(Surv(time) ~ ag), lty=c(2,3))
legend(80, 0.8, c("ag absent", "ag present"), lty=c(2,3))
```

and confidence intervals are obtained easily from the binomial distribution of $\widehat{S}(t)$. For example, the estimated variance is

$$\widehat{S}(t)[1 - \widehat{S}(t)]/n = r(t)[n - r(t)]/n^3 \qquad (12.2)$$

when $r(t)$ is the number of cases still alive (and hence 'at risk') at time t.

This computation introduces the function `survfit` and its associated `plot`, `print` and `summary` methods. It takes a model formula, and if there are factors on the right-hand side, splits the data on those factors, and plots a survivor curve for each factor combination, here just presence or absence of `ag`. (Although the factors can be specified additively, the computation effectively uses their interaction.)

For censored data we have to allow for the decline in the number of cases 'at risk' over time. Let $r(t)$ be the number of cases at risk just before time t, that is those which are in the trial and not yet dead. If we consider a set of intervals $I_i = [t_i, t_{i+1})$ covering $[0, \infty)$, we can estimate the probability p_i of surviving interval I_i as $[r(t_i) - d_i]/r(t_i)$ where d_i is the number of deaths in interval I_i. Then the probability of surviving until t_i is

$$P(T > t_i) = S(t_i) \approx \prod_0^{i-1} p_j \approx \prod_0^{i-1} \frac{r(t_i) - d_i}{r(t_i)}$$

Now let us refine the grid of intervals. Non-unity terms in the product will only appear for intervals in which deaths occur, so the limit becomes

$$\widehat{S}(t) = \prod \frac{r(t_i) - d_i}{r(t_i)}$$

---

[2] Andersen *et al.* (1993, p. 22) indicate that this trial had a sequential stopping rule which invalidates most of the methods used here; it should be seen as illustrative only.

the product being over times at which deaths occur before t (but they could occur simultaneously). This is the Kaplan-Meier estimator. Note that this becomes constant after the largest observed t_i, and for this reason the estimate is only plotted up to the largest t_i. However, the points at the right-hand end of the plot will be very variable, and it may be better to stop plotting when there are still a few individuals at risk.

We can apply similar reasoning to the cumulative hazard

$$H(t_i) \approx \sum_{j \leqslant i} h(t_j)(t_{j+1} - t_j) \approx \sum_{j \leqslant i} \frac{d_j}{r(t_j)}$$

with limit

$$\widehat{H}(t) = \sum \frac{d_j}{r(t_j)} \tag{12.3}$$

again over times at which deaths occur before t. This is the Nelson estimator of the cumulative hazard, and leads to the Altshuler or Fleming-Harrington estimator of the survivor curve

$$\tilde{S}(t) = \exp -\widehat{H}(t) \tag{12.4}$$

The two estimators are related by the approximation $\exp x \approx 1 - x$ for small x, so they will be nearly equal for large risk sets.

Similar arguments to those used to derive the two estimators lead to the standard error formula for the Kaplan-Meier estimator

$$\mathrm{var}\left(\widehat{S}(t)\right) = \widehat{S}(t)^2 \sum \frac{d_j}{r(t_j)[r(t_j) - d_j]} \tag{12.5}$$

often called Greenwood's formula after its version for life tables, and

$$\mathrm{var}\left(\widehat{H}(t)\right) = \sum \frac{d_j}{r(t_j)[r(t_j) - d_j]} \tag{12.6}$$

We leave it to the reader to check that Greenwood's formula reduces to (12.2) in the absence of ties and censoring. Note that if censoring can occur, both the Kaplan-Meier and Nelson estimators are biased; the bias occurs from the inability to give a sensible estimate when the risk set is empty.

Tsiatis (1981) suggested the denominator $r(t_j)^2$ rather than $r(t_j)[r(t_j) - d_j]$ on asymptotic grounds. Both Fleming & Harrington (1991) and Andersen *et al.* (1993) give a rigorous derivation of these formulae (and corrected versions for mixed distributions), as well as calculations of bias and limit theorems which justify asymptotic normality. Klein (1991) discussed the bias and small-sample behaviour of the variance estimators; his conclusions for $\widehat{H}(t)$ are that the bias is negligible and the Tsiatis form of the standard error is accurate (for practical use) provided the expected size of the risk set at t is at least 5. For the Kaplan-Meier estimator Greenwood's formula is preferred, and is accurate enough (but biased downwards) again provided the expected size of the risk set is at least 5.

We can use these formulae to indicate confidence intervals based on asymptotic normality, but we must decide on what scale to compute them. By default the function `survfit` computes confidence intervals on the log survivor (or cumulative hazard) scale, but linear and complementary log-log scales are also available (via the `conf.type` argument). These choices give

$$\widehat{S}(t) \exp\left[\pm k_\alpha \text{ s.e.}(\widehat{H}(t))\right]$$

$$\widehat{S}(t) \left[1 \pm k_\alpha \text{ s.e.}(\widehat{H}(t))\right]$$

$$\exp\left\{\widehat{H}(t) \exp\left[\pm k_\alpha \frac{\text{s.e.}(\widehat{H}(t))}{\widehat{H}(t)}\right]\right\}$$

the last having the advantage of taking values in $(0, 1)$. Bie *et al.* (1987) and Borgan & Liestøl (1990) considered these and an arc-sine transformation; their results indicate that the complementary log-log interval is quite satisfactory for sample sizes as small as 25.

We will not distinguish clearly between log-survivor curves and cumulative hazards, which differ only by sign, yet the natural estimator of the first is the Kaplan-Meier estimator on log scale, and for the second it is the Nelson estimator. This is particularly true for confidence intervals, which we would expect to transform just by a change of sign. Fortunately, practical differences only emerge for very small risk sets, and are then swamped by the very large variability of the estimators.

The function `survfit` also handles censored data, and uses the Kaplan-Meier estimator by default. We can try it on the gehan data:

```
> library(MASS, first=T)
> attach(gehan)
> Surv(time, cens)
 [1]  1   10  22   7   3  32+ 12  23   8  22  17   6   2  16
[15] 11  34+  8  32+ 12  25+  2  11+  5  20+  4  19+ 15   6
[29]  8  17+ 23  35+  5   6  11  13   4   9+  1   6+  8  10+
> plot.factor(gehan)
> plot(log(time) ~ pair)
# product-limit estimators with Greenwood's formula for errors:
> gehan.surv <- survfit(Surv(time, cens) ~ treat, gehan,
         conf.type="log-log")
> summary(gehan.surv)
Call: survfit(formula = Surv(time, cens) ~ treat, data = gehan,
         conf.type = "log-log")
```

treat=6-MP

time	n.risk	n.event	survival	std.err	lower 95% CI	upper 95% CI
6	21	3	0.857	0.0764	0.620	0.952
7	17	1	0.807	0.0869	0.563	0.923
10	15	1	0.753	0.0963	0.503	0.889
13	12	1	0.690	0.1068	0.432	0.849

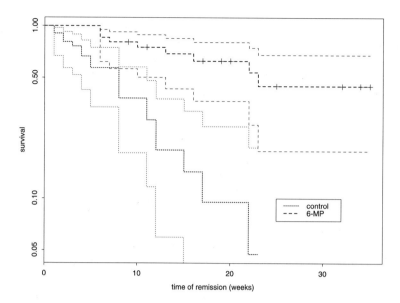

Figure 12.1: Survivor curves (on log scale) for the two groups of the gehan data. The crosses (on the 6-MP curve) represent censoring times. The thicker lines are the estimates, the thinner lines pointwise 95% confidence intervals.

16	11	1	0.627	0.1141	0.368	0.805
22	7	1	0.538	0.1282	0.268	0.747
23	6	1	0.448	0.1346	0.188	0.680

treat=control

time	n.risk	n.event	survival	std.err	lower 95% CI	upper 95% CI
1	21	2	0.9048	0.0641	0.67005	0.975
2	19	2	0.8095	0.0857	0.56891	0.924
3	17	1	0.7619	0.0929	0.51939	0.893
4	16	2	0.6667	0.1029	0.42535	0.825
5	14	2	0.5714	0.1080	0.33798	0.749
8	12	4	0.3810	0.1060	0.18307	0.578
11	8	2	0.2857	0.0986	0.11656	0.482
12	6	2	0.1905	0.0857	0.05948	0.377
15	4	1	0.1429	0.0764	0.03566	0.321
17	3	1	0.0952	0.0641	0.01626	0.261
22	2	1	0.0476	0.0465	0.00332	0.197
23	1	1	0.0000	NA	NA	NA

```
> plot(gehan.surv, conf.int=T, lty=c(3,2), log=T,
      xlab="time of remission (weeks)", ylab="survival")
> plot(gehan.surv, lty=c(3,2), lwd=2, log=T, add=T, cex=2)
> legend(25, 0.1 , c("control","6-MP"), lty=c(2, 3), lwd=2)
```

which calculates and plots (as shown in Figure 12.1) the product-limit estimators for the two groups, giving standard errors calculated using Greenwood's formula. (Confidence intervals are plotted automatically if there is only one group. The `add=T` argument is from the version of `plot.survfit` in our library, so attaching this with `first=T` is *essential*. Whereas there is a `lines` method for survival fits, it is not as versatile.) Other options are available, including `error="tsiatis"` and `type="fleming-harrington"` (which can be abbreviated to the first character). Note that the `plot` method has a `log` argument that plots $\widehat{S}(t)$ on log scale, effectively showing the negative cumulative hazard.

Testing survivor curves

We can test for differences between the groups in the `gehan` example by

```
> survdiff(Surv(time, cens) ~ treat, data=gehan)
                 N Observed Expected (O-E)^2/E (O-E)^2/V
treat=6-MP 21          9     19.3      5.46      16.8
treat=control 21      21     10.7      9.77      16.8

 Chisq= 16.8  on 1 degrees of freedom, p= 4.17e-05
```

This is one of a family of tests with parameter ρ defined by Fleming & Harrington (1981) and Harrington & Fleming (1982). The default $\rho = 0$ corresponds to the *log-rank test*. Suppose t_j are the observed death times. If we condition on the risk set and the number of deaths D_j at time t_j the mean of the number of deaths D_{jk} in group k is clearly $E_{jk} = D_k r_k(t_j)/r(t_j)$ under the null hypothesis (where $r_k(t_j)$ is the number from group k at risk at time j). The statistic used is $(O_k - E_k) = \sum_j \widehat{S}(t_j-)^\rho \left[D_{jk} - E_{jk}\right]$,[3] and from this we compute a statistic $(O - E)^T V^{-1}(O - E)$ with an approximately chi-squared distribution. There are a number of different approximations to the variance matrix V, the one used being the weighted sum over death times of the variance matrices of $D_{jk} - E_{jk}$ computed from the hypergeometric distribution. The sum of $(O_k - E_k)^2/E_k$ provides a conservative approximation to the chi-squared statistic. The final column is $(O - E)^2$ divided by the diagonal of V.

The argument `rho=1` of `survdiff` corresponds approximately to the Peto-Peto modification (Peto & Peto, 1972) of the Wilcoxon test, and is more sensitive to early differences in the survivor curves.

The function `survdiff` can also be used to test whether the data might make come from a specific survivor function S specified by a formula whose right-hand side is an offset giving the predicted survival probability $S(t_j)$ for each case.

A warning: tests of differences between groups are often used inappropriately. The `gehan` dataset has no other covariates, but where there are covariates the differences between the groups may reflect or be masked by differences in the covariates. Thus for the `leuk` dataset

---

[3] $\widehat{S}(t-)$ is the Kaplan-Meier estimate of survival just prior to t, ignoring the grouping.

```
> survdiff(Surv(time) ~ ag, data=leuk)
            N Observed Expected (O-E)^2/E (O-E)^2/V
ag=absent  16       16      9.3      4.83       8.45
ag=present 17       17     23.7      1.90       8.45

 Chisq= 8.4  on 1 degrees of freedom, p= 0.00365
```

is inappropriate as there are differences in distribution of wbc between the two groups. A model is needed to adjust for the covariates.

12.2 Parametric models

Parametric models for survival data have fallen out of fashion with the advent of less parametric approaches such as the Cox proportional hazard models considered in the next section, but they remain a very useful tool, particularly in exploratory work (as usually they can be fitted very much faster than the Cox models).

The simplest parametric model is the exponential distribution with hazard $\lambda_i > 0$. The natural way to relate this to a covariate vector x for the case (including a constant if required) and to satisfy the positivity constraint is to take

$$\log \lambda_i = \beta^T x_i, \qquad \lambda_i = e^{\beta^T x_i}$$

For the Weibull distribution the hazard function is

$$h(t) = \lambda^\alpha \alpha t^{\alpha-1} = \alpha t^{\alpha-1} \exp(\alpha \beta^T x) \tag{12.7}$$

if again make λ an exponential function of the covariates, and so we have the first appearance of the *proportional hazards* model

$$h(t) = h_0(t) \exp \beta^T x \tag{12.8}$$

which we will consider again later. This identification suggests re-parametrising the Weibull by replacing λ^α by λ, but as this just re-scales the coefficients we can move easily from one parametrization to the other. Note that the proportional hazards form (12.8) provides a direct interpretation of the parameters β.

The Weibull is also a member of the class of *accelerated life* models, which have survival time T such that $T \exp \beta^T x$ has a fixed distribution; that is time is speeded up by the factor $\exp \beta^T x$ for an individual with covariate x. This corresponds to replacing t in the survivor function and hazard by $t \exp \beta^T x$, and for models such as the exponential, Weibull and log-logistic with parametric dependence on λt, this corresponds to taking $\lambda = \exp \beta^T x$. For all accelerated-life models we will have

$$\log T = \log T_0 - \beta^T x \tag{12.9}$$

for a random variable T_0 whose distribution does not depend on x, so these are naturally considered as regression models.

For the Weibull the cumulative hazard is linear on a log-log plot, which provides a useful diagnostic aid. For example, for the gehan data (using our version of plot.survfit, so attaching library MASS with first=T is essential):

```
> library(MASS, first=T)
> plot(gehan.surv,  lty=c(3,4), cloglog=T,
      xlab="time of remission (weeks)", ylab="H(t)")
> legend(2, 2 , c("control","6-MP"), lty=c(4, 3))
```

we see excellent agreement with the proportional hazards hypothesis and with a
Weibull baseline (Figure 12.2).

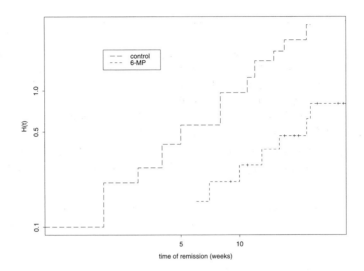

Figure 12.2: A log-log plot of cumulative hazard for the gehan dataset.

The function `survreg` is provided to fit parametric survival models of the
form

$$\log T \sim \beta^T x + \sigma \epsilon \qquad (12.10)$$

This has arguments including

```
survreg(formula, data, subset, na.action,
    link=c("log", "identity"),
    dist=c("extreme", "logistic", "gaussian",
        "exponential", "rayleigh"), fixed)
```

where the first-mentioned of the alternatives are the defaults. (The names can
be abbreviated provided enough is given to uniquely identify them.) The model
formula may include an `offset` term (see page 229). 'Link' here is used in a
different sense to that for GLMs; it is the data which are transformed, not their
means. Thus for the rarely used identity 'link' we have

$$T \sim \beta^T x + \sigma \epsilon$$

obviously with constraints on β. The `distribution` argument specifies the
distribution of ϵ, and σ is known as the *scale*.

With the default log link, the extreme value option corresponds to the model

$$\log T \sim \beta^T x + \sigma \log E$$

for a standard exponential E whereas our Weibull parametrization corresponds to

$$\log T \sim -\log \lambda + \frac{1}{\alpha} \log E$$

Thus `survreg` uses a log-linear Weibull model for $-\log \lambda$ and the scale factor σ estimates $1/\alpha$. (The name 'extreme' comes from the distribution of $\log T$.) The `exponential` distribution comes from fixing $\sigma = \alpha = 1$ and the `rayleigh` distribution from $\sigma = 0.5$; they correspond to specifying `fixed = list(scale=1)` or `0.5`. Note that, confusingly, these two names apply to the distribution of T whereas the first three apply to the distribution of $\log T$. For most purposes it is better to regard the distributions as Weibull, log-logistic, log-normal, exponential and Rayleigh, with a regression for the mean of the log survival time.

We consider exponential analyses, followed by Weibull and log-logistic regression analysis. (The output prior to **S-PLUS 3.4** will be somewhat different.)

```
> options(contrasts=c("contr.treatment", "contr.poly"))
> survreg(Surv(time) ~ ag*log(wbc), leuk, dist="exponential")
   ....
Coefficients:
 (Intercept)       ag log(wbc) ag:log(wbc)
      4.3433 4.1349 -0.15402     -0.32781
   ....
Degrees of Freedom: 33 Total; 29 Residual
-2*Log-Likelihood: 104.55
> summary(survreg(Surv(time) ~ ag + log(wbc), leuk, dist="exp"))
   ....
Coefficients:
               Value Std. Error z value        p
(Intercept)    5.815      1.263    4.60 4.15e-06
         ag    1.018      0.364    2.80 5.14e-03
   log(wbc)   -0.304      0.124   -2.45 1.44e-02

> summary(survreg(Surv(time) ~ ag + log(wbc), leuk)) # Weibull
   ....
Coefficients:
             Value Std. Error z value        p
(Intercept)   5.85      1.323    4.42 9.66e-06
         ag   1.02      0.378    2.70 6.95e-03
   log(wbc) -0.31      0.131   -2.36 1.81e-02

Extreme value distribution: Dispersion (scale) = 1.0407
Degrees of Freedom: 33 Total; 29 Residual
-2*Log-Likelihood: 106
   ....
```

```
> summary(survreg(Surv(time) ~ ag + log(wbc), leuk, dist="log"))
    ....
Coefficients:
              Value Std. Error z value          p
(Intercept)   8.027      1.701    4.72 2.37e-06
         ag   1.155      0.431    2.68 7.30e-03
   log(wbc) -0.609      0.176   -3.47 5.21e-04

Logistic distribution: Dispersion (scale) est = 0.68794
```

The Weibull analysis shows no support for non-exponential shape. For later reference, in the proportional hazards parametrization (12.8) the estimate of the coefficients is $\hat{\beta} = -(5.85, 1.02, -0.310)^T/1.04 = (-5.63, -0.981, 0.298)^T$. The log-logistic distribution, which is an accelerated life model but not a proportional hazards model (in our parametrization), gives a considerably more significant coefficient for log(wbc). Its usual scale parameter τ (as defined on page 343) is estimated as $1/0.688 \approx 1.45$. Solomon (1984) considers fitting proportional-hazard models when an accelerated-life model is appropriate and vice versa, and shows that to first order the coefficients β differ only by a factor of proportionality. Here there is definite evidence of non-proportionality.

We can test for a difference in groups within the Weibull model by the Wald test (the 'z value'). We can perform a likelihood ratio test by the anova method in S-PLUS 3.4 and later.

```
> summary(survreg(Surv(time) ~ log(wbc), leuk))
    ....
Coefficients:
              Value Std. Error z value          p
(Intercept)   7.484      1.376    5.44 5.40e-08
   log(wbc) -0.421      0.144   -2.93 3.41e-03

Extreme value distribution: Dispersion (scale) = 1.1583
Degrees of Freedom: 33 Total; 30 Residual
-2*Log-Likelihood: 113
    ....
> anova(survreg(Surv(time) ~ log(wbc), leuk),
      survreg(Surv(time) ~ ag + log(wbc), leuk))
    ....
          Terms Resid. Df  -2*LL Test Df Deviance Pr(Chi)
1      log(wbc)        30 112.59
2 ag + log(wbc)        29 106.24  +ag  1   6.3572 0.01169
```

The likelihood ratio test statistic is somewhat less significant than the result given by survdiff. More care is needed in earlier versions, as anova would use anova.glm, and the deviances reported are not comparable if the dispersion parameter is allowed to vary. (It seems that the saturated model for the estimated shape parameter is used in the deviance.) S-PLUS 3.3 gives

```
> summary(survreg(Surv(time) ~ log(wbc), leuk, cor=F))
```

```
    . . . .
Coefficients:
              Value Std. Error z value        p
(Intercept)   7.484      1.376    5.44 5.40e-08
   log(wbc)  -0.421      0.144   -2.93 3.41e-03
 Log(scale)   0.147      0.138    1.07 2.85e-01
Extreme value distribution: Dispersion (scale) = 1.1583

     Null Deviance: 45.3 on 32 degrees of freedom
 Residual Deviance: 36.9 on 30 degrees of freedom  (LL= -56.3 )
```

and we can compute the likelihood ratio test from differences in LL, the log-likelihood, which is component $loglik[2]$ of the fit.

If the accelerated-life model holds, $T \exp -\beta^T x$ has the same distribution for all subjects, being standard Weibull, log-logistic and so on. Thus we can get some insight into what the common distribution should be by studying the distribution of $(T_i \exp -\widehat{\beta}^T x_i)$. Another way to look at this is that the residuals from the regression are $\log T_i - \widehat{\beta}^T x_i$ which we have transformed back to the scale of time. For the $leuk$ data we could use, for example,

```
leuk.wei <- survreg(Surv(time) ~ ag + log(wbc), leuk)
ntimes <- leuk$time * exp(-leuk.wei$linear.predictors)
plot(survfit(Surv(ntimes)), log=T)
```

The result (Figure 12.3) is plausibly linear, confirming the suitability of an exponential model. If we wished to test for a general Weibull distribution, we should plot $\log(-\log \widehat{S}(t))$ against $\log t$. (This plot is provided by the $cloglog$ argument to the version of $plot.survfit$ in our library, as we used for Figure 12.2.)

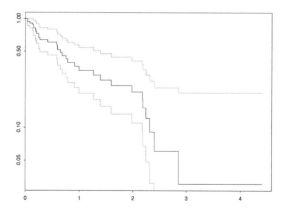

Figure 12.3: A log-log plot of cumulative hazard for the $leuk$ dataset.

Moving on to the $gehan$ dataset, which includes right censoring, we find

```
> survreg(Surv(time, cens) ~ factor(pair) + treat, gehan,
    dist="exp")
```

```
       . . . .
Degrees of Freedom: 42 Total; 20 Residual
-2*Log-Likelihood: 84.228
> summary(survreg(Surv(time, cens) ~ treat, gehan, dist="exp"))
Coefficients:
              Value Std. Error z value        p
(Intercept)   3.69      0.333    11.06 2.00e-28
      treat  -1.53      0.398    -3.83 1.27e-04

Degrees of Freedom: 42 Total; 40 Residual
-2*Log-Likelihood: 98
> summary(survreg(Surv(time, cens) ~ treat, gehan))
Coefficients:
              Value Std. Error z value        p
(Intercept)   3.52      0.252    13.96 2.61e-44
      treat  -1.27      0.311    -4.08 4.51e-05
Extreme value distribution: Dispersion (scale) = 0.73219
```

There is no evidence of close matching of pairs. The difference in log hazard between treatments is $-(-1.27)/0.732 = 1.73$ with a standard error of $0.42 = 0.311/0.732$.

Finally, we consider the motors data, which are analysed by Kalbfleisch & Prentice (1980, §3.8.1). According to Nelson & Hahn (1972), the data were collected to assess survival at $130°C$, for which they found a median of 34 400 hours and a 10 percentile of 17 300 hours.

```
> plot(survfit(Surv(time, cens) ~ factor(temp), motors),
    conf.int=F)
> motor.wei <- survreg(Surv(time, cens) ~ temp, motors)
> summary(motor.wei)
    . . . .
Coefficients:
              Value Std. Error z value        p
(Intercept) 16.3185    0.62296    26.2 3.03e-151
       temp -0.0453    0.00319   -14.2 6.74e-46

Extreme value distribution: Dispersion (scale) = 0.33433
    . . . .
> unlist(predict(motor.wei, data.frame(temp=130), se.fit=T))
  fit.1 se.fit.1 residual.scale df
 10.429  0.65922        3.2033 37

> motor.wei$coef %*% c(1, 130)
       [,1]
[1,] 10.429
> cz <- c(1, 130,0) # compute s.e.
> sqrt(t(cz)%*% motor.wei$var %*% cz)
       [,1]
[1,] 0.222
```

At the time of writing `predict.survreg` was not yet written[4], and the predictions returned by `predict.glm` are not always appropriate (used as class `survreg` inherits from `glm`). The model for $\log T$ at $130°C$ is

$$\log T = 16.32 - 0.0453 \times \texttt{temp} + 0.334 \log E = 10.429(0.22) + 0.334 \log E$$

with a median for $\log T$ of

$$10.429(0.22) + 0.334 \times -0.3665 \approx 10.31(0.22)$$

or a median for T of 30 000 hours with 95% confidence interval (19 300, 46 600) and a 10 percentile for $\log T$ of

$$10.429(0.22) + 0.334 \times -2.2503 \approx 9.68(0.22)$$

or as a 95% confidence interval for the 10 percentile for T, (10 300, 24 800) hours.

Nelson & Hahn worked with $z = 1000/(\texttt{temp} + 273.2)$. We leave the reader to try this; it gives slightly larger quantiles.

12.3   Cox proportional hazards model

Cox (1972) introduced a less parametric approach known as proportional hazards. There is a baseline hazard function $h_0(t)$ which is modified multiplicatively by covariates (including group indicators), so the hazard function for any individual case is

$$h(t) = h_0(t) \exp \boldsymbol{\beta}^T \boldsymbol{x}$$

and the interest is mainly in the proportional factors rather than the baseline hazard. Note that the cumulative hazards will also be proportional, so we can examine the hypothesis by plotting survivor curves for sub-groups on log scale. Later we will allow the covariates to depend on time.

The parameter vector $\boldsymbol{\beta}$ is estimated by maximizing a partial likelihood. Suppose one death occurred at time t_j. Then conditional on this event the probability that case i died is

$$\frac{h_0(t) \exp \boldsymbol{\beta}^T \boldsymbol{x}_i}{\sum_l I(T_l \geqslant t) h_0(t) \exp \boldsymbol{\beta}^T \boldsymbol{x}_l} = \frac{\exp \boldsymbol{\beta}^T \boldsymbol{x}_i}{\sum_l I(T_l \geqslant t) \exp \boldsymbol{\beta}^T \boldsymbol{x}_l} \tag{12.11}$$

which does not depend on the baseline hazard. The partial likelihood for $\boldsymbol{\beta}$ is the product of such terms over all observed deaths, and usually contains most of the information about $\boldsymbol{\beta}$ (the remainder being in the observed times of death). However, we need a further condition on the censoring (Fleming & Harrington, 1991, pp. 138–9) that it is independent and *uninformative* for this to be so; the

---

[4] and had not been for four years.

latter means that the likelihood for censored observations in $[t, t + \Delta t)$ does not depend on β.

The correct treatment of ties causes conceptual difficulties as they are an event of probability zero for continuous distributions. Formally (12.11) may be corrected to include all possible combinations of deaths. As this increases the computational load, it is common to employ the Breslow approximation[5] in which each death is always considered to precede all other events at that time. Let $\tau_i = I(T_i \geqslant t) \exp \beta^T x_i$, and suppose there are d deaths out of m possible events at time t. Breslow's approximation uses the term

$$\prod_{i=1}^{d} \frac{\tau_i}{\sum_1^m \tau_j}$$

in the partial likelihood at time t. Other options are Efron's approximation

$$\prod_{i=1}^{d} \frac{\tau_i}{\sum_1^m \tau_j - \frac{i}{d} \sum_1^d \tau_j}$$

and the 'exact' partial likelihood

$$\prod_{i=1}^{d} \tau_i \Big/ \sum \prod_{k=1}^{d} \tau_{j_k}$$

where the sum is over subsets of $1, \ldots, m$ of size d. One of these terms is selected by the `method` argument of the function `coxph`, with default `efron`.

The baseline cumulative hazard $H_0(t)$ is estimated by rescaling the contributions to the number at risk by $\exp \widehat{\beta}^T x$ in (12.3). Thus in that formula $r(t) = \sum I(T_i \geqslant t) \exp \widehat{\beta}^T x_i$.

The Cox model is easily extended to allow different baseline hazard functions for different groups, and this is automatically done if they declared as `strata`. (The model formula may also include a `offset` term.) For our leukaemia example we have:

```
> library(MASS, first=T)
> attach(leuk)
> leuk.cox <- coxph(Surv(time) ~ ag + log(wbc), leuk)
> summary(leuk.cox)
    ....
            coef exp(coef) se(coef)      z      p
      ag -1.069     0.343    0.429 -2.49 0.0130
log(wbc)  0.368     1.444    0.136  2.70 0.0069

          exp(coef) exp(-coef) lower .95 upper .95
      ag      0.343      2.913     0.148     0.796
log(wbc)      1.444      0.692     1.106     1.886
```

---

[5] first proposed by Peto (1972).

```
Rsquare= 0.377    (max possible= 0.994 )
Likelihood ratio test= 15.6  on 2 df,    p=0.000401
Wald test             = 15.1  on 2 df,    p=0.000537
Efficient score test = 16.5  on 2 df,    p=0.000263

> update(leuk.cox, ~ . -ag)
    ....
Likelihood ratio test=9.19  on 1 df, p=0.00243  n= 33

> leuk.coxs <- coxph(Surv(time) ~ strata(ag) + log(wbc), leuk)
> leuk.coxs
    ....
          coef exp(coef) se(coef)    z      p
log(wbc) 0.391      1.48    0.143 2.74 0.0062
    ....
Likelihood ratio test=7.78  on 1 df, p=0.00529  n= 33

> leuk.coxs1 <- coxph(Surv(time) ~ strata(ag) + log(wbc) +
      ag:log(wbc), leuk)
> leuk.coxs1
    ....
              coef exp(coef) se(coef)     z    p
   log(wbc) 0.183      1.20    0.188 0.978 0.33
ag:log(wbc) 0.456      1.58    0.285 1.598 0.11
    ....
> plot(survfit(Surv(time) ~ ag), lty=2:3, log=T)
> plot(survfit(leuk.coxs), lty=2:3, lwd=3, add=T, log=T)
> legend(80, 0.8, c("ag absent", "ag present"), lty=2:3)
```

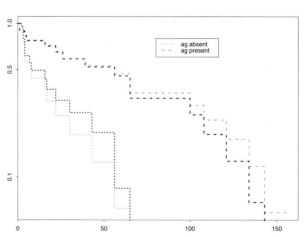

Figure 12.4: Log-survivor curves for the `leuk` dataset. The thick lines are from a Cox model with two strata, the thin lines Kaplan-Meier estimates which ignore the blood counts.

The 'likelihood ratio test' is actually based on (log) partial likelihoods, not the full likelihood, but has similar asymptotic properties. The tests show that there is a significant effect of wbc on survival, but also that there is a significant difference between the two ag groups (although as Figure 12.4 shows, this is less than before adjustment for the effect of wbc).

Note how survfit can take the result of a fit of the proportional hazard model. In the first fit the hazards in the two groups differ only by a factor whereas later they are allowed to have separate baseline hazards (which look very close to proportional). There is marginal evidence for a difference in slope within the two strata. Note how straight the log-survivor functions are in Figure 12.4, confirming the good fit of the exponential model for these data. The Kaplan-Meier survivor curves refer to the populations; those from the coxph fit refer to a patient in the stratum with an average log(wbc) for the whole dataset. This example shows why it is inappropriate just to test (using survdiff) the difference between the two groups: part of the difference is attributable to the lower wbc in the ag absent group.

The test statistics refer to the whole set of covariates. The likelihood ratio test statistic is the change in deviance on fitting the covariates over just the baseline hazard (by strata); the score test is the expansion at the baseline, and so does not need the parameters to be estimated (although this has been done). Thus the score test is still useful if we do not allow iteration:

```
> leuk.cox <- coxph(Surv(time) ~ ag, leuk)
> summary(coxph(Surv(time) ~ log(wbc) +
      offset(predict(leuk.cox, type="lp")), iter.max=0))
   ....
Efficient score test = 7.57  on 1 df,    p=0.00593
```

using the results of a previous fit as a starting point. The R^2 measure quoted by summary.coxph is taken from Nagelkerke (1991).

The general proportional hazards model gives estimated (non-intercept) coefficients $\widehat{\beta} = (-1.07, 0.37)^T$, compared to the Weibull fit of $(-0.981, 0.298)^T$ (on page 353). The log-logistic had coefficients $(-1.155, 0.609)^T$ which under the approximations of Solomon (1984) would be scaled by $\tau/2$ to give $(-0.789, 0.416)^T$ for a Cox proportional-hazards fit if the log-logistic regression model (an accelerated life model) were the true model.

We next consider the Gehan data. We saw before that the pairs have a negligible effect for the exponential model. Here the effect is a little larger, with $P \approx 8\%$. The Gehan data have a large number of (mainly pairwise) ties.

```
> gehan.cox <- coxph(Surv(time, cens) ~ treat, gehan)
> summary(gehan.cox)
      ....
      coef exp(coef) se(coef)    z       p
treat 1.57      4.82    0.412 3.81 0.00014

      exp(coef) exp(-coef) lower .95 upper .95
treat      4.82      0.208      2.15      10.8
```

```
    . . . .
Likelihood ratio test= 16.4  on 1 df,    p=5.26e-05
    . . . .
> coxph(Surv(time, cens) ~ treat, gehan, method="exact")
    . . . .
        coef exp(coef) se(coef)    z        p
treat 1.63      5.09     0.433 3.76 0.00017

Likelihood ratio test=16.2  on 1 df, p=5.54e-05  n= 42

# The next fit is slow
> coxph(Surv(time, cens) ~ treat + factor(pair), gehan,
    method="exact")
    . . . .
Likelihood ratio test=45.5  on 21 df, p=0.00148  n= 42
    . . . .
> 1 - pchisq(45.5 - 16.2, 20)
[1] 0.082018
```

Finally we consider the `motors` data. The exact fit for the motors dataset is much the slowest, as it has large groups of ties.

```
> motor.cox <- coxph(Surv(time, cens) ~ temp, motors)
> motor.cox
    . . . .
        coef exp(coef) se(coef)    z        p
temp 0.0918      1.1    0.0274 3.36 0.00079
    . . . .
> coxph(Surv(time, cens) ~ temp, motors, method="breslow")
    . . . .
        coef exp(coef) se(coef)    z       p
temp 0.0905     1.09    0.0274 3.3 0.00098
    . . . .
> coxph(Surv(time, cens) ~ temp, motors, method="exact")
    . . . .
        coef exp(coef) se(coef)    z        p
temp 0.0947      1.1    0.0274 3.45 0.00056
    . . . .
> plot( survfit(motor.cox, newdata=data.frame(temp=200),
    conf.type="log-log") )
> summary( survfit(motor.cox, newdata=data.frame(temp=130)) )
time n.risk n.event survival  std.err lower 95% CI upper 95% CI
 408     40       4    1.000 0.000254        0.999            1
 504     36       3    1.000 0.000499        0.999            1
1344     28       2    0.999 0.001910        0.995            1
1440     26       1    0.998 0.002698        0.993            1
1764     20       1    0.996 0.005327        0.986            1
2772     19       1    0.994 0.007922        0.978            1
3444     18       1    0.991 0.010676        0.971            1
3542     17       1    0.988 0.013670        0.962            1
```

3780	16	1	0.985 0.016980	0.952	1
4860	15	1	0.981 0.020697	0.941	1
5196	14	1	0.977 0.024947	0.929	1

The function `survfit` has a special method for `coxph` objects that plots the mean and confidence interval of the survivor curve for an average individual (with average values of the covariates). As we see, this can be over-ridden by giving new data, as shown in Figure 12.5. The non-parametric method is unable to extrapolate to $130°C$ as none of the test examples survived long enough to estimate the baseline hazard beyond the last failure at 5196 hours.

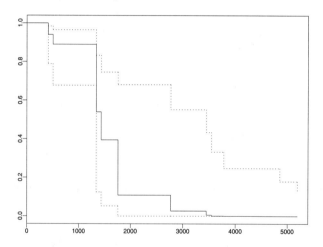

Figure 12.5: The survivor curve for a motor at $200°C$ estimated from a Cox proportional hazards model (solid line) with pointwise 95% confidence intervals (dotted lines).

Residuals

The concept of a residual is a difficult one for binary data, especially as here if the event may not be observed because of censoring. A straightforward possibility is to take

$$r_i = \delta_i - \widehat{H}(t_i)$$

which is known as the *martingale residual* after a derivation from the mathematical theory given by Fleming & Harrington (1991, §4.5). They show that it is appropriate for checking the functional form of the proportional hazards model, for if

$$h(t) = h_0(t)\phi(x^*) \exp \boldsymbol{\beta}^T \boldsymbol{x}$$

for an (unknown) function of a covariate x^*, then

$$E[R \mid X^*] \approx [\phi(X^*) - \overline{\phi}] \sum \delta_i/n$$

and this can be estimated by smoothing a plot of the martingale residuals *vs* x^*, for example using `lowess` or the function `scatter.smooth` based on `loess`.

(The term $\overline{\phi}$ is a complexly weighted mean.) The covariate x^* can be one not included in the model, or one of the terms to check for non-linear effects.

The martingale residuals are the default output of residuals on a coxph fit.

The martingale residuals can have a very skewed distribution, as their maximum value is 1, but they can be arbitrarily negative. The *deviance residuals* are a transformation

$$\text{sign}(r_i)\sqrt{2[-r_i - \delta_i \log(\delta_i - r_i)]}$$

which reduces the skewness, and when squared and summed (approximately) give the deviance. Deviance residuals are best used in plots which will indicate cases not well-fitted by the model.

The *Schoenfeld residuals* (Schoenfeld, 1982) are defined at death times as $x_i - \overline{x}(t_i)$ where $\overline{x}(s)$ is the mean weighted by $\exp \widehat{\beta}^T x$ of the x over only the cases still in the risk set at time s. These residuals form a matrix with one row for each case which died and a column for each covariate. The scaled Schoenfeld residuals (type="scaledsch") are I^{-1} matrix multiplying the Schoenfeld residuals, where I is the (partial) information matrix at the fitted parameters in the Cox model.

The *score residuals* are the terms of efficient score for the partial likelihood, this being a sum over cases of

$$L_i = \big[x_i - \overline{x}(t_i)\big]\delta_i - \int_0^{t_i} \big[x_i(s) - \overline{x}(s)\big]\hat{h}(s)\,\mathrm{d}s$$

Thus the score residuals form an $n \times p$ matrix. They can be used to examine leverage of individual cases by computing (approximately) the change in $\widehat{\beta}$ if the observation was dropped; type="dfbeta" gives this whereas type="dfbetas" scales by the standard errors for the components of $\widehat{\beta}$.

Tests of proportionality of hazards

Once a type of departure from the base model is discovered or suspected, the proportional hazards formulation is usually flexible enough to allow an extended model to be formulated and the significance of the departure tested within the extended model. Nevertheless, some approximations can be useful, and are provided by the function cox.zph for departures of the type

$$\beta(t) = \beta + \theta g(t)$$

for some postulated smooth function g. Grambsch & Therneau (1994) show that the scaled Schoenfeld residuals for case i have, approximately, mean $g(t_i)\theta$ and a computable variance matrix.

The function `cox.zph` has both `print` and `plot` methods. The printed output gives an estimate of the correlation between $g(t_i)$ and the scaled Schoenfeld residuals and a chi-squared test of $\theta = 0$ for each covariate, and an overall chi-squared test. The plot method gives a plot for each covariate, of the scaled Schoenfeld residuals against $g(t)$ with a spline smooth and pointwise confidence bands for the smooth. (Figure 12.8 on page 367 is an example.)

The function g has to be specified. The default in `cox.zph` is $1 - \widehat{S}(t)$ for the Kaplan-Meier estimator, with options for the ranks of the death times, $g \equiv 1$ and $g = \log$ as well as a user-specified function. (The x axis of the plots is labelled by the death times, not $\{g(t_i)\}$.)

12.4   Further examples

VA lung cancer data

S-PLUS supplies the dataset `cancer.vet` on a Veteran's Administration lung cancer trial used by Kalbfleisch & Prentice (1980), but as it has no on-line help, it is not obvious what it is! It is a matrix of 137 cases with right-censored survival time and the covariates

treatment	standard or test
celltype	one of four cell types
Karnofsky score	of performance on scale 0–100, with high values for relatively well patients
diagnosis	time since diagnosis in months at entry to trial
age	in years
therapy	logical for prior therapy

As there are several covariates, we use the Cox model to establish baseline hazards.

```
> library(MASS, first=T)
> VA.temp <- data.frame(cancer.vet)
> dimnames(VA.temp)[[2]] <- c("treat", "cell", "stime",
      "status", "Karn", "diag.time","age","therapy")
> attach(VA.temp)
> VA <- data.frame(stime, status, treat=factor(treat), age,
      Karn, diag.time, cell=factor(cell), prior=factor(therapy))
> detach("VA.temp")
> VA.cox <- coxph(Surv(stime, status) ~ treat + age  + Karn +
      diag.time + cell + prior, VA)
> VA.cox
                 coef exp(coef) se(coef)        z        p
    treat    2.95e-01     1.343  0.20755  1.41945  1.6e-01
      age   -8.71e-03     0.991  0.00930 -0.93612  3.5e-01
     Karn   -3.28e-02     0.968  0.00551 -5.95801  2.6e-09
diag.time    8.18e-05     1.000  0.00914  0.00895  9.9e-01
    cell2    8.62e-01     2.367  0.27528  3.12970  1.7e-03
    cell3    1.20e+00     3.307  0.30092  3.97474  7.0e-05
```

```
    cell4  4.01e-01    1.494  0.28269  1.41955 1.6e-01
    prior  7.16e-02    1.074  0.23231  0.30817 7.6e-01

Likelihood ratio test=62.1  on 8 df, p=1.8e-10  n= 137

> VA.coxs <- coxph(Surv(stime, status) ~ treat + age + Karn +
    diag.time + strata(cell) + prior, VA)
> VA.coxs
               coef exp(coef) se(coef)      z      p
    treat   0.28590     1.331  0.21001  1.361 1.7e-01
      age  -0.01182     0.988  0.00985 -1.201 2.3e-01
     Karn  -0.03826     0.962  0.00593 -6.450 1.1e-10
diag.time  -0.00344     0.997  0.00907 -0.379 7.0e-01
    prior   0.16907     1.184  0.23567  0.717 4.7e-01

Likelihood ratio test=44.3  on 5 df, p=2.04e-08  n= 137

> plot(survfit(VA.coxs), log=T, lty=1:4)
> legend(locator(1), c("squamous", "small", "adeno", "large"),
    lty=1:4)
> plot(survfit(VA.coxs), cloglog=T, lty=1:4)
> attach(VA)
> cKarn <- factor(cut(Karn, 5))
> VA.cox1 <- coxph(Surv(stime, status) ~ strata(cKarn) + cell)
> plot(survfit(VA.cox1), cloglog=T)
> VA.cox2 <- coxph(Surv(stime, status) ~ Karn + strata(cell))
> scatter.smooth(Karn, residuals(VA.cox2))
> detach()
```

Figures 12.6 and 12.7 show some support for proportional hazards between the cell types (except perhaps squamous), and suggest a Weibull or even exponential distribution.

```
> VA.wei <- survreg(Surv(stime, status) ~ treat + age + Karn +
    diag.time + cell + prior, VA)
> summary(VA.wei, cor=F)
    ....
Coefficients:
               Value Std. Error z value       p
(Intercept)  3.262015  0.66253   4.9236 8.50e-07
      treat -0.228523  0.18684  -1.2231 2.21e-01
        age  0.006099  0.00855   0.7131 4.76e-01
       Karn  0.030068  0.00483   6.2281 4.72e-10
  diag.time -0.000469  0.00836  -0.0561 9.55e-01
      cell2 -0.826185  0.24631  -3.3542 7.96e-04
      cell3 -1.132725  0.25760  -4.3973 1.10e-05
      cell4 -0.397681  0.25475  -1.5611 1.19e-01
      prior -0.043898  0.21228  -0.2068 8.36e-01

Extreme value distribution: Dispersion (scale) = 0.92811
```

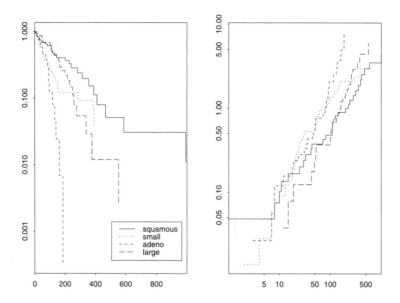

Figure 12.6: Cumulative hazard functions for the cell types in the VA lung cancer trial. The left-hand plot is labelled by survival probability on log scale. The right-hand plot is on log-log scale.

```
Degrees of Freedom: 137 Total; 127 Residual
-2*Log-Likelihood: 392

> VA.exp <- survreg(Surv(stime, status) ~ Karn + cell, VA
    dist="exp")
> summary(VA.exp, cor=F)
    ....
Coefficients:
             Value Std. Error z value        p
(Intercept) 3.4222   0.35463    9.65 4.92e-22
       Karn 0.0297   0.00486    6.11 9.97e-10
      cell2 -0.7102  0.24061   -2.95 3.16e-03
      cell3 -1.0933  0.26863   -4.07 4.70e-05
      cell4 -0.3113  0.26635   -1.17 2.43e-01

Extreme value distribution: Dispersion (scale) fixed at 1
Degrees of Freedom: 137 Total; 132 Residual
-2*Log-Likelihood: 395
```

Note that `scale` does not differ significantly from one, so an exponential distribution is an appropriate summary.

```
> cox.zph(VA.coxs)
              rho  chisq       p
    treat -0.0607  0.545 0.46024
```

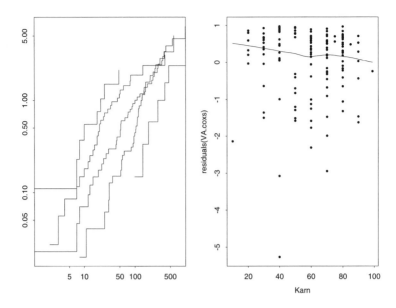

Figure 12.7: Diagnostic plots for the Karnofsky score in the VA lung cancer trial. **Left:** Log-log cumulative hazard plot for 5 groups. **Right:** Martingale residuals vs Karnofsky score, with a smoothed fit.

```
      age  0.1734  4.634 0.03134
     Karn  0.2568  9.146 0.00249
diag.time  0.1542  2.891 0.08909
    prior -0.1574  3.476 0.06226
   GLOBAL      NA 13.488 0.01921
> par(mfrow=c(3,2)); plot(cox.zph(VA.coxs))
```

Closer investigation does show some suggestion of time-varying coefficients in the Cox model. The plot is Figure 12.8. Note that some coefficients which are not significant in the basic model show evidence of varying with time. This suggests that the model with just `Karn` and `cell` may be too simple, and that we need to consider interactions. We automate the search of interactions using `stepAIC`, which has methods for both `coxph` and `survreg` fits. With hindsight, we centre the data.

```
> VA$Karnc <- VA$Karn - 50
> VA.coxc <- update(VA.cox, ~ . - Karn + Karnc)
> VA.cox2 <- stepAIC(VA.coxc, ~ .^2)
> VA.cox2$anova
Initial Model:
Surv(stime, status) ~ treat + age + diag.time + cell + prior
                      + Karnc

Final Model:
```

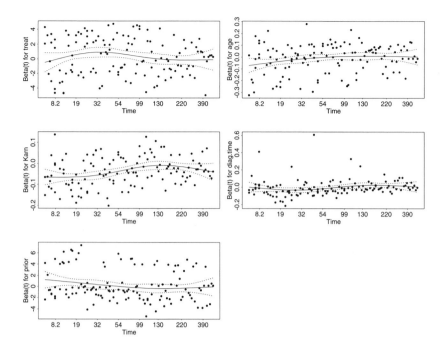

Figure 12.8: Diagnostics plots from cox.zph of the constancy of the coefficients in the proportional hazards model VA.coxs. Each plot is of a component of the Schoenfeld residual against a non-linear scale of time. A spline smoother is shown, together with ±2 standard deviations.

```
Surv(stime, status) ~ treat + diag.time + cell + prior + Karnc
      + prior:Karnc + diag.time:cell + treat:prior + treat:Karnc
```

```
               Step Df Deviance Resid. Df Resid. Dev    AIC
1                                      129     948.79 964.79
2     +prior:Karnc -1   -9.013        128     939.78 957.78
3 +diag.time:cell -3  -11.272        125     928.51 952.51
4             -age  1    0.415        126     928.92 950.92
5     +treat:prior -1   -2.303        125     926.62 950.62
6     +treat:Karnc -1   -2.904        124     923.72 949.72
```

(The 'deviances' here are minus twice log partial likelihoods.) Applying stepAIC to VA.wei leads to the same sequence of steps. As the variables diag.time and Karn are not factors, this will be easier to interpret using nesting:

```
> VA.cox3 <- update(VA.cox2, ~ treat/Karnc + prior*Karnc
      + treat:prior + cell/diag.time)
> VA.cox3
                  coef exp(coef) se(coef)      z       p
      treat  0.8065     2.240   0.27081  2.978 2.9e-03
      prior  0.9191     2.507   0.31568  2.912 3.6e-03
      Karnc -0.0107     0.989   0.00949 -1.129 2.6e-01
```

cell2	1.7068	5.511	0.37233	4.584	4.6e-06
cell3	1.5633	4.775	0.44205	3.536	4.1e-04
cell4	0.7476	2.112	0.48136	1.553	1.2e-01
Karnc %in% treat	-0.0187	0.981	0.01101	-1.695	9.0e-02
prior:Karnc	-0.0481	0.953	0.01281	-3.752	1.8e-04
treat:prior	-0.7264	0.484	0.41833	-1.736	8.3e-02
cell1diag.time	0.0532	1.055	0.01595	3.333	8.6e-04
cell2diag.time	-0.0245	0.976	0.01293	-1.896	5.8e-02
cell3diag.time	0.0161	1.016	0.04137	0.388	7.0e-01
cell4diag.time	0.0150	1.015	0.04033	0.373	7.1e-01

Thus the hazard increases with time since diagnosis in squamous cells, only, and
the effect of the Karnofsky score is only pronounced in the group with prior therapy.
We tried replacing diag.time with a polynomial, with negligible benefit. Using
cox.zph shows a very significant change with time in the coefficients of the
treat*Karn interaction.

```
> cox.zph(VA.cox3)
                    rho      chisq         p
         treat  0.18012  6.10371  0.013490
         prior  0.07197  0.76091  0.383044
         Karnc  0.27220 14.46103  0.000143
         cell2  0.09053  1.31766  0.251013
         cell3  0.06247  0.54793  0.459164
         cell4  0.00528  0.00343  0.953318
Karnc %in% treat -0.20606  7.80427  0.005212
   prior:Karnc -0.04017  0.26806  0.604637
   treat:prior -0.13061  2.33270  0.126682
 cell1diag.time  0.11067  1.62464  0.202446
 cell2diag.time -0.01680  0.04414  0.833596
 cell3diag.time  0.09713  1.10082  0.294086
 cell4diag.time  0.16912  3.16738  0.075123
        GLOBAL       NA 25.52734  0.019661

> par(mfrow=c(2,2))
> plot(cox.zph(VA.cox3), var=c(1,3,7))
```

Stanford heart transplants

This set of data is analysed by Kalbfleisch & Prentice (1980, §5.5.3). (The data
given in Kalbfleisch & Prentice are rounded, but the full data are supplied as
data frame heart (from S-PLUS 3.3 on). It is on survival from early heart
transplant operations at Stanford. The new feature is that patients may change
treatment during the study, moving from the control group to the treatment group at
transplantation, so some of the covariates such as waiting time for a transplant are
time-dependent (in the simplest possible way). Patients who received a transplant
are treated as two cases, before and after the operation, so cases in the transplant
group are in general both right- and left-censored. This is handled by Surv by
supplying entry and exit times. For example, patient 4 has the rows

```
        start  stop event       age         year   surgery transplant
         0.0   36.0    0   -7.73716632  0.49007529        0          0
        36.0   39.0    1   -7.73716632  0.49007529        0          1
```

which show that he waited 36 days for a transplant and then died after 3 days. The proportional hazards model is fitted from this set of cases, but some summaries need to take account of the splitting of patients.

The covariates are age (in years minus 48), year (after 1 Oct 1967) and an indicator for previous surgery. Rather than use the 6 models considered by Kalbfleisch & Prentice, we do our own model selection.

```
> attach(heart)
> coxph(Surv(start, stop, event) ~ transplant*
    (age+surgery+year), heart)
    ....
Likelihood ratio test=18.9  on 7 df, p=0.00852 n= 172
> coxph(Surv(start, stop, event) ~ transplant*(age+year) +
    surgery, heart)
    ....
Likelihood ratio test=18.4  on 6 df, p=0.0053  n= 172
> stan <- coxph(Surv(start, stop, event) ~ transplant*year +
    age + surgery, heart)
> stan
    ....
                  coef exp(coef) se(coef)     z     p
    transplant -0.6213    0.537   0.5311 -1.17 0.240
          year -0.2526    0.777   0.1049 -2.41 0.016
           age  0.0299    1.030   0.0137  2.18 0.029
       surgery -0.6641    0.515   0.3681 -1.80 0.071
transplant:year  0.1974    1.218   0.1395  1.42 0.160

Likelihood ratio test=17.1  on 5 df, p=0.00424  n= 172

> stan1 <- coxph(Surv(start, stop, event) ~ strata(transplant)+
    year + year:transplant + age + surgery, heart)
> plot(survfit(stan1), conf.int=T, log=T, lty=c(1,3))
> legend(locator(1), c("before", "after"), lty=c(1,3))

> plot(year[transplant==0], residuals(stan1, collapse=id),
    xlab = "year", ylab="martingale residual")
> lines(lowess(year[transplant==0],
    residuals(stan1, collapse=id)))
> sresid <- resid(stan1, type="score", collapse=id)
> -100 * sresid %*% stan1$var %*% diag(1/stan1$coef)
```

This analysis suggests that survival rates over the study improved *prior* to transplantation, which Kalbfleisch & Prentice suggest could be due to changes in recruitment. The diagnostic plots of Figure 12.9 show nothing amiss. The collapse argument is needed as those patients who received transplants are treated as two cases, and we need the residual per patient.

Now consider predicting the survival of future patients:

```
# Survivor curve for the "average" subject
> summary(survfit(stan))
# Survivor curve for a subject of age 50 on 1 Oct 1971
#   with prior surgery, transplant after 6 months
#  follow-up for two years
> stan2 <- data.frame(start=c(0,183), stop=c(183,2*365),
      event=c(0,0), year=c(4,4), age=c(50,50), surgery=c(1,1),
      transplant=c(0,1))
> summary(survfit(stan, stan2, individual=T,
      conf.type="log-log"))
```

time	n.risk	n.event	survival	std.err	lower 95% CI	upper 95% CI
....						
165	43	1	0.1107	0.2030	1.29e-05	0.650
186	41	1	0.1012	0.1926	8.53e-06	0.638
188	40	1	0.0923	0.1823	5.54e-06	0.626
207	39	1	0.0838	0.1719	3.51e-06	0.613
219	38	1	0.0759	0.1617	2.20e-06	0.600
263	37	1	0.0685	0.1517	1.35e-06	0.588
285	35	2	0.0548	0.1221	2.13e-06	0.524
308	33	1	0.0486	0.1128	1.24e-06	0.511
334	32	1	0.0431	0.1040	7.08e-07	0.497
340	31	1	0.0380	0.0954	3.97e-07	0.484
343	29	1	0.0331	0.0867	2.10e-07	0.470
584	21	1	0.0268	0.0750	7.30e-08	0.451
675	17	1	0.0205	0.0618	1.80e-08	0.428

The argument `individual=T` is needed to avoid averaging the two cases (which are the same individual). In this case the probability of survival until transplantation is rather low.

Australian AIDS survival

The data on the survival of AIDS patients within Australia are of unusually high quality within that field, and jointly with Dr Patty Solomon we have studied survival up to 1992.[6] There are a large number of difficulties in defining survival from AIDS (acquired immunodeficiency syndrome), in part because as a syndrome its diagnosis is not clear-cut and has almost certainly changed with time. (To avoid any possible confusion, we are studying survival from AIDS and not the HIV infection which is generally accepted as the cause of AIDS.)

The major covariates available were the reported transmission category, and the state or territory within Australia. The AIDS epidemic had started in New South Wales and then spread, so the states will have different profiles of cases in calendar time. A factor which was expected to be important in survival is the widespread availability of zidovudine (AZT) to AIDS patients from mid-1987

---

[6] We are grateful to the Australian National Centre in HIV Epidemiology and Clinical Research for making these data available to us.

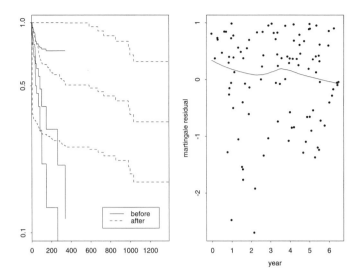

Figure 12.9: Plots for the Stanford heart transplant study. Left: log survivor curves for the two groups. Right: Martingale residuals against calendar time.

which has enhanced survival, and the use of zidovudine for HIV-infected patients from mid-1990, which it was thought might delay the onset of AIDS without necessarily postponing death further.

The transmission categories were:

hs	male homosexual or bisexual contact
hsid	as **hs** and also intravenous drug user
id	female or heterosexual male intravenous drug user
het	heterosexual contact
haem	haemophilia or coagulation disorder
blood	receipt of blood, blood components or tissue
mother	mother with or at risk of HIV infection
other	other or unknown

The data file gave data on all patients whose AIDS status was diagnosed prior to January 1992, with their status then. Since there is a delay in notification of death, some deaths in late 1991 would not have been reported and we adjusted the endpoint of the study to 1 July 1991. A total of 2843 patients were included, of whom about 1770 had died by the end date. The file contained an ID number, the dates of first diagnosis, birth and death (if applicable), as well as the state and the coded transmission category. The following S code was used to translate the file to a suitable format for a time-dependent-covariate analysis, and to combine the states ACT and NSW (as Australian Capital Territory is a small enclave within New South Wales). The code uses loops and took a few minutes, but this needed to be done only once and was still fast compared to the statistical analyses.

As there are a number of patients who are diagnosed at (strictly, after) death, there are a number of zero survivals. The software used to treat these incorrectly, so all deaths were shifted by 0.9 days to occur after other events the same day. Only the file Aids2 is included in our library. To maintain confidentiality the dates have been jittered, and the smallest states combined.

```
Aids <- read.table("aids.dat")
library(date)  # Therneau's library to manipulate dates
attach(Aids)
diag <- mdy.date(diag.m, diag.d, diag.y, F)
dob <- mdy.date(bir.m, bir.d, bir.y, F)
death <- mdy.date(die.m, die.d, die.y)
# the dates were randomly adjusted here
latedeath <- death >= mdy.date(7,1,91)
death[latedeath] <- mdy.date(7,1,91)
status[latedeath] <- "A"
death[is.na(death)] <- mdy.date(7,1,91)
age <- floor((diag-dob)/365.25)
state <- factor(state)
levels(state) <- c("NSW","NSW", "Other", "QLD", "Other",
    "Other", "VIC", "Other")
T.categ <- Aids$T
levels(T.categ) <- c("hs", "hsid", "id", "het", "haem",
    "blood", "mother", "other")
Aids1 <- data.frame(state, sex, diag=as.integer(diag),
    death=as.integer(death), status, T.categ,
    age, row.names=1:length(age))
detach("Aids")
rm(death, diag, status, age, dob, state, T.categ, latedeath)
Aids2 <- Aids1[Aids1$diag < as.integer(mdy.date(7, 1, 91)),]
table(Aids2$state)
  NSW Other QLD VIC
 1780   249 226 588

table(Aids2$T.categ)
   hs hsid id het haem blood mother other
 2465   72 48  41   46    94      7    70

as.integer(c(mdy.date(7,1,87), (mdy.date(7,1,90)),
    mdy.date(12,31,92)))
[1] 10043 11139 12053

time.depend.covar <- function(data) {
    id <- row.names(data)
    n <- length(id)
    ii <- 0
    stop <- start <- late <- T.categ <- state <- age <-
        status <- sex <- numeric(n)
    idno <- character(n)
```

```
events <- c(0, 10043, 11139, 12053)
for (i in 1:n) {
    diag <- data$diag[i]
    death <- data$death[i]
    for (j in 1:(length(events)-1)) {
    crit1 <- events[j]
    crit2 <- events[j+1]
    if( diag < crit2 && death >= crit1 ) {
        ii <- ii+1
        start[ii] <- max(0, crit1-diag)
        stop[ii] <- min(crit2, death+0.9)-diag
        if(death > crit2) status[ii] <- 0
        else status[ii] <- unclass(data$status[i]) - 1 # 0/1
        state[ii] <- data$state[i]
        T.categ[ii] <- data$T.categ[i]
        idno[ii] <- id[i]
        late[ii] <- j - 1
        age[ii] <- data$age[i]
        sex[ii] <- data$sex[i]
    }}
if(i%%100 ==0) print(i)
}
levels(state) <- c("NSW", "Other", "QLD", "VIC")
levels(T.categ) <- c("hs", "hsid", "id", "het", "haem",
    "blood", "mother", "other")
levels(sex) <- c("F", "M")
data.frame(idno, zid=factor(late), start, stop, status,
    state, T.categ, age, sex)
}
Aids3 <- time.depend.covar(Aids2)
```

The factor `zid` indicates whether the patient is likely to have received zidovudine at all, and if so whether it might have been administered during HIV infection.

Our analysis was based on a proportional hazards model which allowed a proportional change in hazard from 1 July 1987 to 30 June 1990 and another from 1 July 1990; the results show a halving of hazard from 1 July 1987 but a nonsignificant change in 1990. This dataset is large, and the analyses take a long time (20 mins) and a lot of memory (up to 10 Mb).

```
> attach(Aids3)
> aids.cox <- coxph(Surv(start, stop, status)
        ~ zid + state + T.categ + sex + age, data=Aids3)
> summary(aids.cox)

            coef exp(coef) se(coef)       z       p
zid1 -0.69087     0.501  0.06578 -10.5034 0.0e+00
zid2 -0.78274     0.457  0.07550 -10.3675 0.0e+00
```

```
     stateOther -0.07246        0.930  0.08964  -0.8083 4.2e-01
       stateQLD  0.18315        1.201  0.08752   2.0927 3.6e-02
       stateVIC  0.00464        1.005  0.06134   0.0756 9.4e-01
  T.categhsid -0.09937          0.905  0.15208  -0.6534 5.1e-01
    T.categid -0.37979          0.684  0.24613  -1.5431 1.2e-01
   T.categhet -0.66592          0.514  0.26457  -2.5170 1.2e-02
  T.categhaem  0.38113          1.464  0.18827   2.0243 4.3e-02
 T.categblood  0.16856          1.184  0.13763   1.2248 2.2e-01
T.categmother  0.44448          1.560  0.58901   0.7546 4.5e-01
 T.categother  0.13156          1.141  0.16380   0.8032 4.2e-01
           sex  0.02421         1.025  0.17557   0.1379 8.9e-01
           age  0.01374         1.014  0.00249   5.5060 3.7e-08

                exp(coef) exp(-coef) lower .95 upper .95
          zid1     0.501     1.995     0.441     0.570
          zid2     0.457     2.187     0.394     0.530
    stateOther     0.930     1.075     0.780     1.109
      stateQLD     1.201     0.833     1.012     1.426
      stateVIC     1.005     0.995     0.891     1.133
   T.categhsid     0.905     1.104     0.672     1.220
     T.categid     0.684     1.462     0.422     1.108
    T.categhet     0.514     1.946     0.306     0.863
   T.categhaem     1.464     0.683     1.012     2.117
  T.categblood     1.184     0.845     0.904     1.550
 T.categmother     1.560     0.641     0.492     4.948
  T.categother     1.141     0.877     0.827     1.572
           sex     1.025     0.976     0.726     1.445
           age     1.014     0.986     1.009     1.019

Rsquare= 0.045   (max possible= 0.998 )
Likelihood ratio test= 185  on 14 df,    p=0
Wald test             = 204  on 14 df,    p=0
Efficient score test = 211  on 14 df,    p=0
```

The effect of sex is nonsignificant, and so dropped in further analyses. There is no detected difference in survival during 1990.

Note that Queensland has a significantly elevated hazard relative to New South Wales (which has over 60% of the cases), and that the intravenous drug users have a longer survival, whereas those infected via blood or blood products have a shorter survival, relative to the first category who form 87% of the cases. We can use stratified Cox models to examine these effects (Figures 12.10 and 12.11).

```
> aids1.cox <- coxph(Surv(start, stop, status)
    ~ zid + strata(state) + T.categ + age, data=Aids3)
> aids1.surv <- survfit(aids1.cox)
> aids1.surv
                 n events mean se(mean) median 0.95LCL 0.95UCL
   state=NSW 1780    1116  639     17.6     481     450     509
 state=Other  249     142  658     42.2     525     436     618
   state=QLD  226     149  519     33.5     439     357     568
```

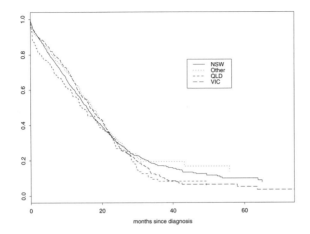

Figure 12.10: Survival of AIDS patients in Australia by state by state.

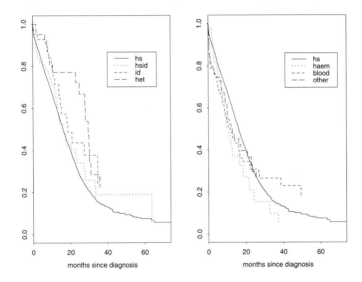

Figure 12.11: Survival of AIDS patients in Australia by transmission category.

```
  state=VIC  588      355    610       26.3     508      424        618
> plot(aids1.surv, mark.time=F, lty=1:4, xscale=365.25/12,
  xlab="months since diagnosis")
> legend(locator(1), levels(state), lty=1:4)

> aids2.cox <- coxph(Surv(start, stop, status)
    ~ zid + state + strata(T.categ) + age, data=Aids3)
> aids2.surv <- survfit(aids2.cox)
> aids2.surv
                  n events mean se(mean) median 0.95LCL 0.95UCL
    T.categ=hs 2465   1533  633     15.6    492   473.9     515
```

T.categ=hsid	72	45	723	86.7	493	396.9	716
T.categ=id	48	19	653	54.3	568	427.9	NA
T.categ=het	40	17	775	57.3	897	753.9	NA
T.categ=haem	46	29	431	53.9	337	209.9	657
T.categ=blood	94	76	583	86.1	358	151.9	666
T.categ=mother	7	3	395	92.6	655	0.9	NA
T.categ=other	70	40	421	40.7	369	24.9	NA

```
> par(mfrow=c(1,2))
> plot(aids2.surv, mark.time=F, lty=1:8, xscale=365.25/12,
    xlab="months since diagnosis", strata=1:4)
> legend(locator(1), levels(T.categ)[1:4], lty=1:4)

> plot(aids2.surv, mark.time=F, lty=1:8, xscale=365.25/12,
    xlab="months since diagnosis", strata=c(1,5,6,8))
> legend(locator(1), levels(T.categ)[c(1,5,6,8)], lty=1:4)
```

The `strata` argument of `plot.survfit` (another of our modifications) allows
us to select strata for the plot.

We now consider the possible non-linearity of log-hazard with age. First we
consider the martingale residual plot (Figure 12.12).

```
cases <- diff(c(0,idno)) != 0
aids.res <- residuals(aids.cox, collapse=idno)
scatter.smooth(age[cases], aids.res, xlab = "age",
    ylab="martingale residual")
```

This shows a slight rise in residual with age over 60, but no obvious effect. The
next step is to replace a linear term in age by a step function, with breaks chosen
from prior experience. The contrast is chosen explicitly to take as base level the
31–40 age group.

```
age2 <- cut(age, c(-1, 15, 30, 40, 50, 60, 100))
c.age <- factor(age2)
levels(c.age) <- c("0-15", "16-30", "31-40",
    "41-50", "51-60", "61+")
table(c.age)
 0-15 16-30 31-40 41-50 51-60 61+
   39  1022  1583   987   269  85
tmp <- diag(6)
dimnames(tmp) <- list(levels(c.age), levels(c.age))
contrasts(c.age) <- tmp[, -3]

summary(coxph(Surv(start, stop, status) ~ zid + state
    + T.categ + c.age, data=Aids3))
    ....
                      coef exp(coef) se(coef)        z         p
    ....
    c.age0-15  2.51e-01    1.285    0.2888  8.67e-01 3.9e-01
    c.age16-30 -1.02e-01    0.903    0.0625 -1.63e+00 1.0e-01
```

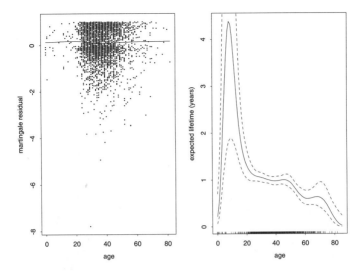

Figure 12.12: Diagnostic plots for age-dependence of AIDS survival in Australia. Left: Martingale residual plot from proportional hazards analysis including a linear effect for age. Right: Predicted survival *vs* age of a NSW hs patient (solid line), with pointwise 95% confidence intervals (dashed lines) and a rug of all observed ages.

```
c.age41-50  8.00e-02    1.083   0.0615   1.30e+00  1.9e-01
c.age51-60  3.72e-01    1.450   0.0972   3.82e+00  1.3e-04
  c.age61+  7.12e-01    2.037   0.1566   4.54e+00  5.5e-06
    ....
Likelihood ratio test= 192  on 17 df,    p=0
    ....
detach("Aids3")
```

Beyond this we could fit a smooth function of age via splines, but to save computational time we defer this to the parametric analysis.

We can also consider a parametric analysis. From the survivor curves the obvious model is the Weibull. Since this is both a proportional hazards model and an accelerated life model, we can include the effect of the introduction of zidovudine by assuming a doubling of survival after July 1987. With 'time' computed on this basis we find

```
make.aidsp <- function(){
    cutoff <- 10043
    btime <- pmin(cutoff, Aids2$death) - pmin(cutoff, Aids2$diag)
    atime <- pmax(cutoff, Aids2$death) - pmax(cutoff, Aids2$diag)
    survtime <- btime + 0.5*atime
    status <- unclass(Aids2$status) - 1
    data.frame(survtime, status, state=Aids2$state,
        T.categ=Aids2$T.categ, age=Aids2$age, sex=Aids2$sex)
}
Aidsp <- make.aidsp()
```

```
attach(Aidsp)
aids.wei <- survreg(Surv(survtime + 0.9, status) ~ state
    + T.categ + sex + age, data=Aidsp)
summary(aids.wei, correlation=F)
    ....
Coefficients:
                 Value Std. Error   z value         p
  (Intercept)  6.41825     0.2098   30.5970 1.34e-205
   stateOther  0.09387     0.0931    1.0079  3.13e-01
     stateQLD -0.18213     0.0913   -1.9956  4.60e-02
     stateVIC -0.00750     0.0637   -0.1177  9.06e-01
 T.categhsid  0.09363     0.1582    0.5918  5.54e-01
   T.categid  0.40132     0.2552    1.5727  1.16e-01
  T.categhet  0.67689     0.2744    2.4667  1.36e-02
T.categhaem -0.34090     0.1956   -1.7429  8.14e-02
T.categblood -0.17336    0.1429   -1.2131  2.25e-01
T.categmother -0.40186    0.6123   -0.6563  5.12e-01
T.categother -0.11279     0.1696   -0.6649  5.06e-01
         sex -0.00426     0.1827   -0.0233  9.81e-01
         age -0.01374     0.0026   -5.2862  1.25e-07

Extreme value distribution: Dispersion (scale) = 1.0405
    ....
```

Note that we still have to avoid zero survival. This shows good agreement with the parameters for the Cox model. The parameter α (the reciprocal of the scale) is close to one. For practical purposes the exponential is a good fit, and the parameters are little changed.

We also considered parametric non-linear functions of age by using a spline function. For useful confidence intervals we include the constant term in the predictions, which are for a NSW hs patient (see Figure 12.12).

```
> summary(survreg(Surv(survtime+0.9, status) ~ state
    + T.categ + age, data=Aidsp), correlation=F)
    ....
Extreme value distribution: Dispersion (scale) = 1.0405
Degrees of Freedom: 2843 Total; 2830 Residual
-2*Log-Likelihood: 7577
> t1 <- bs(0:85, knots = c(15,30,40,50,60))
> aids.bs <- survreg(Surv(survtime+0.9,status) ~ state
    + T.categ + t1[age+1,], data=Aidsp)
> summary(aids.bs, correlation=F)
    ....
Extreme value distribution: Dispersion (scale) = 1.0336
Degrees of Freedom: 2843 Total; 2823 Residual
-2*Log-Likelihood: 7546
> cv <- aids.bs$coef[c(1,12:19)]
> cv.var <- aids.bs$var[c(1,12:19), c(1,12:19)]
> t2 <- cbind(1, t1)
> t3 <- sqrt(diag(t2 %*% cv.var %*% t(t2) ))
```

```
> plot(0:85, exp(t2%*%cv)/365.25, type="l", xlab="age",
    ylab = "expected lifetime (years)")
> lines(0:85, exp(t2%*%cv + 1.96 * t3)/365.25, lty=3)
> lines(0:85, exp(t2%*%cv - 1.96 * t3)/365.25, lty=3)
> rug(age+runif(length(age), -0.5, 0.5), 0.015)
```

12.5 Expected survival rates

In medical applications we may want to compare survival rates to those of a standard population, perhaps to standardize the experience of the population under study. As the survival experience of the general population changes with calendar time, this must be taken into account.

Unfortunately, there are differences between versions in how calendar time is recorded between the versions of the survival analysis functions: the version in S-PLUS uses modified versions of functions from the `chron` library whereas the original `survival4` uses the format of Therneau's library `date` (obtainable from `statlib`). Both record dates in days since 1 Jan 1960, but with class `"dates"` and `"date"`) respectively. For the S-PLUS version the easiest way to specify or print calendar dates is the function `dates`; for datasets such as `aids.dat` with numerical day, month and year data the function `julian` may be useful.

For a cohort study, expected survival is often added to a plot of survivor curves. The function `survexp` is usually used with a formula generated by `ratetable`. The optional argument `times` specifies a vector at which to evaluate survival, by default for all follow times. For example, we could add expected survival for 65-year old US white males to the left plot of Figure 12.9 by

```
year <- dates("7/1/91")
# OR for the statlib version
library(date); year=mdy.date(7,1,1991)

expect <- survexp(~ ratetable(sex=1, year=year, age=65*365.25),
        times = seq(0, 1400, 30), ratetable=survexp.uswhite)
lines(expect$time, expect$surv, lty=4)
```

but as the patients are seriously ill, the comparison is not very useful. As the inbuilt rate tables are in units of days, all of `year`, `age` and `times` must be in days.

Entry and date times can be specified as vectors, when the average survival for the cohort is returned. For individual expected survival, we can use the same form with `cohort=F`, perhaps evaluated at death time.

Some explanation of the averaging used is needed in the cohort case. We can use the cumulative hazard function $H_i(t)$ and survivor function $S_i(t)$ of the exact match (on age and sex) to individual i. There are three possibilities, which differ in the assumptions on what follow-up would have been.

1. The formula has no response. Then the function returns the average of $S_i(t)$. This corresponds to assuming complete follow-up.

2. The death times are given as the response. Then the $H_i(t)$ are averaged over the cases at risk at time t to from a cohort cumulative hazard function and converted to a survivor function.

3. The potential censoring times for each case are given as the response, and conditional=F, when the weights in the cohort cumulative hazard function tion are computed as $S_i(t)I$(potentially in study at t). This corresponds to assuming follow-up until the end of the study.

The first is most useful for forecasting, the other two for comparing with the study outcome. Thus to compare the survival in Figure 12.9 to matched males of the same ages we might use

```
expect <- survexp(stop ~ ratetable(sex=1, year=year*365.25,
    age=(age+48)*365.25),   times = seq(0, 1400, 30),
    ratetable = survexp.uswhite, data = heart,
    subset = diff(c(id, 0)) != 0, cohort = T, conditional = T)
lines(expect$time, expect$surv, lty=4)
```

We do need to extract the second record corresponding to transplanted subjects to get the correct death/censoring time for the cohort matching.

It is possible to use the fit from a coxph model in place of the inbuilt ratetables to compare the present study to an earlier one.

Chapter 13

Multivariate Analysis

Multivariate analysis is concerned with datasets which have more than one response variable for each observational or experimental unit. The datasets can be summarized by data matrices X with n rows and p columns, the rows representing the observations or cases, and the columns the variables. The matrix can be viewed either way, depending whether the main interest is in the relationships between the cases or between the variables. Note that for consistency we represent the variables of a case by the *row* vector x.

The main division in multivariate methods is between those methods which assume a given structure, for example dividing the cases into groups, and those which seek to discover structure from the evidence of the data matrix alone. In pattern-recognition terminology the distinction is between *supervised* and *unsupervised* methods. One of our examples is the (in)famous iris data collected by Anderson (1935) and given and analysed by Fisher (1936). This has 150 cases, which are stated to be 50 of each of the three species *Iris setosa*, *I. virginica* and *I. versicolor*. Each case has four measurements on the length and width of its petals and sepals. *A priori* this is a supervised problem, and the obvious questions are to use measurements on a future case to classify it, and perhaps to ask how the variables vary between the species. (In fact, Fisher (1936) used these data to test a genetic hypothesis which placed *I. versicolor* as a hybrid two-thirds of the way from *I. setosa* to *I. virginica*.) However, the classification of species is uncertain, and similar data have been used to identify species by grouping the cases. (Indeed, Wilson (1982) and McLachlan (1992, §6.9) consider whether the iris data can be split into sub-species.) We end the chapter with a similar example on splitting a species of crab.

Krzanowski (1988) and Mardia, Kent & Bibby (1979) are two general references on multivariate analysis. Our treatment is closer to Krzanowski's. The boundaries of multivariate analysis are rather vague. Many authors would consider MANOVA, the multivariate analysis of variance, as multivariate analysis. There is an S-PLUS function `manova` for MANOVA, but it is much more closely related to `aov` than to the methods of this chapter.

13.1   Graphical methods

The simplest way to examine multivariate data is via a `splom` plot, enhanced to
show the groups as discussed in Chapter 3. A similar effect can be obtained using
the S-PLUS function `brush`,

```
ir <- rbind(iris[,,1], iris[,,2], iris[,,3])
ir.species <- c(rep("s",50), rep("c",50), rep("v",50))
brush(ir)
```

by marking the cases with different symbols (Figure 13.1). Note that coincident
points are not marked. (See Section 3.2 for further details.)

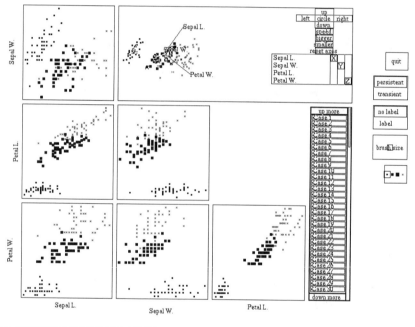

press Button 1 to highlight, Button 2 to downlight

Figure 13.1: brush plot of the `iris` data.

Principal component analysis

Linear methods are the heart of classical multivariate analysis, and depend on
seeking linear combinations of the variables with desirable properties. For the
unsupervised case the main method is *principal component* analysis, which seeks
linear combinations of the columns of X with maximal (or minimal) variance.
Because the variance can be scaled by rescaling the combination, we constrain the
combinations to have unit length.

Let S denote the covariance matrix of the data X, which is defined[1] by

$$nS = (X - n^{-1}\mathbf{1}\mathbf{1}^T X)^T(X - n^{-1}\mathbf{1}\mathbf{1}^T X) = (X^T X - n\overline{x}\overline{x}^T)$$

where $\overline{x} = \mathbf{1}^T X/n$ is the row vector of means of the variables. Then the sample variance of a linear combination xa of a row vector x is $a^T\Sigma a$ and this is to be maximized (or minimized) subject to $\|a\|^2 = a^T a = 1$. Since Σ is a non-negative definite matrix, it has an eigendecomposition

$$\Sigma = C^T\Lambda C$$

where Λ is a diagonal matrix of (non-negative) eigenvalues in decreasing order. Let $b = Ca$, which has the same length as a (since C is orthogonal). The problem is then equivalent to maximizing $b^T\Lambda b = \sum \lambda_i b_i^2$ subject to $\sum b_i^2 = 1$. Clearly the variance is maximized by taking b to be the first unit vector, or equivalently taking a to be the column eigenvector corresponding to the largest eigenvalue of Σ. Taking subsequent eigenvectors gives combinations with as large as possible variance which are uncorrelated with those which have been taken earlier. The ith principal component is then the ith linear combination picked by this procedure. (It is only determined up to a change of sign.)

Another point of view is to seek new variables y_j which are rotations of the old variables to explain best the variation in the dataset. Clearly these new variables should be taken to be the principal components, in order. Suppose we use the first k principal components. Then the subspace they span contains the 'best' k-dimensional view of the data. It both has maximal covariance matrix (both in trace and determinant) and best approximates the original points in the sense of minimizing the sum of squared distance from the points to their projections. The first few principal components are often useful to reveal structure in the data. The principal components corresponding to the smallest eigenvalues are the most nearly constant combinations of the variables, and can also be of interest.

Note that the principal components depend on the scaling of the original variables, and this will be undesirable except perhaps if (as in the iris data) they are in comparable units. (Even in this case, correlations would often be used.) Otherwise it is conventional to take the principal components of the *correlation* matrix, implicitly rescaling all the variables to have unit sample variance.

There are two functions to compute principal components, the S function prcomp (which we do not discuss further) and the S-PLUS function princomp introduced in version 3.2. As princomp returns an object of class "princomp", it will be printed and plotted specially. The argument cor controls whether the covariance or correlation matrix is used (via re-scaling the variables).

```
> ir.pca <- princomp(log(ir), cor=T)
> ir.pca
Standard deviations:
  Comp. 1 Comp. 2 Comp. 3 Comp. 4
```

---

[1] A divisor of $n - 1$ is more conventional, but princomp calls cov.wt, which uses n.

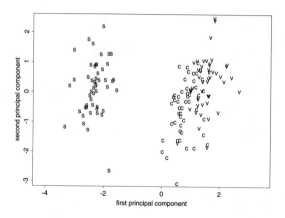

Figure 13.2: First two principal components for the log-transformed `iris` data.

```
  1.7125 0.95238   0.3647 0.16568

   ....
> summary(ir.pca)
Importance of components:
                        Comp. 1 Comp. 2  Comp. 3   Comp. 4
    Standard deviation  1.71246 0.95238 0.364703 0.1656840
Proportion of Variance  0.73313 0.22676 0.033252 0.0068628
 Cumulative Proportion  0.73313 0.95989 0.993137 1.0000000
> plot(ir.pca)
> loadings(ir.pca)
          Comp. 1 Comp. 2 Comp. 3 Comp. 4
Sepal L.   0.504   0.455   0.709   0.191
Sepal W.  -0.302   0.889  -0.331
Petal L.   0.577          -0.219  -0.786
Petal W.   0.567          -0.583   0.580
> ir.pc <- predict(ir.pca)
> eqscplot(ir.pc[,1:2], type="n",
      xlab="first principal component",
      ylab = "second principal component")
> text(ir.pc[,1:2], ir.species)
```

In the terminology of this function, the *loadings* are columns giving the linear combinations a for each principal component, and the *scores* are the data on the principal components. The plot shows the `screeplot`, a barplot of the variances of the principal components labelled by $\sum_{i=1}^{j} \lambda_i / \text{trace}(\Sigma)$. The result of `loadings` is rather deceptive, as small entries are suppressed in printing but will be insignificant only if the correlation matrix is used, and that is *not* the default. The `predict` method rotates to the principal components.

As well as a data matrix `x`, the function `princomp` can accept data via a model formula with an empty left-hand side or as a variance or correlation matrix specified by argument `covlist`, of the form output by `cov.wt` and `cov.mve`. Using the latter is one way to robustify principal component analysis.

Figure 13.2 shows the first two principal components for the `iris` data based on the covariance matrix, revealing the group structure if it had not already been known. *A warning:* principal component analysis will reveal the gross features of the data, which may already be known, and is often best applied to residuals after the known structure has been removed.

There are two books devoted solely to principal components, Jackson (1991) and Jolliffe (1986), which we think over-states its value as a technique. Other projection techniques such as projection pursuit (e.g. Huber, 1985; Friedman, 1987; Jones & Sibson, 1987; Ripley, 1996) choose rotations based on more revealing criteria than variances. (These are not included in S-PLUS, but are available through an S interface to the package XGobi available from `statlib` (see Appendix C) for Unix machines running X11.)

Distance methods

There is a class of methods based on representing the cases in a low-dimensional Euclidean space so that their proximity reflects the similarity of their variables. To do so we have to produce a measure of similarity. The S function `dist` uses one of four distance measures between the points in the p-dimensional space of variables. The default is Euclidean distance; others include `manhattan` (the L_1 distance, summing absolute differences) and `maximum`. Other distance measures, including some appropriate to categorical variables, are discussed in Section 13.2. Distances are often called *dissimilarities*.

The most obvious of the distance methods is *multi-dimensional scaling*, which seeks a configuration in $\mathbb{R}^d$ such that distances between the points best match (in a sense to be defined) those of the distance matrix. Only the classical or metric form of multi-dimensional scaling is implemented in S-PLUS, which is also known as *principal coordinate analysis*. For the `iris` data we can use:

```
ir.scal <- cmdscale(dist(ir), k = 2, eig = T)
ir.scal$points[, 2] <- -ir.scal$points[, 2]
eqscplot(ir.scal$points, type="n")
text(ir.scal$points, ir.species, cex=0.8)
```

where care is taken to ensure correct scaling of the axes (see the top left plot of Figure 13.3). Note that a configuration can be determined only up to translation, rotation and reflection, since Euclidean distance is invariant under the group of rigid motions and reflections. (The plot is reflected to match later ones.) An idea of how good the fit is can be obtained by calculating a measure of 'stress':

```
> distp <- dist(ir)
> dist2 <- dist(ir.scal$points)
> sum((distp - dist2)^2)/sum(distp^2)
[1] 0.001747
```

which shows the fit is good. Using classical multi-dimensional scaling with a Euclidean distance as here is precisely equivalent to plotting the first k principal components (without re-scaling to correlations).

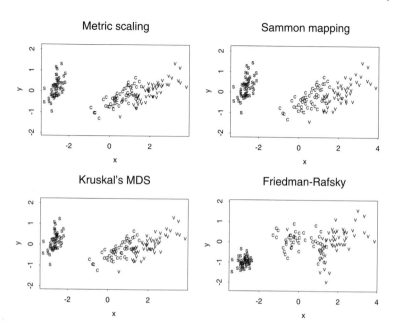

Figure 13.3: Distance-based representations of the `iris` data. The top left plot is by multidimensional scaling, the top right by Sammon's non-linear mapping, the bottom left by Kruskal's isotonic multidimensional scaling and the bottom left via minimum-spanning trees. Note that each is defined up to shifts, rotations and reflections.

A non-metric form of multi-dimensional scaling is Sammon's (1969) non-linear mapping, which given a distance d on n points constructs a k-dimensional configuration with distances $\widetilde{d}$ to minimize a weighted 'stress'

$$E(d, \widetilde{d}) = \frac{1}{\sum_{i \neq j} d_{ij}} \sum_{i \neq j} \frac{(d_{ij} - \widetilde{d}_{ij})^2}{d_{ij}}$$

by an iterative algorithm implemented in our function `sammon`. We have to drop duplicate observations to make sense of $E(d, \widetilde{d})$; running `sammon` will report which observations are duplicates.[2]

```
ir.sam <- sammon(dist(ir[-143,]))
eqscplot(ir.sam$points, type="n")
text(ir.sam$points, label=ir.species[-143], cex=0.8)
```

A more thoroughly non-metric version of goes back to Kruskal and Shepard in the 1960s (see Cox & Cox, 1994; Ripley, 1996). The idea is to choose a

---

[2] In S we would use
`(1:150)[duplicated(do.call("paste", data.frame(ir)))]`

configuration to minimize

$$STRESS^2 = \frac{\sum_{i \neq j} \left[\theta(d_{ij}) - \tilde{d}_{ij}\right]^2}{\sum_{i \neq j} \tilde{d}_{ij}^2}$$

over both the configuration of points and an increasing function θ. Now the location, rotation, reflection and scale of the configuration are all indeterminate. This is implemented in function `isoMDS` which we can use by

```
ir.iso <- isoMDS(dist(ir[-143,]))
eqscplot(ir.iso$points, type="n")
text(ir.iso$points, label=ir.species[-143], cex=0.8)
```

The optimization task is quite difficult and this can be slow.

Minimum-spanning trees can also be used to represent multivariate data in the plane (Friedman & Rafsky, 1981). A minimum spanning tree (MST) is a series of edges which join all points to form a connected graph of minimum total length. (Such trees are usually unique.) The tree is then represented in the plane so that all edges have the right length, and, locally, distances are roughly preserved. It is a less accurate representation than Sammon mapping but much faster (20 times in this example). The function `mstree` includes the Friedman–Rafsky algorithm:

```
> ir.mst <- mstree(ir)
> ir.mst$x <- -ir.mst$x; ir.mst$y <- -ir.mst$y
> eqscplot(ir.mst$y, ir.mst$x, type="n")
> text(ir.mst$y, ir.mst$x, label=ir.species, cex=0.8)
```

where we interchanged the axes only to make comparison easier in Figure 13.3.

Figure 13.4: Isotonic multi-dimensional scaling representation of the `fgl` data. The groups are plotted by the initial letter, except F for window float glass, and N for window non-float glass. Small distances are small on the plot and conversely.

These representations can be quite different in examples such as our dataset `fgl` which do not project well into a small number of dimensions; Figure 13.4 shows a non-metric MDS plot. (We omit one of an identical pair of fragments.)

```
fgl.iso <- isoMDS(dist(as.matrix(fgl[-40, -10])))
eqscplot(fgl.iso$points, type="n", xlab="", ylab="")
text(fgl.iso$points, c("F", "N", "V", "C", "T", "H")
   [fgl$type[-40]], cex=0.6)
```

Biplots

The biplot (Gabriel, 1971) is another method to represent both the cases and variables. We suppose that X has been centred to remove column means. The biplot represents X by two sets of vectors of dimensions n and p producing a rank-2 approximation to X. The best (in the sense of least squares) such approximation is given by replacing Λ in the singular value decomposition of X by D, a diagonal matrix setting $\lambda_3, \ldots$ to zero, so

$$X \approx \widetilde{X} = [\boldsymbol{u}_1 \, \boldsymbol{u}_2] \begin{bmatrix} \lambda_1 & 0 \\ 0 & \lambda_2 \end{bmatrix} \begin{bmatrix} \boldsymbol{v}_1^T \\ \boldsymbol{v}_2^T \end{bmatrix} = GH^T$$

where the diagonal scaling factors can be absorbed into G and H in a number of ways. For example, we could take

$$G = n^{a/2} [\boldsymbol{u}_1 \, \boldsymbol{u}_2] \begin{bmatrix} \lambda_1 & 0 \\ 0 & \lambda_2 \end{bmatrix}^{1-\lambda}, \qquad H = \frac{1}{n^{a/2}} [\boldsymbol{v}_1 \, \boldsymbol{v}_2] \begin{bmatrix} \lambda_1 & 0 \\ 0 & \lambda_2 \end{bmatrix}^{\lambda}$$

The biplot then consists of plotting the $n + p$ 2-dimensional vectors which form the rows of G and H. The interpretation is based on inner products between vectors from the two sets, which give the elements of $\widetilde{X}$. For $\lambda = a = 0$ this is just a plot of the first two principal components and the projections of the variable axes.

The most popular choice is $\lambda = a = 1$ (which Gabriel, 1971, calls the *principal component biplot*). Then G contains the first two principal components scaled to unit variance, so the Euclidean distances between the rows of G represents the Mahalanobis distances (page 397) between the observations and the inner products between the rows of H represent the covariances between the (possibly scaled) variables (Jolliffe, 1986, pp. 77–8); thus the lengths of the vectors represent the standard deviations.

Figure 13.5 shows a biplot with $\lambda = 1$, obtained by[3]

```
library(MASS, first=T)      # corrected biplot.princomp
state <- state.x77[,2:7]; row.names(state) <- state.abb
biplot(princomp(state, cor=T), pc.biplot=T, cex = 0.7, ex=0.8)
```

We specified a rescaling of the original variables to unit variance. (There are additional arguments `scale` which specifies λ and `expand` which specifies a scaling of the rows of H relative to the rows of G, both of which default to 1.)

Gower & Hand (1996) in a book-length discussion of biplots criticize conventional plots such as Figure 13.5. In particular they point out that the axis scales are not at all helpful. Notice that Figure 13.5 has two sets of scales. That on the lower and left axes refers to the values of the rows of G. The upper/right scale is for the values of the rows of H which are shown as arrows. Gower & Hand's preferred style (for $\lambda = 0$) is to omit the external axes and to replace each arrow by a scale for that variable. We leave programming this in S as an exercise for the reader.

---

[3] A corrected version of `biplot.princomp` from our library is used.

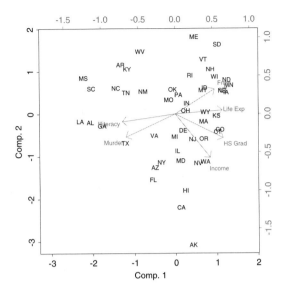

Figure 13.5: Principal component biplot of the part of the `state.x77` data. Distances between states represent Mahalanobis distance, and inner product between variables represent correlations. (The arrows extend 80% of the way along the variable's vector.)

13.2 Cluster analysis

Cluster analysis is concerned with discovering group structure amongst the cases of our n by p matrix. Two general references are Gordon (1981) and Hartigan (1975). Almost all methods are based on a measure of the similarity or dissimilarity between cases. A *dissimilarity coefficient* d is symmetric ($d(A, B) = d(B, A)$), non-negative, and $d(A, A)$ is zero. A similarity coefficient has the scale reversed. Dissimilarities may be a metric

$$d(A, C) \leqslant d(A, B) + d(B, C)$$

or an *ultrametric*

$$d(A, B) \leqslant \max\big(d(A, C), d(B, C)\big)$$

but need not be either. We have already seen several dissimilarities calculated by
`dist`.

Jardine & Sibson (1971) discuss several families of similarity and dissimilarity measures. For categorical variables most dissimilarities are measures of agreement. The *simple matching coefficient* is the proportion of categorical variables on which the cases differ. The *Jaccard coefficient* applies to categorical variables with a preferred level. It is the proportion of such variables with one of the cases at the preferred level in which the cases differ. The `binary` method of `dist` is of this family, being the Jaccard coefficient if all non-zero levels are preferred. Applied to logical variables on two cases it gives the proportion of variables in

which only one is true amongst those which are true on at least one case. There are
many variants of these coefficients; Kaufman & Rousseeuw (1990, §2.5) provide
a readable summary and recommendations.

Ultrametric dissimilarities have the appealing property that they can be represented by a *dendrogram* such as that shown in Figure 13.6, in which the dissimilarity between two cases can be read off from the height at which they join a
single group. Hierarchical clustering methods can be thought of as approximating
a dissimilarity by an ultrametric dissimilarity. Jardine & Sibson argue that one
method, single-link clustering, uniquely has all the desirable properties of a clustering method. This measures distances between clusters by the dissimilarity of
the closest pair, and agglomerates by adding the shortest possible link (i.e. joining
the two closest clusters). Other authors disagree, and Kaufman & Rousseeuw
(1990, §5.2) give a different set of desirable properties leading uniquely to their
preferred method, which views the dissimilarity between clusters as the average of
the dissimilarities between members of those clusters. Another popular method is
complete-linkage, which views the dissimilarity between clusters as the maximum
of the dissimilarities between members.

The S function `hclust` implements these three metrics, selected by its
`method` argument which takes values `compact` (the default, for complete-
linkage), `average` and `connected` (for single-linkage).

The S dataset `swiss.x` gives five measures of socio-economic data on Swiss
provinces about 1888, given by Mosteller & Tukey (1977, pp. 549–551). The data
are percentages, so Euclidean distance is a reasonable choice. We use single-link
clustering:

```
h <- hclust(dist(swiss.x), method="connected")
plclust(h)
cutree(h, 3)
plclust( clorder(h, cutree(h, 3) ))
```

The first plot suggests three main clusters, and the remaining code re-orders the
dendrogram to display (see Figure 13.6) those clusters more clearly. Note that
there are two main groups, with the point 45 well separated from them.

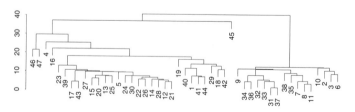

Figure 13.6: Dendrogram for the socio-economic data on Swiss provinces computed by
single-link clustering.

K-means

The K-means clustering algorithm (MacQueen, 1967; Hartigan, 1975; Hartigan & Wong, 1979) chooses a pre-specified number of cluster centres to minimize the within-class sum of squares from those centres. As such it is most appropriate to continuous variables, suitably scaled. The algorithm needs a starting point, so we choose the means of the clusters identified by group-average clustering. The clusters *are* altered (cluster 3 contained just point 45), and are shown in principal-component space in Figure 13.7. (The standard deviations show that a two-dimensional representation is reasonable.)

```
h <- hclust(dist(swiss.x), method="average")
initial <- tapply(swiss.x, list(rep(cutree(h, 3),
    ncol(swiss.x)), col(swiss.x)), mean)
dimnames(initial) <- list(NULL, dimnames(swiss.x)[[2]])
km <- kmeans(swiss.x, initial)
swiss.pca <- princomp(swiss.x)
swiss.pca
Standard deviations:
 Comp. 1 Comp. 2 Comp. 3 Comp. 4 Comp. 5
  42.903  21.202   7.588  3.6879  2.7211
    . . . .
swiss.px <- predict(swiss.pca)
dimnames(km$centers)[[2]] <- dimnames(swiss.x)[[2]]
swiss.centers <- predict(swiss.pca, km$centers)
eqscplot(swiss.px[, 1:2], type="n",
    xlab="first principal component",
    ylab="second principal component")
text(swiss.px[,1:2], km$cluster)
points(swiss.centers[,1:2], pch=3, cex=3)
identify(swiss.px[, 1:2], cex=0.5)
```

Model-based clustering

S-PLUS has functions for "model-based" clustering (Banfield & Raftery, 1993) implemented by the functions `mclust`, `mclass` and `mreloc`. For Figure 13.8 we used

```
h <- mclust(swiss.x, method = "S*")$tree
plclust( clorder(h, cutree(h, 3) ))
```

Note that this works with a data matrix and not with a dissimilarity matrix.

The idea of "model-based" clustering is that the data are independent samples from a series of group populations, but the group labels have been lost. If we knew that the vector γ gave the group labels, and each group had class-conditional pdf $f_i(x; \theta)$, then the likelihood would be

$$\prod_{i=1}^{n} f_{\gamma_i}(x_i; \theta) \qquad (13.1)$$

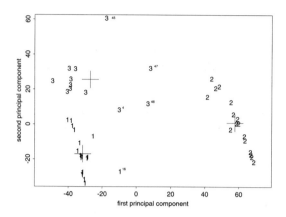

Figure 13.7: The Swiss provinces data plotted on its first two principal components. The labels are the groups assigned by K-means; the crosses denote the group means. Five points are labelled with smaller symbols.

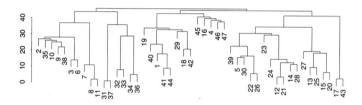

Figure 13.8: Dendrogram for the socio-economic data on Swiss provinces computed by "model-based" clustering.

Since the labels are unknown, these are regarded as parameters in (13.1), and the likelihood maximized over (θ, γ).

Choosing the class-conditional pdfs to be multivariate normal with different means but a common covariance matrix $\Sigma = \sigma^2 I$ leads to the criterion of minimizing the sum of squares to the cluster centre, that is K-means. The other options available are based on multivariate normals with other constraints on the covariance matrices. For example, the default method S* allows clusters of different sizes and orientations but the same pre-specified 'shape' (the ratio of axes of the ellipsoid), and S is similar but constrains to equal size.

The code returns an "approximate weight of evidence" for the number of clusters, which is this case suggests two or three clusters. A further option is "noise" to allow some of the points to come from a homogeneous Poisson process rather than from one of the cluster groups. (It may help to think of this as a $(k + 1)$ st cluster with a very diffuse distribution.)

Note that the hierarchical clustering given by mclust only optimizes the fit criterion in a very crude way. Once the number of clusters is chosen, mreloc can be used to optimize further, although in our example no change occurs. If we allow 'noise' two points are re-allocated:

```
h <- mclust(swiss.x, method = "S*", noise=T)
hclass <- mclass(h, 3)
hclass$class
 [1] 1 2 2 4 1 2 2 2 2 2 2 1 1 1 1 1 1 1 1 1 1 1 1 1 1 1 1 1
[29] 1 1 2 2 2 2 2 2 2 2 1 1 1 1 1 1 4 4 4
mreloc(hclass, swiss.x, method = "S*", noise=T)
 [1] 1 2 2 4 1 2 2 2 2 2 2 1 1 1 1 4 1 1 1 1 1 1 1 1 1 1 1 1
[29] 1 1 2 2 2 2 2 2 2 2 1 1 1 4 1 1 4 4 4
```

Note that this is a random algorithm so the results are not repeatable, and that the class number is the lowest-numbered object in the cluster.

Library `cluster`

The library `cluster` introduced in S-PLUS 3.4 provides an S interface to the FORTRAN clustering routines described in Kaufman & Rousseeuw (1990). These rejoice in the acronymic names of `agnes`, `clara`, `daisy`, `diana`, `fanny`, `mona` and `pam`, and all have `print` and `summary` methods.

The function `daisy` provides a more general way (than `dist`) to compute dissimilarity matrices. The main extension is to variables which are not on interval scale, for example ordinal, logratio and asymmetric binary variables (the latter being treated as in the Jaccard coefficient). The default behaviour of `daisy` is identical to that of `dist`.

The functions `pam`, `clara` and `fanny` are all partitioning methods like K-means; they are described in Kaufman & Rousseeuw (1990) and Ripley (1996, §9.3). Both `pam` and `clara` use the k-medoids criterion of Vinod (1969), which differs from K-means in requiring the cluster centre to be a data point and in using unsquared distances. As this criterion only uses distances between data points, it can also be applied to general dissimilarities. The function `pam` allows a general dissimilarity whereas `clara` works on Euclidean or Manhattan distance between rows of a data matrix, and uses a faster optimization procedure for large datasets.

```
> library(cluster)
> swiss.pam <- pam(swiss.px, 3)
> summary(swiss.pam)
Medoids:
       Comp. 1    Comp. 2 Comp. 3 Comp. 4   Comp. 5
[1,] -29.716   18.22162  1.4265 -1.3206   0.95201
[2,]  58.609    0.56211  2.2320 -4.1778   4.22828
[3,] -28.844  -19.54901  3.1506  2.3870  -2.46842
Clustering vector:
 [1] 1 2 2 1 3 2 2 2 2 2 2 3 3 3 3 3 1 1 1 3 3 3 3 3 3 3 3 3
[29] 1 3 2 2 2 2 2 2 2 2 1 1 1 1 1 1 1 1 1 1
    ....
> eqscplot(swiss.px[, 1:2], type="n",
    xlab="first principal component",
    ylab="second principal component")
> text(swiss.px[,1:2], swiss.pam$clustering)
> points(swiss.pam$medoid[,1:2], pch=3, cex=3)
```

The function `fanny` implements a 'fuzzy' version of the k-medoids criterion. Rather than point i having a membership of just one cluster v, its membership is partitioned amongst clusters as positive weights u_{iv} summing to one. The criterion then is

$$\min_{(u_{iv})} \sum_{v} \frac{\sum_{i,j} u_{iv}^2 u_{jv}^2 \, d_{ij}}{2 \sum_{i} u_{iv}^2}.$$

For our running example we find

```
> fanny(swiss.px, 3)
iterations objective
        16    354.01
Membership coefficients:
          [,1]     [,2]     [,3]
[1,] 0.725016 0.075485 0.199499
[2,] 0.189978 0.643928 0.166094
[3,] 0.191282 0.643596 0.165123
   ....
Closest hard clustering:
 [1] 1 2 2 1 3 2 2 2 2 2 2 3 3 3 3 3 1 1 1 3 3 3 3 3 3 3 3 3
[29] 1 3 2 2 2 2 2 2 2 2 1 1 1 1 1 1 1 1 1
```

Clustering tree of agnes(swiss.x, method = "single")

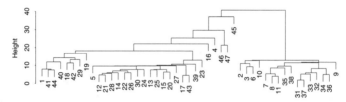

Clustering tree of diana(swiss.x)

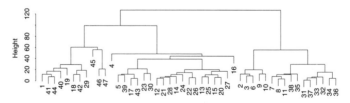

Figure 13.9: Hierarchical clustering of `swiss.x` using library `cluster`.

The remaining methods, `agnes`, `diana` and `mona` are hierarchical clustering methods; `mona` is a specialized method for binary observations only. The function `agnes` is very similar to `hclust` and also implements `average` (its default), `single` (which means `connected`) and `compact` methods of agglomerative clustering. Thus we can use

```
pltree(agnes(swiss.x, method="single"))
```

to obtain[4] a similar dendrogram to Figure 13.6. The generic function `pltree` converts the output of `agnes` into a call to `plclust`.

Function `diana` performs *divisive* clustering, in which the clusters are repeatedly subdivided rather than joined, using the algorithm of Macnaughton-Smith *et al.* (1964). Divisive clustering is an attractive option when a grouping into a few large clusters is of interest. The lower panel of Figure 13.9 was produced by `pltree(diana(swiss.x))`.

13.3 Discriminant analysis

Now suppose that we have a set of g classes, and for each case we know the class (assumed correctly). We can then use the class information to help reveal the structure. Let W denote the within-class covariance matrix, that is the covariance matrix of the variables centred on the class mean, and B denote the between-classes covariance matrix, that is of the predictions by the class means. Let M be the $g \times p$ matrix of class means, and G be the $n \times g$ matrix of class indicator variables (so $g_{ij} = 1$ if and only if case i is assigned to class j). Then the predictions are GM. Let $\bar{x}$ be the means of the variables over the whole sample. Then the sample covariance matrices are

$$W = \frac{(X - GM)^T(X - GM)}{n - g}, \qquad B = \frac{(GM - 1\bar{x})^T(GM - 1\bar{x})}{g - 1} \qquad (13.2)$$

Note that B has rank at most $\min(p, g - 1)$.

Fisher (1936) introduced a linear discriminant analysis seeking a linear combination xa of the variables which has a maximal ratio of the separation of the class means to the within-class variance, that is maximizing the ratio $a^T Ba / a^T Wa$. To compute this, choose a scaling xS of the variables so that they have the identity as their within-group correlation matrix. (One such scaling is take the principal components with respect to W, and re-scale each to unit variance.) On the re-scaled variables the problem is to maximize $a^T Ba$ subject to $\|a\| = 1$, and as we saw before, this is solved by taking a to be the eigenvector of B corresponding to the largest eigenvalue. (Note that re-scaling changes B to $S^T BS$.) The linear combination a is unique up to a change of sign (unless there are multiple eigenvalues, an event of probability zero). The exact multiple of a returned by a program will depend on its definition of the within-class variance matrix. We use the conventional divisor of $n - g$, but divisors of n and $n - 1$ have been used.

As for principal components, we can take further linear components corresponding to the next largest eigenvalues. There will be at most $r = \min(p, g - 1)$ positive eigenvalues. Note that the eigenvalues are the proportions of the between-classes variance explained by the linear combinations, which may help us to choose how many to use. The corresponding transformed variables are called the *linear discriminants* or *canonical variates*. It is often useful to plot the data on the

---

[4] actually, this needs a modification to `pltree.agnes`

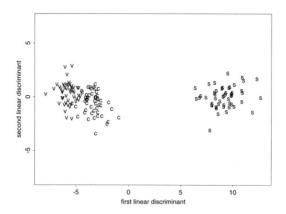

Figure 13.10: The log `iris` data on the first two discriminant axes.

first few linear discriminants (Figure 13.10). Since the within-group covariances should be the identity, we chose an equal-scaled plot.

```
> ir.lda <- lda(log(ir), ir.species)
> ir.lda
Prior probabilities of groups:
      c       s       v
 0.33333 0.33333 0.33333

Group means:
   Sepal L. Sepal W. Petal L. Petal W.
c    1.7773   1.0123  1.44293   0.27093
s    1.6082   1.2259  0.37276  -1.48465
v    1.8807   1.0842  1.70943   0.69675

Coefficients of linear discriminants:
               LD1      LD2
Sepal L.    3.7798  4.27690
Sepal W.    3.9405  6.59422
Petal L.   -9.0240  0.30952
Petal W.   -1.5328 -0.13605

Proportion of trace:
    LD1    LD2
 0.9965 0.0035
> ir.ld <- predict(ir.lda, dimen=2)$x
> eqscplot(ir.ld, type="n", xlab="first linear discriminant",
      ylab="second linear discriminant")
> text(ir.ld, ir.species)
```

This shows that 99.65% of the between-group variance is on the first discriminant axis.

Discrimination for normal populations

An alternative approach to discrimination is *via* probability models. Let π_c denote the prior probabilities of the classes, and $p(x \mid c)$ the densities of distributions of the observations for each class. Then the posterior distribution of the classes after observing x is

$$p(c \mid x) = \frac{\pi_c p(x \mid c)}{p(x)} \propto \pi_c p(x \mid c)$$

and it is fairly simple to show that the allocation rule which makes the smallest expected number of errors chooses the class with maximal $p(c \mid x)$; this is known as the *Bayes rule*.

Now suppose the distribution for class c is multivariate normal with mean μ_c and covariance Σ_c. Then the Bayes rule minimizes

$$\begin{aligned} Q_c &= -2 \log p(x \mid c) - 2 \log \pi_c \\ &= (x - \mu_c) \Sigma_c^{-1} (x - \mu_c)^T + \log |\Sigma_c| - 2 \log \pi_c \end{aligned} \tag{13.3}$$

The first term of (13.3) is the squared *Mahalanobis distance* to the class centre, and can be calculated by the S function `mahalanobis`. The difference between the Q_c for two classes is a quadratic function of x, so the method is known as *quadratic discriminant analysis* and the boundaries of the decision regions are quadratic surfaces in x space. This is implemented by our function `qda`.

Further suppose that the classes have a common covariance matrix Σ. Differences in the Q_c are then *linear* functions of x, and we can maximize $-Q_c/2$ or

$$L_c = x \Sigma^{-1} \mu_c^T - \mu_c \Sigma^{-1} \mu_c^T / 2 + \log \pi_c \tag{13.4}$$

To use (13.3) or (13.4) we have to estimate μ_c and Σ_c or Σ. The 'obvious' estimates are used, the sample mean and covariance matrix within each class, and W for Σ.

How does this relate to Fisher's linear discrimination? The latter gives new variables, the linear discriminants, with unit within-class sample variance, and the differences between the group means lie entirely in the first r variables. Thus on these variables the Mahalanobis distance (with respect to $\widehat{\Sigma} = W$) is just

$$\| x - \mu_c \|^2$$

and only the first r components of the vector depend on c. Similarly, on these variables

$$L_c = x \mu_c^T - \| \mu_c \|^2 / 2 + \log \pi_c$$

and we can work in r dimensions. If there are just two classes, there is a single linear discriminant, and

$$L_2 - L_1 = x(\mu_2 - \mu_1)^T + \text{const}$$

This is an affine function of the linear discriminant, which has coefficient $(\mu_2 - \mu_1)^T$ rescaled to unit length.

We may wish to consider more robust estimates of W (but not B); lda has an argument method = "mve" to use the minimum volume ellipsoid estimate given by cov.mve (see page 266). Other possibilities are described in Ripley (1996, pp. 39–40, 57–8). One is to fit a multivariate t_ν density, which is implemented by setting method="t". This makes a considerable difference for the fgl forensic glass data, as Figure 13.11 shows.

```
par(mfrow=c(1,2))
fgl.lda <- lda(type ~ ., fgl)
fgl.ld <- predict(fgl.lda, dimen=2)$x
eqscplot(fgl.ld[, 2:1], type="n", xlab="LD2", ylab="LD1")
text(fgl.ld[, 2:1],c("F", "N", "V", "C", "T", "H")
    [fgl$type[-40]], cex=0.6)
fgl.rlda <- lda(type ~ ., fgl, method="t")
fgl.rld <- predict(fgl.rlda, dimen=2)$x
eqscplot(fgl.rld[, 2:1], type="n", xlab="LD2", ylab="LD1")
text(fgl.rld[, 2:1],c("F", "N", "V", "C", "T", "H")
    [fgl$type[-40]], cex=0.6)
```

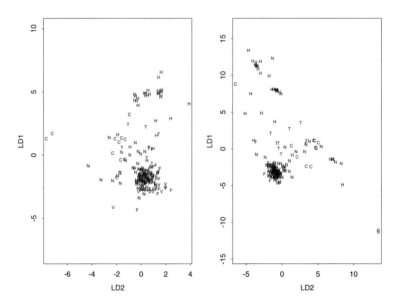

Figure 13.11: The fgl data on the first two discriminant axes. The right-hand plot used robust estimation.

Canonical correlations

Suppose the variables in our data matrix can be split into two sets, and we wish to relate the sets. For definiteness let X be the matrix for the first set of variables, and Y that for the second set. Canonical correlation analysis aims to find linear combinations of each set with maximal correlation. Since we are dealing with correlations, we do not need to impose a fixed scaling on the linear combinations.

Consider first the population version of the problem. The combined matrix $[X\,Y]$ has covariance matrix

$$\Sigma = \begin{bmatrix} \Sigma_{xx} & \Sigma_{xy} \\ \Sigma_{yx} & \Sigma_{yy} \end{bmatrix}$$

and the linear combinations xa and yb (remember our cases are *row vectors*) have correlation

$$\operatorname{corr}(xa, yb) = \frac{a^T \Sigma_{xy} b}{\sqrt{(a^T \Sigma_{xx} a \; b^T \Sigma_{yy} b)}}$$

Now rescale the variables to xS and yT so that they have identity covariance matrix within each set (for example by taking the principal components, which are uncorrelated, and re-scaling each to unit variance). On the re-scaled variables the problem is to maximize $a^T \Sigma_{xy} b$ for $\|a\| = \|b\| = 1$. Let $U\Lambda V^T$ be the singular-value decomposition of Σ_{xy}. Then

$$a^T \Sigma_{xy} b = (U^T a)^T \Lambda (V^T b)$$

which is maximized by taking a and b to be the columns of U and V corresponding to the largest singular value. Following a now familiar pattern we can define further pairs corresponding to the smaller singular values. The correlations which are achieved are equal to the singular values. Note that the problem is unchanged by a change of scale on either a or b, and by a change of sign of both.

The sample version is obtained by replacing the population covariance matrix by its sample estimate. (Here the choice of divisor is irrelevant.) The S function cancor follows precisely the algorithm sketched here, and returns a vector of correlations and matrices of the linear combinations (to be of unit centred length). There are optional parameters for the centring of the cross-covariance matrix.

The dataset swiss.x has 5 columns of percentages. We divide these into social and educational sets:

```
> cancor(swiss.x[,c(1,4,5)]/100, swiss.x[,c(2,3)]/100)
$cor:
[1] 0.76483 0.54349
$xcoef:
          [,1]      [,2]       [,3]
[1,] -0.49049 -0.51314 -0.096209
[2,] -0.14257  0.36880 -0.006300
[3,] -0.63637 -1.37960  4.970178
$ycoef:
          [,1]     [,2]
[1,] 1.766816 -1.8833
[2,] 0.094468  2.1404
$xcenter:
```

```
Agriculture Catholic Infant Mortality
     0.5066   0.41145          0.19943
$ycenter:
Examination Education
    0.16489    0.10979
```

which shows that the first canonical variate is principally on the army draft examination, and the second contrasts the two educational measurements. For neither is the Catholic variable particularly important.

There is a sense in which canonical correlations generalize linear discriminant analysis, for if we perform a canonical correlation analysis on X and G, the matrix of group indicators, we obtain the linear discriminants.

Correspondence analysis

Correspondence analysis is applied to two-way tables of counts, and can be seen as a special case of canonical correlation analysis. Suppose we have an $r \times c$ table N of counts. For example, consider Fisher's (1940) example on colours of eyes and hair of people in Caithness, Scotland:

	fair	red	medium	dark	black
blue	326	38	241	110	3
light	688	116	584	188	4
medium	343	84	909	412	26
dark	98	48	403	681	85

Correspondence analysis seeks 'scores' f and g for the rows and columns which are maximally correlated. Clearly the maximum is one, attained by constant scores, so we seek the largest non-trivial solution. Let X and Y be matrices of the group indicators of the rows and columns respectively. If we ignore means in the variances and correlations, we need to rescale X by $D_r = \sqrt{\mathrm{diag}(n_{i.}/n)}$, and Y by $D_c = \sqrt{\mathrm{diag}(n_{.j}/n)}$, and then take the singular value decomposition $D_r^{-1/2}(X^T Y/n)D_c^{-1/2} = U\Lambda V^T$, so the canonical combinations are the columns of $D_r^{-1/2}U$ and $D_c^{-1/2}V$. Since $X^T Y = N$, we can work directly on the table: see our function corresp.

```
corresp(read.table("Fisher.dat"))
First canonical correlation: 0.44637

Row scores:
    blue     light    medium    dark
-0.89679 -0.98732 0.075306 1.5743

Column scores:
    fair       red    medium    dark   black
-1.2187 -0.52258 -0.094147 1.3189 2.4518
```

13.4 An example: *Leptograpsus variegatus* crabs

Mahon (see Campbell & Mahon, 1974) recorded data on 200 specimens of *Leptograpsus variegatus* crabs on the shore in Western Australia. This occurs in two colour forms, blue and orange, and he collected 50 of each form of each sex and made five physical measurements. These were the carapace (shell) length CL and width CW, the size of the frontal lobe FL and rear width RW, and the body depth BD. Part of the authors' thesis was to establish that the two colour forms were clearly differentiated morphologically, to support classification as two separate species.

We will consider two (fairly realistic) questions:

1. Is there evidence from these morphological data alone of a division into two forms?

2. Can we construct a rule to predict the sex of a future crab of unknown colour form (species)? How accurate do we expect the rule to be?

On the second question, the body depth was measured somewhat differently for the females, so should be excluded from the analysis.

The data are physical measurements, so a sound initial strategy is to work on log scale. This has been done throughout. The data are very highly correlated, and scatterplot matrices and brush plots are none too revealing (try them for yourself).

```
> lcrabs <- log(crabs[,4:8])
> crabs.grp <- c("B", "b", "O", "o")[rep(1:4, rep(50,4))]
> lcrabs.pca <- princomp(lcrabs)
> lcrabs.pc <- predict(lcrabs.pca)
> dimnames(lcrabs.pc) <- list(NULL, paste("PC", 1:5, sep=""))
> lcrabs.pca
Standard deviations:
 Comp. 1  Comp. 2  Comp. 3  Comp. 4   Comp. 5
 0.51664 0.074654 0.047914 0.024804 0.0090522
> loadings(lcrabs.pca)
    Comp. 1 Comp. 2 Comp. 3 Comp. 4 Comp. 5
FL   0.452   0.157   0.438  -0.752   0.114
RW   0.387  -0.911
CL   0.453   0.204  -0.371          -0.784
CW   0.440          -0.672           0.591
BD   0.497   0.315   0.458   0.652   0.136
```

We started by looking at the principal components. (As the data on log scale *are* very comparable, we did not rescale the variables to unit variance.) The first principal component had by far the largest standard deviation (0.52), with coefficients which show it to be a 'size' effect. A plot of the second and third principal components shows an almost total separation into forms (Figure 3.14 and 3.15 on pages 97 and 98) on the third PC, the second PC distinguishing sex. The coefficients of the third PC show that it is contrasting overall size with FL and BD.

To proceed further, for example to do a cluster analysis, we have to remove the dominant effect of size. We used the carapace area as a good measure of size, and divided all measurements by the square root of area. It is also necessary to account for the sex differences, which we can do by analysing each sex separately, or by subtracting the mean for each sex, which we did:

```
cr.scale <- 0.5 * log(crabs$CL * crabs$CW)
slcrabs <- lcrabs - cr.scale
cr.means <- matrix(0, 2, 5)
cr.means[1,] <- apply(slcrabs[crabs$sex=="F",], 2, mean)
cr.means[2,] <- apply(slcrabs[crabs$sex=="M",], 2, mean)
dslcrabs <- slcrabs - cr.means[unclass(crabs$sex),]
lcrabs.sam <- sammon(dist(dslcrabs))
eqscplot(lcrabs.sam$points, type="n", xlab="", ylab="")
text(lcrabs.sam$points, crabs.grp)
```

As the Sammon mapping shows (Figure 13.12), Euclidean distance with this set of variables will be sensible. For this set of data, complete-link clustering makes three errors, and K-means and model-based clustering (with the default criterion S*) one or zero errors, depending on the starting point.

```
crabs.h <- cutree(hclust(dist(dslcrabs)),2)
table(crabs$sp, crabs.h)
      1   2
B   100   0
O     3  97
cr.means[1,] <- apply(dslcrabs[crabs.h==1,], 2, mean)
cr.means[2,] <- apply(dslcrabs[crabs.h==2,], 2, mean)
crabs.km <- kmeans(dslcrabs, cr.means)
table(crabs$sp, crabs.km$cluster)
      1    2
B   99    1
O    0  100
eqscplot(lcrabs.sam$points, type="n", xlab="", ylab="")
text(lcrabs.sam$points, crabs.km$cluster)
table(crabs$sp, mreloc(crabs.h, dslcrabs))
      1    2
B   100   0
O     0  100
```

Discriminant analysis for sex

We noted that BD is measured differently for males and females, so it seemed prudent to omit it from the analysis. To start with, we ignore the differences between the forms. Linear discriminant analysis, for what are highly non-normal populations, finds a variable which is essentially $CL^3RW^{-2}CW^{-1}$, a dimensionally neutral quantity. Six errors are made, all for the blue form:

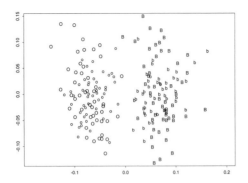

Figure 13.12: Sammon mapping of `crabs` data adjusted for size and sex. Males are coded as capitals, females as lower case, colours as the initial letter of blue or orange.

```
> dcrabs.lda <- lda(crabs$sex ~ FL + RW + CL + CW, lcrabs)
> dcrabs.lda
    ....
Coefficients of linear discriminants:
        LD1
FL   -2.8896
RW  -25.5176
CL   36.3169
CW  -11.8280
dcrabs.pred <- predict(dcrabs.lda)
table(crabs$sex, dcrabs.pred$class)
    F  M
F 97  3
M  3 97
```

It does make sense to take the colour forms into account, especially as the within-group distributions look close to joint normality (look at the data on the linear discriminants). The first two linear discriminants dominate the between-group variation. Figure 13.13 shows the data on those variables.

```
> dcrabs.lda4 <- lda(crabs.grp ~ FL + RW + CL + CW, lcrabs)
Proportion of trace:
   LD1    LD2    LD3
 0.6422 0.3491 0.0087
> dcrabs.pr4 <- predict(dcrabs.lda4, dimen=2)
> dcrabs.pr2 <- dcrabs.pr4$post %*% c(1,1,0,0)
> table(crabs$sex, dcrabs.pr2 > 0.5)
   FALSE TRUE
F     96    4
M      3   97
```

We can not represent all the decision surfaces exactly on a plot. However, using the first two linear discriminants as the data will provide a very good approximation; see Figure 13.13.

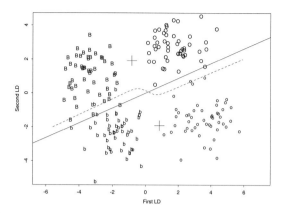

Figure 13.13: Linear discriminants for the `crabs` data. Males are coded as capitals, females as lower case, colours as the initial letter of blue or orange. The crosses are the group means for a linear discriminant for sex (solid line) and the dashed line is the decision boundary for sex based on four groups.

```
cr.t <- dcrabs.pr4$x[,1:2]
eqscplot(cr.t, type="n", xlab="First LD", ylab="Second LD")
text(cr.t, crabs.grp)
perp <- function(x, y) {
    m <- (x+y)/2
    s <- - (x[1] - y[1])/(x[2] - y[2])
    abline(c(m[2] - s*m[1], s))
    invisible()
}
cr.m <- lda(cr.t, crabs$sex)$means
points(cr.m, pch=3, mkh=0.3)
perp(cr.m[1,], cr.m[2,])

cr.lda <- lda(cr.t, crabs.grp)
x <- seq(-6, 6, 0.25)
y <- seq(-2, 2, 0.25)
Xcon <- matrix(c(rep(x,length(y)),
               rep(y, rep(length(x),length(y)))),,2)
cr.pr <- predict(cr.lda, Xcon)$post %*% c(1,1,0,0)
contour(x, y, matrix(cr.pr, length(x), length(y)),
    levels=0.5, labex=0, add=T, lty=3)
```

The reader is invited to try quadratic discrimination on this problem. It performs very marginally better than linear discrimination, not surprisingly since the covariances of the groups appear so similar, as can be seen from the result of

```
for(i in c("O", "o",  "B", "b"))
  print(var(lcrabs[crabs.grp==i, ]))
```

13.5 Factor analysis

We return to discovering structure from the data matrix X alone, without prede-
termined groups. Factor analysis seeks linear combinations xa of the variables,
called *factors*, which represent underlying fundamental quantities of which the
observed variables are expressions. The examples tend to be controversial ones
such as 'intelligence' and 'social deprivation', the idea being that a small number
of factors might explain a large number of measurements in an observational study.

This aim seems close to that of principal component analysis, but the statistical
model differs. For a single common factor f we have

$$x = \mu + \lambda f + u \qquad (13.5)$$

where λ is a vector known as the *loadings* and u is a vector of *unique* (or *specific*)
factors for that observational unit. To help make the model identifiable, we assume
that the factor f has mean zero and variance one, and that u has mean zero and
unknown *diagonal* covariance matrix Ψ. For $k < p$ common factors we have a
vector f of common factors and a loadings matrix Λ, and

$$x = \mu + \Lambda f + u \qquad (13.6)$$

where the components of f have unit variance and are uncorrelated and f and u
are taken to be uncorrelated. Note that *all* the correlations amongst the variables
in x must be explained by the common factors; if we assume joint normality the
observed variables x will be conditionally independent given f.

Principal component analysis also seeks a linear subspace like Λf to explain
the data, but measures the lack of fit by the sum of squares of the u_i. Since
factor analysis allows an arbitrary diagonal covariance matrix Ψ its measure of
fit of the u_i depends on the problem and should be independent of the units of
measurement of the observed variables. (Changing the units of measurement of
the observations does not change the common factors if the loadings and unique
factors are re-expressed in the new units.)

Equation (13.6) and the conditions on f express the covariance matrix Σ of
the data as

$$\Sigma = \Lambda \Lambda^T + \Psi \qquad (13.7)$$

Conversely, if (13.7) holds, there is a k-factor model of the form (13.6). Note that
the common factors $G^T f$ and loadings matrix ΛG give rise to the same model
for Σ, for any $k \times k$ orthogonal matrix G. Choosing an appropriate G is known
as choosing a *rotation*. All we can achieve statistically is to fit the space spanned
by the factors, so choosing a rotation is a way to choose an interpretable basis for
that space. Note that if

$$s = \tfrac{1}{2}p(p+1) - [p(k+1) - \tfrac{1}{2}k(k-1)] = \tfrac{1}{2}(p-k)^2 - \tfrac{1}{2}(p+k) < 0$$

we would expect an infinity of solutions to (13.7). This value is known as the
degrees of freedom, and comes from the number of elements in Σ minus the

number of parameters in Ψ and Λ (taking account of the rotational freedom in Λ since only $\Lambda\Lambda^T$ is determined). Thus it is usual to assume $s \geqslant 0$; for $s = 0$ there may be a unique solution, no solution or an infinity of solutions (Lawley & Maxwell, 1971, pp. 10–11).

The variances of the original variables are decomposed into two parts, the *communality* $h_i^2 = \sum_j \lambda_{ij}^2$ and *uniqueness* ψ_{ii} which is thought of as the 'noise' variance.

Fitting the factor analysis model (13.6) is performed by the S-PLUS function `factanal`. The default method ('principal factor analysis') dates from the days of limited computational power, and is not intrinsically scale invariant—it should not be used. The preferred method is to maximize the likelihood over Λ and Ψ assuming multivariate normality of the factors (f, u), which depends only on the factor space and is scale-invariant. This likelihood can have multiple local maxima; this possibility is usually ignored but `factanal` compares the fit found from five separate starting points. It is possible that the maximum likelihood solution will have some $\widehat{\psi}_{ii} = 0$, so the ith variable lies in the estimated factor space. Opinions differ as to what to do in this case (sometimes known as a *Heywood case*), but often it indicates a lack of data or inadequacy of the factor analysis model. (Bartholomew, 1987, Section 3.6, discusses possible reasons and actions.)

The data matrix X can be specified as the first argument to `factanal` as a matrix or data frame, or via a formula with a null left-hand side[5]. Let us consider the data on Swiss cantons in matrix `swiss.x`.

```
> swiss.FA <- factanal(swiss.x, factors=2, method="mle")
Sums of squares of loadings:
 Factor1 Factor2
  1.9384  1.2923
    ....

Test of the hypothesis that 2 factors are sufficient
versus the alternative that more are required:
The chi square statistic is 2.97 on 1 degree of freedom.
The p-value is 0.0847
    ....
```

The 'Sums of squares of loadings' are the $\sum_i \lambda_{ij}^2$, which do depend on the rotation chosen, although their sum does not. The test statistic is a likelihood ratio test[6] of the fit, and may be used to help select the number of factors; here the fit is marginal with two factors, the maximum possible with five original variables. The `summary` method gives both more and less information:

```
> summary(swiss.FA)
Importance of factors:
```

---

[5] This will not work in earlier versions of S-PLUS (3.2 and 3.3).

[6] with a Bartlett correction: see Bartholomew (1987, p. 46) or Lawley & Maxwell (1971, pp. 35–36). For a Heywood case (as here) Lawley & Maxwell (1971, p. 37) suggest the number of degrees of freedom should be increased by the number of variables with zero uniqueness.

```
                    Factor1 Factor2
      SS loadings 1.93843 1.29230
Proportion Var 0.38769 0.25846
Cumulative Var 0.38769 0.64615
```

The degrees of freedom for the model is 1.

```
Uniquenesses:
  Agriculture Examination Education    Catholic Infant Mortality
      0.40764        0.19008    0.20264 0.00014068                0.96878
```

```
Loadings:
                    Factor1 Factor2
         Agriculture -0.713    0.290
         Examination  0.777   -0.453
           Education  0.893
            Catholic -0.161    0.987
   Infant Mortality           0.170
```

The function loadings gives just the loadings Λ, the smallest numbers in which have been suppressed in the print method. This output is not quite what it appears, as the original variables have been re-scaled to unit variance (with divisor n; equivalently, Σ in (13.7) has been replaced by the the correlation matrix), and so the loading Λ and uniquenesses Ψ refer to the rescaled variables. Bartholomew (1987, p. 49) refers to this as the *standard* or *scale-invariant* form of the parameters Λ and Ψ. The component scale of the returned object relates[7] the output to the original variables.

The scale-invariant output does show that the Catholic variable is very nearly explained by the common factors, and Infant Mortality variable is poorly explained. In fact the uniqueness ψ_{ii} for the Catholic variable is being estimated as zero as tightening the convergence criteria shows:

```
> factanal(swiss.x, factors=2, method="mle",
      control=list(iter.max=100, unique.tol=1e-20))$uniq
  Agriculture Examination Education    Catholic Infant Mortality
      0.40764        0.19008    0.20264 2.8792e-09                0.96878
```

This confirms that the Catholic variable lies in the factor space, so we have a Heywood case. (In this example religion is a plausible candidate for a latent factor.) As the fit is marginal, it is instructive to consider $\Sigma - \hat{\Lambda}\hat{\Lambda}^T - \hat{\Psi}$:

```
> A <- loadings(swiss.FA) %*% t(loadings(swiss.FA)) +
          diag(swiss.FA$uniq)
> round(cor(swiss.x) - A, 3)
              Agriculture Examination Education Catholic Mortality
  Agriculture       0.000      -0.001     0.000        0     -0.145
  Examination      -0.001       0.000     0.000        0      0.001
  Education         0.000       0.000     0.000        0     -0.054
```

---

[7] This is a vector x such that original variable j was *divided* by x_j.

```
         Catholic    0.000      0.000     0.000      0     0.000
         Mortality  -0.145      0.001    -0.054      0     0.000
```

Most of the lack of fit comes from just one correlation.

Note that unlike principal components, common factors are not generated one at a time, and the two-factor space will usually not contain the single-factor space. If we ask for one common factor (the default) rather than two we obtain

```
> swiss.FA1 <- factanal(swiss.x, method="mle")
> swiss.FA1
  ....
Test of the hypothesis that 1 factor is sufficient
versus the alternative that more are required:
The chi square statistic is 17.53 on 5 degrees of freedom.
The p-value is 0.00359
  ....
> summary(swiss.FA1)
  ....
Uniquenesses:
 Agriculture Examination Education Catholic Infant Mortality
     0.52866  2.2139e-06   0.51222  0.67184            0.987

Loadings:
                  Factor1
      Agriculture -0.687
      Examination  1.000
        Education  0.698
         Catholic -0.573
 Infant Mortality -0.114
```

This time the Examination variable is fitted almost exactly. Thus the one-factor solution is the Examination variable, and it is easy to check that this is not in the subspace spanned by the two-factor solution.

It is hard to find examples in the literature for which a factor analysis model fits well: many do not give a measure of fit, or have failed to optimize the likelihood well enough and so failed to detect Heywood cases. We consider an example from Smith & Stanley (1983) as quoted by Bartholomew (1987, pp. 61–65)[8]. Six tests were give to 112 individuals, with covariance matrix

```
          general picture  blocks    maze reading    vocab
 general   24.641   5.991  33.520   6.023  20.755   29.701
 picture    5.991   6.700  18.137   1.782   4.936    7.204
  blocks   33.520  18.137 149.831  19.424  31.430   50.753
    maze    6.023   1.782  19.424  12.711   4.757    9.075
 reading   20.755   4.936  31.430   4.757  52.604   66.762
   vocab   29.701   7.204  50.753   9.075  66.762  135.292
```

---

[8] Bartholomew gives both covariance and correlation matrices, but these are inconsistent. Neither are in the original paper.

The tests were of general intelligence, picture completion, block design, mazes, reading comprehension and vocabulary. Both factanal and princomp can use covariance matrices as input using a covlist argument

```
> ability.cl <- list(cov=ability.cov, center=rep(0,6), n.obs=112)
> ability.FA <- factanal(covlist=ability.cl, method="mle")
> ability.FA
    ....
The chi square statistic is 75.18 on 9 degrees of freedom.
    ....
> ability.FA <- update(ability.FA, factors=2)
> ability.FA
    ....
The chi square statistic is 6.11 on 4 degrees of freedom.
The p-value is 0.191
    ....
> summary(ability.FA)
    ....
Uniquenesses:
 general picture  blocks    maze reading   vocab
 0.45523 0.58933 0.21817 0.76942 0.052463 0.33358

Loadings:
         Factor1 Factor2
general 0.501    0.542
picture 0.158    0.621
 blocks 0.208    0.859
   maze 0.110    0.467
reading 0.957    0.179
  vocab 0.785    0.222
```

Remember that the first variable is a composite measure: it seems that the first factor reflects verbal ability, the second spatial reasoning. The main lack of fit is that the correlation 0.308 between picture and maze is fitted as 0.193.

Factor rotations

There are many criteria for selecting rotations of the factors and loadings matrix; S-PLUS implements 12. There is an auxiliary function rotate which will rotate the fitted Λ according to one of these criteria, which is called via the rotate argument of factanal. The default varimax criterion is to maximize

$$\sum_{i,j}(d_{ij} - \bar{d}_{.j})^2 \qquad \text{where} \qquad d_{ij} = \lambda_{ij}^2 / \sum_j \lambda_{ij}^2 \qquad (13.8)$$

and $\bar{d}_{.j}$ is the mean of the d_{ij}. Thus the varimax criterion maximizes the sum over factors of the variances of the (normalized) squared loadings. The normalizing factors are the communalities which are invariant under orthogonal rotations.

The usual aim of a rotation is to achieve 'simple structure', that is a pattern of loadings which is easy to interpret with a few large and many small coefficients. The effect of normalization is to rescale the variables so the variance explained by the common factors is one for each variable. Normalization makes this rotation criterion scale-invariant; this is not the case for all the criteria, but the **S-PLUS** functions work with the scale-invariant loadings.

Not all the 'rotations' are orthogonal, for example the promax criterion seeks factors (such as arithmetical and verbal reasoning skills in psychology) that might be expected to be correlated. It is constructed by a least-squares fit of Λ to $Q = [|\lambda_{ij}|^4 \text{sign}(\lambda_{ij})]$, and so tends to increase in magnitude large loadings relative to small ones. An initial value of Λ is needed, by default the varimax solution. For our example we have

```
> rotate(swiss.FA, rotation="promax")
Sums of squares of loadings:
[1] 1.9796 1.2511
    ....
Test of the hypothesis that 2 factors are sufficient
versus the alternative that more are required:
The chi square statistic is 2.97 on 1 degree of freedom.
The p-value is 0.0847
    ....
> rotate(loadings(swiss.FA), rotation="promax")
$rmat:
                          [,1]      [,2]
    Agriculture     -0.720923  0.269493
    Examination      0.789990 -0.431096
      Education      0.892821  0.015316
       Catholic     -0.189352  0.981838
Infant Mortality    -0.053304  0.168463
    ....
```

Note that not all rotation methods produce objects of class loadings describing the the rotated factors (the rmat component). so the print method for loadings is not always used, as here. Some care is needed to interpret these *oblique* rotations, as the rotated factors are no longer uncorrelated; for example (13.7) has to modified.

The oblimin criterion is another idea to produce oblique rotations: it minimizes the sum over all pairs of factors of the covariance between the squared loadings for those factors. We can illustrate this on the intelligence test data.

```
> loadings(rotate(ability.FA, rotation="oblimin"))
        Factor1 Factor2
general   0.379   0.513
picture           0.640
 blocks           0.887
   maze           0.483
reading   0.946
  vocab   0.757   0.137
```

```
Component/Factor Correlations:
        Factor1 Factor2
Factor1 1.000   0.356
Factor2 0.356   1.000
```

We can illustrate the oblique rotation graphically; see Figure 13.14.

```
par(pty="s")
L <- loadings(ability.FA)
eqscplot(L, xlim=c(0,1), ylim=c(0,1))
identify(L, dimnames(L)[[1]])
oblirot <- rotate(loadings(ability.FA), rotation="oblimin")
naxes <- solve(oblirot$tmat)
arrows(rep(0,2), rep(0,2), naxes[,1], naxes[,2])
```

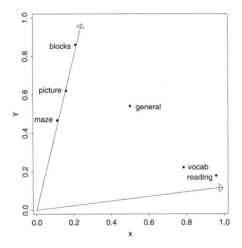

Figure 13.14: The loadings for the intelligence test data after varimax rotation, with the axes for the oblimin rotation shown as arrows.

It is also possible to rotate the loadings from a `princomp` fit, but care is needed as these are not the usual definition (Basilevsky, 1994, p. 258) of loadings for rotation.

Estimating the factor scores

Once factors have been fitted and perhaps interpreted, it may be of interest to estimate the scores of future individuals on the factors. Suppose that the observed vector of observations on a future individual is x_0, and the sample mean is $\bar{x}$. Bartlett suggested the use of (weighted) least squares, that is to regress the observations on the fitted loadings treating the u_i as random $N(0, \widehat{\Psi})$ terms and f as the parameters, giving

$$\widehat{f} = [\widehat{\Lambda}^T \widehat{\Psi}^{-1} \widehat{\Lambda}]^{-1} \widehat{\Lambda}^T \widehat{\Psi}(x_0 - \bar{x}) \qquad (13.9)$$

On the other hand, Thomson noted that if the factors are treated as random variables (as they are in the statistical model),

$$E[f \mid x_0] = \Lambda^T [\Lambda\Lambda^T + \Psi]^{-1}(x_0 - \mu) = \Lambda^T \Sigma^{-1}(x_0 - \bar{x})$$

which suggests the use of

$$\widehat{f} = \widehat{\Lambda}^T \widehat{\Sigma}^{-1}(x_0 - \bar{x}) \tag{13.10}$$

The function `predict.factanal` uses the `type` of `"weighted.ls"` for the Bartlett approach, and `"regression"` for the Thomson approach (its default). The scores for the data are the `scores` component of a `factanal` object, of type specified by the `type` argument to `factanal` (with Thomson scores as the default).

Comparisons with principal component analysis

Despite the many protestations in the literature of a fundamental difference, factor analysis continues to be confused with principal component analysis. We have already noted that selecting the first k principal components fits the model (13.6) with criterion $\sum \|u_i\|^2 = \sum_{i,j} u_{ij}^2$ (page 383). By contrast, maximum-likelihood factor analysis uses the criterion

$$- \operatorname{trace} \Sigma^{-1} S + \log |\Sigma^{-1} S|$$

which matches the observed covariances (or correlations) S to $\Sigma = \Lambda\Lambda^T + \Psi$, and there is no assumption that the specific factors u need be small, just uncorrelated.

Nevertheless, we often find that if the variables have been suitably scaled, for example scaled to unit variance, factor analysis chooses Ψ so that either one (or more) $\widehat{\Psi}_{jj} = 0$ or the $\widehat{\Psi}_{jj}$ are fairly similar and quite small. Then either the factor analysis solution is a subspace containing one or more of the variables or it is likely to be rather similar to the subspace spanned by the first k principal components. (Theoretical support is given by Gower, 1966, and Rao, 1955.) Although in theory the interest in factor analysis is in explaining correlations not variances, this is belied by the output of factor analysis functions (`summary.factanal` indicates the importance of the factors by the proportions of variance explained) and by the way case studies are explained. (See, for example, Sections 8.3 and 8.4 of Reyment & Jöreskog, 1993.)

The fundamental difference is that factor analysis chooses the scaling of the variables via $\widehat{\Psi}$ whereas in principal component analysis the scaling must be chosen by the user. If the user chooses well, there may be little difference in the factors found.

Chapter 14

Tree-based Methods

The use of tree-based models will be relatively unfamiliar to statisticians, although researchers in other fields have found trees to be an attractive way to express knowledge and aid decision-making. Keys such as Figure 14.1 are common in botany and in medical decision-making, and provide a way to encapsulate and structure the knowledge of experts to be used by less-experienced users. Notice how this tree uses both categorical variables and splits on continuous variables.

The automatic construction of decision trees dates from work in the social sciences by Morgan & Sonquist (1963) and Morgan & Messenger (1973). In statistics Breiman *et al.* (1984) had a seminal influence both in bringing the work to the attention of statisticians and in proposing new algorithms for constructing trees. At around the same time decision tree induction was beginning to be used in the field of *machine learning*, notably by Quinlan (1979, 1983, 1986, 1993), and in engineering (Henrichon & Fu, 1969; Sethi & Sarvarayudu, 1982). Whereas there is now an extensive literature in machine learning, further statistical contributions are still sparse. The introduction within S of tree-based models described by Clark & Pregibon (1992) has made the methods much more freely available. Their methods are very much in the spirit of exploratory data analysis, with many functions to investigate trees. (On the other hand, it is not possible to enter trees such as Figure 14.1 without cracking the internal structure. It is a tree, and readers are encouraged to draw it.) Ripley (1996, Chapter 7) gives a comprehensive survey of the subject, with proofs of the theoretical results.

Constructing trees may be seen as a type of variable selection. Questions of interaction between variables are handled automatically, and to a large extent so is monotonic transformation of both the x and y variables. These issues are reduced to which variables to divide on, and how to achieve the split.

Figure 14.1 is a *classification* tree since its endpoint is a factor giving the species. Although this the most common use, it is also possible to have *regression* trees in which each terminal node gives a predicted value, as shown in Figure 14.2 for our dataset cpus .

Much of the machine learning literature is concerned with logical variables and correct decisions. The end point of a tree is a (labelled) partition of the space $\mathcal{X}$ of possible observations. In logical problems it is assumed that there *is* a partition of the space $\mathcal{X}$ which will correctly classify all observations, and the

1.	Leaves subterete to slightly flattened, plant with bulb	2.
	Leaves flat, plant with rhizome	4.
2.	Perianth-tube > 10 mm	**I. × hollandica**
	Perianth-tube < 10 mm	3.
3.	Leaves evergreen	**I. xiphium**
	Leaves dying in winter	**I. latifolia**
4.	Outer tepals bearded	**I. germanica**
	Outer tepals not bearded	5.
5.	Tepals predominately yellow	6.
	Tepals blue, purple, mauve or violet	8.
6.	Leaves evergreen	**I. foetidissima**
	Leaves dying in winter	7.
7.	Inner tepals white	**I. orientalis**
	Tepals yellow all over	**I. pseudocorus**
8.	Leaves evergreen	**I. foetidissima**
	Leaves dying in winter	9.
9.	Stems hollow, perianth-tube 4–7mm	**I. sibirica**
	Stems solid, perianth-tube 7–20mm	10.
10.	Upper part of ovary sterile	11.
	Ovary without sterile apical part	12.
11.	Capsule beak 5–8mm, 1 rib	**I. enstata**
	Capsule beak 8–16mm, 2 ridges	**I. spuria**
12.	Outer tepals glabrous, many seeds	**I. versicolor**
	Outer tepals pubescent, 0–few seeds	**I. × robusta**

Figure 14.1: Key to British species of the genus *Iris*. Simplified from Stace (1991, p. 1140), by omitting parts of his descriptions.

task is to find a tree to describe it succinctly. A famous example of Donald Michie (e.g. Michie, 1989) is whether the space shuttle pilot should use the autolander or land manually (Table 14.1). Some enumeration will show that the decision has been specified for 253 out of the 256 possible observations. Some cases have been specified twice. This body of expert opinion needed to be reduced to a simple decision aid, as shown in Figure 14.3. (Table 14.1 appears to result from a decision tree which differs from Figure 14.3 in reversing the order of two pairs of splits.)

Note that the botanical problem is treated as if it were a logical problem, although there will be occasional specimens which do not meet the specification for their species.

14.1   Partitioning methods

The ideas for classification and regression trees are quite similar, but the terminology differs, so we will consider classification first. Classification trees are more familiar and it is a little easier to justify the tree-construction procedure, so we consider them first.

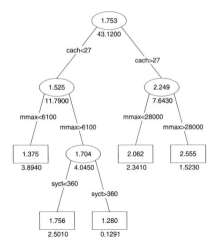

Figure 14.2: A regression tree for the cpu performance data on $\log_{10}$ scale. The value in each node is the prediction for the node, those underneath the nodes indicate the deviance contributions D_i.

Table 14.1: Example decisions for the space shuttle autolander problem.

stability	error	sign	wind	magnitude	visibility	decision
any	any	any	any	any	no	auto
xstab	any	any	any	any	yes	noauto
stab	LX	any	any	any	yes	noauto
stab	XL	any	any	any	yes	noauto
stab	MM	nn	tail	any	yes	noauto
any	any	any	any	Out of range	yes	noauto
stab	SS	any	any	Light	yes	auto
stab	SS	any	any	Medium	yes	auto
stab	SS	any	any	Strong	yes	auto
stab	MM	pp	head	Light	yes	auto
stab	MM	pp	head	Medium	yes	auto
stab	MM	pp	tail	Light	yes	auto
stab	MM	pp	tail	Medium	yes	auto
stab	MM	pp	head	Strong	yes	noauto
stab	MM	pp	tail	Strong	yes	auto

Classification trees

We have already noted that the end-point for a tree is a partition of the space $\mathcal{X}$, and we compare trees by how well that partition corresponds to the correct decision rule for the problem. In logical problems the easiest way to compare partitions is to count the number of errors, or, if we have a prior over the space $\mathcal{X}$, to compute the probability of error.

In statistical problems the distributions of the classes over $\mathcal{X}$ usually overlap,

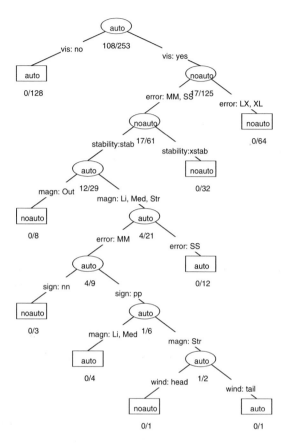

Figure 14.3: Decision tree for shuttle autolander problem. The numbers m/n denote the proportion of training cases reaching that node with the classification in the label.

so there is no partition which completely describes the classes. Then for each cell of the partition there will be a probability distribution over the classes, and the Bayes decision rule will choose the class with highest probability. This corresponds to assessing partitions by the overall probability of misclassification. Of course, in practice we do not have the whole probability structure, but a training set of n classified examples which we assume are an independent random sample. Then we can estimate the misclassification rate by the proportion of the training set which is misclassified.

Almost all current tree-construction methods, including those in S, use a one-step lookahead. That is, they choose the next split in an optimal way, without attempting to optimize the performance of the whole tree. (This avoids a combinatorial explosion over future choices, and is akin to a very simple strategy for playing a game such as chess.) However, by choosing the right measure to optimize at each split, we can ease future splits. It does not seem appropriate to use the misclassification rate to choose the splits.

What class of splits should we allow? Both Breiman *et al.*'s CART methodology and the S methods only allow binary splits, which avoids one difficulty in comparing splits, that of normalization by size. For a continuous variable x_j the allowed splits are of the form $x_j < t$ versus $x_j \geqslant t$. For ordered factors the splits are of the same type. For general factors the levels are divided into two classes. (Note that for L levels there are 2^L possible splits, and if we disallow the empty split and ignore the order, there are still $2^{L-1} - 1$. For ordered factors there are only $L - 1$ possible splits.) Some algorithms, including CART but excluding S, allow linear combination of continuous variables to be split, and Boolean combinations to be formed of binary variables.

The justification for the S methodology is to view the tree as providing a probability model (hence the title 'tree-based models' of Clark & Pregibon, 1992). At each node i of a classification tree we have a probability distribution p_{ik} over the classes. The partition is given by the *leaves* of the tree (also known as terminal nodes). Each case in the training set is assigned to a leaf, and so at each leaf we have a random sample n_{ik} from the multinomial distribution specified by p_{ik}.

We now condition on the observed variables x_i in the training set, and hence we know the numbers n_i of cases assigned to every node of the tree, in particular to the leaves. The conditional likelihood is then proportional to

$$\prod_{\text{cases } j} p_{[j]y_j} = \prod_{\text{leaves } i} \prod_{\text{classes } k} p_{ik}^{n_{ik}}$$

where $[j]$ denotes the leaf assigned to case j. This allows us to define a deviance for the tree as

$$D = \sum_i D_i, \qquad D_i = -2 \sum_k n_{ik} \log p_{ik}$$

as a sum over leaves.

Now consider splitting node s into nodes t and u. This changes the probability model within node s, so the reduction in deviance for the tree is

$$D_s - D_t - D_u = 2 \sum_k \left[n_{tk} \log \frac{p_{tk}}{p_{sk}} + n_{uk} \log \frac{p_{uk}}{p_{sk}} \right]$$

Since we do not know the probabilities, we estimate them from the proportions in the split node, obtaining

$$\hat{p}_{tk} = \frac{n_{tk}}{n_t}, \qquad \hat{p}_{uk} = \frac{n_{uk}}{n_u}, \qquad \hat{p}_{sk} = \frac{n_t \hat{p}_{tk} + n_u \hat{p}_{uk}}{n_s} = \frac{n_{sk}}{n_s}$$

so the reduction in deviance is

$$\begin{aligned}
D_s - D_t - D_u &= 2 \sum_k \left[n_{tk} \log \frac{n_{tk} n_s}{n_{sk} n_t} + n_{uk} \log \frac{n_{uk} n_s}{n_{sk} n_u} \right] \\
&= 2 \Bigg[\sum_k n_{tk} \log n_{tk} + n_{uk} \log n_{uk} - n_{sk} \log n_{sk} \\
&\qquad + n_s \log n_s - n_t \log n_t - n_u \log n_u \Bigg]
\end{aligned}$$

This gives a measure of the value of a split. Note that it is size-biased; there is more value in splitting leaves with large numbers of cases.

The tree construction process takes the maximum reduction in deviance over all allowed splits of all leaves, to choose the next split. (Note that for continuous variates the value depends only on the split of the ranks of the observed values, so we may take a finite set of splits.) The tree construction continues until the number of cases reaching each leaf is small (by default $n_i < 10$ in S) or the leaf is homogeneous enough (by default its deviance is less than 1% of the deviance of the root node in S, which is a size-biased measure). Note that as all leaves not meeting the stopping criterion will eventually be split, an alternative view is to consider splitting any leaf and choose the best allowed split (if any) for that leaf, proceeding until no further splits are allowable.

This justification for the value of a split follows Ciampi *et al.* (1987) and Clark & Pregibon, but differs from most of the literature on tree construction. The more common approach is to define a measure of the impurity of the distribution at a node, and choose the split which most reduces the average impurity. Two common measures are the entropy $\sum p_{ik} \log p_{ik}$ and the Gini index

$$\sum_{j \neq k} p_{ij} p_{ik} = 1 - \sum_k p_{ik}^2$$

As the probabilities are unknown, they are estimated from the node proportions. With the entropy measure, the average impurity differs from D by a constant factor, so the tree construction process is the same, except perhaps for the stopping rule. Breiman *et al.* preferred the Gini index.

Regression trees

The prediction for a regression tree is constant over each cell of the partition of $\mathcal{X}$ induced by the leaves of the tree. The deviance is defined as

$$D = \sum_{\text{cases } j} (y_j - \mu_{[j]})^2$$

and so clearly we should estimate the constant μ_i for leaf i by the mean of the values of the training-set cases assigned to that node. Then the deviance is the sum over leaves of D_i, the corrected sum of squares for cases within that node, and the value of a split is the reduction in the residual sum of squares.

The obvious probability model (and that proposed by Clark & Pregibon) is to take a normal $N(\mu_i, \sigma^2)$ distribution within each leaf. Then D is the usual scaled deviance for a Gaussian GLM. However, the distribution at internal nodes of the tree is then a mixture of normal distributions, and so D_i is only appropriate at the leaves. The tree-construction process has to be seen as a hierarchical refinement of probability models, very similar to forwards variable selection in regression. In contrast, for a classification tree, one probability model can be used throughout the tree construction process.

Missing values

One attraction of tree-based methods is the ease with which missing values can be handled. Consider the botanical key of Figure 14.1. We only need to know about a small subset of the 10 observations to classify any case, and part of the art of constructing such trees is to avoid observations which will be difficult or missing in some of the species (or as in capsules, for some of the cases). A general strategy is to 'drop' a case down the tree as far as it will go. If it reaches a leaf we can predict y for it. Otherwise we use the distribution at the node reached to predict y, as shown in Figure 14.2, which has predictions at all nodes.

An alternative strategy (not implemented in S) is used by many botanical keys and can be seen at nodes 9 and 12 of Figure 14.1. A list of characteristics is given, the most important first, and a decision made from those observations which are available. This is codified in the method of *surrogate splits* in which surrogate rules are available at non-terminal nodes to be used if the splitting variable is unobserved. Another attractive strategy is to split cases with missing values, and pass part of the the case down each branch of the tree (Ripley, 1996, p. 232).

Tree construction in `tree` is based on the cases without any missing observations.

S implementation

The S implementation is based on a class `tree` which has a quite complex internal structure. The tree is a regression tree unless the response variable in the model formula is a factor, and the internal structure differs in the two cases. Beware: it is very easy to code the classes numerically and then fit a regression tree by mistake.

There are serious errors in the implementation prior to S-PLUS 3.4; for earlier versions use

```
library(treefix, first=T)
```

to correct (most of) the problems.

The function `tree` constructs trees, and there are `print`, `summary` and `plot` methods for trees. As our first example consider the computer performance data in our data frame `cpus`. Ein-Dor & Feldmesser (1987) studied data on the performance on a benchmark of a mix of minicomputers and mainframes. The measure was normalized relative to an IBM 370/158-3. There were six machine characteristics, the cycle time (nanoseconds), the cache size (Kb), the main memory size (Kb) and number of channels. (For the latter two there are minimum and maximum possible values; what the actual machine tested had is unspecified.) The original paper gave a linear regression for the square root of performance.

We first consider a tree for the performance and then for log-performance.

```
> cpus.tr <- tree(perf ~ syct+mmin+mmax+cach+chmin+chmax, cpus)
> summary(cpus.tr)
Regression tree:
tree(formula = perf ~ syct + mmin + mmax + cach + chmin + chmax,
     data = cpus)
Variables actually used in tree construction:
[1] "mmax"  "cach"  "chmax"  "mmin"
Number of terminal nodes:  9
Residual mean deviance:  4523 = 904600 / 200
   ....
> print(cpus.tr)
node), split, n, deviance, yval
      * denotes terminal node

 1) root 209 5380000 105.60
   2) mmax<28000 182  585900   60.72
     4) cach<27 141    97850   39.64
       8) mmax<10000 113   36000   32.21 *
       9) mmax>10000 28    30470   69.61 *
     5) cach>27 41  209900 133.20
      10) cach<96.5 34    96490 114.40
        20) mmax<11240 14   12950   72.79 *
        21) mmax>11240 20   42240 143.60 *
      11) cach>96.5 7    43150 224.40 *
   3) mmax>28000 27 1954000 408.30
     6) chmax<59 22  436600 323.20
      12) mmin<12000 15  106500 244.50
        24) cach<56 9    26650 191.60 *
        25) cach>56 6    16700 324.00 *
      13) mmin>12000 7    38480 491.70 *
     7) chmax>59 5  658000 782.60 *
> plot(cpus.tr, type="u");  text(cpus.tr, srt=90)

> cpus.ltr <- tree(log10(perf) ~ syct + mmin + mmax + cach
     + chmin + chmax, cpus)
> summary(cpus.ltr)
Number of terminal nodes:  19
Residual mean deviance:  0.0239 = 4.55 / 190
   ....
> plot(cpus.ltr, type="u");  text(cpus.ltr, srt=90)
```

This output needs some explanation. The model is specified by a model formula
with terms separated by +; interactions make no sense for trees. The first tree
does not use all of the variables, so the summary informs us. (The second tree does
use all 6.) Next we have a compact printout of the tree including the number, sum
of squares and mean at the node. The plots are shown in Figure 14.4. The `type`
parameter is used to reduce over-crowding of the labels; by default the depths
reflect the values of the splits. Plotting uses no labels and allows the tree topology
to be studied: split conditions and the values at the leaves are added by `text`

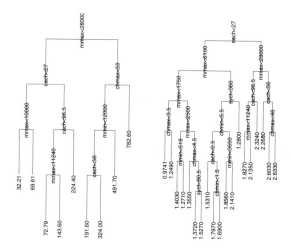

Figure 14.4: Regression trees for the cpu performance data on linear (left) and $\log_{10}$ scale (right).

(which has a number of other options). The elegant plots such as Figure 14.2 are produced directly in POSTSCRIPT by `post.tree`.

For examples of classification trees we use the `iris` data discussed in Chapter 13. We have

```
> ir.species <- factor(c(rep("s",50), rep("c", 50),
    rep("v", 50)))
> ird <- data.frame(rbind(iris[,,1], iris[,,2], iris[,,3]))
> ir.tr <- tree(ir.species ~., ird)
> summary(ir.tr)

Classification tree:
tree(formula = ir.species ~ ., data = ird)
Variables actually used in tree construction:
[1] "Petal.L." "Petal.W." "Sepal.L."
Number of terminal nodes:  6
Residual mean deviance:  0.125 = 18 / 144
Misclassification error rate: 0.0267 = 4 / 150

> ir.tr
node), split, n, deviance, yval, (yprob)
      * denotes terminal node

 1) root 150 330.0 c ( 0.330 0.33 0.330 )
   2) Petal.L.<2.45 50    0.0 s ( 0.000 1.00 0.000 ) *
   3) Petal.L.>2.45 100 140.0 c ( 0.500 0.00 0.500 )
     6) Petal.W.<1.75 54   33.0 c ( 0.910 0.00 0.093 )
      12) Petal.L.<4.95 48    9.7 c ( 0.980 0.00 0.021 )
        24) Sepal.L.<5.15 5    5.0 c ( 0.800 0.00 0.200 ) *
```

```
    25) Sepal.L.>5.15 43    0.0 c ( 1.000 0.00 0.000 ) *
    13) Petal.L.>4.95 6    7.6 v ( 0.330 0.00 0.670 ) *
   7) Petal.W.>1.75 46    9.6 v ( 0.022 0.00 0.980 )
    14) Petal.L.<4.95 6    5.4 v ( 0.170 0.00 0.830 ) *
    15) Petal.L.>4.95 40    0.0 v ( 0.000 0.00 1.000 ) *
```

The (yprob) give the distribution by class within the node. Note how the second split on Petal length occurs twice, and that splitting node 12 is attempting to classify one case of *I. virginica* without success. Thus viewed as a decision tree we would want to snip off nodes 24, 25, 14, 15. We can do so interactively:

```
> plot(ir.tr)
> text(ir.tr, all=T)
> ir.tr1 <- snip.tree(ir.tr)
node number:  12
   tree deviance =  18.05
   subtree deviance =  22.77
node number:  7
   tree deviance =  22.77
   subtree deviance =  26.99
> ir.tr1
node), split, n, deviance, yval, (yprob)
      * denotes terminal node

 1) root 150 330.0 c ( 0.330 0.33 0.330 )
   2) Petal.L.<2.45 50    0.0 s ( 0.000 1.00 0.000 ) *
   3) Petal.L.>2.45 100 140.0 c ( 0.500 0.00 0.500 )
     6) Petal.W.<1.75 54  33.0 c ( 0.910 0.00 0.093 )
      12) Petal.L.<4.95 48    9.7 c ( 0.980 0.00 0.021 ) *
      13) Petal.L.>4.95 6    7.6 v ( 0.330 0.00 0.670 ) *
     7) Petal.W.>1.75 46    9.6 v ( 0.022 0.00 0.980 ) *
> summary(ir.tr1)

Classification tree:
   ....
Number of terminal nodes:  4
Residual mean deviance:  0.185 = 27 / 146
Misclassification error rate: 0.0267 = 4 / 150

par(pty="s")
plot(ird[, 3],ird[, 4], type="n",
   xlab="petal length", ylab="petal width")
text(ird[, 3], ird[, 4], as.character(ir.species))
par(cex=2)
partition.tree(ir.tr1, add=T)
par(cex=1)
```

where we clicked twice in succession with mouse button 1 on each of nodes 12 and 7 to remove their subtrees, then with button 2 to quit.

The decision region is now entirely in the petal length – petal width space, so we can show it by partition.tree. This example shows the limitations

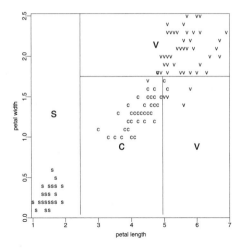

Figure 14.5: Partition for the `iris` data induced by the snipped tree.

of the one-step-ahead tree construction, for Weiss & Kapouleas (1989) used a
rule-induction program to find the set of rules

> If `Petal length` < 3 then *I. setosa.*
> If `Petal length` > 4.9 or `Petal width` > 1.6 then *I. virginica.*
> Otherwise *I. versicolor.*

Compare this with Figure 14.5, which makes one more error.

For the forensic glass dataset `fgl` which has 6 classes we can use

```
> fgl.tr <- tree(type ~ ., fgl)
> summary(fgl.tr)
Classification tree:
tree(formula = type ~ ., data = fgl)
Number of terminal nodes:  24
Residual mean deviance:  0.649 = 123 / 190
Misclassification error rate: 0.15 = 32 / 214
> plot(fgl.tr);   text(fgl.tr, all=T, cex=0.5)
> fgl.tr1 <- snip.tree(fgl.tr)
> tree.screens()
> plot(fgl.tr1)
> tile.tree(fgl.tr1, fgl$type)
> close.screen(all = T)
```

Once more snipping is required.

Fine control

Missing values are handled by the argument `na.action` of `tree`. The default is
`na.fail`, to abort if missing values are found; an alternative is `na.omit` which
omits all cases with a missing observation. The `weights` and `subset` arguments
are available as in all model-fitting functions.

The predict method predict.tree can be used to predict future cases. This automatically handles missing values[1] by dropping cases down the tree until a NA is encountered or a leaf is reached. With option split=T it splits cases as described in Ripley (1996, p. 232)

There are three control arguments specified under tree.control, but which may be passed directly to tree. The mindev controls the threshold for splitting a node. The parameters minsize and mincut control the size thresholds; minsize is the threshold for a node size, so nodes of size minsize or larger are candidates for a split. Daughter nodes must exceed mincut for a split to be allowed. The defaults are minsize = 10 and mincut = 5.

To make a tree fit exactly, the help page recommends the use of mindev = 0 and minsize = 2. However, for the shuttle data this will split an already pure node. A small value such as mindev = 1e-6 is safer. For example, for the shuttle data:

```
> shuttle.tr <- tree(use ~ ., shuttle, subset=1:253,
                       mindev=1e-6, minsize=2)
> shuttle.tr
node), split, n, deviance, yval, (yprob)
      * denotes terminal node

  1) root 253 350.0  auto ( 0.57 0.43 )
    2) vis: no 128    0.0  auto ( 1.00 0.00 ) *
    3) vis: yes 125  99.0  noauto ( 0.14 0.86 )
      6) error: MM, SS 61  72.0  noauto ( 0.28 0.72 )
       12) stability:stab 29  39.0  auto ( 0.59 0.41 )
          24) magn: Out 8    0.0  noauto ( 0.00 1.00 ) *
          25) magn: Light, Med, Str 21  20.0  auto ( 0.81 0.19 )
            50) error: MM 9  12.0  auto ( 0.56 0.44 )
             100) sign: nn 3    0.0  noauto ( 0.00 1.00 ) *
             101) sign: pp 6    5.4  auto ( 0.83 0.17 )
               202) magn: Light, Med 4  0.0  auto ( 1.00 0.00 ) *
               203) magn: Strong 2    2.8  auto ( 0.50 0.50 )
                 406) wind: head 1    0.0  noauto ( 0.00 1.00 ) *
                 407) wind: tail 1    0.0  auto ( 1.00 0.00 ) *
            51) error: SS 12    0.0  auto ( 1.00 0.00 ) *
       13) stability:xstab 32    0.0  noauto ( 0.00 1.00 ) *
      7) error: LX, XL 64    0.0  noauto ( 0.00 1.00 ) *
> post.tree(shuttle.tr)
> shuttle1 <- shuttle[254:256, ]  # 3 missing cases
  stability error sign  wind   magn  vis
1      stab    MM   nn  head  Light  yes
2      stab    MM   nn  head Medium  yes
3      stab    MM   nn  head Strong  yes
> predict(shuttle.tr, shuttle1)
    auto  noauto
```

---

[1] Prior to S-PLUS 3.4 another method was used which will give incorrect results if any continuous variable used in the tree has missing values. This is corrected in the treefix library.

254	0	1
255	0	1
256	0	1

14.2   Cutting trees down to size

With 'noisy' data, that is when the distributions for the classes overlap, it is quite possible to grow a tree which fits the training set well, but which has adapted too well to features of that subset of $\mathcal{X}$. Similarly, regression trees can be too elaborate and over-fit the training data. We need an analogue of variable selection in regression.

The established methodology is tree cost-complexity *pruning*, first introduced by Breiman *et al.* (1984). They considered rooted subtrees of the tree $\mathcal{T}$ grown by the construction algorithm, that is the possible results of snipping off terminal subtrees on $\mathcal{T}$. The pruning process chooses one of the rooted subtrees. Let R_i be a measure evaluated at the leaves, such as the deviance or the number of errors, and let R be the value for the tree, the sum over the leaves of R_i. Let the size of the tree be the number of leaves. Then Breiman *et al.* showed that the set of rooted subtrees of $\mathcal{T}$ which minimize the cost-complexity measure

$$R_\alpha = R + \alpha\,\text{size}$$

is itself nested. That is, as we increase α we can find the optimal trees by a sequence of snip operations on the current tree (just like pruning a real tree). This produces a sequence of trees from the size of $\mathcal{T}$ down to just the root node, but it may prune more than one node at a time. (Short proofs of these assertions are given by Ripley, 1996, Chapter 7. The tree $\mathcal{T}$ is not necessarily optimal for $\alpha = 0$, as we shall see.)

Pruning is implemented in the function `prune.tree`, which can be asked for the trees for one or more values of α (its argument k) or for a tree of a particular size (its argument `best`). (It is possible that there is not an optimal true of the requested size, in which case the next larger tree is returned.) The default measure is the deviance, but for classification trees one can specify `method="misclass"` to use the error count.[2]

Let us prune the cpus tree. Plotting the output of `prune.tree`, the deviance against size, allows us to choose a likely break point (Figure 14.6):

```
> plot(prune.tree(cpus.ltr))
> cpus.ltr1 <- prune.tree(cpus.ltr, best=8)
> plot(cpus.ltr1);   text(cpus.ltr1)
```

Note that size 9 does not appear in the tree sequence.

One way to choose the parameter α is to consider pruning as a method of variable selection. Akaike's information criterion (AIC) penalizes minus twice

---

[2] Unfortunately the version supplied in S-PLUS 3.0 to 3.3 contains several errors. Our library `treefix` contains a replacement function.

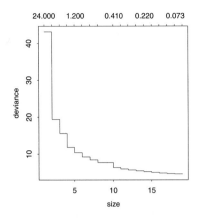
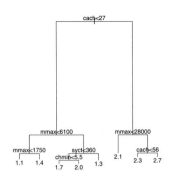

Figure 14.6: Pruning cpus.ltr. (Left) deviance *vs* size. The top axis is of α. (Right) the pruned tree of size 8.

log-likelihood by twice the number of parameters. For classification trees with K classes choosing $\alpha = 2(K-1)$ will find the rooted subtree with minimum AIC. For regression trees one approximation (Mallows' C_p) to AIC is to replace the minus twice log-likelihood by the residual sum of squares divided by $\hat{\sigma}^2$ (our best estimate of σ^2), so we could take $\alpha = 2\hat{\sigma}^2$. One way to select $\hat{\sigma}^2$ would be from the fit of the full tree model. In our example this suggests $\alpha = 2 \times 0.02394$, which makes no change. Other criteria suggest that AIC and C_p tend to over-fit and choose larger constants in the range 2–6. (Note that counting parameters in this way does not take the selection of the splits into account.)

For the classification tree on the fgl data this suggests $\alpha = 2 \times (6-1) = 10$, and we have

```
> summary(prune.tree(fgl.tr, k=10))
Classification tree:
snip.tree(tree = fgl.tr, nodes = c(11, 10, 15, 108, 109, 12, 26))
Variables actually used in tree construction:
[1] "Mg" "Na" "Al" "Fe" "Ba" "RI" "K"  "Si"
Number of terminal nodes:  13
Residual mean deviance:  0.961 = 193 / 201
Misclassification error rate: 0.182 = 39 / 214
```

Cross-validation

We really need a better way to choose the degree of pruning. If a separate validation set is available, we can predict on that set, and compute the deviance versus α for the pruned trees. This will often have a minimum, and we can choose the smallest tree whose deviance is close to the minimum.

If no validation set is available we can make one by splitting the training set. Suppose we split the training set into 10 (roughly) equally sized parts. We can then use 9 to grow the tree and test it on the tenth. This can be done in 10 ways, and we can average the results. This is done by the function `cv.tree`. Note that as ten trees must be grown, the process can be slow, and that the averaging is done for fixed α and not fixed tree size.

```
set.seed(123)
plot(cv.tree(cpus.ltr,, prune.tree))
post.tree(prune.tree(cpus.ltr, best=4))
```

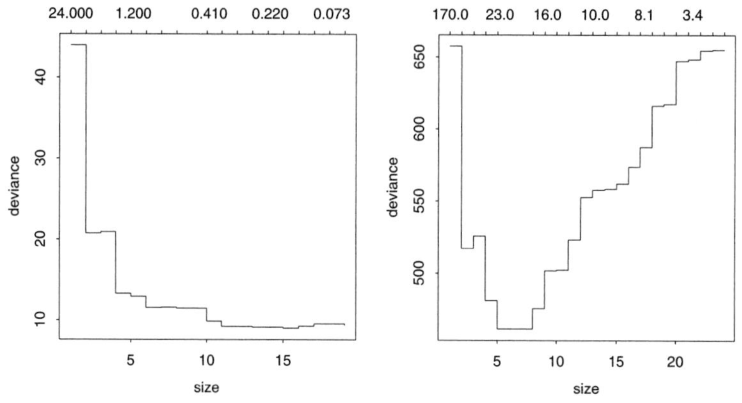

Figure 14.7: Cross-validation plots for pruning (left) `cpus.ltr` and (right) `fgl.tr`.

The answers can be far from convincing as Figure 14.7 shows. The algorithm randomly divides the training set, and sometimes a second attempt will give a better answer. It may be worthwhile if CPU time allows to average over several random divisions.

Let us prune the tree grown on the `fgl` dataset.

```
set.seed(123)
fgl.cv <- cv.tree(fgl.tr,, prune.tree)
for(i in 2:5)  fgl.cv$dev <- fgl.cv$dev +
    cv.tree(fgl.tr,, prune.tree)$dev
fgl.cv$dev <- fgl.cv$dev/5
plot(fgl.cv)
misclass.tree(fgl.tr)
[1] 32
misclass.tree(prune.tree(fgl.tr, best=5))
[1] 65
```

These re-substitution error counts are optimistically biased.

There is a difficulty with cross-validating the deviance; if at some leaf a class occurs in the test set but not in the training set the deviance will be infinity. (If this occurs at the root node, the deviance will be infinite for all prunings.) To avoid this, zero probabilities are replaced by a very small probability (10^{-3} by default).

Error-rate pruning

Pruning on error-rate is attractive, as it will remove terminal splits which have the same class for each leaf, and allows the use of cross-validation to estimate the error rate. Note that `cv.tree` passes its ... argument only to some of its calls to the pruning function. We can work around this by replacing `prune.tree` by the function `prune.misclass`:

```
> set.seed(123)
> fgl.cv <- cv.tree(fgl.tr,, prune.misclass)
> for(i in 2:5)  fgl.cv$dev <- fgl.cv$dev +
      cv.tree(fgl.tr,, prune.misclass)$dev
> fgl.cv$dev <- fgl.cv$dev/5
> fgl.cv
$size:
 [1] 24 18 16 12 11  9  6  5  4  3  1
$dev:
 [1]  73.6  74.2  74.0  72.6  73.0  73.4  78.8  89.0  90.4  92.6
[11] 146.4
> plot(fgl.cv)
> prune.misclass(fgl.tr)
$size:
 [1] 24 18 16 12 11  9  6  5  4  3  1
$dev:
 [1]  32  32  33  37  39  44  58  65  73  84 138
      . . . .
```

which suggests that a pruned tree of size 5 suffices. Note that in this example cross-validation gives a 35% higher estimate of the error rate.

We said on page 426 that we should take the smallest tree whose performance is close to the minimum. Breiman *et al.* (1984) suggest the *one s.e. rule* by taking the smallest pruned tree whose error rate is within one standard deviation of the minimum. Under cross-validation the number of errors made is approximately Poisson, so in this example we would seek an answer within 9 of the minimum.

14.3   Low birth weights revisited

We return to the example on low birth weights studied in Sections 7.2 and 11.1. As it has a binary response, it is fitted as a classification tree using the binomial log-likelihood. We can continue to use AIC by pruning with $\alpha = 2$. Note that the AIC for this tree, $153.5 + 2 \times 19 = 191.5$, is smaller than for the best logistic regression model of Section 7.2.

```
> bwt.tr <- tree(low ~ ., bwt)
> summary(bwt.tr)

Classification tree:
tree(formula = low ~ ., data = bwt)
```

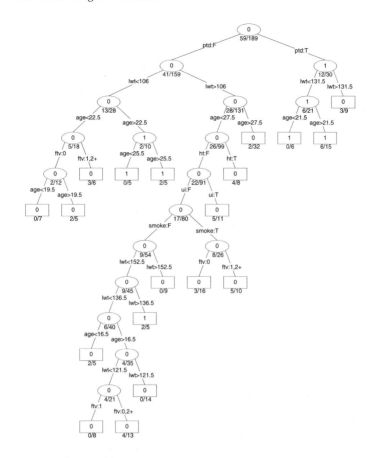

Figure 14.8: AIC-pruned tree for low birth-weight data.

```
Number of terminal nodes:  28
Residual mean deviance:  0.907 = 146 / 161
Misclassification error rate: 0.217 = 41 / 189

> bwt.tr1 <- prune.tree(bwt.tr, k=2)
> summary(bwt.tr1)

Classification tree:
Variables actually used in tree construction:
[1] "ptd"   "lwt"   "age"   "ftv"   "ht"     "ui"     "smoke"
Number of terminal nodes:  19
Residual mean deviance:  0.903 = 153 / 170
Misclassification error rate: 0.228 = 43 / 189
> prune.misclass(bwt.tr1)
$size:
[1] 19 11  5  2  1
$dev:
```

```
[1] 43 43 44 53 59
> bwt.tr3 <- prune.misclass(bwt.tr1, best=5)
```

The AIC-pruned tree is shown in Figure 14.8. This can be pruned by a further 8
nodes without changing its predictions at all, and to size 5 (as shown in Figure 14.9)
at the price of one additional misclassification on the training set.

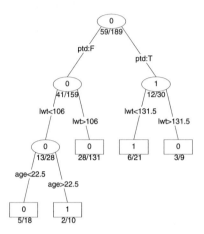

Figure 14.9: Error-rate pruned version of Figure 14.8.

Finally, we consider cross-validated error-rate pruning:

```
> set.seed(123)
> bwt.cv <- cv.tree(bwt.tr,, prune.misclass)
> for(i in 2:5)  bwt.cv$dev <- bwt.cv$dev +
      cv.tree(bwt.tr,, prune.misclass)$dev
> bwt.cv$dev <- bwt.cv$dev/5
> bwt.cv
$size:
[1] 28 16  8  5  2  1
$dev:
[1] 73.6 74.6 74.2 73.0 65.4 63.8
```

which once again shows a much higher error rate estimate, and suggests no split
at all!

For a binary classification such as this one it can be useful to use one of the
features of `text.tree`:

```
plot(bwt.tr3)
text(bwt.tr3, label="1")
```

which labels the leaves by the estimated probability of a low birth weight. Unfortunately this option is not provided for `post.tree`.

Chapter 15

Time Series

There are now a large number of books on time series. Our philosophy and notation are close to those of the applied book by Diggle (1990) (from which some of our examples are taken). Brockwell & Davis (1991) and Priestley (1981) provide more theoretical treatments, and Bloomfield (1976) and Priestley are particularly thorough on spectral analysis. Brockwell & Davis (1996) is an excellent low-level introduction to the theory.

Functions for time series have been included in S for some years, and further time-series support was one of the earliest enhancements of S-PLUS. In the current system regularly spaced time series are of class `rts`, and are created by the function `rts`. (We will not cover the classes `cts` for series of dates and `its` for irregularly spaced series.)

Our first running example is `lh`, a series of 48 observations at 10 minute intervals on luteinizing hormone levels for a human female taken from Diggle (1990). This was created by

```
> lh <- rts(scan(n=48))
2.4 2.4 2.4 2.2 2.1 1.5 2.3 2.3 2.5 2.0 1.9 1.7
2.2 1.8 3.2 3.2 2.7 2.2 2.2 1.9 1.9 1.8 2.7 3.0
2.3 2.0 2.0 2.9 2.9 2.7 2.7 2.3 2.6 2.4 1.8 1.7
1.5 1.4 2.1 3.3 3.5 3.5 3.1 2.6 2.1 3.4 3.0 2.9
```

Printing it gives

```
> lh
 1: 2.4 2.4 2.4 2.2 2.1 1.5 2.3 2.3 2.5 2.0 1.9 1.7 2.2 1.8
15: 3.2 3.2 2.7 2.2 2.2 1.9 1.9 1.8 2.7 3.0 2.3 2.0 2.0 2.9
29: 2.9 2.7 2.7 2.3 2.6 2.4 1.8 1.7 1.5 1.4 2.1 3.3 3.5 3.5
43: 3.1 2.6 2.1 3.4 3.0
  start deltat frequency
      1      1         1
```

which shows the attribute vector `tspar` of the class `rts`, which is used for plotting and other computations. The components are the `start`, the label for the first observation, `deltat` (Δt), the increment between observations, and `frequency`, the reciprocal of `deltat`. Note that the final index can be deduced from the attributes and length. Any of `start`, `deltat`, `frequency` and `end` can be specified in the call to `rts`, provided they are specified consistently. In this example the units are known, and we can also specify this:

```
> lh <- rts(scan(n=48), start=1, deltat=1, units="10mins")
    ....
> lh
 1: 2.4 2.4 2.4 2.2 2.1 1.5 2.3 2.3 2.5 2.0 1.9 1.7 2.2 1.8
15: 3.2 3.2 2.7 2.2 2.2 1.9 1.9 1.8 2.7 3.0 2.3 2.0 2.0 2.9
29: 2.9 2.7 2.7 2.3 2.6 2.4 1.8 1.7 1.5 1.4 2.1 3.3 3.5 3.5
43: 3.1 2.6 2.1 3.4 3.0 2.9
 start deltat frequency
     1      1         1
 Time units :   10mins
```

Our second example is a seasonal series. Our dataset `deaths` gives monthly deaths in the UK from a set of common lung diseases for the years 1974 to 1979, from Diggle (1990). This was read into S by

```
> deaths <- rts(scan(n=72), start=1974, frequency=12,
    units="months")
3035 2552 2704 2554 2014 1655 1721 1524 1596 2074 2199 2512
2933 2889 2938 2497 1870 1726 1607 1545 1396 1787 2076 2837
2787 3891 3179 2011 1636 1580 1489 1300 1356 1653 2013 2823
3102 2294 2385 2444 1748 1554 1498 1361 1346 1564 1640 2293
2815 3137 2679 1969 1870 1633 1529 1366 1357 1570 1535 2491
3084 2605 2573 2143 1693 1504 1461 1354 1333 1492 1781 1915
> deaths
        Jan  Feb  Mar  Apr  May  Jun  Jul  Aug  Sep  Oct  Nov
1974: 3035 2552 2704 2554 2014 1655 1721 1524 1596 2074 2199
1975: 2933 2889 2938 2497 1870 1726 1607 1545 1396 1787 2076
1976: 2787 3891 3179 2011 1636 1580 1489 1300 1356 1653 2013
1977: 3102 2294 2385 2444 1748 1554 1498 1361 1346 1564 1640
1978: 2815 3137 2679 1969 1870 1633 1529 1366 1357 1570 1535
1979: 3084 2605 2573 2143 1693 1504 1461 1354 1333 1492 1781
    ....
 start   deltat frequency
 1974 0.083333        12
 Time units :   months
```

Note how the specification of `units="months"` has triggered a special form of labelling of the print. Quarterly data (with both `frequency=4` and `units="quarters"`) are also treated specially.

There is a series of functions to extract aspects of the time base:

```
> tspar(deaths)
 start   deltat frequency
 1974 0.083333        12
attr(, "units"):
[1] "months"
> start(deaths)
```

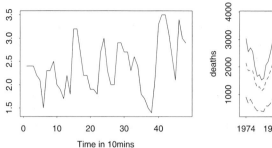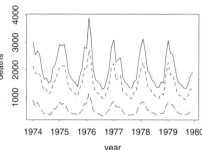

Figure 15.1: Plots by `ts.plot` of `lh` and the three series on deaths by lung diseases. In the right-hand plot the dashed series is for males, the long dashed series for females and the solid line for the total.

```
[1] 1974
> end(deaths)
[1] 1979.9
> frequency(deaths)
 frequency
        12
> units(deaths)
[1] "months"
> cycle(deaths)
      Jan Feb Mar Apr May Jun Jul Aug Sep Oct Nov Dec
1974:   1   2   3   4   5   6   7   8   9  10  11  12
1975:   1   2   3   4   5   6   7   8   9  10  11  12
1976:   1   2   3   4   5   6   7   8   9  10  11  12
1977:   1   2   3   4   5   6   7   8   9  10  11  12
1978:   1   2   3   4   5   6   7   8   9  10  11  12
1979:   1   2   3   4   5   6   7   8   9  10  11  12
 start   deltat frequency
  1974 0.083333        12
Time units :  months
```

Time series can be plotted by `plot`, but the functions `ts.plot`, `ts.lines` and `ts.points` are provided for time-series objects. All can plot several related series together. For example, the `deaths` series is the sum of two series `mdeaths` and `fdeaths` for males and females. Figure 15.1 was created by

```
> par(mfrow = c(2,2))
> ts.plot(lh)
> ts.plot(deaths, mdeaths, fdeaths, lty=c(1,3,4), xlab="year",
    ylab="deaths")
```

The functions `ts.union` and `ts.intersect` bind together multiple time series. The time axes are aligned and only observations at times which appear in all the series are retained with `ts.intersect`; with `ts.union` the combined series covers the whole range of the components, possibly as NA values. The result is a matrix with `tspar` attributes set, or a data frame if argument `dframe=T` is set. We discuss most of the methodology for multiple time series in Section 15.4.

The function `window` extracts a sub-series of a single or multiple time series, by specifying `start` and/or `end`.

The function `lag` shifts the time axis of a series back by k positions, default one. Thus `lag(deaths, k=3)` is the series of deaths shifted one quarter into the past. This can cause confusion, as most people think of lags as shifting time and not the series: that is the current value of a series lagged by one year is last year's, not next year's.

The function `diff` takes the difference between a series and its lagged values, and so returns a series of length $n - k$ with values lost from the beginning (if $k > 0$) or end. (The argument `lag` specifies k and defaults to one. Note that the lag is used in the usual sense here, so `diff(deaths, lag=3)` is equal to `deaths - lag(deaths, k=-3)`!) The function `diff` has an argument `differences` which causes the operation to be iterated. For later use, we denote the dth difference of series X_t by $\nabla^d X_t$, and the dth difference at lag s by $\nabla_s^d X_t$.

The function `aggregate aggregate.ts` can be used to change the frequency of the time base. For example to obtain quarterly sums or annual means of `deaths`:

```
> aggregate(deaths, 4, sum)
           1    2    3    4
1974: 8291 6223 4841 6785
    ....
> aggregate(deaths, 1, mean)
1974: 2178.3 2175.1 2143.2 1935.8 1995.9 1911.5
```

Each of the functions `lag`, `diff` and `aggregate` can also be applied to multiple time series objects formed by `ts.union` or `ts.intersect`.

15.1 Second-order summaries

The theory for time series is based on the assumption of second-order stationarity after removing any trends (which will include seasonal trends). Thus second moments are particularly important in the practical analysis of time series. We assume that the series X_t runs throughout time, but is observed only for $t = 1, \ldots, n$. We will use the notations X_t and $X(t)$ interchangeably. The series will have a mean μ, often taken to be zero, and the covariance and correlation

$$\gamma_t = \text{cov}\,(X_{t+\tau}, X_\tau), \qquad \rho_t = \text{corr}\,(X_{t+\tau}, X_\tau)$$

do not depend on τ. The covariance is estimated for $t > 0$ from the $n-t$ observed pairs $(X_{1+t}, X_1), \ldots, (X_n, X_{n-t})$. If we just take the standard correlation or covariance of these pairs we will use different estimates of the mean and variance for each of the subseries $X_{1+t}, \ldots, X_n$ and $X_1, \ldots, X_{n-t}$, whereas under our assumption of second-order stationarity these have the same mean and variance. This suggests the estimators

$$c_t = \frac{1}{n} \sum_{s=\max(1,-t)}^{\min(n-t,n)} [X_{s+t} - \overline{X}][X_s - \overline{X}], \qquad r_t = \frac{c_t}{c_0}$$

Note that we use divisor n even though there are $n - |t|$ terms. This is to ensure that the sequence (c_t) is the covariance sequence of some second-order stationary time series.[1] Note that all of γ, ρ, c, r are symmetric functions ($\gamma_{-t} = \gamma_t$ and so on).

The function `acf` computes and by default plots the sequences (c_t) and (r_t), known as the *autocovariance* and *autocorrelation* functions. The argument `type` controls which is used, and defaults to the correlation.

Our definitions are easily extended to several time series observed over the same interval. Let

$$\gamma_{ij}(t) = \mathrm{cov}\,(X_i(t + \tau), X_j(\tau))$$

$$c_{ij}(t) = \frac{1}{n} \sum_{s=\max(1,-t)}^{\min(n-t,n)} [X_i(s + t) - \overline{X_i}][X_j(s) - \overline{X_j}]$$

which are not symmetric in t for $i \neq j$. These forms are used by `acf` for multiple time series:

```
acf(lh)
acf(lh, type="covariance")
acf(deaths)
acf(ts.union(mdeaths, fdeaths))
```

The `type` may be abbreviated in any unique way, for example `cov`. The output is shown in Figures 15.2 and 15.3. Note that approximate 95% confidence limits are shown for the autocorrelation plots; these are for an independent series for which $\rho_t = I(t = 0)$. As with a time series *a priori* one is expecting autocorrelation, these limits must be viewed with caution. In particular, if any ρ_t is non-zero, all the limits are invalid.

Note that for a series with a non-unit frequency such as `deaths` the lags are expressed in the basic time unit, here years. The function `acf` chooses the number of lags to plot unless this is specified by the argument `lag.max`. Plotting can be suppressed by setting argument `plot=F`. The function returns a list which can be plotted subsequently by `acf.plot`.

The plots of the `deaths` series show the pattern typical of seasonal series, and the autocorrelations do not damp down for large lags. Note how one of the cross-series is only plotted for negative lags. We have $c_{ji}(t) = c_{ij}(-t)$, so the cross terms are needed for all lags, whereas the terms for a single series are symmetric about 0. The labels are confusing: the plot in row 2 column 1 shows c_{12} for negative lags, a reflection of the plot of c_{21} for positive lags.

---

[1] That is, the covariance sequence is positive-definite.

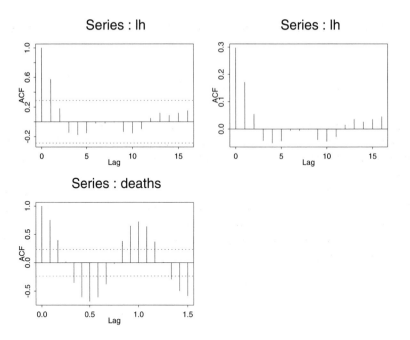

Figure 15.2: acf plots for the series lh and deaths . The top row shows the autocorrelation (left) and autocovariance (right).

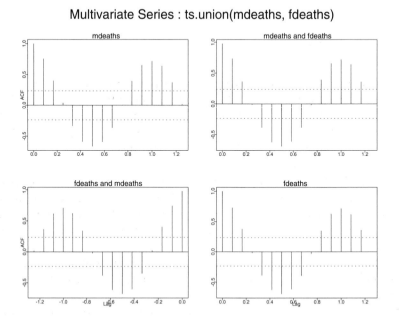

Figure 15.3: Autocorrelation plots for the multiple time series of male and female deaths.

Spectral analysis

The spectral approach to second-order properties is better able to separate short-term and seasonal effects, and also has a sampling theory which is easier to use for non-independent series.

We will only give a brief treatment; extensive accounts are given by Bloomfield (1976) and Priestley (1981). Be warned that accounts differ in their choices of where to put the constants in spectral analysis; we have tried to follow S-PLUS as far as possible.

The covariance sequence of a second-order stationary time series can always be expressed as

$$\gamma_t = \frac{1}{2\pi} \int_{-\pi}^{\pi} e^{i\omega t} \, \mathrm{d}F(\omega)$$

for the *spectrum* F, a finite measure on $(-\pi, \pi]$. Under mild conditions which exclude purely periodic components of the series, the measure has a density known as the *spectral density* f, so

$$\gamma_t = \frac{1}{2\pi} \int_{-\pi}^{\pi} e^{i\omega t} f(\omega) \, \mathrm{d}\omega = \int_{-1/2}^{1/2} e^{i\omega_f t} f(\omega_f) \, \mathrm{d}\omega_f \tag{15.1}$$

where in the first form the frequency ω is in units of radians/time and in the second form ω_f is in the units of cycles/time, and in both cases time is measured in the units of Δt. If the time series object has a `frequency` greater than one and time is measured in the base units, the spectral density will be divided by `frequency`.

The Fourier integral can be inverted to give

$$f(\omega) = \sum_{-\infty}^{\infty} \gamma_t e^{-i\omega t} = \gamma_0 \left[1 + 2 \sum_{1}^{\infty} \rho_t \cos(\omega t) \right] \tag{15.2}$$

By the symmetry of γ_t, $f(-\omega) = f(\omega)$, and we need only consider f on $(0, \pi)$. Equations (15.1) and (15.2) are the first place the differing constants appear. Bloomfield and Brockwell & Davis omit the factor $1/2\pi$ in (15.1) which therefore appears in (15.2).

The basic tool in estimating the spectral density is the *periodogram*. For a frequency ω we effectively compute the squared correlation between the series and the sine/cosine waves of frequency ω by

$$I(\omega) = \left| \sum_{t=1}^{n} e^{-i\omega t} X_t \right|^2 / n = \frac{1}{n} \left[\left\{ \sum_{t=1}^{n} X_t \sin(\omega t) \right\}^2 + \left\{ \sum_{t=1}^{n} X_t \cos(\omega t) \right\}^2 \right] \tag{15.3}$$

Frequency 0 corresponds to the mean, which is normally removed. The frequency π corresponds to a cosine series of alternating ± 1 with no sine series. Bloomfield (but not Brockwell & Davis) has a factor $1/2\pi$ in the definition of the periodogram. S-PLUS appears to divide by the `frequency` to match its view of the spectral density.

The periodogram is related to the autocovariance function by

$$I(\omega) = \sum_{-\infty}^{\infty} c_t e^{-i\omega t} = c_0 \left[1 + 2 \sum_{1}^{\infty} r_t \cos(\omega t) \right]$$

$$c_t = \frac{1}{2\pi} \int_{-\pi}^{\pi} e^{i\omega t} I(\omega) \, d\omega$$

and so conveys the same information. However, each form makes some of that information easier to interpret.

Asymptotic theory shows that $I(\omega) \sim f(\omega)E$ where E has a standard exponential distribution, except for $\omega = 0$ and $\omega = \pi$. Thus if $I(\omega)$ is plotted on log scale, the variation about the spectral density is the same for all $\omega \in (0, \pi)$ and is given by a Gumbel distribution (for that is the distribution of $\log E$). Further, $I(\omega_1)$ and $I(\omega_2)$ will be asymptotically independent at distinct frequencies. Indeed if ω_k is a *Fourier frequency* of the form $\omega_k = 2\pi k/n$, then the periodogram at two Fourier frequencies will be approximately independent for large n. Thus although the periodogram itself does not provide a consistent estimator of the spectral density, if we assume that the latter is smooth, we can average over adjacent independently distributed periodogram ordinates and obtain a much less variable estimate of $f(\omega)$. A kernel smoother is used of the form

$$\hat{f}(\omega) = \frac{1}{h} \int K\left(\frac{\lambda - \omega}{h}\right) I(\omega) \, d\lambda$$

$$\approx \frac{2\pi}{nh} \sum_k K\left(\frac{\omega_k - \omega}{h}\right) I(\omega_k) = \sum_k g_k I(\omega_k)$$

for a probability density K. The parameter h controls the degree of smoothing. To see its effect we approximate the mean and variance of $\hat{f}(\omega)$:

$$\mathrm{var}\left(\hat{f}(\omega)\right) \approx \sum_k g_k^2 f(\omega_k)^2 \approx f(\omega)^2 \sum_k g_k^2 \approx \frac{2\pi}{nh} f(\omega)^2 \int K(x)^2 \, dx$$

$$E\left(\hat{f}(\omega)\right) \approx \sum_k g_k f(\omega_k) \approx f(\omega) + \frac{f''(\omega)}{2} \sum_k g_k (\omega_k - \omega)^2$$

$$\mathrm{bias}\left(\hat{f}(\omega)\right) \approx \frac{f''(\omega)}{2} h^2 \int x^2 K(x) \, dx$$

so as h increases the variance decreases but the bias increases. We see that the ratio of the variance to the squared mean is approximately $g^2 = \sum_k g_k^2$. If $\hat{f}(\omega)$ had a distribution proportional to a χ_ν^2, this ratio would be $2/\nu$, so $2/g^2$ is referred to as the equivalent degrees of freedom. Bloomfield and S-PLUS refer to $\sqrt{2 \, \mathrm{bias}\left(\hat{f}(\omega)\right)/f''(\omega)}$ as the *bandwidth*, which is proportional to h.

To understand these quantities, consider a simple moving average over $2m+1$ Fourier frequencies centred on a Fourier frequency ω. Then the variance is $f(\omega)^2/(2m + 1)$ and the equivalent degrees of freedom is $2(2m + 1)$, as we would expect on averaging $2m + 1$ exponential (or χ_2^2) variates. The bandwidth is approximately

$$\frac{(2m+1)2\pi}{n} \frac{1}{\sqrt{12}}$$

and the first factor is the width of the window in frequency space. (Since S-PLUS works in cycles rather than radians, the bandwidth is about $(2m + 1)/n\sqrt{12}$ frequency on its scale.) The bandwidth is thus a measure of the size of the smoothing window, but rather smaller than the effective width.

The workhorse function for spectral analysis is `spectrum`, which with its default options computes and plots the periodogram on log scale. The function `spectrum` calls `spec.pgram` to do most of the work. (Note: `spectrum` by default removes a linear trend from the series before estimating the spectral density.) For our examples we can use:

```
par(mfrow=c(2,2))
spectrum(lh)
spectrum(deaths)
```

with the result shown in Figure 15.4.

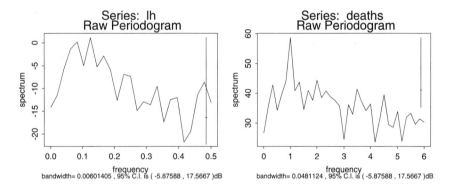

Figure 15.4: Periodogram plots for `lh` and `deaths`.

Note how elaborately labelled the figures are. The plots are on log scale, in units of *decibels*, that is the plot is of $10 \log_{10} I(\omega)$. The function `spec.pgram` returns the bandwidth and (equivalent) degrees of freedom as components `bandwidth` and `df`.

The function `spectrum` also produces smoothed plots, using repeated smoothing with modified Daniell smoothers (Bloomfield, 1976), which are moving averages giving half weight to the end values of the span. Trial-and-error is needed to choose the spans (Figures 15.5 and 15.6):

```
par(mfrow=c(2,2))
spectrum(lh)
spectrum(lh, spans=3)
spectrum(lh, spans=c(3,3))
```

```
spectrum(lh, spans=c(3,5))

spectrum(deaths)
spectrum(deaths, spans=c(3,3))
spectrum(deaths, spans=c(3,5))
spectrum(deaths, spans=c(5,7))
```

The spans should be odd integers, and it helps to produce a smooth plot if they are different and at least two are used. The width of the centre mark on the 95% confidence interval indicator indicates the bandwidth.

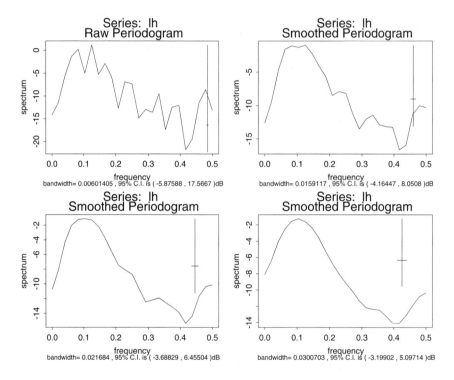

Figure 15.5: Spectral density estimates for lh.

The periodogram has other uses. If there are periodic components in the series the distribution theory given above does not apply, but there will be peaks in the plotted periodogram. Smoothing will reduce those peaks, but they can be seen quite clearly by plotting the *cumulative periodogram*

$$U(\omega) = \sum_{0 < \omega_k \leqslant \omega} I(\omega_k) \Big/ \sum_{1}^{\lfloor n/2 \rfloor} I(\omega_k)$$

against ω. The cumulative periodogram is also very useful as a test of whether a particular spectral density is appropriate, as if we replace $I(\omega)$ by $I(\omega)/f(\omega)$, $U(\omega)$ should be a straight line. Further, asymptotically, the maximum deviation

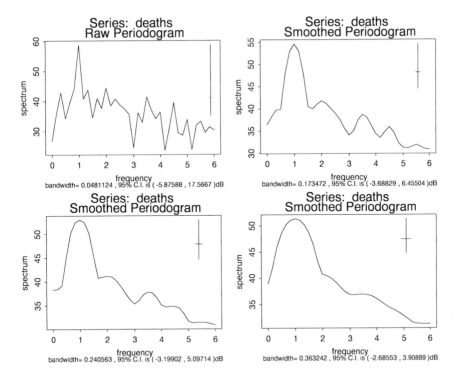

Figure 15.6: Spectral density estimates for `deaths`.

from that straight line has a distribution given by that of the Kolmogorov-Smirnov statistic, with a 95% limit approximately $1.358/[\sqrt{m} + 0.11 + 0.12/\sqrt{m}]$ where $m = \lfloor n/2 \rfloor$ is the number of Fourier frequencies included. This is particularly useful for a residual series with f constant, in testing if the series is uncorrelated.

The distribution theory can be made more accurate, and the peaks made sharper, by *tapering* the de-meaned series (Bloomfield, 1976). The magnitude of the first α and last α of the series is tapered down towards zero by a cosine bell, that is X_t is replaced by

$$X'_t = \begin{cases} (1 - \cos \frac{\pi(t-0.5)}{\alpha n})X_t & t \leqslant \alpha n \\ X_t & \alpha n < t < (1-\alpha)n \\ (1 - \cos \frac{\pi(n-t+0.5)}{\alpha n})X_t & t \geqslant (1-\alpha)n \end{cases}$$

The proportion α is controlled by the parameter `taper` of `spec.pgram`, and defaults to 10%. It should rarely need to be altered. (The taper function `spec.taper` can be called directly if needed.) Tapering does increase the variance of the periodogram and hence the spectral density estimate, by about 12% for the default taper, but it will decrease the bias near peaks very markedly, if those peaks are not at Fourier frequencies.

We cannot show this directly using `spectrum`, since that plots frequency zero, and as the taper is reduced, the mean of the tapered series and hence $I(0)$ goes to zero. We need to use a taper for the plot used to set the scale:

```
spectrum(deaths)
deaths.spc <- spec.pgram(deaths, taper=0)
lines(deaths.spc$freq, deaths.spc$spec, lty=3)
```

In this example tapering does not help resolve the peaks, but they are at Fourier frequencies. S-PLUS does not supply a function for the cumulative periodogram, but we can write one using the fft function to compute a discrete Fourier transform.

```
cpgram <- function(ts, taper=0.1,
    main=paste("Series: ", deparse(substitute(ts))) )
{
    x <- as.vector(ts)
    x <- x[!is.na(x)]
    x <- spec.taper(scale(x, T, F), p=taper)
    y <- Mod(fft(x))^2/length(x)
    y[1] <- 0
    n <- length(x)
    x <- (0:(n/2))*frequency(ts)/n
    if(length(x)%%2==0) {
        n <- length(x)-1
        y <- y[1:n]
        x <- x[1:n]
    } else y <- y[1:length(x)]
    xm <- frequency(ts)/2
    mp <- length(x)-1
    crit <- 1.358/(sqrt(mp)+0.12+0.11/sqrt(mp))
    oldpty <- par()$pty
    par(pty="s")
    plot(x, cumsum(y)/sum(y), type="s", xlim=c(0, xm),
        ylim=c(0, 1), xaxs="i", yaxs="i", xlab="frequency",
        ylab="", pty="s")
    lines(c(0, xm*(1-crit)), c(crit, 1))
    lines(c(xm*crit, xm), c(0, 1-crit))
    title(main = main)
    invisible(par(pty=oldpty))
}
```

Dropping missing values helps when using the function with residual series, which start with missing values. The results for our examples are shown in Figure 15.7, with 95% confidence bands.

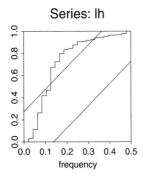   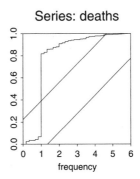

Figure 15.7: Cumulative periodogram plots for `lh` and `deaths`.

15.2   ARIMA models

In the late 1960s Box and Jenkins advocated a methodology for time series based on finite-parameter models for the second-order properties, so this approach is often named after them. Let ϵ_t denote a series of uncorrelated random variables with mean zero and variance σ^2. A moving average process of order q (MA(q)) is defined by

$$X_t = \sum_{0}^{q} \beta_j \epsilon_{t-j} \tag{15.4}$$

an autoregressive process of order p (AR(p)) is defined by

$$X_t = \sum_{1}^{p} \alpha_i X_{t-i} + \epsilon_t \tag{15.5}$$

and an ARMA(p, q) process is defined by

$$X_t = \sum_{1}^{p} \alpha_i X_{t-i} + \sum_{0}^{q} \beta_j \epsilon_{t-j} \tag{15.6}$$

(S-PLUS reverses the sign of the MA coefficients for $j > 0$.)

Note that we do not need both σ^2 and β_0 for a MA(q) process, and we will take $\beta_0 = 1$. Some authors put the regression terms of (15.5) and (15.6) on the left-hand side and reverse the sign of α_i. Any of these processes can be given mean μ by adding μ to each observation.

An ARIMA(p, d, q) process (where the I stands for integrated) is a process whose dth difference $\nabla^d X$ is an ARMA(p, q) process.

Equation (15.4) will always define a second-order stationary time series, but (15.5) and (15.6) need not. They need the condition that all the (complex) roots of the polynomial

$$\phi_\alpha(z) = 1 - \alpha_1 z - \cdots - \alpha_p z^p$$

lie outside the unit disc. (The function `polyroot` can be used to check this.) However, there are in general 2^q sets of coefficients in (15.4) which give the same second-order properties, and it is conventional to take the set with roots of

$$\phi_\beta(z) = 1 + \beta_1 z + \cdots + \beta_q z^q$$

outside the unit disc. Let B be the backshift or lag operator defined by $BX_t = X_{t-1}$. Then we conventionally write an ARMA process as

$$\phi_\alpha(B)X = \phi_\beta(B)\epsilon \tag{15.7}$$

The function `arima.sim` simulates an ARIMA process. Simple usage is of the form

```
ts.sim <- arima.sim(list(order=c(1,1,0), ar=0.7), n=200)
```

which generates a series whose first differences follow an AR(1) process.

Model identification

A lot of attention has been paid to *identifying* ARMA models, that is choosing plausible values of p and q by looking at the second-order properties. Much of the literature is reviewed by de Gooijer *et al.* (1985). Nowadays it is computationally feasible to fit all plausible models and choose on the basis of their goodness of fit, but some simple diagnostics are still useful. For an MA(q) process we have

$$\gamma_k = \sigma^2 \sum_{i=0}^{q-|k|} \beta_i \beta_{i+|k|}$$

which is zero for $|k| > q$, and this may be discernible from plots of the ACF. For an AR(p) process the population autocovariances are generally all non-zero, but they satisfy the Yule–Walker equations

$$\rho_k = \sum_1^p \alpha_i \rho_{k-i}, \qquad k > 0 \tag{15.8}$$

This motivates the *partial autocorrelation function*. The partial correlation between X_s and X_{s+t} is the correlation after regression on $X_{s+1}, \ldots, X_{s+t-1}$, and is zero for $t > p$ for an AR(p) process. The PACF can be estimated by solving the Yule–Walker equations (15.8) with $p = t$ and ρ replaced by r, and is given by the `type="partial"` option of `acf`:

```
acf(lh, type="partial")
acf(deaths, type="partial")
```

as shown in Figure 15.8. These are short series, so no definitive pattern emerges, but `lh` might be fitted well by an AR(1) or perhaps an AR(3) process.

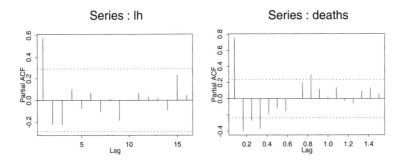

Figure 15.8: Partial autocorrelation plots for the series `lh` and `deaths`.

Model fitting

Selection amongst ARMA processes can be done by Akaike's information criterion
(AIC) which penalizes the deviance by twice the number of parameters; the model
with the smallest AIC is chosen. (All likelihoods considered assume a Gaussian
distribution for the time series.) Fitting can be done by the functions `ar` or
`arima.mle`:

```
> lh.ar1 <- ar(lh, F, 1)
> cpgram(lh.ar1$resid, main="AR(1) fit to lh")
> lh.ar <- ar(lh, order.max=9)
> lh.ar$order
[1] 3
> lh.ar$aic
 [1] 18.30668  0.99567  0.53802  0.00000  1.49036  3.21280
 [7]  4.99323  6.46950  8.46258  8.74120
> cpgram(lh.ar$resid, main="AR(3) fit to lh")
> lh1 <- lh - mean(lh)
> lh.arima1 <- arima.mle(lh1, model=list(order=c(1,0,0)),
      n.cond=3)
> arima.diag(lh.arima1)
> lh.arima3 <- arima.mle(lh1, model=list(order=c(3,0,0)),
      n.cond=3)
> arima.diag(lh.arima3)
> lh.arima11 <- arima.mle(lh1, model=list(order=c(1,0,1)),
      n.cond=3)
> arima.diag(lh.arima11)
```

This first fits an AR(1) process and obtains, after removing the mean,

$$X_t = 0.576 X_{t-1} + \epsilon_t$$

with $\sigma^2 = 0.208$. It then uses AIC to choose the order amongst AR processes
and selects $p = 3$ and fits

$$X_t = 0.653 X_{t-1} - 0.064 X_{t-2} - 0.227 X_{t-3} + \epsilon_t$$

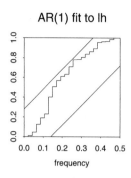

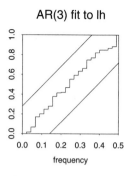

Figure 15.9: Cumulative periodogram plots for residuals of AR models fitted to `lh`.

with $\sigma^2 = 0.196$ and AIC is reduced by 0.996. (For `ar` the component `aic` is the excess over the best fitting model, and it starts from $p = 0$.) The function `ar` by default fits the model by solving the Yule–Walker equations (15.8) with ρ replaced by r. An alternative is to use `method="burg"`.

The function `arima.mle` fits by maximum likelihood and does not include a mean. Confusingly, the `loglik` component returned by `arima.mle` is a measure of the deviance (minus twice the log-likelihood). Equally confusingly, the likelihood considered is not the full likelihood but a likelihood conditional on a set of starting values for the AR and difference terms, so the AIC given cannot be compared between models unless the component `n.cond` is the same.[2] The conditional likelihood conditions on $p + d$ starting values for a non-seasonal series, or `n.cond` if this is larger. The fitted models are

$$X_t = 0.586(0.121)X_{t-1} + \epsilon_t$$

with $\sigma^2 = 0.211$ and AIC $= 59.61$,

$$X_t = 0.658(0.145)X_{t-1} - 0.066(0.174)X_{t-2} - 0.234(0.145)X_{t-3} + \epsilon_t$$

with $\sigma^2 = 0.190$ and AIC $= 59.09$ and

$$X_t = 0.463(0.218)X_{t-1} + \epsilon_t + 0.200(0.241)\epsilon_{t-1}$$

with $\sigma^2 = 0.205$ and AIC $= 60.41$, which shows the MA term is not worthwhile.

The diagnostic plots are shown in Figures 15.9 and 15.10. The cumulative periodograms of the residuals show that the AR(1) process has not removed all the correlation. The bottom panel of the `arima.diag` plots the P-value for the Box & Pierce (1970) *portmanteau test*

$$Q_K = n \sum_1^K c_k^2 \tag{15.9}$$

---

[2] This is both unnecessary and unfortunate, as the full likelihood could be computed quite easily; see Brockwell & Davis (1991, §8.7).

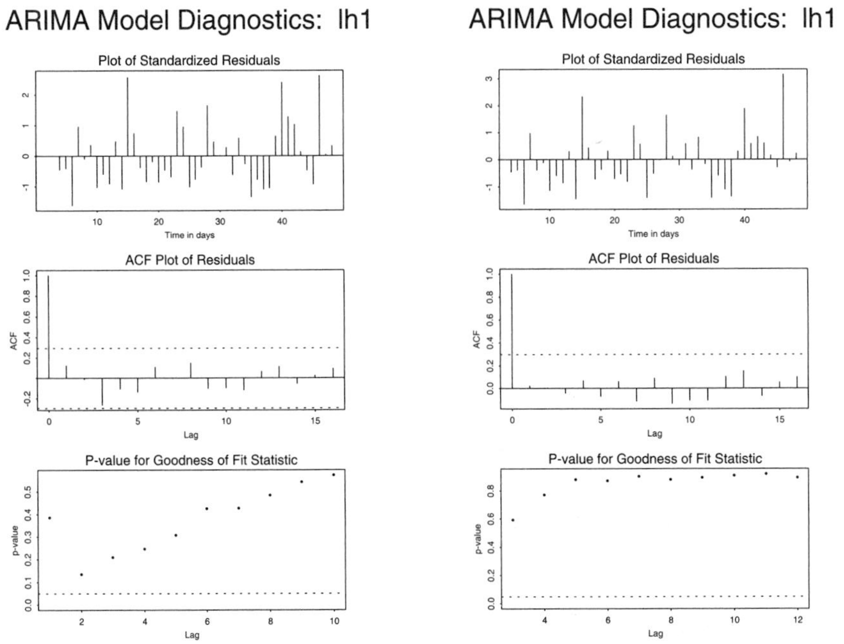

ARIMA Model Diagnostics: lh1   ARIMA Model Diagnostics: lh1

Figure 15.10: Diagnostic plots for ARIMA models fitted to `lh`. (Left) AR(1) and (right) AR(3).

applied to the residuals (and not the Ljung-Box test as a comment in the function suggests). Here the maximum K is set by the parameter `gof.lag` (which defaults to 10) plus $p + q$. Note that although an AR(3) model fits better according to AIC, this is not clear-cut from the diagnostic plots. Although the AIC is smaller, a formal test of the difference in deviances, 4.52, using a χ_2^2 distribution is not significant.

   The function `arima.diag` can produce (standardized) residuals from a fit by `arima.mle`, by setting `plot=F`, `acf.resid=F`, `gof.lag=0` and `resid=T` or `std.resid=T`.

   The function `arima.mle` can also include differencing and so fit an ARIMA model (the middle integer in `order` is d).

Forecasting

Forecasting is relatively straightforward using the function `arima.forecast`:

```
lh.fore <- arima.forecast(lh1, n=12, model=lh.arima3$model)
lh.fore$mean <- lh.fore$mean + mean(lh)
ts.plot(lh, lh.fore$mean, lh.fore$mean+2*lh.fore$std.err,
    lh.fore$mean-2*lh.fore$std.err)
```

(see Figure 15.11) but the standard errors do not include the effect of estimating the mean and the parameters of the ARIMA model.

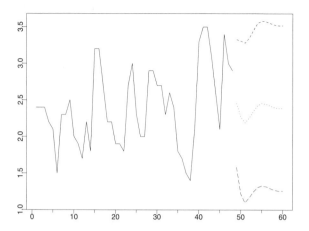

Figure 15.11: Forecasts for 12 periods (of 10 mins) ahead for the series `lh`. The dashed curves are approximate pointwise 95% confidence intervals.

Spectral densities via AR processes

The spectral density of an ARMA process has a simple form: it is given by

$$f(\omega) = \sigma^2 \left| \frac{1 + \sum_s \beta_s e^{-is\omega}}{1 - \sum_t \alpha_t e^{-it\omega}} \right|^2 \qquad (15.10)$$

and so we can estimate the spectral density by substituting parameter estimates in (15.10). It is most usual to fit high-order AR models, both because they can be fitted rapidly, and since they can produce peaks in the spectral density estimate by small values of $|1 - \sum \alpha_t e^{-it\omega}|$ (which correspond to nearly non-stationary fitted models since there must be roots of $\phi_\alpha(z)$ near $e^{-i\omega}$).

This procedure is implemented by `spectrum` with `method="ar"`, which calls `spec.ar`. Although popular because it often produces visually pleasing spectral density estimates, it is not recommended (e.g. Thomson, 1990).

Regression terms

The `arima` family of functions can also handle regressions with ARIMA residual processes, that is models of the form

$$X_t = \sum \gamma_i Z_t^{(i)} + \eta_t, \qquad \phi_\alpha(B)\Delta^d \eta_t = \phi_\beta(B)\epsilon \qquad (15.11)$$

for one or more external time series $z^{(i)}$. This can be specified for simulation to `arima.sim`, the parameters γ estimated by `arima.mle`, and forecasts computed by `arima.forecast` (provided forecasts of the external series are available). Again, the variability of the parameter estimates $\widehat{\gamma}$ is not taken into account in the computed prediction standard errors.

15.3   Seasonality

For a seasonal series there are two possible approaches. One is to decompose the series, usually into a trend, a seasonal component and a residual, and to apply non-seasonal methods to the residual component. The other is to model all the aspects simultaneously.

Decompositions

Two decomposition algorithms are available. The function `sabl` dates from 1982 and is being superseded by the function `stl`. Both are fairly complex, and the on-line documentation should be consulted for full details and references (principally Cleveland *et al.*, 1990).

The function `stl` can extract a strictly periodic component plus a remainder:

```
deaths.stl <- stl(deaths, "periodic")
ts.plot(deaths, deaths.stl$sea, deaths.stl$rem)
```

as shown in Figure 15.12.

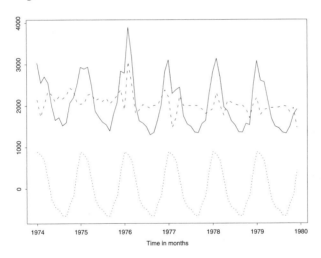

Figure 15.12: `stl` decomposition for the `deaths` series (solid line). The dotted series is the seasonal component, the dashed series the remainder.

The function `monthplot` plots the seasonal component of a series decomposed by `sabl` or `stl`.

We now return to complete the analysis of the `deaths` series by analysing the non-seasonal component. The results are shown in Figure 15.13.

```
> dsd <- deaths.stl$rem
> ts.plot(dsd)
> acf(dsd)
> acf(dsd, type="partial")
```

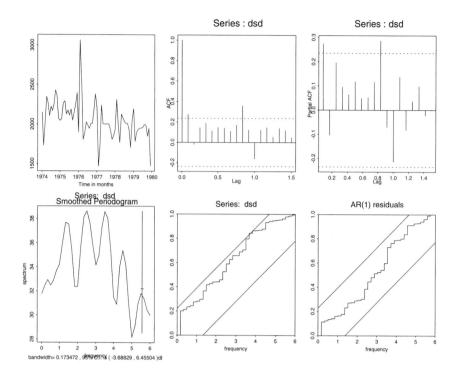

Figure 15.13: Diagnostics for an AR(1) fit to the remainder of an `stl` decomposition of the `deaths` series.

```
> spectrum(dsd, span=c(3,3))
> cpgram(dsd)
> dsd.ar <- ar(dsd)
> dsd.ar$order
[1] 1
> dsd.ar$aic
 [1]  3.64856  0.00000  1.22644  0.40857  1.75586  3.46936
  ....
> dsd.ar$ar
[1,] 0.27469
> cpgram(dsd.ar$resid, main="AR(1) residuals")
> dsd.rar <- ar.gm(dsd)
> dsd.rar$ar
  ....
[1] 0.41493
```

The large jump in the cumulative periodogram at the lowest (non-zero) Fourier frequency is caused by the downward trend in the series. The spectrum has dips at the integers since we have removed the seasonal component and hence all of the frequency and its multiples. (The dip at 1 is obscured by the peak at the Fourier frequency to its left; see the cumulative periodogram.) The plot of the remainder series shows exceptional values for February–March 1976 and 1977. As there are only 6 cycles, the seasonal pattern is difficult to establish at all precisely.

The robust AR-fitting function `ar.gm` helps to overcome the effect of outliers such as February 1976, and produces a considerably higher AR coefficient. The values for February 1976 and 1977 are heavily down-weighted. Unfortunately its output does not mesh well with the other time-series functions.

Seasonal ARIMA models

The function `diff` allows differences at lags greater than one, so for a monthly series the difference at lag 12 is the difference from this time last year. Let s denote the period, often 12. We can then consider ARIMA models for the sub-series sampled s apart, for example for all Januaries. This corresponds to replacing B by B^s in the definition (15.7). Thus an ARIMA $(P, D, Q)_s$ process is a seasonal version of an ARIMA process. However, we may include both seasonal and non-seasonal terms, obtaining a process of the form

$$\Phi_{AR}(B)\Phi_{SAR}(B^s)Y = \Phi_{MA}(B)\Phi_{SMA}(B^s)\epsilon, \qquad Y = (I-B)^d(I-B^s)^D X$$

If we expand this out, we see that it is an ARMA $(p + sP, q + sQ)$ model for Y_t, but parametrized in a special way with large numbers of zero coefficients. It can still be fitted as an ARMA process, and `arima.mle` can handle models specified in this form, by specifying extra terms in the argument `model` with the period set. (Examples follow.)

To identify a suitable model we first look at the seasonally differenced series. Figure 15.14 suggests that this may be over-differencing, but that the non-seasonal term should be an AR(2).

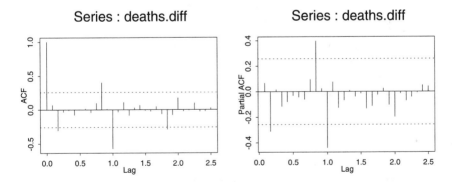

Figure 15.14: Autocorrelation and partial autocorrelation plots for the seasonally differenced `deaths` series. The negative values at lag 12 suggest over-differencing.

```
> deaths.diff <- diff(deaths, 12)
> acf(deaths.diff, 30)
> acf(deaths.diff, 30, type="partial")
```

```
> ar(deaths.diff)
$order:
[1] 12
    ....
$aic:
 [1]   7.8143   9.5471   5.4082   7.3929   8.5839 10.1979 12.1388
 [8]  14.0201 15.7926 17.2504   8.9905 10.9557   0.0000   1.6472
[15]   2.6845   4.4097   6.4047   8.3152
# this suggests the seasonal effect is still present.
> deaths.arima1 <- arima.mle(deaths, model=list(
      list(order=c(2,0,0)), list(order=c(0,1,0), period=12)) )
> deaths.arima1$aic
[1] 847.41
> deaths.arima1$model[[1]]$ar   # the non-seasonal part
[1]   0.12301 -0.30522
> sqrt(diag(deaths.arima1$var.coef))
[1] 0.12504 0.12504
> arima.diag(deaths.arima1, gof.lag=24)
# suggests need a seasonal AR term
> deaths1 <- deaths - mean(deaths)
> deaths.arima2 <- arima.mle(deaths1, model=list(
      list(order=c(2,0,0)),  list(order=c(1,0,0), period=12)) )
> deaths.arima2$aic
[1] 845.38
> deaths.arima2$model[[1]]$ar # non-seasonal part
[1]   0.21601 -0.25356
> deaths.arima2$model[[2]]$ar # seasonal part
[1] 0.82943
> sqrt(diag(deaths.arima2$var.coef))
[1] 0.12702 0.12702 0.07335
> arima.diag(deaths.arima2, gof.lag=24)
> cpgram(arima.diag(deaths.arima2, plot=F, resid=T)$resid)
> deaths.arima3 <- arima.mle(deaths, model=list(
      list(order=c(2,0,0)), list(order=c(1,1,0), period=12)) )
> deaths.arima3$aic   # not comparable to those above
[1] 638.21
> deaths.arima3$model[[1]]$ar
[1]   0.41212 -0.26938
> deaths.arima3$model[[2]]$ar
[1] -0.7269
> sqrt(diag(deaths.arima3$var.coef))
[1] 0.14199 0.14199 0.10125
> arima.diag(deaths.arima3, gof.lag=24)
> arima.mle(deaths1, model=list(list(order=c(2,0,0)),
      list(order=c(1,0,0), period=12)), n.cond=26 )$aic
[1] 664.14
> deaths.arima4 <- arima.mle(deaths1, model=list(
      list(order=c(2,0,0)), list(order=c(2,0,0), period=12)) )
> deaths.arima4$aic
[1] 634.07
```

```
> deaths.arima4$model[[1]]$ar
[1]   0.47821 -0.27873
> deaths.arima4$model[[2]]$ar
[1]   0.17346 0.63474
> sqrt(diag(deaths.arima4$var.coef))
[1]   0.14160 0.14160 0.11393 0.11393
```

The AR-fitting suggests a model of order 12 (of up to 16) which indicates that seasonal effects are still present. The diagnostics from the ARIMA($(2, 0, 0) \times (0, 1, 0)_{12}$) model suggest problems at lag 12. Dropping the differencing in favour of a seasonal AR term gives an AIC which favours the second model. However, the diagnostics suggest that there is still seasonal structure in the residuals, so we include differencing and a seasonal AR term. This gives a seasonal model of the form

$$(I + 0.727B^s)(I - B^s)X = (I - 0.273B^s - 0.727B^{2s})X = \epsilon$$

which suggests a seasonal AR(2) term might be more appropriate.

The best fit found is an ARIMA($(2, 0, 0) \times (2, 0, 0)_{12}$) model, but this does condition on the first 26 observations, including the exceptional value for February 1976 (see Figure 15.1). Fitting the ARIMA($(2, 0, 0) \times (1, 0, 0)_{12}$) model to the same series shows that the seasonal AR term does help the fit considerably.

We consider another example on monthly accidental deaths in the USA 1973–8, from Brockwell & Davis (1991) contained in our dataset accdeaths:

```
> dacc <- diff(accdeaths, 12)
> ts.plot(dacc)
> acf(dacc, 30)
> acf(dacc, 30, "partial")
> ddacc <- diff(dacc)
> ts.plot(ddacc)
> acf(ddacc, 30)
> acf(ddacc, 30, "partial")
> ddacc.1 <- arima.mle(ddacc-mean(ddacc),
       model=list(list(order=c(0,0,1)),
       list(order=c(0,0,1), period=12)))
$model[[1]]$ma:
[1] 0.48834
$model[[2]]$ma:
[1] 0.58534
$aic:
[1] 852.72
$loglik:
[1] 848.72
$sigma2:
[1] 94629
> sqrt(diag(ddacc.1$var.coef))
[1] 0.11361 0.10556
> ddacc.2 <- arima.mle(ddacc-mean(ddacc),
```

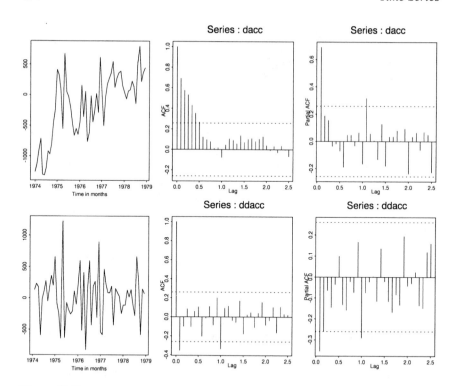

Figure 15.15: Seasonally differenced (top row) and then differenced (bottom row) versions of the accidental deaths series `accdeath` with ACF and PACF plots.

```
          model=list(order=c(0,0,13),
          ma.opt=c(T,F,F,F,F,T,F,F,F,F,F,T,T)),
          max.iter=50, max.fcal=100)
$model$ma:
 [1]  0.60784  0.00000  0.00000  0.00000  0.00000  0.41119
 [7]  0.00000  0.00000  0.00000  0.00000  0.00000  0.67693
[13] -0.47260
$aic:
[1] 869.85
$loglik:
[1] 843.85
$sigma2:
[1] 70540
> sqrt(diag(ddacc.2$var.coef))
 [1] 0.11473 0.10798 0.10798 0.10798 0.10798 0.10798 0.12052
 [8] 0.10798 0.10798 0.10798 0.10798 0.10798 0.11473
```

The plots (Figure 15.15) suggest the use of $\nabla\nabla_{12}X$, and this has a non-zero mean. The first model fitted is

$$\nabla\nabla_{12}X = 28.83 + (1 - 0.488B)(1 - 0.585B^{12})\epsilon$$

and the second model comes from selecting promising non-zero terms in a general MA(13) process, as

$$\nabla\nabla_{12}X = 28.83 + (1 - 0.608B - 0.411B^6 - 0.677B^{12} + 0.473B^{13})\epsilon$$

Note that the AIC is wrong; it should be 851.85 as there are parameters set to zero (although this does not allow for selection). This fit illustrates the ability to constrain coefficients in an ARIMA fit. That standard errors are returned for zero parameters suggests that the standard errors are wrong. Standard likelihood theory suggests deleting rows from the inverse of the information matrix:

```
> dd.VI <- solve(ddacc.2$var.coef)
> sqrt(diag(
     solve(dd.VI[ddacc.2$model$ma.opt,ddacc.2$model$ma.opt])
     ))
[1] 0.096691 0.085779 0.094782 0.095964
```

which shows the power of the S language.

Trading days

Economic and financial series can be affected by the number of trading days in the month or quarter in question. The function `arima.td` computes a multiple time series with seven series giving the number of days in the months and the differences between the number of Saturdays, Sundays, Mondays, Tuesdays, Wednesdays and Thursdays, and the number of Fridays in the month or quarter. This can be used as a regression term in an ARIMA model. It also has other uses. For the `deaths` series we might consider dividing by the number of days in the month to find daily death rates; we already have enough problems with February!

15.4   Multiple time series

The second-order time-domain properties of multiple time series were covered in Section 15.1. The function `ar` will fit AR models to multiple time series, but ARIMA fitting is confined to univariate series. Let X_t denote a multiple time series, and ϵ_t a correlated sequence of identically distributed random variables. Then a vector AR(p) process is of the form

$$X_t = \sum_i^p A_i X_{t-i} + \epsilon_t$$

for matrices A_i. Further, the components of ϵ_t may be correlated, so we will assume that this has covariance matrix Σ. Again there is a condition on the coefficients, that

$$\det\left[I - \sum_1^p A_i z^i\right] \neq 0 \text{ for all } |z| \leqslant 1$$

The parameters can be estimated by solving the multiple version of the Yule–Walker equations (Brockwell & Davis, 1991, §11.5), and this is used by `ar.yw`, the function called by `ar`. (The other method, `ar.burg`, also handles multiple series.)

Spectral analysis for multiple time series

The definitions of the spectral density can easily be extended to a pair of series. The cross-covariance is expressed by

$$\gamma_{ij}(t) = \frac{1}{2\pi} \int_{-\pi}^{\pi} e^{i\omega t} \, dF_{ij}(\omega)$$

for a finite complex measure on $(-\pi, \pi]$, which will often have a density f_{ij} so that

$$\gamma_{ij}(t) = \frac{1}{2\pi} \int_{-\pi}^{\pi} e^{i\omega t} f_{ij}(\omega) \, d\omega$$

and

$$f_{ij}(\omega) = \sum_{-\infty}^{\infty} \gamma_{ij}(t) e^{-i\omega t}$$

Note that since $\gamma_{ij}(t)$ is not necessarily symmetric, the sign of the frequency becomes important, and f_{ij} is complex. Conventionally it is written as $c_{ij}(\omega) - i\, q_{ij}(\omega)$ where c is the *co-spectrum* and q is the *quadrature spectrum*. Alternatively we can consider the amplitude $a_{ij}(\omega)$ and phase $\phi_{ij}(\omega)$ of $f_{ij}(\omega)$. Rather than use the amplitude directly, it is usual to work with the *coherence*

$$b_{ij}(\omega) = \frac{a_{ij}(\omega)}{\sqrt{f_{ii}(\omega) f_{jj}(\omega)}}$$

which lies between zero and one.

The *cross-periodogram* is

$$I_{ij}(\omega) = \left[\sum_{s=1}^{n} e^{-i\omega s} X_i(s) \sum_{t=1}^{n} e^{i\omega t} X_j(t) \right] \Big/ n$$

and is a complex quantity. It is useless as an estimator of the amplitude spectrum, since if we define

$$J_i(\omega) = \sum_{s=1}^{n} e^{-i\omega s} X_i(s)$$

then

$$|I_{ij}(\omega)| \Big/ \sqrt{I_{ii}(\omega) I_{jj}(\omega)} = |J_i(\omega) J_j(\omega)^*| \big/ |J_i(\omega)| \, |J_j(\omega)| = 1$$

but smoothed versions can provide sensible estimators of both the coherence and phase.

The function `spec.pgram` will compute the coherence and phase spectra given a multiple time series. The results are shown in Figure 15.16.

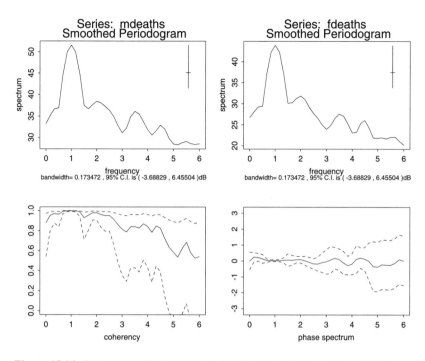

Figure 15.16: Coherence and phase spectra for the two deaths series, with 95% pointwise confidence intervals.

```
spectrum(mdeaths, spans=c(3,3))
spectrum(fdeaths, spans=c(3,3))
mfdeaths.spc <- spec.pgram(ts.union(mdeaths, fdeaths),
    spans=c(3,3))
plot(mfdeaths.spc$freq, mfdeaths.spc$coh, type="l",
    ylim=c(0,1), xlab="coherency", ylab="")
gg <- 2/mfdeaths.spc$df
se <- sqrt(gg/2)
lines(mfdeaths.spc$freq, tanh(atanh(mfdeaths.spc$coh) +
    1.96*se), lty=3)
lines(mfdeaths.spc$freq, tanh(atanh(mfdeaths.spc$coh) -
    1.96*se), lty=3)
plot(mfdeaths.spc$freq, mfdeaths.spc$phase, type="l",
    ylim=c(-pi, pi), xlab="phase spectrum", ylab="")
cl <- asin( pmin( 0.9999, qt(0.95, 2/gg-2)*
    sqrt(gg*(mfdeaths.spc$coh^{-2} - 1)/(2*(1-gg)) ) ) )
lines(mfdeaths.spc$freq, mfdeaths.spc$phase + cl, lty=3)
lines(mfdeaths.spc$freq, mfdeaths.spc$phase - cl, lty=3)
```

These confidence intervals follow Bloomfield (1976, §8.5). At the frequency of 1/year there is a strong signal common to both series, so the coherence is high and both coherence and phase are determined very precisely. At high frequencies there is little information, and the phase cannot be fixed at all precisely.

It is helpful to consider what happens if the series are not aligned:

```
mfdeaths.spc <- spec.pgram(ts.union(mdeaths, lag(fdeaths, 4)),
    spans=c(3,3))
plot(mfdeaths.spc$freq, mfdeaths.spc$coh, type="l",
    ylim=c(0,1), xlab="coherency", ylab="")
gg <- 2/mfdeaths.spc$df
se <- sqrt(gg/2)
lines(mfdeaths.spc$freq, tanh(atanh(mfdeaths.spc$coh) +
    1.96*se), lty=3)
lines(mfdeaths.spc$freq, tanh(atanh(mfdeaths.spc$coh) -
    1.96*se), lty=3)
phase <- (mfdeaths.spc$phase + pi)%%(2*pi) - pi
plot(mfdeaths.spc$freq, phase, type="l",
    ylim=c(-pi, pi), xlab="phase spectrum", ylab="")
cl <- asin( pmin( 0.9999, qt(0.95, 2/gg-2)*
    sqrt(gg*(mfdeaths.spc$coh^{-2} - 1)/(2*(1-gg)) ) ) )
lines(mfdeaths.spc$freq, phase + cl, lty=3)
lines(mfdeaths.spc$freq, phase - cl, lty=3)
```

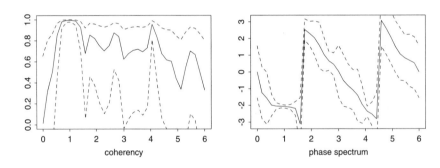

Figure 15.17: Coherence and phase spectra for the re-aligned deaths series, with 95% pointwise confidence intervals.

The results are shown in Figure 15.17. The phase has an added component of slope $2\pi * 4$, since if $X_2(t) = X_1(t - \tau)$,

$$\gamma_{12}(t) = \gamma_{11}(t + \tau), \qquad f_{11}(\omega) = f_{11}(\omega)e^{-i\tau\omega}$$

For more than two series we can consider all the pairwise coherence and phase spectra, which are returned by spec.pgram.

15.5   Nottingham temperature data

We now consider a substantial example. The data are mean monthly air temperatures ($°F$) at Nottingham Castle for the months January 1920 – December

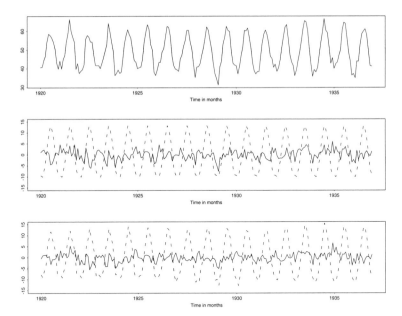

Figure 15.18: Plots of the first 17 years of the `nottem` dataset. Top is the data, middle the `stl` decomposition with a seasonal periodic component and bottom the `stl` decomposition with a 'local' seasonal component.

1939, from *'Meteorology of Nottingham'*, in *City Engineer and Surveyor*. They also occur in Anderson (1976). We will use the years 1920–1936 to forecast the years 1937–1939 and compare with the recorded temperatures. The data are series `nottem` in our library.

```
nott <- window(nottem, end=c(1936,12))
ts.plot(nott)
nott.stl <- stl(nott, "period")
ts.plot(nott.stl$rem-49, nott.stl$sea,
    ylim = c(-15, 15), lty=c(1,3))
nott.stl <- stl(nott, 5)
ts.plot(nott.stl$rem-49, nott.stl$sea,
    ylim = c(-15, 15), lty=c(1,3))
boxplot(split(nott, cycle(nott)), names=month.abb)
```

Figures 15.18 and 15.19 show clearly that February 1929 is an outlier. It *is* correct—it was an exceptionally cold month in England. The `stl` plots show that the seasonal pattern is fairly stable over time. Since the value for February 1929 will distort the fitting process, we altered it to a low value for February of 35. We first model the remainder series:

```
> nott[110] <- 35
> nott.stl <- stl(nott, "period")
> nott1 <- nott.stl$rem - mean(nott.stl$rem)
```

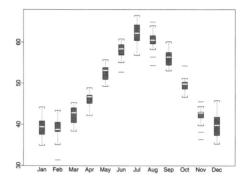

Figure 15.19: Monthly boxplots of the first 17 years of the `nottem` dataset.

```
> acf(nott1)
> acf(nott1,, "partial")
> cpgram(nott1)
> ar(nott1)$aic
 [1] 13.67432  0.00000  0.11133  2.07849  3.40381  5.40125
> plot(0:23, ar(nott1)$aic, xlab="order", ylab="AIC",
      main="AIC for AR(p)")
> nott1.ar1 <- arima.mle(nott1, model=list(order=c(1,0,0)))
> nott1.ar1$model$ar:
[1] 0.27255
> sqrt(nott1.ar1$var.coef)
ar(1) 0.067529
> nott1.fore <- arima.forecast(nott1, n=36,
      model=nott1.ar1$model)
> nott1.fore$mean <- nott1.fore$mean + mean(nott.stl$rem) +
                        as.vector(nott.stl$sea[1:36])
> ts.plot(window(nottem, 1937), nott1.fore$mean,
      nott1.fore$mean+2*nott1.fore$std.err,
      nott1.fore$mean-2*nott1.fore$std.err, lty=c(3,1,2,2))
> title("via Seasonal Decomposition")
```

(see Figures 15.20 and 15.21) all of which suggest an AR(1) model. (Remember that a seasonal term has been removed, so we expect negative correlation at lag 12.) The confidence intervals in Figure 15.21 for this method ignore the variability of the seasonal terms. We can easily make a rough adjustment. Each seasonal term is approximately the mean of 17 approximately independent observations (since 0.2725516^{12} is negligible). Those observations have variance about 5.05 about the seasonal term, so the seasonal term has standard error about $\sqrt{5.05/17} = 0.55$, compared to the 2.25 for the forecast. The effect of estimating the seasonal terms is in this case negligible. (Note that the forecast errors are correlated with errors in the seasonal terms.)

We now move to the Box-Jenkins methodology of using differencing:

```
> acf(diff(nott,12), 30)
```

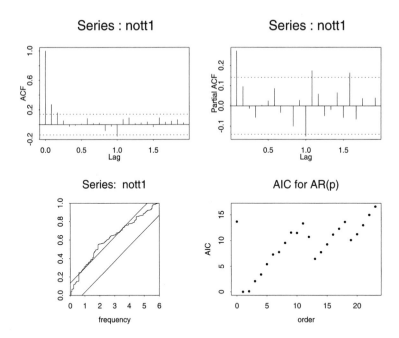

Figure 15.20: Summaries for the remainder series of the `nottem` dataset.

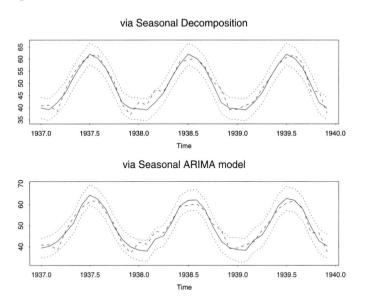

Figure 15.21: Forecasts (solid), true values (dashed) and approximate 95% confidence intervals for the `nottem` series. The upper plot is via a seasonal decomposition, the lower plot via a seasonal ARIMA model.

```
> acf(diff(nott,12), 30, "partial")
> cpgram(diff(nott,12))
> nott.arima1 <- arima.mle(nott,
      model=list(list(order=c(1,0,0)), list(order=c(2,1,0),
      period=12)))
> nott.arima1
$model[[1]]$ar:
[1] 0.32425
$model[[2]]$ar:
[1] -0.87576 -0.31311
> sqrt(diag(nott.arima1$var.coef))
[1] 0.073201 0.073491 0.073491
> arima.diag(nott.arima1, gof.lag=24)
> nott.fore <- arima.forecast(nott, n=36,
      model=nott.arima1$model)
> ts.plot(window(nottem, 1937), nott.fore$mean,
      nott.fore$mean+2*nott.fore$std.err,
      nott.fore$mean-2*nott.fore$std.err, lty=c(3,1,2,2))
> title("via Seasonal ARIMA model")
```

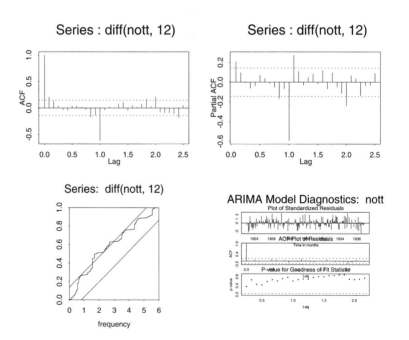

Figure 15.22: Seasonal ARIMA modelling of the `nottem` series. The top row shows the ACF and partial ACF of the yearly differences, which suggest an AR model with seasonal and non-seasonal terms. The bottom row shows the cumulative periodogram of the differenced series and the output of `arima.diag` for the fitted model.

(see Figures 15.21 and 15.22) which produces slightly wider confidence intervals, but happens to fit the true values slightly better. The fitted model is

$$(I + 0.875B^s + 0.313B^{2s})(I - B^s)(I - 0.324B)X = \epsilon$$

15.6   Regression with autocorrelated errors

We touched briefly on the use of regression terms with the functions `arima.mle` and `arima.forecast` on page 448. In this section we consider a number of ways to use S-PLUS facilities to study regression with autocorrelated errors. They are most pertinent when the regression rather than time-series prediction is of primary interest.

Our main example is taken from Reynolds (1994). She describes a small part of a study of the long-term temperature dynamics of beaver (*Castor canadensis*) in north-central Wisconsin. Body temperature was measured by telemetry every 10 minutes for four females, but data from a one period of less than a day for each of two animals is used there (and here). Columns indicate the day (December 12-13, 1990 and November 3-4, 1990 for the two examples), time (hhmm on 24 hour clock), temperature ($^{\circ}C$) and a binary index of activity (0 = animal inside retreat; 1 = animal outside retreat).

These data are available on the diskette of Lange *et al.* (1994) and at `statlib`. Figure 15.23 shows the two series. The first series has a missing observation (at 22:20), and this and the pattern of activity suggests that it is easier to start with beaver 2.

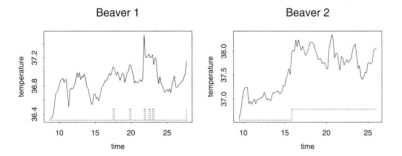

Figure 15.23: Plots of temperature (solid) and activity (dashed) for two beavers. The time is shown in hours since midnight of the first day of observation.

```
attach(beav1)
beav1$hours <- 24*(day-346) + trunc(time/100) + (time%%100)/60
detach(); attach(beav2)
beav2$hours <- 24*(day-307) + trunc(time/100) + (time%%100)/60
detach()
par(mfrow=c(2,2))
```

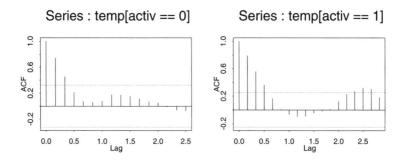

Figure 15.24: ACF of the beaver 2 temperature series before and after activity begins.

```
plot(beav1$hours, beav1$temp, type="l", xlab="time",
    ylab="temperature", main="Beaver 1")
usr <- par("usr"); usr[3:4] <- c(-0.2, 8); par(usr=usr)
lines(beav1$hours, beav1$activ, type="s", lty=2)
plot(beav2$hours, beav2$temp, type="l", xlab="time",
    ylab="temperature", main="Beaver 2")
usr <- par("usr"); usr[3:4] <- c(-0.2, 8); par(usr=usr)
lines(beav2$hours, beav2$activ, type="s", lty=2)
```

Beaver 2

Looking at the series before and after activity begins suggests a moderate amount
of autocorrelation, confirmed by the plots in Figure 15.24.

```
attach(beav2)
temp <- rts(temp, start=8+2/3, frequency=6, units="hours")
activ <- rts(activ, start=8+2/3, frequency=6, units="hours")
acf(temp[activ==0]); acf(temp[activ==1]) # also look at PACFs
ar(temp[activ==0]); ar(temp[activ==1])
```

Fitting an $AR(p)$ model to each part of the series selects $AR(1)$ models with
coefficients 0.74 and 0.79, so a common $AR(1)$ model for the residual series
looks plausible.

We begin by fitting by `arima.mle` a common $AR(1)$ model by `arima.mle`
to find the baseline deviance of -122.86. Adding a term for activity gives
a deviance of -140.78, and coefficients of 37.284 ($°C$) for the mean whilst
inactive and and 0.584 for the change in mean due to activity. The AR coefficient
is estimated as $0.8255(0.056)$. We might consider that we need a smoother
transition in temperature at a change in activity, but as there is only one transition
we cannot learn much.

```
arima.mle(temp, xreg=rep(1, length(temp)), model=list(ar=0.75))
arima.mle(temp, xreg=cbind(1, activ), model=list(ar=0.75))
```

We can test for diurnal variation by adding a sine-wave term. This reduces the
deviance to -142.55, giving an increase in AIC for two further parameters.

```
dreg <- cbind(sin(2*pi*hours/24), cos(2*pi*hours/24))
arima.mle(temp, xreg=cbind(1, activ,dreg), model=list(ar=0.75))
```

How can we find standard errors for the regression parameter estimates? Unfortunately, `arima.mle` does not give them. We could use a plot of the log-likelihood to find a confidence region. Or we could use first principles and compute $(X^T \widehat{\Sigma}^{-1} X)^{-1}$, assuming that the covariance structure $\widehat{\Sigma}$ of the fitted AR model is the true structure. However, the most promising idea is to use *pre-whitening*. Our model is

$$y = X\beta + \eta, \qquad (I - \alpha B)\eta = \epsilon,$$

and so

$$(I - \alpha B)y = (I - \alpha B)X\beta + \epsilon$$

which we can fit by least-squares, using $(I - \widehat{\alpha} B)X$ as the regressors. This is sometimes known as the Cochrane–Orcutt scheme (Cochrane & Orcutt, 1949), and gives standard errors of 0.096 and 0.098. The mean when active is estimated as $37.868(0.076)$. These standard errors ignored the estimation of the AR model but are often reliable enough.

```
alpha <- 0.8255
stemp <- temp - alpha*lag(temp, -1)
X <- cbind(1, activ); sX <- X[-1, ] - alpha*X[-100, ]
beav2.ls <- lm(stemp ~ -1 + sX)
beav2.sls <- summary(beav2.ls)
Coefficients:
          Value Std. Error t value Pr(>|t|)
     sX 37.284    0.096    389.071   0.000
sXactiv  0.584    0.098      5.934   0.000
sqrt(t(c(1,1)) %*% beav2.sls$cov %*% c(1,1)) * beav2.sls$sigma
        [,1]
[1,] 0.076229
plot(hours[-1], residuals(beav2.ls))
detach(); rm(temp, activ)
```

Looking at the residuals from this regression compares each observation with the prediction from the current activity and the immediate past observation. No particular pattern emerges.

Using `lme`

We can use the function `lme` from library `nlme` which has the ability to include serial correlations within a cluster (and we take the one animal as one cluster).

```
> library(nlme)  # may be needed
> beav2.lme <- lme(temp ~ activ, cluster = ~rep(1,100),
      data=beav2, serial.structure="ar1", est.method="ML")
> summary(beav2.lme)
    ....
Cluster Residual Variance: 0.063888
```

```
Serial Correlation Parameter(s): 0.87315
                Value Approx. Std.Error z ratio(C)
(Intercept) 37.19193              0.11198    332.1320
      activ   0.61421              0.10763      5.7067
```

(This is slow – around 30 seconds.) There are some end effects due to the sharp initial rise in temperature:

```
summary(lme(temp ~ activ, cluster = ~rep(1,95),
    data=beav2[6:100,], serial.structure="ar1", est.method="ML"))
    ....
Cluster Residual Variance: 0.047885
Serial Correlation Parameter(s): 0.838
                Value Approx. Std.Error z ratio(C)
(Intercept) 37.24998              0.095304    390.8556
      activ   0.60281              0.098259      6.1349
```

and REML estimates of the standard errors are substantially larger.

Beaver 1

Applying the same ideas to beaver 2, we can select an initial covariance model based on the observations before the first activity at 17:30. The autocorrelations again suggest an $AR(1)$ model, whose coefficient is fitted as 0.82. We included as regressors activity now and 10, 20 and 30 minutes ago. This gives a mean when inactive of $36.859(0.032)$.

```
attach(beav1)
temp <- rts(c(temp[1:82], NA, temp[83:114]), start=9.5,
            frequency=6, units="hours")
activ <- rts(c(activ[1:82], NA, activ[83:114]), start=9.5,
            frequency=6, units="hours")
acf(temp[1:53]) # and also type="partial"
ar(temp[1:53])

act <- c(rep(0, 10), activ)
X <- cbind(1, act=act[11:125], act1 = act[10:124],
        act2 = act[9:123], act3 = act[8:122])
arima.mle(temp, xreg=X, model=list(ar=0.82))
$model$ar:
[1] 0.8025

alpha <- 0.80
stemp <- temp - alpha*lag(temp, -1)
sX <- X[-1, ] - alpha * X[-115,]
beav1.ls <- lm(stemp ~ -1 + sX, na.action=na.omit)
summary(beav1.ls, cor=F)
Coefficients:
        Value Std. Error t value Pr(>|t|)
    sX  36.856    0.039    939.833   0.000
```

```
sXact    0.254    0.039    6.464    0.000
sXact1   0.171    0.051    3.352    0.001
sXact2   0.162    0.051    3.148    0.002
sXact3   0.105    0.043    2.448    0.016
detach(); rm(temp, activ)
```

All the terms are significant, and adding earlier activity does not improve the model. (Note that we need to be careful with missing values here: which rows contain missing values depend on which terms are included in the model.)

Our analysis shows that there is a difference in temperature between activity and inactivity, and that temperature may build up gradually with activity (which seems physiologically reasonable). A *caveat* is that the apparent outlier at 21:50, at the start of a period of activity, contributes considerably to this conclusion.

Exercises

15.6.1 Use the information gained in the analysis of beav1 to refine the analysis for beav2.

15.6.2 If you have access to the S-PLUS module S+SPATIALSTATS, consider how to apply the spatial linear model function slm to this problem.

15.6.3 Consider the problem of estimating the effect of seat belt legislation on road accident casualties in the UK considered by Harvey & Durbin (1986). The data (from Harvey (1989)) are in the series drivers.

15.7   Other time-series functions

S-PLUS has a number of time-series functions which are used less frequently and we have not yet discussed. This section is only cursory.

Many of the other functions implement various aspects of filtering, that is converting one times series into another while emphasising some features and de-emphasising others. A linear filter is of the form

$$Y_t = \sum_j a_j X_{t-j}$$

which is implemented by the function filter. The coefficients are supplied, and it is assumed that they are non-zero only for $j \geqslant 0$ (sides=1) or $-m \leqslant j \leqslant m$ (sides=2 , the default). A linear filter affects the spectrum by

$$f_Y(\omega) = \left| \sum a_s e^{-is\omega} \right|^2 f_X(\omega)$$

and filters are often described by aspects of the gain function $| \sum a_s e^{-is\omega} |$. Kernel smoothers such as ksmooth are linear filters when applied to regularly-spaced time series.

Another way to define a linear filter is recursively (as in exponential smoothing), and this can be done by `filter`, using

$$Y_t = \sum_{s=1}^{\ell} a_s Y_{t-s}$$

in which case ℓ initial values must be specified by the argument `init`.

Converting an ARIMA process to the innovations process ϵ is one sort of recursive filtering, implemented by the function `arima.filt`.

A large number of smoothing operations such as `lowess` can be regarded as filters, but they are non-linear. The functions `acm.filt`, `acm.ave` and `acm.smo` provide filters resistant to outliers.

Complex demodulation is a technique to extract approximately periodic components from a time series. It is discussed in detail by Bloomfield (1976, Chapter 7) and implemented by the function `demod`.

Some time series exhibit correlations which never decay exponentially, as they would for an ARMA process. One way to model these phenomena is fractional differencing (Brockwell & Davis, 1991, §13.2). Suppose we expand ∇^d by a binomial expansion:

$$\nabla^d = \sum_{j=0}^{\infty} \frac{\Gamma(j-d)}{\Gamma(j+1)\Gamma(-d)} B^j$$

and use the right-hand side as the definition for non-integer d. This will only make sense if the series defining $\nabla^d X_t$ is mean-square convergent. A fractional ARIMA process is defined for $d \in (-0.5, 0.5)$ by the assumption that $\nabla^d X_t$ is an ARMA(p, q) process, so

$$\phi(B)\nabla^d X = \theta(B)\epsilon, \qquad \text{so} \qquad \phi(B)X = \theta(B)\nabla^{-d}\epsilon$$

and we can consider it also as an ARMA(p, q) process with fractionally integrated noise. The spectral density is of the form

$$f(\omega) = \sigma^s \left| \frac{\theta(e^{-i\omega})}{\phi(e^{-i\omega})} \right|^2 \times |1 - e^{-i\omega}|^{-2d}$$

and the behaviour as ω^{-2d} at the origin will help identify the presence of fractional differencing.

The functions `arima.fracdiff` and `arima.fracdiff.sim` implement fractionally-differenced ARIMA processes.

Chapter 16

Spatial Statistics

Spatial statistics is a recent and graphical subject which is ideally suited to implementation in S; S itself includes one spatial interpolation method, akima, and loess which can be used for two-dimensional smoothing, but the specialist methods of spatial statistics have been added and are given in our library spatial. The main references for spatial statistics are Ripley (1981, 1988), Diggle (1983), Upton & Fingleton (1985) and Cressie (1991). Not surprisingly, our notation is closest to that of Ripley (1981).

The S-PLUS module[1] S+SPATIALSTATS provides more comprehensive (and more polished) facilities for spatial statistics than those provided in our library spatial. Details of how to work through our examples in that module may be found in the on-line complements to this book. (See the preface.)

16.1 Spatial interpolation and smoothing

We provide three examples of datasets for spatial interpolation. The dataset topo contains 52 measurements of topographic height (in feet) within a square of side 310 feet (labelled in 50 feet units). The datasets shkap and npr1 are measurements on oil fields in the (then) USSR and in the USA. Both contain permeability measurements (a measure of the ease of oil flow in the rock) and npr1 also has porosity (the volumetric proportion of the rock which is pore space).

Suppose we are given n observations $Z(x_i)$ and we wish to map the process $Z(x)$ within a region D. (The sample points x_i are usually, but not always, within D.) Although our treatment will be quite general, our S code assumes D to be a two-dimensional region, which covers the majority of examples. There are however applications to the terrestrial sphere and in three dimensions in mineral and oil applications.

---

[1] S-PLUS modules are additional-cost products; contact your S-PLUS distributor for details.

Trend surfaces

One of the earliest methods was fitting *trend surfaces*, polynomial regression surfaces of the form

$$f((x,y)) = \sum_{r+s \leqslant p} a_{rs} x^r y^s \qquad (16.1)$$

where the parameter p is the order of the surface. There are $P = (p+1)(p+2)/2$ coefficients. Originally (16.1) was fitted by least squares, and could for example be fitted using lm with poly which will give polynomials in one or more variables. There will however be difficulties in prediction, and predict.gam must be used to ensure that the correct orthogonal polynomials are generated. This is rather inefficient in applications such as ours in which the number of points at which prediction is needed may far exceed n. Our function surf.ls implicitly rescales x and y to $[-1, 1]$, which ensures that the first few polynomials are far from collinear. We show some low-order trend surfaces for the topo dataset in Figure 16.1, generated by:

```
library(spatial)
topo.ls <- surf.ls(2, topo)
trsurf <- con2tr(trmat(topo.ls, 0, 6.5, 0, 6.5, 30))
c2 <- contourplot(z ~ x * y, trsurf, at = seq(600, 1000, 25),
    xlab = list("Degree=2", cex=1.5), aspect=1, ylab="",
    panel = function(x, y, subscripts, ...) {
        panel.contourplot(x, y, subscripts, ...)
        panel.xyplot(topo$x,topo$y, cex=0.5)
    }
)
topo.ls <- surf.ls(3, topo)
 . . . .
topo.ls <- surf.ls(4, topo)
 . . . .
topo.ls <- surf.ls(6, topo)
 . . . .
print(c2, position=c(0.0, 0.48, 0.6, 1.02), more=T)
print(c3, position=c(0.32, 0.48, 0.92, 1.02), more=T)
print(c4, position=c(0.0, 0.0, 0.6, 0.55), more=T)
print(c6, position=c(0.32, 0.0, 0.92, 0.55))
```

Figure 16.1 shows trend surfaces for the topo dataset. The highest degree, 6, has 28 coefficients fitted from 52 points. The higher-order surfaces begin to show the difficulties of fitting by polynomials in two or more dimensions, when inevitably extrapolation is needed at the edges.

There are several other ways to show trend surfaces in S-PLUS. Figure 16.2 shows a greyscale plot from levelplot and a perspective plot from wireframe. They were generated by

```
topo.ls <- surf.ls(4, topo)
trsurf <- trmat(topo.ls, 0, 6.5, 0, 6.5, 30)
trsurf[c("x", "y")] <- expand.grid(x=trsurf$x, y=trsurf$y)
```

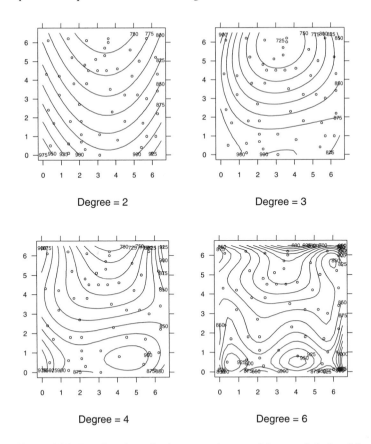

Figure 16.1: Trend surfaces for the `topo` dataset, of degrees 2, 3, 4 and 6.

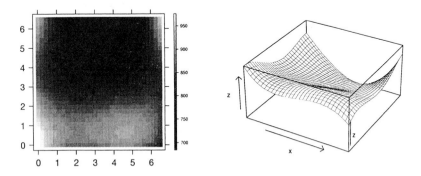

Figure 16.2: The quartic trend surfaces for the `topo` dataset.

```
plt1 <- levelplot(z ~ x * y, trsurf, aspect=1,
            at = seq(650, 1000, 10),   xlab = "", ylab = "")
plt2 <- wireframe(z ~ x * y, trsurf, aspect=c(1, 0.5),
            screen = list(z = -30, x = -60))
```

```
print(plt1, position = c(0, 0, 0.5, 1), more=T)
print(plt2, position = c(0.45, 0, 1, 1))
```

One difficulty with fitting trend surfaces is that in most applications the ob-servations are not regularly spaced, and sometimes they are most dense where the surface is high (for example in mineral prospecting). This makes it important to take the spatial correlation of the errors into consideration. We thus suppose that

$$Z(\boldsymbol{x}) = \boldsymbol{f}(\boldsymbol{x})^T \boldsymbol{\beta} + \epsilon(\boldsymbol{x})$$

for a parametrized trend term such as (16.1) and a zero-mean spatial stochastic process $\epsilon(\boldsymbol{x})$ of errors. We assume that $\epsilon(\boldsymbol{x})$ possesses second moments, and has covariance matrix

$$C(\boldsymbol{x}, \boldsymbol{y}) = \mathrm{cov}\left(\epsilon(\boldsymbol{x}), \epsilon(\boldsymbol{y})\right)$$

(this assumption will be relaxed slightly later). Then the natural way to estimate $\boldsymbol{\beta}$ is by *generalized least squares*, that is to minimize

$$[Z(\boldsymbol{x}_i) - \boldsymbol{f}(\boldsymbol{x}_i)^T \boldsymbol{\beta}]^T [C(\boldsymbol{x}_i, \boldsymbol{x}_j)]^{-1} [Z(\boldsymbol{x}_i) - \boldsymbol{f}(\boldsymbol{x}_i)^T \boldsymbol{\beta}]$$

We need some simplified notation. Let $\boldsymbol{Z} = F\boldsymbol{\beta} + \boldsymbol{\epsilon}$ where

$$F = \begin{bmatrix} \boldsymbol{f}(\boldsymbol{x}_1)^T \\ \vdots \\ \boldsymbol{f}(\boldsymbol{x}_n)^T \end{bmatrix}, \qquad \boldsymbol{Z} = \begin{bmatrix} Z(\boldsymbol{x}_1) \\ \vdots \\ Z(\boldsymbol{x}_n) \end{bmatrix}, \qquad \boldsymbol{\epsilon} = \begin{bmatrix} \epsilon(\boldsymbol{x}_1) \\ \vdots \\ \epsilon(\boldsymbol{x}_n) \end{bmatrix}$$

and let $K = [C(\boldsymbol{x}_i, \boldsymbol{x}_j)]$. We assume that K is of full rank. Then the problem is to minimize

$$[\boldsymbol{Z} - F\boldsymbol{\beta}]^T K^{-1} [\boldsymbol{Z} - F\boldsymbol{\beta}] \tag{16.2}$$

The Choleski decomposition (Golub & Van Loan, 1989; Nash, 1990) finds a lower-triangular matrix L such that $K = LL^T$. (The S function chol is unusual in working with $U = L^T$.) Then minimizing (16.2) is equivalent to

$$\min_{\boldsymbol{\beta}} \| L^{-1} [\boldsymbol{Z} - F\boldsymbol{\beta}] \|^2$$

which reduces the problem to one of ordinary least squares. To solve this we use the QR decomposition (Golub & Van Loan, 1989) of $L^{-1}F$ as

$$QL^{-1}F = \begin{bmatrix} R \\ 0 \end{bmatrix}$$

for an orthogonal matrix Q and upper-triangular $P \times P$ matrix R. Write

$$QL^{-1}\boldsymbol{Z} = \begin{bmatrix} \boldsymbol{Y}_1 \\ \boldsymbol{Y}_2 \end{bmatrix}$$

as the upper P and lower $n - P$ rows. Then $\hat{\boldsymbol{\beta}}$ solves

$$R\hat{\boldsymbol{\beta}} = \boldsymbol{Y}_1$$

which is easy to compute as R is triangular.

Trend surfaces for the topo data fitted by generalized least squares are shown later (Figure 16.5), where we discuss the choice of the covariance function C.

Local trend surfaces

We have commented on the difficulties of using polynomials as global surfaces. There are two ways to make their effect local. The first is to fit a polynomial surface for each predicted point, using only the nearby data points. The function loess is of this class, and provides a wide range of options. By default it fits a quadratic surface by weighted least squares, the weights ensuring that 'local' data points are most influential. We will only give details for the span parameter α less than one. Let $q = \lfloor \alpha n \rfloor$, and let δ denote the Euclidean distance to the q th nearest point to x. Then the weights are

$$w_i = \left[1 - \left(\frac{d(x, x_i)}{\delta} \right)^3 \right]_+^3$$

for the observation at x_i. ($[\quad]_+$ denotes the positive part.) Full details of loess are given by Cleveland *et al.* (1992). For our example we have:

```
topo.loess <- loess(z ~ x * y, topo, degree=2, span = 0.25,
    normalize=T)
topo.plt <- expand.grid(x=seq(0, 6.5, 0.1), y=seq(0, 6.5, 0.1))
topo.plt$pred <- as.vector(predict(topo.loess, topo.plt))
contourplot(pred ~ x * y, topo.plt, aspect=1,
    at = seq(700, 1000, 25),
    xlab = list("Loess degree=2", cex=1.5), ylab="",
    panel = function(x, y, subscripts, ...) {
        panel.contourplot(x, y, subscripts, ...)
        panel.xyplot(topo$x,topo$y, cex=0.5)
    }
)
topo.plt1 <- expand.grid(x=seq(0, 6.5, 0.2), y=seq(0, 6.5, 0.2))
topo.lo <- predict(topo.loess, topo.plt1, se.fit=T)
topo.plt1$se <- topo.lo$se.fit
contourplot(se ~ x * y, topo.plt1, at = seq(5, 25, 5),
    . . . .
topo.loess <- loess(z ~ x * y, topo, degree=1, span = 0.25,
    normalize=T)
    . . . .
```

We turn normalization off to use Euclidean distance on unscaled variables. Note that the predictions from loess are confined to the range of the data in each of the x and y directions even though we requested them to cover the square: this is a side-effect of the algorithms used. The standard-error calculations are slow and memory-intensive, so we reduce the grid resolution.

Although loess allows a wide range of smoothing via its parameter span, it is designed for exploratory work and has no way to choose the smoothness except to 'look good'.

The Dirichlet tessellation of a set of points is the set of *tiles*, each of which is associated with a data point, and is the set of points nearer to that data point

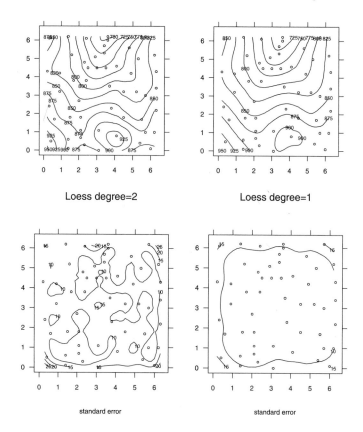

Figure 16.3: `loess` surfaces and prediction standard errors for the `topo` dataset.

than any other. There is an associated triangulation, the Delaunay triangulation, in which data points are connected by an edge of the triangulation if and only if their Dirichlet tiles share an edge. (Algorithms and examples are given in Ripley (1981, §4.3). There is S and FORTRAN code in library `delaunay` available from `statlib`.) Akima's (1978) fitting method fits a fifth-order trend surface within each triangle of the Delaunay triangulation; details are given in Ripley (1981, §4.3). The S implementation is the function `interp`; Akima's example is in datasets `akima.x`, `akima.y` and `akima.z`. The method is forced to interpolate the data, and has no flexibility at all to choose the smoothness of the surface. The arguments `ncp` and `extrap` control details of the method: see the on-line help for details. For Figure 16.4 we used

```
topo.akima <- con2tr(interp(topo$x, topo$y, topo$z))
contourplot(z ~ x * y, topo.akima, aspect=1,
    at = seq(700, 1000, 25),
    xlab = list("interp default", cex=1.5), ylab="",
    panel = function(x, y, subscripts, ...) {
        panel.contourplot(x, y, subscripts, ...)
```

```
      panel.xyplot(topo$x,topo$y, cex=0.5)
   }
)

topo.mar <- list(x = seq(0, 6.5, 0.1), y=seq(0, 6.5, 0.1))
topo.akima <- con2tr(interp(topo$x, topo$y, topo$z,
   topo.mar$x, topo.mar$y, ncp=4, extrap=T))
contourplot(z ~ x * y, topo.akima, aspect=1,
   at = seq(700, 1000, 25),
   xlab = list("interp", cex=1.5), ylab="",
   panel = function(x, y, subscripts, ...) {
      panel.contourplot(x, y, subscripts, ...)
      panel.xyplot(topo$x,topo$y, cex=0.5)
   }
)
```

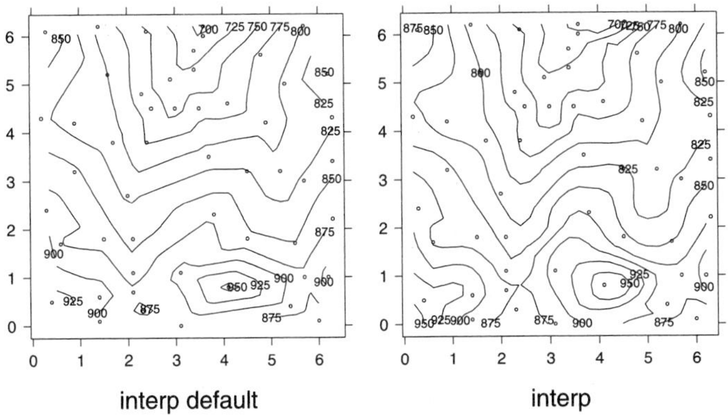

Figure 16.4: interp surfaces for the topo dataset.

16.2 Kriging

Kriging is the name of a technique developed by Matheron in the early 1960s for mining applications which has been independently discovered many times. Journel & Huijbregts (1978) give a comprehensive guide to its application in the mining industry. In its full form, *universal kriging*, it amounts to fitting a process of the form

$$Z(x) = f(x)^T \beta + \epsilon(x)$$

by generalized least squares, predicting the value at x of both terms and taking their sum. Thus it differs from trend-surface prediction which predicts $\epsilon(x)$ by zero. In what is most commonly termed *kriging*, the trend surface is of degree zero, that is a constant.

Our derivation of the predictions is given by Ripley (1981, pp. 48–50). Let $k(x) = [C(x, x_i)]$. The computational steps are

1. Form $K = [C(\boldsymbol{x}_i, \boldsymbol{y}_i)]$, with Cholesky decomposition L.

2. Form F and $\boldsymbol{Z}$.

3. Minimize $\|L^{-1}\boldsymbol{Z} - L^{-1}F\beta\|^2$, reducing $L^{-1}F$ to R.

4. Form $\boldsymbol{W} = \boldsymbol{Z} - F\hat{\beta}$, and $\boldsymbol{y}$ such that $L(L^T y) = \boldsymbol{W}$.

5. Predict $Z(\boldsymbol{x})$ by $\hat{Z}(\boldsymbol{x}) = \boldsymbol{y}^T k(\boldsymbol{x}) + \boldsymbol{f}(\boldsymbol{x})^T \hat{\beta}$, with error variance given by $C(\boldsymbol{x}, \boldsymbol{x}) - \|\boldsymbol{e}\|^2 + \|\boldsymbol{g}\|^2$ where

$$Le = k(\boldsymbol{x}), \qquad R^T g = \boldsymbol{f}(\boldsymbol{x}) - (L^{-1}F)^T e.$$

This recipe involves only linear algebra and so can be implemented in S, but our C version is about 10 times faster. For the `topo` data we have (Figure 16.5):

```
topo.ls <- surf.ls(2, topo)
trsurf <- con2tr(trmat(topo.ls, 0, 6.5, 0, 6.5, 30))
    ....
topo.gls <- surf.gls(2, expcov, topo, d=0.7)
trsurf <- con2tr(trmat(topo.ls, 0, 6.5, 0, 6.5, 30))
    ....
prsurf <- con2tr(prmat(topo.gls, 0, 6.5, 0, 6.5, 50))
    ....
sesurf <- con2tr(semat(topo.gls, 0, 6.5, 0, 6.5, 30))
    ....
```

Covariance estimation

To use either generalized least squares or kriging we have to know the covariance function C. We will assume that

$$C(\boldsymbol{x}, \boldsymbol{y}) = c(d(\boldsymbol{x}, \boldsymbol{y})) \tag{16.3}$$

where $d()$ is Euclidean distance. (An extension known as *geometric anisotropy* can be incorporated by re-scaling the variables, as we did for the Mahalanobis distance in Chapter 13.) We can compute a *correlogram* by dividing the distance into a number of bins and finding the covariance between pairs whose distance falls into that bin, then dividing by the overall variance.

Choosing the covariance is very much an iterative process, as we need the covariance of the residuals, and the fitting of the trend surface by generalized least squares depends on the assumed form of the covariance function. Further, as we have residuals their covariance function is a biased estimator of c. In practice it is important to get the form right for small distances, for which the bias is least.

Although $c(0)$ must be one, there is no reason why $c(0+)$ should not be less than one. This is known in the kriging literature as a *nugget effect* since it could arise from a very short-range component of the process $Z(\boldsymbol{x})$. A more general explanation is measurement error. In any case, if there is a nugget effect, the

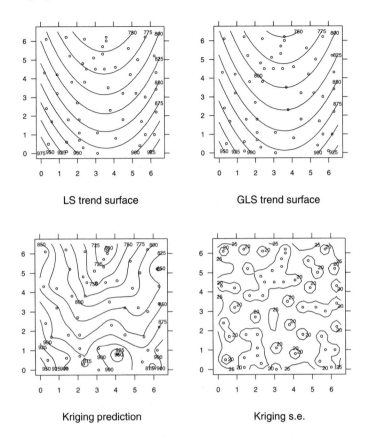

Figure 16.5: Trend surfaces by least squares and generalized least squares, and a kriged surface and standard error of prediction, for the `topo` dataset.

predicted surface will have spikes at the data points, and so effectively will not interpolate but smooth.

The kriging literature tends to work with the *variogram* rather than the covariance function. More properly termed the semi-variogram, this is defined by

$$V(\boldsymbol{x}, \boldsymbol{y}) = \frac{1}{2} E[Z(\boldsymbol{x}) - Z(\boldsymbol{y})]^2$$

and is related to C by

$$V(\boldsymbol{x}, \boldsymbol{y}) = \frac{1}{2}[C(\boldsymbol{x}, \boldsymbol{x}) + C(\boldsymbol{y}, \boldsymbol{y})] - C(\boldsymbol{x}, \boldsymbol{y}) = c(0) - c(d(\boldsymbol{x}, \boldsymbol{y}))$$

under our assumption (16.3). However, since different variance estimates will be used in different bins, the empirical versions will not be so exactly related. Much heat and little light emerges from discussions of their comparison.

There are a number of standard forms of covariance functions which are commonly used. A nugget effect can be added to each. The exponential covariance has

$$c(r) = \sigma^2 \exp{-r/d}$$

the so-called Gaussian covariance is

$$c(r) = \sigma^2 \exp{-(r/d)^2}$$

and the spherical covariance is in two dimensions

$$c(r) = \sigma^2 \left[1 - \frac{2}{\pi} \left(\frac{r}{d} \sqrt{1 - \frac{r^2}{d^2}} + \sin^{-1} \frac{r}{d} \right) \right]$$

and in three dimensions (but also valid as a covariance function in two)

$$c(r) = \sigma^2 \left[1 - \frac{3r}{2d} + \frac{r^3}{2d^3} \right]$$

for $r \leqslant d$ and zero for $r > d$. Note that this is genuinely local, since points at a greater distance than d from x are given zero weight at step 5 (although they do affect the trend surface).

We promised to relax the assumption of second-order stationarity slightly. As we only need to predict residuals, we only need a covariance to exist in the space of linear combinations $\sum a_i Z(x_i)$ which are orthogonal to the trend surface. For degree 0, this corresponds to combinations with sum zero. It is possible that the variogram is finite, without the covariance existing, and there are extensions to more general trend surfaces given by Matheron (1973) and reproduced by Cressie (1991, §5.4). In particular, we can always add a constant to c without affecting the predictions (except perhaps numerically). Thus if the variogram v is specified, we work with covariance function $c = \text{const} - v$ for a suitably large constant. The main advantage is in allowing us to use certain functional forms which do not correspond to covariances, such as

$$v(d) = d^\alpha, 0 \leqslant \alpha < 2 \quad \text{or} \quad d^3 - \alpha d$$

The variogram $d^2 \log d$ corresponds to a thin-plate spline in $\mathbb{R}^2$ (see Wahba, 1990 and the review in Cressie, 1991, §3.4.5).

Our functions `correlogram` and `variogram` allow the empirical correlogram and variogram to be plotted and functions `expcov`, `gaucov` and `sphercov` compute the exponential, Gaussian and spherical covariance functions (the latter in two and three dimensions) and can be used as arguments to `surf.gls`. For our running example we have

```
topo.kr <- surf.ls(2, topo)
correlogram(topo.kr, 25)
d <- seq(0, 7, 0.1)
lines(d, expcov(d, 0.7))
variogram(topo.kr, 25)
```

see Figure 16.6. We then consider fits by generalized least squares.

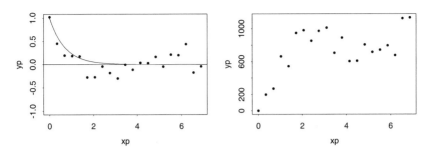

Figure 16.6: Correlogram (left) and variogram (right) for the residuals of topo dataset from a least-squares quadratic trend surface.

```
topo.kr <- surf.gls(2, expcov, topo, d=0.7)
correlogram(topo.kr, 25)
lines(d, expcov(d, 0.7))
lines(d, gaucov(d, 1.0, 0.3), lty=3) # try nugget effect
topo.kr <- surf.gls(2, gaucov, topo, d=1, alph=0.3)
prsurf <- con2tr(prmat(topo.kr, 0, 6.5, 0, 6.5, 50))
contourplot(z ~ x * y, prsurf, at = seq(600, 1000, 25),
    aspect=1, xlab = "", ylab="",
    panel = function(x, y, subscripts, ...) {
        panel.contourplot(x, y, subscripts, ...)
        panel.xyplot(topo$x,topo$y, cex=0.5)
    }
)
sesurf <- con2tr(semat(topo.kr, 0, 6.5, 0, 6.5, 30))
contourplot(z ~ x * y, sesurf, at = c(15, 20, 25),
    aspect=1, xlab = "", ylab="",
    panel = function(x, y, subscripts, ...) {
        panel.contourplot(x, y, subscripts, ...)
        panel.xyplot(topo$x,topo$y, cex=0.5)
    }
)

topo.kr <- surf.ls(0, topo)
correlogram(topo.kr, 25)
lines(d, gaucov(d, 2, 0.05))
topo.kr <- surf.gls(0, gaucov, topo, d=2, alph=0.05, nx=10000)
prsurf <- con2tr(prmat(topo.kr, 0, 6.5, 0, 6.5, 50))
contourplot(z ~ x * y, prsurf, at = seq(600, 1000, 25),
    aspect=1, xlab = "", ylab="",
    panel = function(x, y, subscripts, ...) {
        panel.contourplot(x, y, subscripts, ...)
        panel.xyplot(topo$x,topo$y, cex=0.5)
    }
)
sesurf <- con2tr(semat(topo.kr, 0, 6.5, 0, 6.5, 30))
contourplot(z ~ x * y, sesurf, at = c(15, 20, 25),
```

```
            aspect=1, xlab = "", ylab="",
            panel = function(x, y, subscripts, ...) {
                panel.contourplot(x, y, subscripts, ...)
                panel.xyplot(topo$x,topo$y, cex=0.5)
            }
    )
```

We first fit a quadratic surface by least squares, then try one plausible covariance function (Figure 16.8). Re-fitting by generalized least squares suggests this function and another with a nugget effect, and we predict the surface from both. The first was shown in Figure 16.5, the second in Figure 16.7. We also consider not using a trend surface but a longer-range covariance function, also shown in Figure 16.7. (The small nugget effect is to ensure numerical stability as without it the matrix K is very ill-conditioned; the correlations at short distances are very near one. We increased nx for a more accurate look-up table of covariances.)

16.3   Point process analysis

A spatial point pattern is a collection of n points within a region $D \subset \mathbb{R}^2$. The number of points is thought of as random, and the points are considered to be generated by a stationary isotropic point process in $\mathbb{R}^2$. (This means that there is no preferred origin or orientation of the pattern.) For such patterns probably the most useful summaries of the process are the first and second moments of the counts $N(A)$ of the numbers of points within a set $A \subset D$. The first moment can be specified by a single number, the *intensity* λ giving the expected number of points per unit area, obviously estimated by n/a where a denotes the area of D.

The second moment can be specified by Ripley's K function. For example, $\lambda K(t)$ is the expected number of points within distance t of a point of the pattern. The benchmark of complete randomness is the Poisson process, for which $K(t) = \pi t^2$, the area of the search region for the points. Values larger than this indicate clustering on that distance scale, and smaller values indicate regularity. This suggests working with $L(t) = \sqrt{K(t)/\pi}$, which will be linear for a Poisson process.

We only have a single pattern from which to estimate K or L. The definition in the previous paragraph suggests an estimator of $\lambda K(t)$; average over all points of the pattern the number seen within distance t of that point. This would be valid but for the fact that some of the points will be outside D and so invisible. There are a number of edge-corrections available, but that of Ripley (1976) is both simple to compute and rather efficient. This considers a circle centred on the point x and passing through another point y. If the circle lies entirely within D, the point is counted once. If a proportion $p(x, y)$ of the circle lies within D, the point is counted as $1/p$ points. (We may want to put a limit on for small p, to reduce the variance at the expense of some bias.) This gives an estimator $\lambda \widehat{K}(t)$ which is unbiased for t up to the circumradius of D (so that it is possible

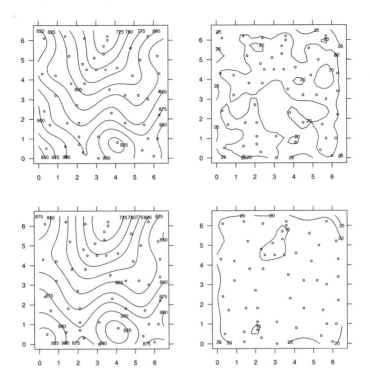

Figure 16.7: Two more kriged surfaces and standard errors of prediction for the topo dataset. The top row uses a quadratic trend surface and a nugget effect. The bottom row is without a trend surface.

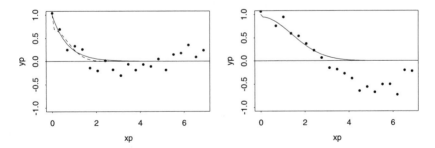

Figure 16.8: Correlograms for the topo dataset. (Left) residuals from quadratic trend surface showing exponential covariance (solid) and Gaussian covariance (dashed). (Right) raw data with fitted Gaussian covariance function.

to observe two points $2t$ apart). Since we do not know λ, we estimate it by $\hat{\lambda} = n/a$. Finally

$$\widehat{K}(t) = \frac{a}{n^2} \sum_{\boldsymbol{x} \in D, d(\boldsymbol{y},\boldsymbol{x}) \leqslant t} \frac{1}{p(\boldsymbol{x}, \boldsymbol{y})}$$

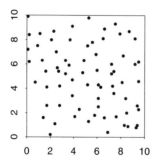
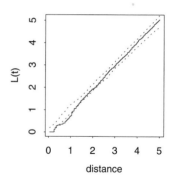

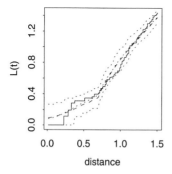

Figure 16.9: The Swedish pines dataset from Ripley (1981), with two plots of $L(t)$. That at the upper right shows the envelope of 100 binomial simulations, that at the lower left the average and the envelope (dotted) of 100 simulations of a Strauss process with $c = 0.2$ and $R = 0.7$. Also shown (dashed) is the average for $c = 0.15$. All units are in metres.

and obviously we estimate $L(t)$ by $\sqrt{\widehat{K}(t)/\pi}$. We find that on square-root scale the variance of the estimator varies little with t.

Our first example is the Swedish pines data of Ripley (1981, §8.6). This records 72 trees within a 10-metre square. Figure 16.9 shows that $\widehat{L}$ is not straight, and comparison with simulations from a binomial process (a Poisson process conditioned on $N(D) = n$, so n independently uniformly distributed points within D) shows that the lack of straightness is significant. The upper two panels of Figure 16.9 were produced by the following code:

```
library(spatial)
pines <- ppinit("pines.dat")
par(mfrow=c(2,2), pty="s")
plot(pines, xlim=c(0,10), ylim=c(0,10), xlab="", ylab="",
    xaxs="i", yaxs="i")
plot(Kfn(pines,5), type="s", xlab="distance", ylab="L(t)")
lims <- Kenvl(5, 100, Psim(72))
```

```
lines(lims$x, lims$l, lty=2)
lines(lims$x, lims$u, lty=2)
```

The function `ppinit` reads the data from the file and also the coordinates of a rectangular domain D. The latter can be reset, or set up for simulations, by the function `ppregion`. (It *must* be set for each session.) The function `Kfn` returns an estimate of $L(t)$ and other useful information for plotting, for distances up to its second argument `fs` (for full-scale).

The functions `Kaver` and `Kenvl` return the average and, for `Kenvl`, also the extremes of K-functions (on L scale) for a series of simulations. The function `Psim(n)` simulates the binomial process on n points within the domain D which has already been set.

Alternative processes

We need to consider alternative point processes to the Poisson. One of the most useful for regular point patterns is the so-called Strauss process, which is simulated by `Strauss(n, c, r)`. This has a density of n points proportional to

$$c^{\text{number of } R\text{-close pairs}}$$

and so has $K(t) < \pi t^2$ for $t \leqslant R$ (and up to about $2R$). For $c = 0$ we have a 'hard-core' process which never generates pairs closer than R and so can be envisaged as laying down the centres of non-overlapping discs of diameter $r = R$.

Figure 16.9 also shows the average and envelope of the L-plots for a Strauss process fitted to the pines data by Ripley (1981). There the parameters were chosen by trial-and-error based on a knowledge of how the L-plot changed with (c, R). Ripley (1988) considers the estimation of c for known R by the pseudo-likelihood. This is done by our function `pplik` and returns an estimate of about $c = 0.15$ ('about' since it uses numerical integration). As Figure 16.9 shows, the difference between $c = 0.2$ and $c = 0.15$ is small. We used the following code:

```
ppregion(pines)
plot(Kfn(pines,1.5), type="s", xlab="distance", ylab="L(t)")
lims <- Kenvl(1.5, 100, Strauss(72, 0.2, 0.7))
lines(lims$x, lims$a, lty=2)
lines(lims$x, lims$l, lty=2)
lines(lims$x, lims$u, lty=2)
pplik(pines, 0.7)
lines(Kaver(1.5, 100, Strauss(72, 0.15, 0.7)), lty=3)
```

which took about 15 seconds.

The theory is given by Ripley (1988, p. 67). For a point $\xi \in D$ let $t(\xi)$ denote the number of points of the pattern within distance t of ξ. Then the pseudo-likelihood estimator solves

$$\frac{\int_D t(\xi) c^{t(\xi)} \, d\xi}{\int_D c^t(\xi) \, d\xi} = \frac{\#(R\text{-close pairs})}{n} = \frac{n\widehat{K}(R)}{a}$$

and the left-hand side is an increasing function of c. The function pplik uses the S-PLUS function uniroot to find a solution in the range $(0, 1]$.

Other processes for which simulation functions are provided are the binomial process (Psim(n)) and Matérn's sequential spatial inhibition process (SSI(n, r)) which sequentially lays down centres of discs of radius r that do not overlap existing discs.

Chapter 17

Classification

Classification is an increasingly important application of modern methods in statistics. In the statistical literature the word is used in two distinct senses. The entry (Hartigan, 1982) in the original Encyclopedia of Statistical Sciences uses the sense of *cluster analysis* discussed in Chapter 13, and this is the main business of the International Federation of Classification Societies. Modern usage is leaning to the other meaning (Ripley, 1997) of allocating future cases to one of g classes. This is similar to *discriminant analysis*, but we are not interested in the differences between the classes *per se*. Medical diagnosis is an archetypal classification problem in the modern sense. (The older statistical literature sometimes refers to this as *allocation*.)

Both senses of classification form the subject of *pattern recognition*. Cluster analysis is *unsupervised* pattern recognition, since the groups are *a priori* unknown, whereas diagnosis is *supervised*.

S-PLUS contains several tools for classification problems which have been covered in earlier chapters. We have seen linear and quadratic discriminant analysis and classification trees. Neural networks have provided the impetus for much of the current interest in classification, and other modern regression techniques can also be used. In this chapter we consider how to use these tools for some practical classification problems. The background (and these examples) are discussed further in Ripley (1996).

17.1 Theory

In the terminology of pattern recognition the given cases with their classifications are known as the *training set*, and future cases form the *test set*. Our primary measure of success is the error (or misclassification) rate. Note that we would obtain (possibly seriously) biased estimates by re-classifying the training set, but that the error rate on a test set randomly chosen from the whole population will be an unbiased estimator.

It may be helpful to know the type of errors made. A *confusion matrix* gives the number of cases with true class i classified as of class j. In some problems some errors are considered to be worse than others, so we assign costs L_{ij} to allocating a case of class i to class j. Then we will be interested in the average error cost rather than the error rate.

It is fairly easy to show (Ripley, 1996, p. 19) that the average error cost is minimized by the *Bayes rule* which is to allocate to the class c minimizing $\sum_i L_{ic} p(i \mid x)$ where $p(i \mid x)$ is the posterior distribution of the classes after observing x. If the costs of all errors are the same, this rule amounts to choosing the class c with the largest posterior probability $p(c \mid x)$. The minimum average cost is known as the *Bayes risk*.

We saw in Section 13.3 how $p(c \mid x)$ can be computed for normal populations, and how estimating the Bayes rule with equal error costs leads to linear and quadratic discriminant analysis. As our functions `predict.lda` and `predict.qda` return posterior probabilities, they can also be used for classification with error costs.

The posterior probabilities $p(c \mid x)$ may also be estimated directly. For just two classes we can model $p(1 \mid x)$ using a logistic regression, fitted by `glm`. For more than two classes we need a multiple logistic model: it may be possible to fit this using a surrogate log-linear Poisson GLM model, but using the `multinom` function in library `nnet` will usually be faster and easier.

Classification trees model the $p(c \mid x)$ directly by a special multiple logistic model, one in which the right-hand side is a single factor specifying which leaf the case will be assigned to by the tree. Again, since the posterior probabilities are given by the `predict` method it is easy to estimate the Bayes rule for unequal error costs.

Predictive and 'plug-in' rules

In the last few paragraphs we skated over an important point. To find the Bayes rule we need to know the posterior probabilities $p(c \mid x)$. Since these are unknown we use an explicit or implicit parametric family $p(c \mid x; \theta)$. In the methods considered so far we act as if $p(c \mid x; \hat{\theta})$ are the actual posterior probabilities, where $\hat{\theta}$ is an estimate computed from the training set $\mathcal{T}$, often by maximizing some appropriate likelihood. This is known as the 'plug-in' rule. However, the 'correct' estimate of $p(c \mid x)$ is (Ripley, 1996, §2.4) to use the *predictive* estimates

$$\tilde{p}(c \mid x) = P(c = c \mid X = x, \mathcal{T}) = \int p(c \mid x; \theta) p(\theta \mid \mathcal{T}) \, d\theta \qquad (17.1)$$

If we are very sure of our estimate $\hat{\theta}$ there will be little difference between $p(c \mid x; \hat{\theta})$ and $\tilde{p}(c \mid x)$; otherwise the predictive estimate will normally be less extreme (not as near 0 or 1). The 'plug-in' estimate ignores the uncertainty in the parameter estimate $\hat{\theta}$ which the predictive estimate takes into account.

It is not often possible to perform the integration in (17.1) analytically, but it *is* possible for linear and quadratic discrimination with appropriate 'vague' priors on θ (Aitchison & Dunsmore, 1975; Geisser, 1993; Ripley, 1996). This estimate is implemented by `method = "predictive"` of the `predict` methods for our functions lda and qda. Often the differences are small, especially for linear discrimination, *provided* there is enough data for a good estimate of the variance matrices. When there is not, Moran & Murphy (1979) argue that considerable improvement can be obtained by using an unbiased estimator of $\log p(x \mid c)$, implemented by the argument method = "debiased".

Non-parametric rules

There are a number of non-parametric classifiers based on non-parametric estimates of the class densities or of the log posterior. Library class implements the k–nearest neighbour classifier and related methods (Devijver & Kittler, 1982; Ripley, 1996) and learning vector quantization (Kohonen, 1990, 1995; Ripley, 1996). These are all based on finding the k nearest examples in some reference set, and taking a majority vote amongst the classes of these k examples, or, equivalently, estimating the posterior probabilities $p(c \mid x)$ by the proportions of the classes amongst the k examples.

The methods differ in their choice of reference set. The k–nearest neighbour methods use the whole training set or an edited subset. Learning vector quantization is similar to K-means in selecting points in the space other than the training set examples to summarize the training set, but unlike K-means it takes the classes of the examples into account.

These methods almost always measure 'nearest' by Euclidean distance.

17.2 A simple example

We will illustrate these methods by a small example taken from Aitchison & Dunsmore (1975, Tables 11.1–3) and used for the same purpose by Ripley (1996). The data are on diagnostic tests on patients with Cushing's syndrome, a hypersensitive disorder associated with over-secretion of cortisol by the adrenal gland. This dataset has three recognized types of the syndrome represented as a, b, c. (These encode 'adenoma', 'bilateral hyperplasia' and 'carcinoma', and represent the underlying cause of over-secretion. This can only be determined histopathologically.) The observations are urinary excretion rates (mg/24h) of the steroid metabolites tetrahydrocortisone and pregnanetriol, and are considered on log scale.

There are six patients of unknown type (marked u), one of whom was later found to be of a fourth type, and another was measured faultily.

Figure 17.1 shows the classifications produced by lda and the various options of quadratic discriminant analysis. This was produced by

```
predplot <- function(object, main="", len=100, ...)
{
    plot(Cushings[,1], Cushings[,2], log="xy", type="n",
        xlab="Tetrahydrocortisone", ylab = "Pregnanetriol", main)
    text(Cushings[1:21,1], Cushings[1:21,2],
        as.character(tp))
```

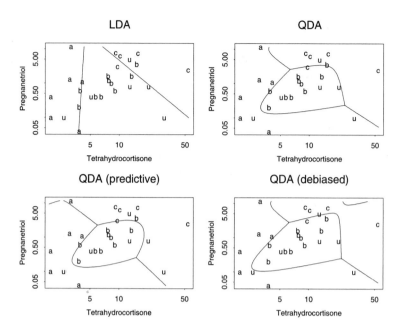

Figure 17.1: Linear and quadratic discriminant analysis applied to the Cushing's syndrome data.

```
        text(Cushings[22:27,1], Cushings[22:27,2], "u")
        xp <- seq(0.6, 4.0, length=len)
        yp <- seq(-3.25, 2.45, length=len)
        cushT <- expand.grid(Tetrahydrocortisone=xp,
          Pregnanetriol=yp)
        Z <- predict(object, cushT, ...); zp <- unclass(Z$class)
        zp <- Z$post[,3] - pmax(Z$post[,2], Z$post[,1])
        contour(xp/log(10), yp/log(10), matrix(zp, len),
          add=T, levels=0, labex=0)
        zp <- Z$post[,1] - pmax(Z$post[,2], Z$post[,3])
        contour(xp/log(10), yp/log(10), matrix(zp, len),
          add=T, levels=0, labex=0)
        invisible()
    }
    cush <- log(as.matrix(Cushings[, -3]))
    tp <- factor(Cushings$Type[1:21])
    cush.lda <- lda(cush[1:21,], tp); predplot(cush.lda, "LDA")
    cush.qda <- qda(cush[1:21,], tp); predplot(cush.qda, "QDA")
    predplot(cush.qda, "QDA (predictive)", method="predictive")
    predplot(cush.qda, "QDA (debiased)", method="debiased")
```

(This code is quite memory-intensive (10Mb), so readers may wish to reduce `len` here and below.)

We can contrast these with logistic discrimination performed by

```
library(nnet)
Cf <- data.frame(tp = tp,
   Tetrahydrocortisone = log(Cushings[1:21,1]),
   Pregnanetriol = log(Cushings[1:21,2]) )
cush.multinom <- multinom(tp ~ Tetrahydrocortisone
   + Pregnanetriol, Cf, maxit=250)
xp <- seq(0.6, 4.0, length=100); np <- length(xp)
yp <- seq(-3.25, 2.45, length=100)
cushT <- expand.grid(Tetrahydrocortisone=xp,
      Pregnanetriol=yp)
Z <- predict(cush.multinom, cushT, type="probs")
plot(Cushings[,1], Cushings[,2], log="xy", type="n",
  xlab="Tetrahydrocortisone", ylab = "Pregnanetriol")
text(Cushings[1:21,1], Cushings[1:21,2], as.character(tp))
text(Cushings[22:27,1], Cushings[22:27,2], "u")
zp <- Z[,3] - pmax(Z[,2], Z[,1])
contour(xp/log(10), yp/log(10), matrix(zp, np),
   add=T, levels=0, labex=0)
zp <- Z[,1] - pmax(Z[,2], Z[,3])
contour(xp/log(10), yp/log(10), matrix(zp, np),
   add=T, levels=0, labex=0)
```

When, as here, the classes have quite different variance matrices, linear and logistic discrimination can give quite different answers (compare Figures 17.1 and 17.2).

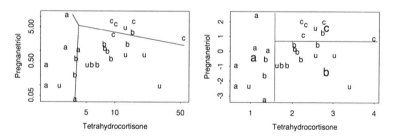

Figure 17.2: Logistic regression and classification trees applied to the Cushing's syndrome data.

For classification trees we can use

```
cush.tr <- tree(tp ~ Tetrahydrocortisone + Pregnanetriol, Cf)
plot(cush[,1], cush[,2], type="n",
  xlab="Tetrahydrocortisone", ylab = "Pregnanetriol")
text(cush[1:21,1], cush[1:21,2], as.character(tp))
text(cush[22:27,1], cush[22:27,2], "u")
par(cex=1.5); partition.tree(cush.tr, add=T); par(cex=1)
```

With such a small dataset we make no attempt to refine the size of the tree, shown in Figure 17.2.

Neural networks

Neural networks provide a flexible non-linear extension of multiple logistic regression, as we saw in Section 11.4. We can consider them for this example by the following code.[1]

```
library(nnet)
cush <- cush[1:21,]; tpi <- class.ind(tp)

plt <- function(...) {
    plot(Cushings[,1], Cushings[,2], log="xy", type="n",
    xlab="Tetrahydrocortisone", ylab = "Pregnanetriol", ...)
    for(il in 1:4) {
        set <- Cushings$Type==levels(Cushings$Type)[il]
        text(Cushings[set, 1], Cushings[set, 2],
            as.character(Cushings$Type[set]), col = 2 + il) }
}

plt.bndry <- function(size=0, decay=0, ...)
{
    cush.nn <- nnet(cush, tpi, skip=T, softmax=T, size=size,
        decay=decay, maxit=1000)
    invisible(b1(predict(cush.nn, cushT), ...))
}

b1 <- function(Z, ...)
{
    zp <- Z[,3] - pmax(Z[,2], Z[,1])
    contour(xp/log(10), yp/log(10), matrix(zp, np),
        add=T, levels=0, labex=0, ...)
    zp <- Z[,1] - pmax(Z[,3], Z[,2])
    contour(xp/log(10), yp/log(10), matrix(zp, np),
        add=T, levels=0, labex=0, ...)
}
par(mfrow=c(2,2))
plt("Size = 2")
set.seed(1); plt.bndry(size=2, col=2)
set.seed(3); plt.bndry(size=2, col=3); plt.bndry(size=2, col=4)

plt("Size = 2, lambda = 0.001")
set.seed(1); plt.bndry(size=2, decay=0.001, col=2)
set.seed(2); plt.bndry(size=0, decay=0.001, col=4)

plt("Size = 2, lambda = 0.01")
set.seed(1); plt.bndry(size=2, decay=0.01, col=2)
set.seed(2); plt.bndry(size=2, decay=0.01, col=4)

plt("Size = 5, 20  lambda = 0.01")
set.seed(2); plt.bndry(size=5, decay=0.01, col=1)
set.seed(2); plt.bndry(size=20, decay=0.01, col=2)
```

---

[1] The colours are set for a Trellis device.

If this proves to be too slow, reduce `maxit=1000` to 200 or less.

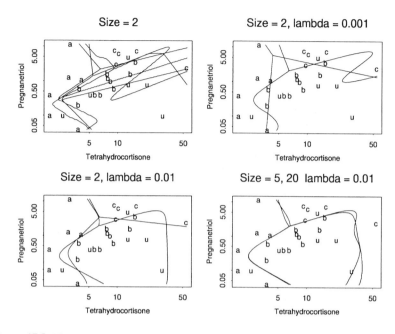

Figure 17.3: Neural networks applied to the Cushing's syndrome data. Each panel shows the fits from two or three local maxima of the (penalized) log-likelihood.

The results are shown in Figure 17.3. We see that in all cases there are multiple local maxima of the likelihood.

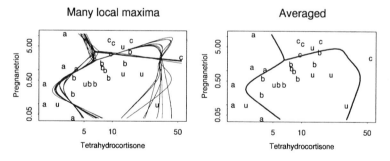

Figure 17.4: Neural networks applied to the Cushing's syndrome data. The right panel shows many local minima and their average (as the thick line).

Once we have a penalty, the choice of the number of hidden units is often not critical (see Figure 17.3). The spirit of the predictive approach is to average the predicted $p(c \mid x)$ over the local maxima. A simple average will often suffice:

```
plt("Many local maxima")
Z <- matrix(0, nrow(cushT), ncol(tpi))
```

```
for(iter in 1:20) {
    set.seed(iter)
    cush.nn <- nnet(cush, tpi,  skip=T, softmax=T, size=3,
        decay=0.01, maxit=1000, trace=F)
    Z <- Z + predict(cush.nn, cushT)
    cat("final value", format(round(cush.nn$value,3)), "\n")
    b1(predict(cush.nn, cushT), col=2, lwd=0.5)
}
plt("Averaged")
b1(Z, lwd=3)
```

but more sophisticated integration can be done (Ripley, 1996, §5.5). Note that there are two quite different types of local maxima occurring here, and some local maxima occur several times (up to convergence tolerances).

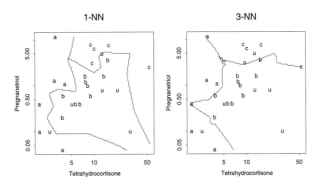

Figure 17.5: k-nearest neighbours applied to the Cushing's syndrome data.

Nonparametric methods

The simplest nonparametric method is k-nearest neighbours. We use Euclidean distance on the logged covariates, rather arbitrarily treating them equally.

```
library(class)
par(pty="s", mfrow=c(1,2))
plot(Cushings[,1], Cushings[,2], log="xy", type="n",
    xlab="Tetrahydrocortisone", ylab = "Pregnanetriol", "1-NN")
text(Cushings[1:21,1], Cushings[1:21,2], as.character(tp))
text(Cushings[22:27,1], Cushings[22:27,2], "u")
Z <- knn(scale(cush, F, c(3.4, 5.7)),
        scale(cushT, F, c(3.4, 5.7)), tp)
contour(xp/log(10), yp/log(10), matrix(as.numeric(Z=="a"), np),
        add=T, levels=0.5, labex=0)
contour(xp/log(10), yp/log(10), matrix(as.numeric(Z=="c"), np),
        add=T, levels=0.5, labex=0)
plot(Cushings[,1], Cushings[,2], log="xy", type="n",
    xlab="Tetrahydrocortisone", ylab = "Pregnanetriol", "3-NN")
```

```
text(Cushings[1:21,1], Cushings[1:21,2], as.character(tp))
text(Cushings[22:27,1], Cushings[22:27,2], "u")
Z <- knn(scale(cush, F, c(3.4, 5.7)),
         scale(cushT, F, c(3.4, 5.7)), tp, k=3)
contour(xp/log(10), yp/log(10), matrix(as.numeric(Z=="a"), np),
        add=T, levels=0.5, labex=0)
contour(xp/log(10), yp/log(10), matrix(as.numeric(Z=="c"), np),
        add=T, levels=0.5, labex=0)
```

This dataset is too small to try the editing and LVQ methods.

17.3   Forensic glass

The forensic glass dataset fgl has 214 points from six classes with nine measurements, and provides a fairly stiff test of classification methods. As we have seen (Figures 3.18 on page 101, 5.4 on page 173, 13.4 on page 387 and 13.11 on page 398) the types of glass do not form compact well-separated groupings, and the marginal distributions are far from normal. There are some small classes (with 9, 13 and 17 examples), so we cannot use quadratic discriminant analysis.

We will assess their performance by 10-fold cross-validation, using the same random partition for all the methods. Logistic regression provides a suitable benchmark (as is often the case).

```
set.seed(123); rand <- sample (10, 214, replace=T)
con<- function(x,y)
{
    tab <- table(x,y)
    print(tab)
    diag(tab) <- 0
    cat("error rate = ", round(100*sum(tab)/length(x),2),"%\n")
    invisible()
}
CVtest <- function(fitfn, predfn, ...)
{
  res <- fgl$type
  for (i in sort(unique(rand))) {
     cat("fold ",i,"\n", sep="")
     learn <- fitfn(rand != i, ...)
     res[rand == i] <- predfn(learn, rand==i)
  }
  res
}
res.multinom <- CVtest(
    function(x, ...) multinom(type ~ ., fgl[x,], ...),
    function(obj, x) predict(obj, fgl[x, ],type="class"),
    maxit=1000, trace=F )

> con(fgl$type, res.multinom)
```

```
       WinF WinNF Veh Con Tabl Head
  WinF   44    20   4   0    2    0
 WinNF   20    50   0   3    2    1
   Veh    9     5   3   0    0    0
   Con    0     4   0   8    0    1
  Tabl    0     2   0   0    4    3
  Head    1     1   0   3    1   23
error rate =   38.79 %

res.lda <- CVtest(
   function(x, ...) lda(type ~ ., fgl[x, ], ...),
   function(obj, x) predict(obj, fgl[x, ])$class )
> con(fgl$type, res.lda)
    ....
error rate =   37.38 %
```

Using predictive LDA gives the same error rate.

Figure 13.4 on page 387 suggests that nearest neighbour methods might work well, and the 1–nearest neighbour classifier is (to date) unbeatable in this problem. We can estimate a lower bound for the Bayes risk as 10% by the method of Ripley (1996, pp. 196–7).

```
library(class)
fgl0 <- fgl[ ,-10] # drop type
{ res <- fgl$type
  for (i in sort(unique(rand))) {
     cat("fold ",i,"\n", sep="")
     sub <- rand == i
     res[sub] <- knn(fgl0[!sub, ], fgl0[sub,], fgl$type[!sub],
                     k=1)
  }
  res } -> res.knn1
> con(fgl$type, res.knn1)
       WinF WinNF Veh Con Tabl Head
  WinF   59     6   5   0    0    0
 WinNF   12    57   3   3    1    0
   Veh    2     4  11   0    0    0
   Con    0     2   0   8    1    2
  Tabl    1     0   0   1    6    1
  Head    0     4   1   1    1   22
error rate =   23.83 %
> res.lb <- knn(fgl0, fgl0, fgl$type, k=3, prob=T, use.all=F)
> table(attr(res.lb, "prob"))
 0.333333 0.666667    1
       10       64 140
1/3 * (64/214) = 0.099688
```

We saw in Chapter 14 that we could fit a classification tree of size about 6 to this dataset. We need to cross-validate over the choice of tree size, which does vary by group from 4 to 11. (The design of cv.tree and model.frame.tree is not helpful here; we need to keep x visible.)

```
res.tree <- CVtest(
    function(x, ...) {
        tr <- tree(type ~ ., fgl[x,], ...)
        assign("x", x, f=1)
        r <- cv.tree(tr,, prune.misclass)
        rmin <- max(seq(along=r$dev)[r$dev < min(r$dev)+9])
        cat("size chosen was", r$size[rmin], "\n")
        prune.misclass(tr, k = r$k[rmin])
    },
    function(obj, x)
        predict(obj, fgl[x, ], type="class")
)
con(fgl$type, res.tree)
        WinF WinNF Veh Con Tabl Head
WinF     56    13   1   0    0    0
WinNF    24    38   2   7    4    1
Veh      11     5   1   0    0    0
Con       0     3   0   9    0    1
Tabl      0     1   0   1    4    3
Head      1     3   0   3    0   22
error rate =  39.25 %
```

In fact in this example we do better (34.6%) by not pruning at all.

We leave to the reader to try neural networks and LVQ methods: some help is given in the on-line complement for this chapter.

Calibration plots

One measure that a suitable model for $p(c \mid x)$ has been found is that the predicted probabilities are *well calibrated*, that is that a fraction of about p of the events we predict with probability p actually occur. Methods for testing calibration of probability forecasts have been developed in connection with weather forecasts (Dawid, 1982, 1986).

For the forensic glass example we are making six probability forecasts for each case, one for each class. To ensure that they are genuine forecasts, we should use the cross-validation procedure. A minor change to the code gives the probability predictions:

```
CVprobs <- function(fitfn, predfn, ...)
{
    res <- matrix(, 214, 6)
    for (i in sort(unique(rand))) {
        cat("fold ",i,"\n", sep="")
        learn <- fitfn(rand != i, ...)
        res[rand == i,] <- predfn(learn, rand==i)
    }
    res
}
probs.multinom <- CVprobs(
```

```
function(x, ...) multinom(type ~ ., fgl[x,], ...),
function(obj, x) predict(obj, fgl[x, ],type="probs"),
maxit=1000, trace=F )
```

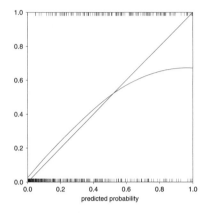

Figure 17.6: Calibration plot for multiple logistic fit to the `fgl` data.

We can plot these and smooth them by

```
probs.yes <- as.vector(class.ind(fgl$type))
probs <- as.vector(probs.multinom)
par(pty="s")
plot(c(0,1), c(0,1), type="n", xlab="predicted probability",
    ylab="", xaxs="i", yaxs="i", las=1)
rug(probs[probs.yes==0], 0.02, side=1, lwd=0.5)
rug(probs[probs.yes==1], 0.02, side=3, lwd=0.5)
abline(0,1)
newp <- seq(0, 1, length=100)
lines(newp, predict(loess(probs.yes ~ probs, span=1), newp))
```

A method with an adaptive bandwidth such as `loess` is needed here, as the distribution of points along the x axis can be very much more uneven than in this example. The result is shown in Figure 17.6. This plot does show substantial overconfidence in the predictions, especially at probabilities close to one. Indeed, only 22/64 of the events predicted with probability greater than 0.9 occurred. (The underlying cause is the multimodal nature of some of the underlying class distributions.)

Where calibration plots are not straight, the best solution is to find a better model. Sometimes the overconfidence is minor, and mainly attributable to the use of plug-in rather than predictive estimates. Then the plot can be used to adjust the probabilities (which may need then further adjustment to sum to one for more than two classes).

Appendix A

Datasets and Software

The software and datasets used in this book are available over the World Wide Web. Point your browser at

    http://www.stats.ox.ac.uk/pub/MASS2/sites.html

to obtain a current list of sites; please use a site near you. We expect this list to include

    http://www.stats.ox.ac.uk/pub/MASS2
    http://arcola.stats.adelaide.edu.au/pub/MASS2
    http://lib.stat.cmu.edu/S/MASS2
    http://franz.stat.wisc.edu/pub/MASS2

The online instructions tell you how to install the software under both Unix and Windows. Note that S-PLUS is a commercial system, and is *not* included. In case of difficulty in accessing the software please email

    MASS@stats.ox.ac.uk

The datasets and software remain the property of the authors or of the original source, but may be freely distributed provided the source is acknowledged.

The on-line complements as well as printable versions of the on-line help for our software and answers to selected exercises are available at these sites.

A.1 Libraries

Our software is packaged as four library sections, plus section `treefix` for S-PLUS 3.2 and 3.3 and section `helpfix` for Windows.

MASS

This contains all the datasets and a number of S functions, as well as number of other datasets which we have used in learning or teaching.

nnet

Software for feed-forward neural networks with a single hidden layer and for multinomial log-linear models.

spatial

Software for spatial smoothing and the analysis of spatial point patterns. This directory contains a number of datasets of point patterns, described in the text file PP.files (PP.fil under Windows).

class

Functions for nonparametric classification, by k-nearest neighbours and learning vector quantization.

Using our libraries

Under Windows you can make the S objects in a library available by

```
library(name)
```

The help file for a library can be browsed (in the usual Windows help browser) by

```
help(library="name")
```

whether or not the `library` call has been used.

Under Unix the libraries may have been installed as system libraries or private libraries; if the latter you need first to issue the command

```
assign("lib.loc", "path to private library directory", w=0)
```

to say where the libraries are. Then S objects in a library are made available by

```
library(name)
```

and

```
library(help=name)
```

gives a short description of the library and a listing of its contents. After attaching the library, help on its objects can be obtained in the same ways as for system objects (page 6).

How to use libraries is considered in detail in Appendix C.

A.2   Caveat

These datasets and software are provided in good faith, but none of the authors, publishers or distributors warrant their accuracy nor can be held responsible for the consequences of their use.

We have tested the software as widely as we are able but it is inevitable that system dependencies will arise. We are unlikely to be in a position to assist with such problems.

Appendix B

Common S-PLUS Functions

This Appendix contains the call sequences and short descriptions of about 300 commonly used S-PLUS functions, with cross references to descriptions in the main text. Note that the call sequences have been abbreviated to include only the most commonly used arguments and are possibly subject to change; for full current details see the online help.

Functions marked by † are from our libraries.

```
abline(a, b)
abline(coef)
abline(reg)
abline(h=, v=)
```
> Page 75. Adds a line to a plot, often a regression line.

```
ace(x, y, wt, monotone=NULL, linear=NULL)
```
> Page 335. Performs a form of nonlinear regression which nonparametrically transforms both the y and the x variables to produce an additive model. The transformations are chosen to maximize the correlation between the transformed y and the sum of the transformed x's.

```
acf(x, lag.max, type="correlation", plot=T)
```
> Page 435. Estimates and plots autocovariance, autocorrelation or partial autocorrelation function.

```
aggregate(x, ...)
```
> Page 124. Splits up data by factors and computes summary for each subset.

```
aggregate(x, nf=1, fun=sum, ndeltat=1)
```
> Page 434. For a time-series first argument, returns a shorter time series with observations that are the result of fun applied to a partition of the original time series.

```
all(..., na.rm=F)
```
> Page 114. Gives a single logical value that is the 'and' of the logical values that are input.

```
anova(object, ...)
```
> Page 193. Produces an object of class "anova" from a fit of a statistical model, usually an analysis of variance or deviance.

```
any(..., na.rm=F)
```
> Page 114. Gives a single logical value that is the 'or' of the logical values that are input.

`aov(formula, data)`

> Page 191. Compute the analysis of variance for the specified model.

`apply(X, MARGIN, FUN, ...)`

> Page 118. Returns a vector or array by applying a specified function FUN to sections of an array.

`ar(x, aic=T, order.max, method="yule-walker")`

> Page 445. Fits a model of the form $x_t = \alpha_1 x_{t-1} + \cdots + \alpha_p x_{t-p} + e(t)$.

`arg.dialog(x)`

> Page 30. (Windows) Pops up a dialog the argument names and their defaults for a function, and allows them to be changed and the function call submitted.

`args(x)`

> Page 29. Displays the argument names and their defaults for a function.

`arima.diag(z, gof.lag=1)`

> Page 446. Computes diagnostics for an ARIMA model. The diagnostics include the autocorrelation function of the residuals, the standardized residuals, and the portmanteau goodness-of-fit test statistic.

`arima.forecast(x, model, n, xreg, reg.coef)`

> Page 447. Forecasts a univariate time series using an ARIMA model. Also estimates standard errors.

`arima.mle(x, model, n.cond, xreg)`

> Page 445. Fit a univariate ARIMA model by Gaussian maximum (conditional) likelihood, conditioning on at least the first n.cond observations.

`arima.sim(model, n=100, xreg, reg.coef)`

> Page 444. Simulates a univariate ARIMA time series. By default the innovations are Gaussian.

`array(data=NA, dim, dimnames=NULL)`

> Page 49. Creates an array. Arrays are matrices and higher-dimensional generalizations of matrices.

`assign(x, value, frame, where, ...)`

> Page 155. Assigns a value to a name. The location of the assignment can either be a specific frame (useful when writing functions) or a data directory. This is not often used directly except to write to frame 0, the session frame.

`attach(what=NULL, pos=2, name)`

> Page 46. Adds a directory, data frame or list to the S search path.

`attr(x, which)`

> Page 22. Returns a specific attribute of an object. This function can also be used to create a new attribute.

`attributes(x)`

> Returns or changes all of the attributes of an object.

`avas(x, y, wt, monotone, linear)`

> Page 335. Computes a form of nonlinear regression which transforms both the dependent and independent variables to produce an additive model with constant residual variance.

`axis(side, at, labels=T, ticks=T)`

> Page 80. Adds an axis to the current plot. The side, positioning of tick marks, labels and other options can be specified.

`backsolve(r, x)`
> Page 55. Solves a system of linear equations when the matrix representing the system is upper triangular. The function `solve.upper` is essentially the same.

`biplot.default(obs, bivars, var.axes=T)`
> Page 388. Produces a plot in which both the observations and the variables are represented in a two dimensional space.

† `boxcox(model, lambda, plotit=T, interp, ...)`
> Page 216. Computes and plots log-likelihoods for the Box-Cox transformation $y^\lambda - 1$.

`boxplot(...)`
> Page 172. Produces side by side boxplots from one or more vectors.

`browser(object, ...)`
> Page 145. Allows the user to inspect the contents of an object or function frame.

`brush(x, collab, rowlab, hist=F, spin=T)`
> Pages 76 and 382. Creates, for certain graphics devices, a matrix of all two-dimensional scatterplots of multivariate data plus optional histograms and three-dimensional spinning plot, all of which may have points highlighted interactively.

`bs(x, df, knots, degree=3, intercept=FALSE)`
> Page 324. Generate a B-spline basis matrix for polynomial splines.

`bwplot(formula, box.ratio=1...)`
> Page 93. Produces boxplot(s) from a Trellis formula.

`c(...)`
> Pages 21, 24 and 42. Concatenates objects into a vector or a list. The result is a list if one or more of the objects is a list, and a vector otherwise.

`cancor(x, y, xcenter, ycenter)`
> Page 399. Performs canonical correlation analysis to find maximally correlated linear relationships between two groups of multivariate data. By default the data are centred using means.

`cat(..., file, sep=" ", fill=F)`
> Page 62. Coerces arguments to mode character and then prints to standard output, or to a specified file.

`cbind(...)`
> Page 36. Returns a matrix that is pieced together from several vectors and/or matrices.

`chol(x, pivot=F)`
> Page 55. Returns an upper-triangular matrix U which is the Choleski decomposition of a matrix, that is $X = U^T U$. See also `Choleski` for Matrices.

`close.screen(n, all=F)`
> Page 79. Closes one or all the screens of a split screen.

`cloud(formula, ...)`
> Page 106. Trellis plot of a 3D point cloud.

`cmdscale(d, k=2, eig=F, add=F)`
> Page 385. Performs classical (metric) multi-dimensional scaling to represent data in a low-dimensional Euclidean space. (This is also known as principal coordinates analysis.)

```
coef(object)
coefficients(object)
```
Page 193. Extract the coefficients from a fitted model.

```
contourplot(formula, at, cuts=7, ...)
```
Page 96. Produce a contour plot specified by a Trellis formula.

```
contrasts(x)
```
Page 200. Returns the matrix which is the `contrasts` attribute of an object.

```
cor(x, y=x, trim=0)
```
Page 168. Returns the correlation matrix of a data matrix, or correlations between matrices or vectors. A trimming fraction may be specified.

† `corresp(table)`

Page 400. Correspondence analysis assigns scores to the rows and columns of a two-way table in such a way that they are maximally correlated amongst non-constant scores.

```
cov.mve(x, cor=F, print=T)
```
Page 266. Find robust multivariate estimates of location and covariance matrix from a data matrix via the minimum volume ellipsoid estimate. The result uses the mve to reject points before returning estimates based on the cleaned data.

```
cov.wt(x, wt=rep(1, nrow(x)), cor=F, center=T)
```
Page 168. Find weighted estimates of mean and covariance matrix from a data matrix, and optionally (cor=T) of the correlation matrix.

```
coxph(formula, data, subset, iter.max=10,
    method=c("efron","breslow","exact"))
```
Page 357. Fit a Cox proportional hazards model in survival analysis. Time-dependent variables, time-dependent strata, multiple events per subject, and other extensions are incorporated using the counting process formulation.

† `cpgram(ts, taper=0.1, main)`

Page 442. computes and plots a cumulative periodogram of a time series

```
cut(x, breaks, labels)
```
Page 324. Creates a category object by dividing continuous data into intervals. Either specific cut points or the number of equal width intervals can be specified. Warning: this creates intervals of the form $(a, b]$, that is the cutpoint is in the interval to its left.

```
cutree(tree, k=0, h=0)
```
Page 390. An auxiliary function for cluster analysis. It returns a vector of group numbers for the observations that were clustered. Either the number of groups desired or a clustering height may be specified.

```
cv.tree(object, rand, FUN=shrink.tree)
```
Page 427. Cross-validates the tree sequence obtained by either shrinking or pruning a tree.

```
data.dump(list, file="dumpdata")
```
Page 62. Creates a file containing an ASCII representation of the list of objects that are named for use with `data.restore` .

```
data.frame(...)
```
Page 28. Constructs a data frame object from vectors, matrices and other data frames.

```
data.restore(file, print=F)
```
Page 62. Puts data objects into the local directory that were put into a file with `data.dump` .

`debugger(data=last.dump)`
> Page 145. Allows the user to browse in the function frames after an error has occurred, provided `options(error=dump.frames)` is in use.

`density(x, n=50, window="g")`
> Page 179. Kernel density estimate of univariate data. Options include the choice of the window to use and the number of points at which to estimate the density.

`densityplot(formula, cut, from, n=50, to, width, window="g")`
> Page 179. Trellis plot using `density` to produce density estimate(s) of univariate data.

`deparse(expr, short=F)`
> Page 128. Returns a vector of character strings representing the expression.

`deriv(expr, namevec, function.arg)`
`deriv3(expr, namevec, function.arg, hessian = T)`
> Pages 271 and 282. function which takes the first partial derivative (`deriv`) or first and second partial derivatives (`deriv3`) of a S-PLUS expression with respect to a given variable or variables.

`detach(what=2, save=T)`
> Page 46. Removes a database from the search path. The database may be specified by a number, by name, or with a logical vector.

`deviance(object, ...)`
> Page 193. Returns the deviance (twice log-likelihood for full model minus this model) or weighted residual sum of squares of the fitted model object.

`dev.ask(ask=T)`
> Page 71. Forces every graphics device to pause before starting a new plot; press the Return key to draw the next plot.

`dev.copy(device, ..., which)`
> Page 72. Starts a graphics device given by the `device` argument or uses an active device given by `which` (default, the next on the list). It copies the current graph onto the new graphics device and sets this graphics device as the current device.

`dev.cur()`
> Page 71. Returns the number and name of the graphics device currently accepting graphics commands.

`dev.list()`
> Page 71. Returns the number and names of all active devices.

`dev.off(which=dev.cur())`
> Page 71. Shuts down the device specified by the `which` argument and returns the number of the current device after it has been shut down.

`dev.print(device=postscript, ...)`
> Page 72. Starts a graphics device given by the `device` and copies the graph from the current graphics device onto this new graphics device then turns off the new graphics device.

`dev.set(which)`
> Page 71. Returns the number and name of the (newly) current graphics device. (This may not be the one you asked for if the argument `which` does not refer to an active device.)

`diag(x)`
> Page 54. Either creates a diagonal matrix or returns the diagonal of a matrix.

```
diff(x, lag=1, differences=1)
```
> Page 434. Returns a time series or other object which is the result of differencing the input.

```
dim(x)
```
> Page 23. Returns or changes the dim attribute, which describes the dimensions of a matrix, data frame or array.

```
dimnames(x)
```
> Page 50. Returns or changes the dimnames attribute of an array. This is a list of the same length as the dimension of the array, giving names (possibly NULL) for each dimension.

```
dist(x, metric="euclidean")
```
> Page 385. Returns a distance structure that represents all of the pairwise distances between objects in the data. The choices for the metric are "euclidean", "maximum", "manhattan", and "binary".

```
dotplot(formula, ...)
```
> Page 100. Trellis plot of univariate data. for editing.

```
dump(list, fileout="dumpdata", full.precision=T)
```
> Page 62. Creates an ASCII file representing a group of S objects, suitable for editing.

```
eigen(x, symmetric=all(x==t(x)), only.values=F)
```
> Page 55. Returns the eigenvalues and (if desired) eigenvectors of a square matrix.

† ```eqscplot(x, y, tol=0.04, ...)```
> Page 76. Plots with geometrically equal scales, that is 1cm represents the same units on each axis.

```
Error(terms)
```
> Page 301. Used on the right-hand side of a model formula to enclose terms, separated by +, which are to be treated as random effects.

```
exists(name, where, ...)
```
> Page 155. Searches for an S object and returns a logical stating whether the object exists in the search path or not. The position of the object and/or its mode can be specified.

```
expand.grid(...)
```
> Page 77. Creates a data frame containing all combinations of its arguments, vectors, factors or lists of these. It is most often used to make regular grids for prediction.

```
factanal(x, factors=1, method="principal")
```
> Page 406. Perform factor analysis on multivariate data x. The number of factors is specified by factors; the method should be set to "mle". Instead of x, the covariance matrix can be specified via covlist.

```
factor(x, levels, labels, exclude=NA)
```
> Page 25. A factor is a character vector with class attribute "factor" and a levels attribute which determines what character strings may be included in the vector. The function factor creates a factor object out of data and allows the levels attribute to be set.

```
fft(z, inverse=F)
```
> Page 442. Performs the fast Fourier transform on a vector or array of numeric or complex values.

```
fitted(object)
fitted.values(object)
```
> Page 193. Extracts the fitted values from a fitted model.

```
fix(x, file=tempfile("fix"), window=F)
```
> Page 140. Invokes an editor on an S object and overwrites it with the edited version.

```
format(x, ...)
```
> Page 63. Coerces to character strings using a common format. The function is useful for building custom output displays, and is often used in conjunction with cat to print such displays.

```
gam(formula, family=gaussian, data, weights, subset)
```
> Page 327. Fits a generalized additive model.

```
get(name, where, ...)
```
> Page 46. Searches for an S object and returns the object. The position of the object and/or its mode can be specified.

```
glm(formula, family=gaussian, data, weights, subset)
```
> Page 227. Fits a generalized linear model.

```
graphics.off()
```
> Page 71. Shuts down all of the active graphics devices.

```
hclust(dist, method="compact", sim=)
```
> Page 390. Performs hierarchical clustering on a distance or similarity structure. Choices of method are "compact" (also called complete linkage), "average", and "connected" (also called single link).

```
help(name="help", offline=F)
```
> Page 6. Shows on-line help documentation, or prints it out.

```
help.off()
```
> Page 7. Shuts down the S-PLUS help system.

```
help.start(gui)
```
> Page 6. Starts the window system for help in Unix versions of S-PLUS. This system allows you to search for help files by topic and to display them.

```
hist(x, nclass, breaks, plot=T, probability=F, ...)
```
> Page 168. Plots a histogram.

```
histogram(formula, breaks, endpoints, nint, type="percent", ...)
```
> Page 169. Trellis plot of one or more 'histograms', actually frequency counts in bins.

† `histplot(formula, nbins, h, x0, breaks, prob=T, ...)`
> Page 170. Trellis plot of one or more histograms.

```
history(pattern=".", max=10, menu=T, reverse=T)
```
> Page 66. Retrieves recent S expressions, optionally restricting the search to those expressions matching a specified pattern.

```
identify(x, y, labels=seq(along=x), n=length(x))
```
> Page 80. Interactively identify points in a plot on a suitable graphics device.

```
ifelse(test, yes, no)
```
> Page 114. Vector version of 'if then else'.

```
inspect(x)
```
> Page 141. An interactive debugging facility with breakpoints, tracking reports, and interactive examination or modification of S-PLUS objects.

`integrate(f, lower, upper, max.subdiv=100)`
>Approximates the integral of a real-valued function over a given interval, and estimates the absolute error in the approximation.

`interp(x, y, z, xo, yo, ncp=0, extrap=F)`
>Page 474. Interpolates the value of the third variable onto an evenly spaced grid of the first two variables, by Akima's method.

`is.na(x)`
>Page 35. Returns an object similar to the input which is filled with logicals denoting whether the corresponding element of the input is 'NA'.

† `isoMDS(d, y=cmdscale(d), k=2, niter=50, trace=T)`
>Page 387. Computes Kruskal's form of non-metric multidimensional scaling.

† `Kaver(fs, nsim, ...)`
>Page 483. Computes and averages `nsim` simulations of Ripley's K function.

† `kde2d(x, y, h, n=25, lims=c(range(x), range(y)))`
>Page 184. Two-dimensional kernel density estimation with an axis-aligned bivariate normal kernel, evaluated on a square grid.

† `Kenvl(fs, nsim, ...)`
>Page 483. Computes and finds the envelope (pointwise max and min) and average of `nsim` simulations of Ripley's K function.

† `Kfn(pp, fs, k=100)`
>Page 483. Computes Ripley's K function, actually $L = \sqrt{K/\pi}$.

`kmeans(x, centers, iter.max=10)`
>Page 391. Performs K-means clustering, using the number of initial centres as the number of groups.

`ksmooth(x, y=NULL, kernel="box", bandwidth=0.5)`
>Page 327. Estimates a probability density or performs scatterplot smoothing using kernel estimates.

`l1fit(x, y, intercept=T, print=T)`
>Page 257. Performs an L_1 regression.

`lag(x, k=1)`
>Page 434. Returns a time series like the input but shifted in time.

`lapply(X, FUN, ...)`
>Page 121. Returns a list which is the result of a function that is applied to each element of a list,

† `lda(formula, data, prior=proportions)`
`lda(x, grouping, prior=proportions)`
>Page 396. Computes linear discriminant analysis specified either by a formula or a data matrix and vector or factor of groups.

`length(x)`
>Page 22. Returns the length of the object (such as a vector or list).

`levelplot(formula, at, cuts=7, ...)`
>Page 96. Produce a greylevel or pseudo-colour plot specified by a Trellis formula.

`library(section, first=F, help=NULL, lib.loc)`
>Page 517. Attaches a section of an S library. A library section is a way of packaging a collection of S functions, datasets, documentation, and compiled code.

`lines(x, y, type="l")`
> Page 75. Adds lines to the current plot. The type of line can be specified as well as other graphical parameters.

`list(...)`
> Page 24. Creates a list from the specified objects.

`lm(formula, data, weights, subset, ...)`
> Page 192. Fits a linear model.

`lme(fixed, random=fixed, cluster, data, start, ...)`
> Page 306. Fits a linear model with (possibly) both fixed and random effects.

`lm.influence(lm)`
> Page 205. Computes statistics used in measuring the influence of the observations on the original fit of a linear model.

`lmsreg(x, y, intercept=T)`
> Page 262. Returns a resistant regression estimate that minimizes the median of the squared residuals.

`lo(..., span=0.5, degree=1)`
> Page 327. Allows the user to specify a loess fit in the formula for a generalized additive model.

`loadings(x)`
> Page 383. Returns the loadings component of an object, usually from `princomp` or `factanal`.

`location.m(x, location, scale, psi.fun="bisquare", parameters)`
> Page 252. Robust M-estimator of location.

`locator(n=500, type="n")`
> Page 80. Returns the coordinates specified interactively on a plot. Points and/or lines may be added to the plot.

`loess(formula, data, weights, subset, span=0.75, degree=2,`
`    family=c("gaussian", "symmetric"))`
> Page 473. Fits a local regression (loess) model.

`loglin(table, margin, start)`
> Page 240. Estimates test statistics and parameter values for a log-linear analysis of a multidimensional contingency table.

`lowess(x, y, f=2/3, iter=3, delta=.01*range(x))`
> Page 326. Gives a robust, local smooth of scatterplot data. Among other options is the fraction `f` of data smoothed at each point.

`ltsreg(x, y, intercept=T)`
> Page 262. Returns a resistant regression estimate that minimizes the sum of the smallest half of the squared residuals.

`mad(y, center=median(y), constant=1.4826, na.rm=F, low=F)`
> Page 249. Returns a robust scale estimate of the data.

`mahalanobis(x, center, cov, inverted=F)`
> Page 397. Returns a vector of the Mahalanobis distances for the rows of a data matrix.

`masked(where=1)`
> Page 46. Reports objects with common names on different databases.

`matrix(data=NA, nrow, ncol, byrow=F, dimnames=NULL)`
> Page 49. Creates a matrix.

`mclass(m, n.clust, aux=NULL)`

Page 391. Uses the output from `mclust` to determine the classification corresponding to a given number of clusters.

`mclust(x, method="S*", noise=F)`

Page 391. Performs hierarchical clustering via a wide range of clustering options, calculates a Bayesian criterion for choosing the number of clusters, and optionally allows for noise or 'outliers'.

`mean(x, trim=0, na.rm=F)`

Page 32. Computes a mean, possibly after removing missing values and trimming a fraction from each end of the data range.

`median(x)`

Page 249. Computes a median.

`misclass.tree(tree, detail=F)`

Page 427. Returns the number of misclassification errors for a classification tree.

`missing(name)`

Page 129. Returns a logical value that indicates whether or not `name` was supplied as an argument to the function in which `missing` is called.

`mreloc(classification, x, method="S*", noise=F)`

Page 391. Performs iterative relocation for a given clustering criterion and classification. See also `mclust`.

`ms(formula, data, start, control, trace=F)`

Page 286. Minimizes a sum of nonlinear functions over parameters in a data frame.

`mstree(x, plane=T)`

Page 387. Returns a vector or a list which gives the minimal spanning tree and, optionally, multivariate planing information.

† `multinom(formula, data, weights, subset, Hess=F, summ=0, ...)`

Page 242. Fits multinomial log-linear models via neural networks.

`names(x)`

Page 23. Returns or changes the `"names"` attribute of an object, usually a list or a vector.

`nlme(fixed, random=fixed, cluster, data, start, ...)`

Page 314. Fits a non-linear model with (possibly) both fixed and random effects.

`nlminb(start, objective, gradient=NULL, hessian=NULL, scale=1,`
`    control=NULL, lower=-Inf, upper=Inf)`

Page 292. Minimizer for smooth nonlinear functions subject to box-constrained parameters.

`nlregb(nres, start, residuals, jacobian=NULL, scale=NULL,`
`    control=NULL, lower=-Inf, upper=Inf, ...)`

Page 284. Fits a nonlinear regression model via least squares subject to box-constrained parameters.

`nls(formula, data, start, control, trace=F)`

Page 269. Fits a nonlinear regression model via least squares.

† `nnet(x, y, weights, size, Wts, linout=F, entropy=F, softmax=F,`
`    skip=F, rang=0.7, decay=0, maxit=100, trace=T)`

Page 340. Fits a single-hidden-layer neural network.

† `nnet.Hess(net, x, y)`
> Page 340. Computes Hessian of the objective function at the fitted values of a nnet object.

`ns(x, df, knots, intercept=F)`
> Page 324. Generates a basis matrix for natural cubic splines.

`numeric(n)`
> Page 30. Allocates space for a vector of length `n`.

`objects(where=1, frame=NULL, pattern)`
> Page 45. Returns a vector of character strings which are the names of S objects in a position on the search path, or in a memory frame.

`optimize(f, interval)`
> Page 286. Approximates a local optimum of a continuous univariate function within a given interval.

`options(...)`
> Page 64. Provides a means to control some of the behaviour of S such as the number of digits printed, the maximum number of characters to place on a line, and strategies for memory management.

`order(..., na.last=T)`
> Page 53. Returns an integer vector containing the permutation that will sort the input into ascending order.

`ordered(x, levels, labels, exclude=NA)`
> Page 26. Returns an object of class `"ordered"` which is an ordered factor.

`outer(X, Y, FUN="*", ...)`
> Page 54. Performs an outer product operation given two arrays (or vectors) and, optionally, a function.

`par(...)`
> Page 82. Provides control over the action of the graphics device—layout, colour and so on.

`paste(..., sep=" ", collapse=NULL)`
> Page 42. Returns a vector of character strings which is the result of pasting corresponding elements of the input vectors together. If the `collapse` argument is used, a single string is returned.

`plclust(tree)`
> Page 390. Creates a plot of a clustering tree given a structure produced by `hclust` or `mclust`.

`plot(x, ...)`
> Page 73. Creates a plot on the current graphics device. A generic function which can plot many types of object.

`points(x, y, type="p", ...)`
> Page 75. Adds points to the current plot. The type of point can be specified as well as other graphical parameters.

`poly(x, ...)`
> Page 470. Returns a matrix of orthonormal polynomials, which represents a basis for polynomial regression.

`polygon(x, y, density=-1, angle=45, border=T)`
> Page 70. Adds a polygon with the specified vertices to the current plot.

`polyroot(z)`
> Page 444. Finds all roots of a polynomial with real or complex coefficients.

```
postscript(file="", width, height, onefile=T)
```
Page 73. Allows graphics to be produced for a POSTSCRIPT printer.

```
post.tree(tree, digits=.Options$digits - 3, pretty=0,
     pointsize=12)
```
Page 421. Generates a POSTSCRIPT presentation plot of a tree object.

† `ppinit(file)`
Page 483. Loads and initializes a spatial point process.

† `pplik(pp, R, ng=25)`
Page 483. Computes the pseudo-likelihood estimate of the parameter c of a Strauss point process.

```
ppreg(x, y, min.term, max.term=min.term, xpred=NULL, optlevel=2,
     bass=0, span="cv")
```
Page 332. Computes projection pursuit regression, an exploratory nonlinear regression method that models y as a sum of nonparametric functions of projections of the x variables.

† `ppregion(xl=0, xu=1, yl=0, yu=1)`
Page 483. Sets the rectangular domain $(xl, xu) \times (yl, yu)$ for a spatial point process.

```
predict(object, newdata, type, se.fit=F)
```
Page 193. Returns predictions from a fitted model.

```
princomp(x, covlist, scores=T, cor=F, subset=T)
```
Page 383. Finds principal components for multivariate data. The argument `cor` determines if the scaling to a correlation matrix takes place. The covariance matrix can be specified directly via `covlist`, for example the result of `cov.mve`.

```
print(x, ...)
```
Prints the input objects.

```
print.trellis(x, position, split, more=F)
```
Page 105. Prints Trellis objects.

† `prmat(obj, xl, xu, yl, yu, n)`
Page 476. Computes matrix of prediction in kriging over a grid.

```
profile(fitted, which, maxpts, ...)
```
Page 276. Explores the behavior of the objective function near the solution— currently has methods for `ms` and `nls`.

```
prune.misclass(tree, k=NULL, best=NULL, newdata)
```
Page 428. A version of `prune.tree` that correctly computes the `"misclass"` method.

```
prune.tree(tree, k=NULL, best=NULL, method))
```
Page 425. Determines a nested sequence of subtrees of the supplied tree by recursively "snipping" off the least important splits, based upon the cost-complexity measure. If `k` is supplied, the optimal subtree for that penalty is returned, and if `best` is supplied it determines the tree size.

† `Psim(n)`
Page 483. Simulates a binomial spatial point process, that is a Poisson process conditioned on n points.

```
q()
```
Page 5. Terminates the current S session.

† qda(formula, data, prior=proportions)
† qda(x, grouping, prior=proportions)
> Page 397. Computes the quadratic discriminant analysis. Specified either by a formula or a matrix and a grouping variable.

qqline(x, ...)
> Page 165. Fits and plots a line through the quartiles of a normal QQ-plot.

qqnorm(x, ...)
> Page 165. A normal QQ-plot.

qqplot(x, y, plot=T)
> Page 164. QQ-plot of two data vectors (of the same length).

qr(x)
> Page 56. Returns a representation of the QR decomposition of a rectangular matrix x. The functions qr.Q, qr.R and qr.X operate on the result to give the components and to reconstruct x. qr.coef(qr, y), qr.fitted(qr, y), qr.resid(qr, y), qr.qty(qr, y) and qr.qy(qr, y) apply the QR decomposition to the least squares fit of y by x.

quantile(x, probs=seq(0,1,.25), na.rm=F)
> Page 164. Returns a vector of the desired quantiles of a data vector.

raov(formula, data, ...)
> Page 297. Performs an analysis of variance including the estimation of variance components for models with only random effects.

rbind(...)
> Page 36. Returns a matrix that is pieced together from several vectors and/or matrices.

read.table(file, header, sep, row.names, col.names, as.is=F)
> Page 37. Reads in a file in table format and creates a data frame with the same number of rows as there are lines in the file, and the same number of variables as there are fields in the file.

rep(x, times, length)
> Page 33. Replicates the input either a certain number of times or to a certain length.

resid(object)
residuals(object)
> Page 193. Extracts the residuals from a fitted model.

† rlm(formula, data, weights, subset, na.action,k=1.345, ...)
> Page 260. Fit robustly a multiple linear regression.

rm(..., list=NULL)
> Page 47. Removes objects from the working directory.

rotate.default(x, rotation=NULL, orthogonal=T, parameters=NULL, normalize=T)
> Page 409. Computes rotations, the possible rotations include procrustes, promax, the oblimin family, and the orthomax family. There are methods for factanal and princomp.

rreg(x, y, w=rep(1,nrow(x)), int=T, method=wt.default)
> Page 259. Robust regression by an M-estimator.

rts(x=NA, start=1, deltat=1, frequency=1)
> Page 431. Defines a univariate or multivariate regularly spaced time series.

rug(x, ticksize, side=1, lwd=0.1)
> Page 336. Adds a rug (a series of ticks indicating the data locations) to a plot. The default `lwd` gives too thin a linewidth on most phototypesetters – we have used 0.5.

s(x, df=4, spar=0)
> Page 327. Specifies a smooth function in the formula for a generalized additive model.

† sammon(d, y=cmdscale(d), k=2, trace=T, ...)
> Page 386. Computes Sammon's non-linear mapping, one form of non-metric multidimensional scaling.

sample(x, size, replace=F, prob)
> Page 166. Produces a vector of length size of objects randomly chosen from the population.

sapply(X, FUN, ..., simplify=T)
> Page 121. Returns a vector, matrix, or list as the result of applying a function to an S-PLUS object, usually a list.

scale(x, center=T, scale=T)
> Centres and then scales the columns of a matrix. By default each column in the result has mean 0 and sample standard deviation 1.

scale.tau(y, center=median(y), tuning=1.95)
> Page 252. Returns a robust scale estimate of the data.

scan(file="", what=numeric(), n, sep, multi.line=F, ...)
> Page 38. Reads data from a text file or interactively from standard input. Options are available to control how the file is read and the structure of the data object.

screen(n, new=T)
> Page 79. Changes the current screen in a split screen.

search()
> Page 44. Returns the current search path.

segments(x1, y1, x2, y2)
> Page 88. Adds line segments to the current plot.

† semat(obj, xl, xu, yl, yu, n, se)
> Page 476. Computes a matrix of standard errors of prediction in kriging over a grid.

seq(...)
> Page 33. Creates a vector of evenly spaced numbers. The beginning, end, spacing and length of the sequence can be specified.

set.seed(i)
> Page 166. Puts the random number generator into one of $i = 1, \ldots, 1000$ reproducible states.

sink(file, command, append=F)
> Page 62. Directs the output to a file rather than to the terminal, or back if `file` is missing.

solve(A, b)
> Page 55. Solves the linear equations $Ax = b$, or inverts A if b is omitted. It has methods `solve.upper`, `solve.qr` and many more in library `Matrix`.

sort(x, partial=NULL, na.last=NA)
> Page 52. Returns a vector which is a sorted version of the input. By default missing values are deleted.

`sort.list(x, partial=NULL, na.last=T)`
>   Page 53. Returns a vector of integers giving the order of the data.

`source(file, local=F)`
>   Page 61. Takes input from the specified file.

`spec.ar(ar.list, n.freq, frequency, plot=F)`
>   Page 448. Computes the spectrum of a time series using the results of an autoregressive fit.

`spec.pgram(x, spans=1, taper=0.1, pad=0, detrend=T, demean=F)`
>   Page 439. Estimates the spectrum of a time series by smoothing the periodogram, and optionally plots the spectral estimate.

`spec.taper(x, p=0.1)`
>   Page 441. Tapers a time series by means of a split-cosine-bell window.

`spectrum(x, method="pgram", plot=T, ...)`
>   Page 439. Estimates and plots time-series spectra using either a smoothed periodogram or a fitted autoregression.

`spin(x, collab, highlight=rep(F, nrow(x)))`
>   Page 77. Rotates multivariate data so that the three dimensional structure of the variables chosen can be perceived.

`split(data, group)`
>   Page 459. Returns a list in which each component is a vector of values from data that correspond to a unique value in group.

`split.screen(figs, screen, erase=T)`
>   Page 78. Splits a graphics display into multiple screens.

`splom(formula, ...)`
>   Page 91. A trellis plot of one or more scatterplot matrices. The formula is of the form ~ x | g1 * g2 * ..., where x is a numeric matrix or a data frame.

† `SSI(n, r)`
>   Page 484. Simulates a spatial sequential inhibition point process, This has n points laid down successively, each at least distance r from its predecessors.

`stem(x, nl, scale, twodig=F, fence=2, head=T, depth=F)`
>   Page 170. Prints a stem-and-leaf plot.

`step(object, scope, scale, direction, trace=T, keep, steps)`
>   Page 220. Performs stepwise model selection by approximate AIC for glm-like models.

† `stepAIC(object, scope, scale, direction, trace=T, keep, steps, screen)`
>   Page 237. Performs stepwise model selection by exact AIC.

`stepfun(x, y, type="left")`
>   Page 80. Computes a step function from (x, y) points sorted into increasing order on x. The default type gives the left endpoint of the step; the alternative is "right".

`stl(time.series, ...)`
>   Page 449. Decomposes a time series into frequency components of variation by a sequence of loess smoothings.

`stop(message="")`
>   Page 129. Issues a message and exits from the current function.

`Strauss(n, c=0, r)`

> Page 483. Simulates a Strauss spatial point process with inhibition distance r and inhibition strength c.

`stripplot(formula, jitter=F, ...)`

> Page 99. Trellis plot of univariate data. for editing.

`substitute(expr, frame)`

> Page 128. Returns an object like `expr` but with substitutions made relative to `frame`.

`summary(object, ...)`

> Provides a summary of an object.

`supsmu(x, y, wt=rep(1,length(y)), span="cv", periodic=F, bass=0)`

> Page 327. Returns a list containing x and y components that are a smoothed version of the input data. This algorithm is designed to be fast, and by default uses cross-validation to pick the span.

`surf.gls(np, covmod, x, y, z, nx, ...)`

> Page 478. Fits a spatial trend surface by generalized least squares.

`surf.ls(np, x, y, z)`

> Page 470. Fits a spatial trend surface by least squares.

`Surv(time, event)`
`Surv(time, time2, event)`

> Page 344. Computes a regression target for censored data for use in survival analysis.

`survdiff(formula, data, rho=0, subset)`

> Page 349. Tests if there is a difference between two or more survival curves using the G_ρ family of tests, or for a single curve against a known alternative.

`survexp(formula, data, times, cohort=T, conditional=T, ...)`

> Page 379. Returns the expected survival of a cohort of subjects or of each individual subject.

`survfit( object, data, weights, subset, newdata, individual=F,`
`    conf.int=.95, se.fit=T)`

> Page 345. Computes an estimate of a survival curve or computes the predicted survivor function for a cox proportional hazards model.

`survreg(formula, data, subset, na.action, link=c("log","identity"),`
`    dist=c("extreme", "logistic", "gaussian", "exponential"))`

> Page 351. Regression of censored survival data on explanatory variables.

`sweep(A, MARGIN, STATS, FUN="-", ...)`

> Page 119. Returns an array like the input A with STATS "swept" out over the margins specified. Used to remove means, rescale to unit variances, and so on.

`svd(x, nu=min(nrow(x),ncol(x)), nv=min(nrow(x),ncol(x)))`

> Page 56. Returns a list containing the singular value decomposition of the input.

`symbols(x, y, circles=, squares=, rectangles=, stars=,`
`    thermometers=, boxplots=, add=F, inches=T)`

> Page 75. Puts a symbol on a plot at each of the specified locations. The symbols can be circles, squares, rectangles, star plots, thermometers, or boxplots.

`synchronize(database)`

> Page 47. Forces consistency between the evaluator and the data directories, frame 0, etc. Used to write out current values, or to update the current session's view of attached databases if these have been updated.

t(x)
>   Page 35. Matrix transpose.

table(...)
>   Page 26. Returns a contingency table (array) with one dimension per argument.

tapply(X, INDICES, FUN, ...)
>   Page 120. Applies a function to each cell of a ragged array.

text(x, ...)
text.default(x, y, labels=seq(along=x))
>   Page 75. Places text at the given positions in the current plot.

title(main, sub, xlab, ylab, axes=F)
>   Page 80. Adds titles to the current plot.

traceback()
>   Page 141. Prints the calls in the process of evaluation at the time of an error.

tree(formula)
>   Page 419. Grows a tree object from a specified formula and data.

† trmat(obj, xl, xu, yl, yu, n)
>   Page 470. Computes a fitted spatial trend surface over a grid of points.

† truehist(data, nbins, h, x0, breaks, prob=T, ...)
>   Page 170. Plots a true histogram.

ts.intersect(..., dframe=F)
>   Page 433. Binds several time series together into a single multivariate time series whose time domain is the intersection of the time domains of the component series.

ts.plot(..., type="l")
>   Page 433. Plots one or more time series on the current graphics device.

ts.union(..., dframe=F)
>   Page 433. Binds several time series together into a single multivariate time series whose time domain is the union of the time domains of the component series.

twoway(x, trim=.5, iter=6, eps, print=F)
>   Page 255. Returns a list containing estimated row and column effects as well as a grand effect and the residuals. The default is to give estimates from a median polish.

uniroot(f, interval)
>   Page 286. Approximates a root or zero of a continuous univariate function given an interval for which the function has opposite signs at the endpoints.

update(object, formula, ..., evaluate=T, class)
>   Page 193. Allows a new model to be created from an old model by providing only those arguments that need to be changed.

var(x, y=x)
>   Page 32. Returns the variance of a vector or the covariance matrix of a data matrix or two vectors.

varcomp(formula, data, method="minque0")
>   Page 297. Returns an object of class "varcomp" which provides estimates of variance components, coefficients and the random variables.

† vcov(x, ...)
>   Page 228. Returns the variance-covariance matrix of the coefficients of a fitted model.

```
warning(message="")
```
> Page 128. Issues a message from a function.

```
window(x, start=start(x), end=end(x))
```
> Page 434. Returns a time series with a new start and/or end.

```
wireframe(formula, at, drape=F, ...)
```
> Page 96. Trellis plot of a 3D wireframe surface with optional colour drape.

```
write(x, file="data", ncolumns, append=F)
```
> Page 61. Writes the contents of the data to a file in ASCII format.

```
write.table(data, file="", sep=",", dimnames.write=T)
```
> Page 61. Writes a data frame or matrix to a file in ASCII format.

```
xyplot(formula, ...)
```
> Page 96. One or more scatterplots, in Trellis graphics.

Appendix C

Using S-PLUS Libraries

A library in S-PLUS is a convenient way to package together S objects for a common purpose, and to allow these to extend the system. A library section is a directory containing a .Data subdirectory, which may contain a .Help subdirectory. (Under Windows it has a _Data directory, and perhaps a _Help subdirectory or a .hlp file.) The directory should also contain a README file describing its contents, and may also contain object modules for dynamic loading.

The structure of a library section is the same as that of a working directory, but the library function makes libraries much more convenient to use. Conventionally libraries are stored in a standard place, the subdirectory library of the main S-PLUS directory. Which library sections are available can be found by the library command with no arguments; for example on one of our systems this gives:

```
> library()
Library "/usr/local/splus/library"
The following sections are available in the library
```

SECTION	BRIEF DESCRIPTION
chron	Functions to handle dates and times.
cluster	Modern methods for cluster analysis.
Defunct	Some functions that are no longer supported in S-PLUS.
demo	Demo of S-PLUS. Do not attach directly. See help("demo").
examples	Functions and objects from The New S Language.
external	Handle external (large) objects.
image	Display images.
maps	Display of maps with projections.
mathematica	Interface to Mathematica system.
Matrix	New Matrix class functions for numerical linear algebra
progdraw	Sdraw example from Programmer's Manual.
progexam	Examples from Programmer's Manual.
semantics	Functions from chapter 11 of The New S Language
trellis	New graphics functions

517

```
ash                    Scott's ASH functions
bootstrap              Efron & Tibshirani's bootstrap functions
class                  classification tools
haerdle                'Smoothing Techniques' by W. Haerdle
MASS                   main library of Venables & Ripley
nnet                   neural networks functions
spatial                spatial statistics library
xgobi                  support functions for XGobi
```

Further information (the contents of the README file) on any section is given by

```
library(help=section_name)
```

and the library section itself is made available by

```
library(section_name)
```

This has two actions. It attaches the .Data subdirectory of the section at the end of the search path (having checked that it has not already been attached), and executes the function .First.lib if one exists within that .Data subdirectory.

Sometimes it is necessary to have functions in a library which will replace standard system functions (for example to correct bugs or to extend their functionality). This can be done by attaching the library as the second dictionary on the search path with

```
library(section_name, first=T)
```

Of course, attaching other dictionaries with attach or other libraries with first=T will push previously attached libraries down the search path.

Private libraries

So far we have only considered system-wide library sections installed under the main S-PLUS directory, which usually requires privileged access to the operating system. It is also possible to use a private library, by giving library the argument lib.loc or by assigning the object lib.loc in the current session dictionary (frame 0). This should be a vector of directory names which are searched in order for library sections before the system-wide library. For example, on another of our systems we get

```
> assign(where=0, "lib.loc", "/users/ripley/S/library")
> library()
Library "/users/ripley/S/library"
The following sections are available in the library:

SECTION           BRIEF DESCRIPTION

MASS              main library
nnet              neural nets
```

```
spatial           spatial statistics
class             classification

Library "/packages/splus3.4/library"
The following sections are available in the library:

SECTION           BRIEF DESCRIPTION

chron             Functions to handle dates and times.
     ....
```

Because `lib.loc` is local to the session, it must be assigned for each session. The `.First` function is often a convenient place to do so (see page 64); see its help page for other ways using `.First.local` or the S_FIRST environment variable.

C.1   Sources of libraries

Many S-PLUS users have generously collected together their functions and datasets together into libraries and made them publicly available. An archive of sources for library sections is maintained at Carnegie-Mellon University as a service to the statistical profession by Mike Meyer. The World Wide Web address is

> http://lib.stat.cmu.edu/S/

There is a mirror of `statlib` in the UK at

> http://www.hensa.ac.uk/ftp/mirrors/statlib/

Amongst the libraries available from `statlib` are the following which have been mentioned elsewhere in this book:

ash	1D, 2D and 3D ASH code from Scott (1992)
bootstrap.funs	Bootstrap functions from Efron & Tibshirani (1993)
class	S functions for classification
delaunay	Dirichlet tessellation and Delaunay triangulation
haerdle	S and C code and datasets from Härdle (1991)
KernSmooth	Kernel density estimation and smoothing
mda	Mixture and flexible discriminant analysis
nlme	linear and non-linear mixed effect models
postscriptfonts	Additional capabilities for postscript under Unix
pspline	Penalized splines, for estimating derivatives of a smooth fit
robeth	ROBETH code for use with Marazzi (1993)
xgobi	XGobi dynamic graphics package

There are a number of other archives, current details of which may be found in *F*requently *A*sked *Q*uestions list (with answers) from the S-news mailing list available in various formats at

```
http://www.stat.math.ethz.ch/S-FAQ/S-faq_toc.html (on-line)
ftp://ftp.stat.math.ethz.ch/pub/Doc/ (text, DVI, ps, texinfo)
```

The convention is to distribute libraries as 'shar' archives; these are text files which when used as scripts for the Bourne shell sh unpack to give all the files needed for a library section. Check files such as Install for installation instructions; these usually involve editing the Makefile and typing make.

Several of these libraries are available pre-packaged for Windows users in .zip archives; check the WWW addresses

```
http://lib.stat.cmu.edu/DOS/S
http://www.stats.ox.ac.uk/pub/SWin
```

which also point to tools to help Windows users access 'shar' archives intended for Unix. Beware that most libraries packaged for S-PLUS 3.x are incompatible with S-PLUS 4.0 since the format of compiled C or FORTRAN code has changed.

Users are encouraged to share their own efforts; Mike Meyer welcomes submissions (see the file submissions from S). Submitters should try to be aware of the potential differences between different S-PLUS platforms.

C.2   Creating a library section

This is a topic covered in detail in the on-line complements. If the library uses no compiled C or FORTRAN code the process is fairly simple.

1. Create a directory in the library with the section name, and create .Data and .Data/.Help subdirectories (_Data and _Data_Help in Windows).

2. Using this as the working directory, run S-PLUS and create the desired objects.

3. Create help files using prompt (see page 146).

4. Create a README file (README.TXT in Windows) describing briefly the purpose of the library and with one-line descriptions of the public objects.

5. If any start-up actions are needed, put these in a .First.lib function.

References

Abbey, S. (1988) Robust measures and the estimator limit. *Geostandards Newsletter* **12**, 241–248.

Abramowitz, M. and Stegun, I. A. (1965) *Handbook of Mathematical Functions with Formulas, Graphs and Mathematical Tables*. New York: Dover.

Aitchison, J. (1986) *The Statistical Analysis of Compositional Data*. London: Chapman and Hall.

Aitchison, J. and Dunsmore, I. R. (1975) *Statistical Prediction Analysis*. Cambridge: Cambridge University Press.

Aitkin, M. (1978) The analysis of unbalanced cross classifications (with discussion). *Journal of the Royal Statistical Society A* **141**, 195–223.

Akaike, H. (1974) A new look at statistical model identification. *IEEE Transactions on Automatic Control* **AU–19**, 716–722.

Akima, H. (1978) A method of bivariate interpolation and smooth surface fitting for irregularly distributed data points. *ACM Transactions on Mathematical Software* **4**, 148–159.

Analytical Methods Committee (1987) Recommendations for the conduct and interpretation of co-operative trials. *The Analyst* **112**, 679–686.

Analytical Methods Committee (1989a) Robust statistics — how not to reject outliers. Part 1. Basic concepts. *The Analyst* **114**, 1693–1697.

Analytical Methods Committee (1989b) Robust statistics — how not to reject outliers. Part 2. Inter-laboratory trials. *The Analyst* **114**, 1699–1702.

Andersen, P. K., Borgan, Ø., Gill, R. D. and Keiding, N. (1993) *Statistical Models based on Counting Processes*. New York: Springer-Verlag.

Anderson, E. (1935) The irises of the Gaspe peninsula. *Bulletin of the American Iris Society* **59**, 2–5.

Anderson, E. and ten others (1995) *LAPACK User's Guide*. Second Edition. Philadelphia: SIAM.

Anderson, O. D. (1976) *Time Series Analysis and Forecasting. The Box-Jenkins Approach*. London: Butterworths.

Atkinson, A. C. (1985) *Plots, Transformations and Regression*. Oxford: Oxford University Press.

Atkinson, A. C. (1986) Comment: Aspects of diagnostic regression analysis. *Statistical Science* **1**, 397–402.

Atkinson, A. C. (1988) Transformations unmasked. *Technometrics* **30**, 311–318.

Azzalini, A. and Bowman, A. W. (1990) A look at some data on the Old Faithful geyser. *Applied Statistics* **39**, 357–365.

Banfield, J. D. and Raftery, A. E. (1993) Model-based Gaussian and non-Gaussian clustering. *Biometrics* **49**, 803–821.

Barnett, V. and Lewis, T. (1985) *Outliers in Statistical Data.* Second Edition. Chichester: John Wiley and Sons.

Bartholomew, D. J. (1987) *Latent Variable Analysis and Factor Analysis.* London: Griffin.

Basilevsky, A. (1994) *Statistical Factor Analysis and Related Methods.* New York: John Wiley and Sons.

Bassett, G. W. and Koenker, R. (1978) Asymptotic theory of least absolute error regression. *Journal of the American Statistical Association* **73**, 618–622.

Bates, D. M. and Chambers, J. M. (1992) Nonlinear models. Chapter 10 of Chambers & Hastie (1992).

Bates, D. M. and Watts, D. G. (1980) Relative curvature measures of nonlinearity (with discussion). *Journal of the Royal Statistical Society, Series B* **42**, 1–25.

Bates, D. M. and Watts, D. G. (1988) *Nonlinear Regression Analysis and its Applications.* New York: John Wiley and Sons.

Beale, E. M. L. (1960) Confidence intervals in non-linear estimation (with discussion). *Journal of the Royal Statistical Society B* **22**, 41–88.

Becker, R. A. (1994) A brief history of S. In *Computational Statistics: Papers Collected on the Occasion of the 25th Conference on Statistical Computing at Schloss Reisenburg*, eds P. Dirschedl and R. Osterman, pp. 81–110. Heidelberg: Physica-Verlag.

Becker, R. A. and Chambers, J. M. (1988) Auditing of data analyses. *SIAM Journal of Scientific and Statistical Computing* **9**, 747–760.

Becker, R. A., Chambers, J. M. and Wilks, A. R. (1988) *The NEW S Language.* New York: Chapman and Hall. (Formerly Monterey: Wadsworth and Brooks/Cole.).

Bie, O., Borgan, Ø. and Liestøl, K. (1987) Confidence intervals and confidence bands for the cumulative hazard rate function and their small sample properties. *Scandinavian Journal of Statistics* **14**, 221–233.

Bishop, C. M. (1995) *Neural Networks for Pattern Recognition.* Oxford: Clarendon Press.

Bishop, Y. M. M., Fienberg, S. E. and Holland, P. W. (1975) *Discrete Multivariate Analysis.* Cambridge, MA.: MIT Press.

Bloomfield, P. (1976) *Fourier Analysis of Time Series: An Introduction.* New York: John Wiley and Sons.

Bloomfield, P. and Steiger, W. L. (1983) *Least Absolute Deviations: Theory, Applications, and Algorithms.* Boston: Birkhauser.

Borgan, Ø. and Liestøl, K. (1990) A note on confidence intervals and bands for the survival function based on transformations. *Scandinavian Journal of Statistics* **17**, 35–41.

Box, G. E. P. and Cox, D. R. (1964) An analysis of transformations (with discussion). *Journal of the Royal Statistical Society B* **26**, 211–252.

Box, G. E. P., Hunter, W. G. and Hunter, J. S. (1978) *Statistics for Experimenters.* New York: John Wiley and Sons.

Box, G. E. P. and Pierce, D. A. (1970) Distribution of residual autocorrelations in autoregressive-integrated moving average time series models. *Journal of the American Statistical Association* **65**, 1509–1526.

Breiman, L. and Friedman, J. H. (1985) Estimating optimal transformations for multiple regression and correlations (with discussion). *Journal of the American Statistical Association* **80**, 580–619.

Breiman, L., Friedman, J. H., Olshen, R. A. and Stone, C. J. (1984) *Classification and Regression Trees*. Monterey: Wadsworth and Brooks/Cole.

Brent, R. (1973) *Algorithms for Minimization without Derivatives*. Englewood Cliffs, NJ: Prentice-Hall.

Brockwell, P. J. and Davis, R. A. (1991) *Time Series: Theory and Methods*. Second Edition. New York: Springer-Verlag.

Brockwell, P. J. and Davis, R. A. (1996) *Introduction to Time Series and Forecasting*. New York: Springer.

Brownlee, K. A. (1965) *Statistical Theory and Methodology in Science and Engineering*. Second Edition. New York: John Wiley and Sons.

Campbell, N. A. and Mahon, R. J. (1974) A multivariate study of variation in two species of rock crab of genus *Leptograpsus*. *Australian Journal of Zoology* **22**, 417–425.

Cao, R., Cuevas, A. and González-Manteiga, W. (1994) A comparative study of several smoothing methods in density estimation. *Computational Statistics and Data Analysis* **17**, 153–176.

Chambers, J. M. and Hastie, T. J. eds (1992) *Statistical Models in S*. New York: Chapman and Hall. (Formerly Monterey: Wadsworth and Brooks/Cole.)

Ciampi, A., Chang, C.-H., Hogg, S. and McKinney, S. (1987) Recursive partitioning: a versatile method for exploratory data analysis in biostatistics. In *Biostatistics*, eds I. B. McNeil and G. J. Umphrey, pp. 23–50. New York: Reidel.

Clark, L. A. and Pregibon, D. (1992) Tree-based models. Chapter 9 of Chambers & Hastie (1992).

Cleveland, R. B., Cleveland, W. S., McRae, J. E. and Terpenning, I. (1990) STL: A seasonal-trend decomposition procedure based on loess (with discussion). *Journal of Official Statistics* **6**, 3–73.

Cleveland, W. S. (1993) *Visualizing Data*. Summit, NJ: Hobart Press.

Cleveland, W. S., Grosse, E. and Shyu, W. M. (1992) Local regression models. Chapter 8 of Chambers & Hastie (1992).

Cochrane, D. and Orcutt, G. H. (1949) Application of least-squares regression to relationships containing autocorrelated error terms. *Journal of the American Statistical Association* **44**, 32–61.

Collett, D. (1991) *Modelling Binary Data*. London: Chapman and Hall.

Collett, D. (1994) *Modelling Survival Data in Medical Research*. London: Chapman & Hall.

Copas, J. B. (1988) Binary regression models for contaminated data (with discussion). *Journal of the Royal Statistical Society series B* **50**, 225–266.

Cox, D. R. (1972) Regression models and life-tables (with discussion). *Journal of the Royal Statistical Society B* **34**, 187–220.

Cox, D. R. and Oakes, D. (1984) *Analysis of Survival Data*. London: Chapman and Hall.

Cox, D. R. and Snell, E. J. (1989) *The Analysis of Binary Data*. Second Edition. London: Chapman and Hall.

Cox, T. F. and Cox, M. A. A. (1994) *Multidimensional Scaling*. London: Chapman & Hall.

Cressie, N. A. C. (1991) *Statistics for Spatial Data*. New York: John Wiley and Sons.

Cybenko, G. (1989) Approximation by superpositions of a sigmoidal function. *Mathematics of Controls, Signals, and Systems* **2**, 303–314.

Daniel, C. and Wood, F. S. (1980) *Fitting Equations to Data*. Second Edition. New York: John Wiley and Sons.

Darroch, J. N. and Ratcliff, D. (1972) Generalized iterative scaling for log-linear models. *Annals of Mathematical Statistics* **43**, 1470–1480.

Davidian, M. and Giltinan, D. M. (1995) *Nonlinear Models for Repeated Measurement Data*. London: Chapman & Hall.

Davies, P. L. (1993) Aspects of robust linear regression. *Annals of Statistics* **21**, 1843–1899.

Davison, A. C. and Hinkley, D. V. (1997) *Bootstrap Methods and Their Application*. Cambridge: Cambridge University Press.

Davison, A. C. and Snell, E. J. (1991) Residuals and diagnostics. Chapter 4 of Hinkley *et al.* (1991).

Dawid, A. P. (1982) The well-calibrated Bayesian (with discussion). *Journal of the American Statistical Association* **77**, 605–613.

Dawid, A. P. (1986) Probability forecasting. In *Encyclopedia of Statistical Sciences*, eds S. Kotz, N. L. Johnson and C. B. Read, volume 7, pp. 210–218. New York: John Wiley and Sons.

Deming, W. E. and Stephan, F. F. (1940) On a least-squares adjustment of a sampled frequency table when the expected marginal totals are known. *Annals of Mathematical Statistics* **11**, 427–444.

Devijver, P. A. and Kittler, J. V. (1982) *Pattern Recognition: A Statistical Approach*. Englewood Cliffs, NJ: Prentice-Hall.

Diaconis, P. and Shahshahani, M. (1984) On non-linear functions of linear combinations. *SIAM Journal of Scientific and Statistical Computing* **5**, 175–191.

Diggle, P. J. (1983) *Statistical Analysis of Spatial Point Patterns*. London: Academic Press.

Diggle, P. J. (1990) *Time Series: A Biostatistical Introduction*. Oxford: Oxford University Press.

Diggle, P. J., Liang, K.-Y. and Zeger, S. L. (1994) *Analysis of Longitudinal Data*. Oxford: Clarendon Press.

Dixon, W. J. (1960) Simplified estimation for censored normal samples. *Annals of Mathematical Statistics* **31**, 385–391.

Dodge, Y. (1985) *Analysis of Experiments With Missing Data*. New York: John Wiley and Sons.

Duda, R. O. and Hart, P. E. (1973) *Pattern Classification and Scene Analysis*. New York: John Wiley and Sons.

Efron, B. (1982) *The Jackknife, the Bootstrap, and Other Resampling Plans*. Philadelphia: Society for Industrial and Applied Mathematics.

Efron, B. and Hinkley, D. V. (1978) Assessing the accuracy of the maximum likelihood estimator: Observed versus expected Fisher information (with discussion). *Biometrika* **65**, 457–487.

Efron, B. and Tibshirani, R. (1993) *An Introduction to the Bootstrap*. New York: Chapman and Hall.

Ein-Dor, P. and Feldmesser, J. (1987) Attributes of the performance of central processing units: A relative performance prediction model. *Communications of the ACM* **30**, 308–317.

Emerson, J. D. and Hoaglin, D. C. (1983) Analysis of two-way tables by medians. In Hoaglin *et al.* (1983), pp. 165–210.

Emerson, J. D. and Wong, G. Y. (1985) Resistant non-additive fits for two-way tables. In Hoaglin *et al.* (1985), pp. 67–124.

Everitt, B. S. and Hand, D. J. (1981) *Finite Mixture Distributions*. London: Chapman and Hall.

Feigl, P. and Zelen, M. (1965) Estimation of exponential survival probabilities with concomitant information. *Biometrics* **21**, 826–838.

Firth, D. (1991) Generalized linear models. Chapter 3 of Hinkley *et al.* (1991).

Fisher, R. A. (1925) Theory of statistical estimation. *Proceedings of the Cambridge Philosophical Society* **22**, 700–725.

Fisher, R. A. (1936) The use of multiple measurements in taxonomic problems. *Annals of Eugenics (London)* **7**, 179–188.

Fisher, R. A. (1940) The precision of discriminant functions. *Annals of Eugenics (London)* **10**, 422–429.

Fisher, R. A. (1947) The analysis of covariance method for the relation between a part and the whole. *Biometrics* **3**, 65–68.

Fleming, T. R. and Harrington, D. P. (1981) A class of hypothesis tests for one and two sample censored survival data. *Communications in Statistics* **A10**(8), 763–794.

Fleming, T. R. and Harrington, D. P. (1991) *Counting Processes and Survival Analysis*. New York: John Wiley and Sons.

Freedman, D. and Diaconis, P. (1981) On the histogram as a density estimator: L_2 theory. *Zeitschrift für Wahrscheinlichkeitstheorie und verwandte Gebiete* **57**, 453–476.

Friedman, J. H. (1987) Exploratory projection pursuit. *Journal of the American Statistical Association* **82**, 249–266.

Friedman, J. H. (1991) Multivariate adaptive regression splines (with discussion). *Annals of Statistics* **19**, 1–141.

Friedman, J. H. and Rafsky, L. C. (1981) Graphics for the multivariate two-sample problem (with discussion). *Journal of the American Statistical Association* **76**, 277–295.

Friedman, J. H. and Stuetzle, W. (1981) Projection pursuit regression. *Journal of the American Statistical Association* **76**, 817–823.

Funahashi, K. (1989) On the approximate realization of continuous mappings by neural networks. *Neural Networks* **2**, 183–192.

Gabriel, K. R. (1971) The biplot graphical display of matrices with application to principal component analysis. *Biometrika* **58**, 453–467.

Gehan, E. A. (1965) A generalized Wilcoxon test for comparing arbitrarily singly-censored samples. *Biometrika* **52**, 203–223.

Geisser, S. (1993) *Predictive Inference: An Introduction*. New York: Chapman & Hall.

Golub, G. H. and Van Loan, C. F. (1989) *Matrix Computations*. Second Edition. Baltimore: Johns Hopkins University Press.

Goodall, C. (1983) M-estimators of location: An outline of the theory. In Hoaglin *et al.* (1983), pp. 339–403.

Goodman, L. A. (1978) *Analyzing Qualitative/Categorical Data: Log-Linear Models and Latent-Structure Analysis*. Cambridge, MA: Abt Books.

de Gooijer, J. G., Abraham, B., Gould, A. and Robinson, L. (1985) Methods for determining the order of an autoregressive-moving average process: A survey. *International Statistical Review* **53**, 301–329.

Gordon, A. D. (1981) *Classification: Methods for the Exploratory Analysis of Multivariate Data*. London: Chapman and Hall.

Gower, J. C. (1966) Some distance properties of latent roots and vector methods used in multivariate analysis. *Biometrika* **53**, 325–338.

Gower, J. C. and Hand, D. J. (1996) *Biplots*. London: Chapman & Hall.

Grambsch, P. and Therneau, T. M. (1994) Proportional hazards tests and diagnostics based on weighted residuals. *Biometrika* **81**, 515–526.

Green, P. J. and Silverman, B. W. (1994) *Nonparametric Regression and Generalized Linear Models. A Roughness Penalty Approach*. London: Chapman & Hall.

Haberman, S. J. (1978) *Analysis of Qualitative Data. Volume 1: Introductory Topics*. New York: Academic Press.

Haberman, S. J. (1979) *Analysis of Qualitative Data. Volume 2: New Developments*. New York: Academic Press.

Hampel, F. R., Ronchetti, E. M., Rousseeuw, P. J. and Stahel, W. A. (1986) *Robust Statistics. The Approach Based on Influence Functions*. New York: John Wiley and Sons.

Hand, D. J., Daly, F., McConway, K., Lunn, D. and Ostrowski, E. eds (1993) *A Handbook of Small Data Sets*. London: Chapman & Hall.

Härdle, W. (1991) *Smoothing Techniques with Implementation in S*. New York: Springer-Verlag.

Harrington, D. P. and Fleming, T. R. (1982) A class of rank test procedures for censored survival data. *Biometrika* **69**, 553–566.

Hartigan, J. A. (1975) *Clustering Algorithms*. New York: John Wiley and Sons.

Hartigan, J. A. (1982) Classification. In *Encyclopedia of Statistical Sciences*, eds S. Kotz, N. L. Johnson and C. B. Read, volume 2, pp. 1–10. New York: John Wiley and Sons.

Hartigan, J. A. and Wong, M. A. (1979) A K-means clustering algorithm. *Applied Statistics* **28**, 100–108.

Harvey, A. C. (1989) *Forecasting, Structural Time Series Models and the Kalman Filter*. Cambridge University Press.

Harvey, A. C. and Durbin, J. (1986) The effects of seat belt legislation on British road casualties: A case study in structural time series modelling (with discussion). *Journal of the Royal Statistical Society series A* **149**, 187–227.

Hastie, T. J. and Tibshirani, R. J. (1990) *Generalized Additive Models*. London: Chapman and Hall.

Hauck, Jr., W. W. and Donner, A. (1977) Wald's test as applied to hypotheses in logit analysis. *Journal of the American Statistical Association* **72**, 851–853.

Heiberger, R. M. (1989) *Computation for the Analysis of Designed Experiments*. New York: John Wiley and Sons.

Heiberger, R. M. and Becker, R. A. (1992) Design of an S function for robust regression using iteratively reweighted least squares. *Journal of Computational and Graphical Statistics* **1**, 181–196.

Henrichon, Jr., E. G. and Fu, K.-S. (1969) A nonparametric partitioning procedure for pattern classification. *IEEE Transactions on Computers* **18**, 614–624.

Hertz, J., Krogh, A. and Palmer, R. G. (1991) *Introduction to the Theory of Neural Computation*. Redwood City, CA: Addison-Wesley.

Hettmansperger, T. P. and Sheather, S. J. (1992) A cautionary note on the method of least median squares. *American Statistician* **46**, 79–83.

Hinkley, D. V., Reid, N. and Snell, E. J. eds (1991) *Statistical Theory and Modelling. In honour of Sir David Cox, FRS*. London: Chapman and Hall.

Hoaglin, D. C., Mosteller, F. and Tukey, J. W. eds (1983) *Understanding Robust and Exploratory Data Analysis*. New York: John Wiley and Sons.

Hoaglin, D. C., Mosteller, F. and Tukey, J. W. eds (1985) *Exploring Data Tables, Trends and Shapes*. New York: John Wiley and Sons.

Hoaglin, D. C., Mosteller, F. and Tukey, J. W. eds (1991) *Fundamentals of Exploratory Analysis of Variance*. New York: John Wiley and Sons.

Hornik, K., Stinchcombe, M. and White, H. (1989) Multilayer feedforward networks are universal approximators. *Neural Networks* **2**, 359–366.

Hosmer, D. W. and Lemeshow, S. (1989) *Applied Logistic Regression*. New York: John Wiley and Sons.

Huber, P. J. (1967) The behavior of maximum likelihood estimates under nonstandard conditions. In *Proceedings of the Fifth Berkeley Symposium on Mathematical Statistics and Probability*, eds L. M. Le Cam and J. Neyman, volume 1, pp. 221–233. Berkeley, CA: University of California Press.

Huber, P. J. (1981) *Robust Statistics*. New York: John Wiley and Sons.

Huber, P. J. (1985) Projection pursuit (with discussion). *Annals of Statistics* **13**, 435–525.

Huet, S., Bouvier, A., Gruet, M.-A. and Jolivet, E. (1996) *Statistical Tools for Nonlinear Regression. A Practical Guide with S-PLUS Examples*. New York: Springer-Verlag.

Iglewicz, B. (1983) Robust scale estimators and confidence intervals for location. In Hoaglin *et al.* (1983), pp. 405–431.

Ingrassia, S. (1992) A comparison between the simulated annealing and the EM algorithms in normal mixture decompositions. *Statistics and Computing* **2**, 203–211.

Jackson, J. E. (1991) *A User's Guide to Principal Components*. New York: John Wiley and Sons.

Jardine, N. and Sibson, R. (1971) *Mathematical Taxonomy*. London: John Wiley and Sons.

John, P. W. M. (1971) *Statistical Design and Analysis of Experiments*. New York: Macmillan.

Jolliffe, I. T. (1986) *Principal Component Analysis*. New York: Springer-Verlag.

Jones, M. C., Marron, J. S. and Sheather, S. J. (1996) A brief survey of bandwidth selection for density estimation. *Journal of the American Statistical Association* **91**, 401–407.

Jones, M. C. and Sibson, R. (1987) What is projection pursuit? (with discussion). *Journal of the Royal Statistical Society A* **150**, 1–36.

Journel, A. G. and Huijbregts, C. J. (1978) *Mining Geostatistics*. London: Academic Press.

Kalbfleisch, J. D. and Prentice, R. L. (1980) *The Statistical Analysis of Failure Time Data*. New York: John Wiley and Sons.

Kaufman, L. and Rousseeuw, P. J. (1990) *Finding Groups in Data. An Introduction to Cluster Analysis*. New York: John Wiley and Sons.

Kent, J. T., Tyler, D. E. and Vardi, Y. (1994) A curious likelihood identity for the multivariate t-distribution. *Communications in Statistics—Simulation and Computation* **23**, 441–453.

Klein, J. P. (1991) Small sample moments of some estimators of the variance of the Kaplan-Meier and Nelson-Aalen estimators. *Scandinavian Journal of Statistics* **18**, 333–340.

Knuth, D. E. (1968) *The Art of Computer Programming, Volume 1: Fundamental Algorithms*. Reading, MA: Addison-Wesley.

Kohonen, T. (1990) The self-organizing map. *Proceedings IEEE* **78**, 1464–1480.

Kohonen, T. (1995) *Self-Organizing Maps*. Berlin: Springer-Verlag.

Krzanowski, W. J. (1988) *Principles of Multivariate Analysis. A User's Perspective*. Oxford: Oxford University Press.

Laird, N. M. and Ware, J. H. (1982) Random-effects models for longitudinal data. *Biometrics* **38**, 963–974.

Lange, N., Ryan, L., Billard, L., Brillinger, D., Conquest, L. and Greenhouse, J. eds (1994) *Case Studies in Biometry*. New York: John Wiley and Sons.

Lauritzen, S. L. (1996) *Graphical Models*. Oxford: Clarendon Press.

Lawless, J. F. (1982) *Statistical Models and Methods for Lifetime Data*. New York: John Wiley and Sons.

Lawless, J. F. (1987) Negative binomial and mixed poisson regression. *Canadian Journal of Statistics* **15**, 209–225.

Lawley, D. N. and Maxwell, A. E. (1971) *Factor Analysis as a Statistical Method*. Second Edition. London: Butterworths.

Lindstrom, M. J. and Bates, D. M. (1990) Nonlinear mixed effects models for repeated measures data. *Biometrics* **46**, 673–687.

Ludbrook, J. (1994) Repeated measurements and multiple comparisons in cardiovascular research. *Cardiovascular Research* **28**, 303–311.

Macnaughton-Smith, P., Williams, W. T., Dale, M. B. and Mockett, L. G. (1964) Dissimilarity analysis: a new technique of hierarchical sub-division. *Nature* **202**, 1034–1035.

MacQueen, J. (1967) Some methods for classification and analysis of multivariate observations. in. In *Proceedings of the Fifth Berkeley Symposium on Mathematical Statistics and Probability*, eds L. M. Le Cam and J. Neyman, volume 1, pp. 281–297. Berkeley, CA: University of California Press.

Mandel, J. (1969) A method of fitting empirical surfaces to physical or chemical data. *Technometrics* **11**, 411–429.

Marazzi, A. (1993) *Algorithms, Routines and S Functions for Robust Statistics*. Pacific Grove, CA: Wadsworth and Brooks/Cole.

Mardia, K. V., Kent, J. T. and Bibby, J. M. (1979) *Multivariate Analysis*. London: Academic Press.

Matheron, G. (1973) The intrinsic random functions and their applications. *Advances in Applied Probability* **5**, 439–468.

McCullagh, P. and Nelder, J. A. (1989) *Generalized Linear Models*. Second Edition. London: Chapman and Hall.

McCulloch, W. S. and Pitts, W. (1943) A logical calculus of ideas immanent in neural activity. *Bulletin of Mathematical Biophysics* **5**, 115–133.

McLachlan, G. J. (1992) *Discriminant Analysis and Statistical Pattern Recognition*. New York: John Wiley and Sons.

McLachlan, G. J. and Basford, K. E. (1988) *Mixture Models: Inference and Applications to Clustering*. New York: Marcel Dekker.

Michie, D. (1989) Problems of computer-aided concept formation. In *Applications of Expert Systems 2*, ed. J. R. Quinlan, pp. 310–333. Glasgow: Turing Institute Press / Addison-Wesley.

Miller, R. G. (1981) *Survival Analysis*. New York: John Wiley and Sons.

Moran, M. A. and Murphy, B. J. (1979) A closer look at two alternative methods of statistical discrimination. *Applied Statistics* **28**, 223–232.

Morgan, J. N. and Messenger, R. C. (1973) THAID: *a Sequential Search Program for the Analysis of Nominal Scale Dependent Variables*. Survey Research Center, Institute for Social Research, University of Michigan.

Morgan, J. N. and Sonquist, J. A. (1963) Problems in the analysis of survey data, and a proposal. *Journal of the American Statistical Association* **58**, 415–434.

Mosteller, F. and Tukey, J. W. (1977) *Data Analysis and Regression*. Reading, MA: Addison-Wesley.

Nagelkerke, N. J. D. (1991) A note on a general definition of the coefficient of determination. *Biometrika* **78**, 691–692.

Narula, S. C. and Wellington, J. F. (1982) The minimum sum of absolute errors regression: A state of the art survey. *International Statistical Review* **50**, 317–326.

Nash, J. C. (1990) *Compact Numerical Methods for Computers. Linear Algebra and Function Minimization*. Second Edition. Bristol: Adam Hilger.

Nelson, W. D. and Hahn, G. J. (1972) Linear estimation of a regression relationship from censored data. Part 1 – simple methods and their application (with discussion). *Technometrics* **14**, 247–276.

Park, B.-U. and Turlach, B. A. (1992) Practical performance of several data-driven bandwidth selectors (with discussion). *Computational Statistics* **7**, 251–285.

Peto, R. (1972) Contribution to the discussion of the paper by D. R. Cox. *Journal of the Royal Statistical Society B* **34**, 205–207.

Peto, R. and Peto, J. (1972) Asymptotically efficient rank invariant test procedures (with discussion). *Journal of the Royal Statistical Society A* **135**, 185–206.

Plackett, R. L. (1974) *The Analysis of Categorical Data*. London: Griffin.

Prater, N. H. (1956) Estimate gasoline yields from crudes. *Petroleum Refiner* **35**, 236–238.

Priestley, M. B. (1981) *Spectral Analysis and Time Series*. London: Academic Press.

Pringle, R. M. and Rayner, A. A. (1971) *Generalized Inverse Matrices with Applications to Statistics*. London: Griffin.

Quinlan, J. R. (1979) Discovering rules by induction from large collections of examples. In *Expert Systems in the Microelectronic Age*, ed. D. Michie. Edinburgh: Edinburgh University Press.

Quinlan, J. R. (1983) Learning efficient classification procedures and their application to chess end-games. In *Machine Learning*, eds R. S. Michalski, J. G. Carbonell and T. M. Mitchell, pp. 463–482. Palo Alto: Tioga.

Quinlan, J. R. (1986) Induction of decision trees. *Machine Learning* **1**, 81–106.

Quinlan, J. R. (1993) *C4.5: Programs for Machine Learning*. San Mateo, CA: Morgan Kaufmann.

Rao, C. R. (1955) Estimation and tests of significance in factor analysis. *Psychometrika* **20**, 93–111.

Rao, C. R. (1971a) Estimation of variance and covariance components—MINQUE theory. *Journal of Multivariate Analysis* **1**, 257–275.

Rao, C. R. (1971b) Minimum variance quadratic unbiased estimation of variance components. *Journal of Multivariate Analysis* **1**, 445–456.

Rao, C. R. (1973) *Linear Statistical Inference and its Applications*. Second Edition. New York: John Wiley and Sons.

Rao, C. R. and Kleffe, J. (1988) *Estimation of Variance Components and Applications*. Amsterdam: North-Holland.

Rao, C. R. and Mitra, S. K. (1971) *Generalized Inverse of Matrices and its Applications*. New York: John Wiley and Sons.

Ratkowsky, D. A. (1983) *Nonlinear Regression Modeling: A Unified Practical Approach*. New York: Marcel Dekker.

Ratkowsky, D. A. (1990) *Handbook of Nonlinear Regression Models*. New York: Marcel Dekker.

Redner, R. A. and Walker, H. F. (1984) Mixture densities, maximum likelihood and the EM algorithm. *SIAM Review* **26**, 195–239.

Reyment, R. and Jöreskog, K. G. (1993) *Applied Factor Analysis in the Natural Sciences*. Cambridge: Cambridge University Press.

Reynolds, P. S. (1994) Time-series analyses of beaver body temperatures. Chapter 11 of Lange *et al.* (1994).

Ries, P. N. and Smith, H. (1963) The use of chi-square for preference testing in multidimensional problems. *Chemical Engineering Progress* **59**, 39–43.

Ripley, B. D. (1976) The second-order analysis of stationary point processes. *Journal of Applied Probability* **13**, 255–266.

Ripley, B. D. (1981) *Spatial Statistics*. New York: John Wiley and Sons.

Ripley, B. D. (1987) *Stochastic Simulation*. New York: John Wiley and Sons.

Ripley, B. D. (1988) *Statistical Inference for Spatial Processes*. Cambridge: Cambridge University Press.

Ripley, B. D. (1993) Statistical aspects of neural networks. In *Networks and Chaos – Statistical and Probabilistic Aspects*, eds O. E. Barndorff-Nielsen, J. L. Jensen and W. S. Kendall, pp. 40–123. London: Chapman and Hall.

Ripley, B. D. (1994) Neural networks and flexible regression and discrimination. In *Statistics and Images 2*, ed. K. V. Mardia, volume 2 of *Advances in Applied Statistics*, pp. 39–57. Abingdon: Carfax.

Ripley, B. D. (1996) *Pattern Recognition and Neural Networks*. Cambridge: Cambridge University Press.

Ripley, B. D. (1997) Classification (update). In *Encyclopedia of Statistical Sciences*, eds S. Kotz, C. B. Read and D. L. Banks, volume Update 1. New York: John Wiley and Sons, pp. 110–116.

Robinson, G. K. (1991) That BLUP is a good thing: The estimation of random effects (with discussion). *Statistical Science* **6**, 15–51.

Rosenberg, P. S. and Gail, M. H. (1991) Backcalculation of flexible linear models of the Human Immunodeficiency Virus infection curve. *Applied Statistics* **40**, 269–282.

Ross, G. J. S. (1970) The efficient use of function minimization in non-linear maximum-likelihood estimation. *Applied Statistics* **19**, 205–221.

Ross, G. J. S. (1990) *Nonlinear Estimation*. New York: Springer-Verlag.

Rousseeuw, P. J. and Leroy, A. M. (1987) *Robust Regression and Outlier Detection*. New York: John Wiley and Sons.

Rousseeuw, P. J. and van Zomeren, B. C. (1990) Unmasking multivariate outliers and leverage points (with discussion). *Journal of the American Statistical Association* **85**, 633–651.

Sammon, J. W. (1969) A non-linear mapping for data structure analysis. *IEEE Transactions on Computers* **C-18**, 401–409.

Santer, T. J. and Duffy, D. E. (1989) *The Statistical Analysis of Discrete Data*. New York: Springer-Verlag.

Scheffé, H. (1959) *The Analysis of Variance*. New York: John Wiley and Sons.

Schoenfeld, D. (1982) Partial residuals for the proportional hazards model. *Biometrika* **69**, 239–241.

Schwarz, G. (1978) Estimating the dimension of a model. *Annals of Statistics* **6**, 461–464.

Scott, D. W. (1979) On optimal and data-based histograms. *Biometrika* **66**, 605–610.

Scott, D. W. (1992) *Multivariate Density Estimation. Theory, Practice, and Visualization*. New York: John Wiley and Sons.

Seber, G. A. F. and Wild, C. J. (1989) *Nonlinear Regression*. New York: John Wiley and Sons.

Sethi, I. K. and Sarvarayudu, G. P. R. (1982) Hierarchical classifier design using mutual information. *IEEE Transactions on Pattern Analysis and Machine Intelligence* **4**, 441–445.

Sheather, S. J. and Jones, M. C. (1991) A reliable data-based bandwidth selection method for kernel density estimation. *Journal of the Royal Statistical Society B* **53**, 683–690.

Silverman, B. W. (1985) Some aspects of the spline smoothing approach to non-parametric regression curve fitting (with discussion). *Journal of the Royal Statistical Society B* **47**, 1–52.

Silverman, B. W. (1986) *Density Estimation for Statistics and Data Analysis*. London: Chapman and Hall.

Smith, G. A. and Stanley, G. (1983) Clocking g: relating intelligence and measures of timed performance. *Intelligence* **7**, 353–368.

Solomon, P. J. (1984) Effect of misspecification of regression models in the analysis of survival data. *Biometrika* **71**, 291–298.

Stace, C. (1991) *New Flora of the British Isles*. Cambridge: Cambridge University Press.

Staudte, R. G. and Sheather, S. J. (1990) *Robust Estimation and Testing*. New York: John Wiley and Sons.

Stevens, W. L. (1948) Statistical analysis of a non-orthogonal tri-factorial experiment. *Biometrika* **35**, 346–367.

Stone, M. (1974) Cross-validatory choice and assessment of statistical predictions (with discussion). *Journal of the Royal Statistical Society B* **36**, 111–147.

Street, J. O., Carroll, R. J. and Ruppert, D. (1988) A note on computing robust regression estimates via iteratively reweighted least squares. *American Statistician* **42**, 152–154.

Thisted, R. A. (1988) *Elements of Statistical Computing. Numerical Computation*. New York: Chapman and Hall.

Thomson, D. J. (1990) Time series analysis of Holocene climate data. *Philosophical Transactions of the Royal Society A* **330**, 601–616.

Tibshirani, R. (1988) Estimating transformations for regression via additivity and variance stabilization. *Journal of the American Statistical Association* **83**, 394–405.

Titterington, D. M., Smith, A. F. M. and Makov, U. E. (1985) *Statistical Analysis of Finite Mixture Distributions*. Chichester: John Wiley and Sons.

Tsiatis, A. A. (1981) A large sample study of Cox's regression model. *Annals of Statistics* **9**, 93–108.

Tukey, J. W. (1960) A survey of sampling from contaminated distributions. In *Contributions to Probability and Statistics*, eds I. Olkin, S. Ghurye, W. Hoeffding, W. Madow and H. Mann, pp. 448–485. Stanford: Stanford University Press.

Upton, G. J. G. and Fingleton, B. J. (1985) *Spatial Data Analysis by Example*. Volume 1. Chichester: John Wiley and Sons.

Velleman, P. F. and Hoaglin, D. C. (1981) *Applications, Basics, and Computing of Exploratory Data Analysis*. Boston: Duxbury.

Vinod, H. (1969) Integer programming and the theory of grouping. *Journal of the American Statistical Association* **64**, 506–517.

Wahba, G. (1990) *Spline Models for Observational Data*. Philadelphia: SIAM.

Wahba, G., Wang, Y., Gu, C., Klein, R. and Klein, B. (1995) Smoothing spline ANOVA for exponential families, with application to the Wisconsin epidemiological study of diabetic retinopathy. *Annals of Statistics* **23**, 1865–1895.

Wand, M. P. and Jones, M. C. (1995) *Kernel Smoothing*. Chapman & Hall.

Weisberg, S. (1985) *Applied Linear Regression*. Second Edition. New York: John Wiley and Sons.

Weiss, S. M. and Kapouleas, I. (1989) An empirical comparison of pattern recognition, neural nets and machine learning classification methods. In *Proc. 11th International Joint Conference on Artificial Intelligence, Detroit, 1989*, pp. 781–787.

White, H. (1982) Maximum likelihood estimation of mis-specified models. *Econometrika* **50**, 1–25.

Whittaker, J. (1990) *Graphical Models in Applied Multivariate Statistics*. Chichester: John Wiley and Sons.

Wilkinson, G. N. and Rogers, C. E. (1973) Symbolic description of factorial models for analysis of variance. *Applied Statistics* **22**, 392–399.

Williams, E. J. (1959) *Regression Analysis*. New York: John Wiley and Sons.

Wilson, S. R. (1982) Sound and exploratory data analysis. In *COMPSTAT 1982, Proceedings in Computational Statistics*, eds H. Caussinus, P. Ettinger and R. Tamassone, pp. 447–450. Vienna: Physica-Verlag.

Wood, L. A. and Martin, G. M. (1964) Compressibility of natural rubber at pressures below 500 kg/cm^2. *Journal of Research National Bureau of Standards* **68A**, 259–268.

Yates, F. (1935) Complex experiments. *Journal of the Royal Statistical Society (Supplement)* **2**, 181–247.

Yates, F. (1937) *The Design and Analysis of Factorial Experiments*. Technical communication No. 35. Harpenden, England: Imperial Bureau of Soil Science.

Index

Entries in this font are names of S objects. Page numbers in **bold** are to the most comprehensive treatment of the topic. Appendix B provides an annotated index to common functions, indexed here in *italics*.

!, 6
!=, 34
+, 5, 20
->, 21
., 196
..., 19, 29, 129, 131
.First, 64, 519
.First.lib, 518, 520
.First.local, 519
.Last, 65
.Last.value, 148
.Options, 64, 154
.Random.seed, 166
<, 34
<-, 20
<<-, 155
<=, 34
==, 34
>, xvii, 3, 5, 20, 34
>=, 34
[, 21
[[, 6, 24
#, 20
$, xvii
%*%, 36
%/%, 31
%%, 31
%o%, 54
&, 34, 114
&&, 114
^, 30
|, 34, 114
||, 114

A-estimator, 252
abbey, *see* Datasets
abbreviate, 43
abbreviation, 24, 29, 43
    of arguments, 130
abline, 12, 13, 70, **75**, *499*
abs, 31
accdeaths, *see* Datasets
accelerated life, 345, 350

ace, 335, *499*
acf, **435**, 444, *499*
acf.plot, 435
acm.ave, 468
acm.filt, 468
acm.smo, 468
acos, 31
add1, 218
additive models, 323
    fitting, 327
    generalized, 330–331
aggregate, 124, *499*
aggregate.ts, 434, *499*
agnes, 393–395
AIC, **221**, 227, 236, 425, 428, 445, 446, 455
AIDS, 294, *see also* Dataset Aids
Akaike's information criterion, *see* AIC
akima, 469
alias, 211, 213
all, 35, **114**, *499*
all.equal, 35, 114
allocated, 157
allocation, 485
alternating conditional expectation, 334
Altshuler's estimator, 346
analysis of deviance, 226
analysis of variance, 191, 193, 195, 301
    mixed models, 298
    multistratum, 301
    random effects, 297
animals, *see* Datasets
annihilator, 198
annotation, 112
anova, **193**, 195, 218, 228, 232, 244, 353, *499*
any, 35, **114**, *499*
aov, 17, **191**, 212, 216, 219, 301, *500*
aov.genyates, 212
aperm, 119
append, 49
apply, 118, 119, 121, 280, *500*
approx, 326
ar, **445**, 446, 455, 456, *500*

`ar.burg`, 456
`ar.gm`, 451
`ar.yw`, 456
`arg.dialog`, 30, *500*
`args`, 29, *500*
arguments, 29
    "...", 129
    abbreviating, 115, 130
    default values, 29
    matching, 130–131
ARIMA models, 443–448
    filtering by, 468
    fitting, 445
    forecasting, 447
    fractional, 468
    identification, 444
    regression terms, 448
    robust fitting, 451
    seasonal, 451–455
`arima.diag`, 446, 447, *500*
`arima.filt`, 468
`arima.forecast`, 447, 448, 463, *500*
`arima.fracdiff`, 468
`arima.fracdiff.sim`, 468
`arima.mle`, **445**, 446–448, 451, 463–465, *500*
`arima.sim`, **444**, 448, *500*
`arima.td`, 455
`array`, 49, *500*
arrays, 23
    arithmetic on, 52
    dimensions, 49
    dropping dimensions, 51
    index order, 50
    indexing, 49, 50
        by a matrix, 50
    ragged, 120
`arrows`, 70, 75
`as.character`, 28
`as.data.frame`, 58
`as.data.frame.array`, 108
`as.data.frame.ts`, 108
`as.matrix`, 28, 54, **58**
`as.vector`, 28, 43
`as.xxx`, 28
`asin`, 31
`assign`, **155**, 161, *500*, 518
assignment, 3, 20
asymptotic relative efficiency, 248
`atan`, 31
`attach`, 12, **46**, *500*, 518
`attr`, *500*
attributes, 22, 126
`attributes`, *500*
audit trails, 66
Australian AIDS survival, *see* Dataset `Aids`
autocorrelation function, 435

partial, 444
autocovariance function, 435
autoregressive process, 443
`avas`, 335, *500*
`axis`, 70, **80**, 88, *500*

backfitting, 327
`backsolve`, **55**, 60, *501*
bandwidth, 179, 181, 438
`banking`, 112
`barchart`, 92, **93**
barcharts, 73
`barplot`, 70, **73**
BATCH, 67, 159
Bayes risk, 486
Bayes rule, 397, 486
`bcv`, 183
binomial, 128, 224, 230
binomial process, 482
`biplot`, 388, *501*
biplots, 388
`birthwt`, *see* Datasets
bootstrap, **186**, 281
`bootstrap`, 188, 281
boundary kernels, 184
box-and-whisker plots, 93
Box-Cox transformations, 154, 215
Box-Jenkins' methodology, 443
`boxcox`, **216**, *501*
`boxplot`, 172, *501*
boxplots, 93, 99, 108, 172
`break`, 116
breakdown point, 247
`browser`, 140, 145, 146, *501*
`brush`, 13, 15, 70, 72, 76, 77, 382, *501*
`bs`, 209, **324**, 378, *501*
`budworm`, *see* Datasets
`butterfly`, 17
`bwplot`, 92, **93**, 99, 108, 172, *501*
`by`, 124, 125

C, 200
`c`, 21, 24, 25, 42, *501*
calibration plot, 495–496
`cancer.vet`, *see* Datasets
`cancor`, 399, *501*
canonical correlation, 398
canonical link, 225
canonical variates, 395
CART, 417
`cat`, **62**, 63, 117, *501*
`cats`, *see* Datasets
`cbind`, 28, **36**, 60, *501*
`cdf.compare`, 177
ceiling, 31, 34
censoring, 344
    right, 344, 354

uninformative, 356
`cex`, 75, 79
character size (on plots), 75
character strings, 21
`charmatch`, 44, 131
`chem`, *see* Datasets
chi-square test, 177
chi-squared statistic, 117, 229
`chisq.gof`, 177
`chol`, **55**, 472, *501*
`Choleski`, 60
Choleski decomposition, 55, 60, 472
`chull`, 11
`clara`, 393
`class`, 133
classification, 485–496
    non-parametric, 487
classification trees, 413, 415–418, 486, 489,
    494
clipboard, 38, 73
`close.screen`, **79**, *501*
`cloud`, 92, **106**, *501*
cluster analysis, 389–395, 485
    fuzzy, 394
    K-means, 391
    model-based, 391
`cmdscale`, 385, *501*
co-spectrum, 456
Cochrane–Orcutt scheme, 465
`coef`, **193**, 228, 274, 287, 308, *502*
`coefficients`, 193
coercion, 28
coherence, 456
    confidence intervals for, 457
`col`, 51
`color.key`, 92, 97
command-line editor, 7, 9
commands history, 66
comments, 20
communality, 406, 409
complex demodulation, 468
complex numbers, 30
`con2tr`, 77
concatenate, *see* c
condition numbers, 60
conditional execution, 113
conditioning plots, 103, 109, 110
confounding, 210
confusion matrix, 485
`contour`, 12, 70, **77**, 78, 264, 280
`contourplot`, 77, 78, 92, **96**, *502*
`contr.helmert`, 200
`contr.poly`, 200
`contr.sdif`, 202
`contr.sum`, 200
`contr.treatment`, 200
contrast matrix, 198, 200, 201

Helmert, 199
`contrasts`, 200, 376, *502*
control structures, 113–115
`coop`, *see* Datasets
`cor`, 168, *502*
`cor.test`, 176, 177
correlation, 168
    canonical, 398
correlogram, 476, 478
`correlogram`, 478
`corresp`, 400, *502*
correspondence analysis, 400
`cos`, 31
`cosh`, 31
`count.fields`, 39
`cov.mve`, **266**, 384, 398, *502*
`cov.trob`, 266
`cov.wt`, **168**, 383, 384, *502*
covariance matrix, *see* variance matrix
covariances
    exponential, 477
    Gaussian, 477
    spatial, 476–480
    spherical, 477
covariates
    time-dependent, 368, 371
Cox proportional hazards model, 356–361
`cox.zph`, 362, 363
`coxph`, 357, 359, 361, 362, 380, *502*
`cpgram`, 442, *502*
`cpus`, *see* Datasets
`crabs`, *see* Datasets
cross-periodogram, 456
cross-validation, 182, 326, 426, 493–495
`crossprod`, 54
`crosstabs`, 120
`cts`, 431
`cummax`, 31, 159
`cummin`, 31, 159
`cumprod`, 31, 159
`cumsum`, 31, 159
cumulative periodogram, 440
Cushing's syndrome, 487–493, *see also* Datasets
`Cushings`, *see* Datasets
`cut`, 27, **324**, 376, *502*
`cutree`, 390, *502*
`cv.tree`, 427, 494, *502*

D, 288
`daisy`, 393
data
    file names, 38
    reading, 36–41
        from keyboard, 38
    two-dimensional, 184
    univariate summaries, 168
    writing, 61

data frames, 27–28
    as databases, 46
    converting to/from matrices, 58
    dropping to vectors, 51
    joining, 28
    reading, 36
    writing, 61
data.class, 135
data.dump, 62, *502*
data.frame, 12, **28**, *502*
data.matrix, 54, **58**
data.restore, 62, *502*
database, 46
databases, 153
Datasets
    abbey, 171, 253
    accdeaths, 453
    Aids, 370
    animals, 125
    birthwt, 234, 330, 428
    budworm, 230
    cancer.vet, 363
    cats, 191, 284
    chem, 169, 171, 253
    coop, 297
    cpus, 413, 419, 427
    crabs, 97, 98, 124, 125, 401–404
    Cushings, 487–492
    deaths, 73, 93, 108, 432–435, 449, 455
    detergent, 239
    faithful, 169, 178, 181, 184, 287
    fdeaths, 93, 108, 433
    fgl, 100, 101, 387, 398, 423, 426–428,
        493, 496
    gehan, 344, 347–351, 354
    geyser, 169, 291
    heart, 369, 379
    hills, 13, 15, 16, 76, 205, 264
    iris, 108, 118, 119, 383, 385, 396, 421
    leuk, 344, 352, 357
    lh, 431
    mammals, 125, 335, 336
    mcycle, 324
    mdeaths, 93, 108, 433
    menarche, 234
    michelson, 16
    motors, 345, 355, 360
    mydata, 21–23, 52, 58
    nottem, 458, 459
    npr1, 469
    oats, 300, 309, 311
    painters, 27, 61, 427
    petrol, 304, 321
    phones, 256
    pines, 482
    quine, 120, 214, 243, 244
    Rabbit, 315, 322

    rock, 328–330, 332–334, 336, 340, 341
    shkap, 469
    shoes, 172–174, 176, 189
    shuttle, 414, 424
    Sitka, 312, 314, 322
    stack.loss, 262
    stack.x, 262
    state.x77, 389
    stormer, 94, 278, 279, 283
    swiss.df, 86, 103
    swiss.fertility, 170
    swiss.x, 390, 394, 399, 406
    topo, 77, 78, 96, 111, 469, 470, 472, 476
    tperm, 169, 179
    wtloss, 208, 267, 273, 277, 284
dates, 379
dbwrite, 161
deaths, *see* Datasets
debugger, 140, **145**, *503*
debugging, 139–145
decibels, 439
decision trees, 413
decomposition
    Choleski, 55, 60, 472
    eigen-, 55, 60
    LU, 60
    QR, 56, 60, 472
    Schur, 60
    singular value, 56, 60
degrees of freedom, equivalent, 326, 333,
        438
Delaunay triangulation, 474, 519
demod, 468
dendrogram, 390
density, **179**, 181, *503*
density estimation, 177–186
    bandwidth selection, 182
    end effects, 184
    kernel, 179
    MISE, 182
    parametric, 289
    WARP, 179, 183
densityplot, 92, **179**, 181, *503*
deparse, 128, *503*
deriv, **271**, 282, 288, 319
deriv3, **282**, 288, *503*
design, 213
designed experiments, 209
    generating, 212–214
det, 60
detach, 13, **46**, *503*
detergent, *see* Datasets
determinant, 57, 60
dev.ask, 71, *503*
dev.copy, 72, *503*
dev.cur, 71, 130, *503*
dev.list, 71, *503*

dev.next, 71
dev.off, 71, *503*
dev.prev, 71
dev.print, 72, *503*
dev.set, 71, *503*
deviance, 226, 274, 279, 417, 418, 446
    scaled, 226
deviance, **193**, 228, *503*
device
    graphics, *see* graphics device
Devices, 71
dget, 147
diag, 50, **54**, *503*
Diagonal, 59
diagonal (of) matrix, 54
diana, 393–395
dictionary, 46
diff, **434**, 451, *504*
digamma function, 151
dim, **23**, 49, 51, 52, *504*
dimnames, **50**, 58, *504*
Dirichlet tessellation, 473, 519
discriminant analysis, 395–398, 402–404, 485
    linear, 395, 486
    quadratic, 397, 486
dissimilarities, 385, 389
    ultrametric, 390
dissimilarity matrix, 393
dist, **385**, 389, *504*
distance methods, 385
distributions
    Gumbel, 438
    log-logistic, 343, 350
    negative binomial, 242
    Poisson
        zero-truncated, 116
    Rayleigh, 352
    table of, 164
    univariate, 163
    Weibull, 343, 350
dmvnorm, 164
dnorm, 288
do.call, **123**, 124, 181, 386
dos, 149
dos.time, 149
dotchart, 70
dotplot, 92, **100**, *504*
dput, 147
drop, 51
drop1, 218, 220
dummy.coef, 198
dump, 20, **62**, 140, *504*
dump.calls, 146
dump.frames, 146
dumpdata, 62
duplicated, 32, 48

ed, 65, 140

editing, 140
    command-line, 7, 9
eigen, **55**, 60, *504*
eigenvalues, 55
entropy, 418
environment, 1
    customizing, 63–65
EPSF, 73
eqscplot, 76, *504*
equal.count, 92, **103**
Error, **301**, *504*
errors
    finding, 139–145
eruptions, *see* Dataset faithful
estimation
    generalized least squares, 472
    least squares, 269
    maximum likelihood, 116, 269, 286, 339
    pseudo-likelihood, 483
eval, 153
evaluation, lazy, 127
execution, conditional, 113
exists, 122, 155, *504*
exit actions, 131
exp, 31
expand.grid, **77**, 123, 212, 239, 264, 280, 473, *504*
expcov, 478
expected survival, 379–380
experiments
    designed, 209
    generating designs, 212–214
    split-plot, 300
expn, 271
expression, 3, 20
    arithmetical, 30
    logical, 34
expression, 153

F, 21, 34
fac.design, 212, 214
faces, 70
factanal, **406**, 409, *504*, 507, 511
factor, 25, *504*
factor analysis, 405–412
    rotation, 405, 409
    scores, 411
factorial designs, 209
    fractional, 212
factors, 25–27
    functions of, 120
    matrix subscripts from, 240
    ordered, 26, 103
    response, 238
    splitting by, *see* split
    stimulus, 238
faithful, *see* Datasets

family, 228, 233, 242
fanny, 393, 394
fdeaths, *see* Datasets
fft, 442, *504*
fgl, *see* Datasets
fig, 83, 84
figures
    multiple, 78
    in Trellis, 105
file.exists, 147, 149
filter, 467, 468
find, 45
finishing, 5, 9
Fisher information, 285
Fisher scoring, 226
fitted, 13, **193**, 230, 274, 287, 299, 309, *505*
fitted values, 193
    from random effect models, 299
fix, 65, **140**, 155, *505*
fixed effects, 304, 314
fixed.effects, 308, 318
Fleming-Harrington estimator, 346
floor, 31
For, 116, **159**
for, 115–117, 158
forecasting, 447
forensic glass, 100, 493–496, *see also* Dataset fgl
format, 63, *505*
formula, 213
Fourier frequency, 438
fpl, 317
fractionate, 214
fractions, 201
frame, 70, **78**
frames
    0 and 1, 154, 156
    evaluation, 153
frequency.polygon, 177
function, 125
functions
    anonymous, 54
    calling, 29
    debugging, 139–145
    editing, 140
    finishing, 64
    generic, 133
    group method, 139
    method, 134
    of a matrix, 55
    recursive, 149
    replacement, 22
    startup, 64
    value of, 126
    writing, 3, 125

gam, 323, **327**, 333, *505*

gam.plot, 336
Gamma, 224
gamma, 31
gamma function, 31, 151
    log, *see* lgamma
gaucov, 478
gaussian, 224, 229, 230
gehan, 359, *see* Datasets
generalized additive models, 330–331
generalized inverse, 56
generalized least squares, 472, 475, 476, 480
generalized linear models, 223–245, 295
    algorithms, 225
    analysis of deviance, 226
    family, 224
    linear predictor, 223
    scale parameter, 223
    score function, 224
    variance function, 225
generic functions, 133
geometric anisotropy, 476
get, **46**, 154, 155, 161, *505*
geyser, *see* Datasets
Gini index, 418
ginv, 56
glm, **227**, 228–230, 242, 244, 259, 293, 486, *505*
glm.links, 243
glm.nb, 244
glm.variances, 243
graphical design, 69
graphics
    object-oriented editable, 108–112
graphics device, 69
    multiple, 71
graphics.off, 71, *505*
graphsheet, 71, 77, 108
Greenwood's formula, 346
grep, 43, 45
grids, 81
group method function, 139
growth curve
    exponential, 314
    logistic, 316

hardcopy, 72
hat matrix, 204
Hauck–Donner phenomenon, 237
hazard function, 343
    cumulative, 343, 357
hclust, 390, *505*
heart, *see* Datasets
heart transplants, *see* Dataset heart
Helmert contrast matrices, 199
help

on-line, 2, 6, 10
     preparing, 145
help, 6, 10, *505*
help.off, 7, *505*
help.start, 6, *505*
Hessian, 291, 340
Heywood case, 406
hills, *see* Datasets
hist, 11, 30, 70, **168**, *505*
hist.FD, 169
hist.scott, 169
hist2d, 12, 70, **184**
histogram, 92, 168, **169**, *505*
histograms, 168–170
     average shifted *(ASH)*, 178
     two-dimensional, 184
history, 66, *505*
histplot, 170, *505*
hpgl, 71
hplj, 71
huber, 253
Huber's proposal 2, 251
hubers, 253

I(), 28
identify, 13, 70, **80**, 90, *505*
Identity, 59
if, 113, 114
ifelse, **114**, 158, *505*
Im, 30
image, 12, 70, **77**
information
     Fisher, 285
     observed, 285, 293
inner product, 36
input, 36, 61
     from a file, 61
inspect, 140–142, 144, 145, *505*
installation of software, 497
integrate, 150, 151, *506*
integration
     numerical, 150
interaction.plot, 70
interp, 474, *506*
interpolation, 469–476
interrupt, 5, 9
inverse.gaussian, 224
invisible, 136, 156
Iris, key to British species, 414
iris, *see* Datasets
iris4d, 71
is.na, 35, *506*
is.random, 298
is.xxx, 28
isoMDS, 387, *506*
iterative proportional scaling, 225, 240
its, 431

Jaccard coefficient, 389, 393
jackknife, 188
julian, 379

K function, 480
K-means clustering, 391
k-medoids, 393
Kaplan-Meier estimator, 346
Kaver, 483, *506*
kde2d, 184, *506*
keep.order, 240
Kenvl, 483, *506*
kernel density estimation
     bandwidth, 181
     bandwidth selection, 182
     boundary kernels, 184
kernel density estimator, 179
kernel smoother, 326
key, 92, **107**
Kfn, 483, *506*
kmeans, **391**, *506*
knots, 324
kriging, 475–476
     universal, 475
kruskal.test, 177
ks.gof, 177
ksmooth, **327**, 467, *506*

l1fit, 257, *506*
lag, 434, *506*
LAPACK library, 58
lapply, 118, **121**, 158, *506*
last.dump, 145, 146
layout
     crossed and nested, 202
     one-way, 197, 198
     two-way, 202
     unbalanced, 214
lazy evaluation, 127
lda, **396**, 398, 487, *506*
leaps, 221
learning vector quantization, 487
legend, 81, 82, 107, 111
legend, 70, 81, 107
length, 22, *506*
letters, 22, 48
leuk, *see* Datasets
levelplot, 92, 96, 470, *506*
levels, 27
levels of a factor, 25
leverage, 204, 229
lgamma, 31
lh, *see* Datasets
libraries, 517
     creating, 520
     private, 518
     sources of, 519

library
  ash, 184
  bootstrap, 188
  chron, 379
  class, 487, 498
  cluster, 393, 394
  date, 379
  delaunay, 474
  haerdle, 170, 327
  helpfix, 10, 497
  KernSmooth, 186, 327
  MASS, 11, 45, 56, 120, 151, 170, 183,
      191, 201, 214, 216, 228, 233, 234,
      237, 244, 273, 274, 276, 282–284,
      288, 297, 350, 497
  Matrix, 58
  mda, 341
  nlme, 272, 273, 304, 317, 465
  nlmedata, 273, 317
  nls2, 267
  nnet, 242, 340, 486, 498
  postscriptfonts, 82
  pspline, 327
  robeth, 247
  spatial, 158, 469, 498
  survival4, 343
  treefix, 46, 419, 424, 425, 497
  trellis, 89
library, *506*, **517**, 518
likelihood
  partial, 356
  profile, 215, 216
  pseudo-, 483
  quasi-, 226
limits.bca, 189
limits.emp, 189
linear regression, 488
linear discriminant analysis, 395, 486
  debiased, 487
  predictive, 486
linear discriminants, 395
linear equations
  solution of, 55
linear mixed effects models, 304–313
linear models, 4, 191–304
  diagnostics, 204–208
  formulae, 196–204
  leverage, 204
  prediction, 208
  rank-deficient, 216
  selection, 217–222
  transformations, 215
  weighted, 192
linear predictor, 223
lines
  staircase, 80
lines, 12, 70, **75**, 80, 349, *507*

link function, 223
list, 24, *507*
lists, 3, 24–25
  converting to vectors, 24
  functions of, 120
lm, 12, 191, **192**, 216, 219, 228, 260, 327,
    329, 470, *507*
lm.influence, **205**, *507*
lme, **306**, 309, 314, 315, 465, *507*
lmsreg, **262**, 263, *507*
lo, **327**, 333, *507*
loadings, 405–407, 409, 410
  rotation of, 409
loadings, **383**, 407, 410, *507*
location.m, **252**, 253, *507*
locator, 16, 70, 79, **80**, *507*
loess, 12, 110, 263, 264, 323, 326, 361,
    469, **473**, 496, *507*
loess.smooth, 326
log, 31
log-logistic distribution, 343
log10, 31
logistic regression, 293, 330, 486, 493
  multiple, 486
logistic response function, 316
loglin, 240, *507*
logtrans, 217
longitudinal data, 304
loops, 115–118, 158
lower case, 148
lower.tri, 51, 159
lowess, **326**, 361, 468, *507*
ltsreg, 13, 256, **262**, 263, *507*
lu, 60
lung cancer data, *see* Dataset cancer.vet

M-estimation, 250
  regression, 257
machine learning, 413
mad, 68, **249**, 252, 253, *507*
mahalanobis, 397, *507*
Mahalanobis distance, 397
mai, 83, 84
make.call, 288
make.family, 242
make.fields, 40
make.groups, 108
Mallow's C_p, 425
mammals, *see* Datasets
MANOVA, 381
manova, 381
mar, 83, 84
masked, 46, *507*
mat2tr, 77
match, 44, 131
match.arg, 131
matplot, 70, **87**

Matrix, 58
matrix, 3, 23, 58–60
    arithmetic on, 52
    converting to/from data frame, 58
    determinant, 57
    diagonal, 54
    dropping dimensions, 51
    eigenvalues of, 55
    elementary operations, 35
    functions of, 55
    generalized inverse of, 56
    index order, 50
    indexing, 50
      by a matrix, 50
    inversion, 55
    joining, 36
    multiplication, 36
    number of cols, 35
    number of rows, 35
    operators, 54
    outer product, 54
    QR decomposition, 56
    singular value decomposition, 56
    subscripting, 59
    trace, 57
    transpose, 35
    writing, 61
matrix, 23, **49**, 50, *507*
Matrix library, 58–60
    Choleski decomposition, 60
matrix subscripts, 50
    for factors, 240
Matrix.class, 59, 60
max, 31
maximum likelihood estimation, 116, 269, 286
    residual, 299
McCulloch–Pitts model, 339
mclass, 391, *508*
mclust, 391, 392, *508*
mcycle, *see* Datasets
mdeaths, *see* Datasets
mean, 168, 248
    trimmed, 249
mean, **32**, 118, 249, 253, *508*
median, 168, 172, 187, 248, 249
    distribution of, 187
    polish, 254, 255
median, 53, **249**, 253, *508*
memory
    limited, 157–162
    usage, 157
memory.size, 157
menarche, *see* Datasets
merge, 124, 125
method dispatch, 134
method functions, 134

mex, 76, 79, 83, 84
mfcol, 78
mfrow, 78, 84
michelson, *see* Datasets
Michelson experiment, 16
min, 31
minimization, 286–295
minimum-spanning trees, 387
MINQUE, 299
misclass.tree, 427, *508*
missing, 129, *508*
missing values, 35, 168, 192, 419, *see also*
    NA
    in model formulae, 42
mixed effects models
    linear, 304–313
    non-linear, 314–322
mixture models, 287
mkh, 75
mode, 22
model formulae, 41–42, 229
    for linear models, 196–204
    for lme, 306
    for multistratum models, 301
    for nlme, 314
    for non-linear regressions, 269
    for princomp, 384
    for trees, 420
    for Trellis, 91
    in survival analysis, 344, 345, 357, 379
    order of terms in, 240
model matrix, 196–198, 202, 203, 275
model.frame.tree, 494
model.matrix, 275, 332
model.tables, 211, 302, 303
mona, 393, 394
monthplot, 449
motif, 71–73, 77, 91
motors, *see* Datasets
moving average process, 443
mreloc, 391, 392, *508*
ms, **286**, 287, 289, 294, *508*, 510
MS-DOS, 3, 10, 41, 44, 148, 149
mstree, 387, *508*
mtext, 70, **81**
multi-dimensional scaling, 385, 386
multinom, 242, 486, *508*
multistratum models, 299–304, 306, 309
multivariate analysis, 381–412
mvrnorm, 164

NA, **35**, 37, 53
na.action, 41, 42
na.fail, 35, 42
na.gam.replace, 35
na.omit, 35
names, **23**, 24, 48, *508*

nchar, 42
nclass.FD, 169
nclass.scott, 169
ncol, 35, 51
nearest neighbour classifier, 487, 494
neg.bin, 228, **243**, 244
negative binomial distribution, 242
negative.binomial, 244
neural networks, 337–341, 490–492
next, 116
NextMethod, 136, 138
nlme, **314**, 317–319, *508*
nlminb, **292**, 294, *508*
nlregb, **284**, 286, *508*
nls, 228, **269**, 272, 274, 286, 317–319, *508*, 510
nlsList, 317
nnet, 340, *508*
nnet.Hess, 340, *509*
nnls.fit, 284
non-linear mixed effects models, 314–322
non-linear models, 267–286
    curvature diagnostics, 282
    fitting, 268
        with first derivatives, 271
        linear parameters, 275
        method functions, 274
        objects, 274
        self-starting, 272–273
norm, 60
nottem, *see* Datasets
npr1, *see* Datasets
nroff, 146
nrow, 35, 51
ns, 209, **324**, *509*
ntrellis, 89
nugget effect, 476, 480
Null, 202
numeric, **30**, *509*

oa.design, 214
oats, *see* Datasets
objdiff, 148
object browser, 109
object-oriented programming, 133
objects
    editing, 140
    finding, 44–47
    masked, 46
    names, 19
    reading, 62
    transferring, 62
    writing, 62
objects, 11, **45**, 155, *509*
oblimin rotation, 410
offset, **229**, 349
oma, 84

omd, 84
omi, 84
on.exit, 132, 148
one-way layout, 197, 198, 202
openlook, 71–73, 77
operating system, calls to, 147
operator
    arithmetic, 30
    assignment, 3, 20
    colon, 21, 32
    exponentiation, 30
    integer divide, 31
    logical, 22, 34
    matrix, 54
    matrix multiplication, 36
    modulo, 31
    outer product, 54
    precedence, 32
    remainder, 31
optimize, 286, *509*
options, xvii, 62, 63, **64**, 132, 140, 154, *509*
order, 53, *509*
ordered, **26**, 27, *509*
outer, **54**, 158, 159, 280, *509*
outliers, 247, 299, 459
    down-weighting, 248
output, 61
    to a file, 61

painters, *see* Datasets
pairs, 70, 112, 237, 283
pam, 393
panel function, 93
par, 70, **82**, *509*
parameters
    figure, 82
    graphics, 82
    high-level, 85
    layout, 82, 83
    linear, 268
partial likelihood, 356
partition.tree, 422
partitioning methods, 414
paste, 29, **42**, 62, *509*
pattern recognition, 485–496
PCA, *see* principal components analysis
pdf.graph, 71
periodogram, 437
    cross-, 456
    cumulative, 440–442
permeability, 328, 469
permutation tests, 189
persp, 12, 70, **77**, 333
persp.setup, 70
perspp, 70
petrol, *see* Datasets
phones, *see* Datasets

pi, 20
pie, 70
piechart, 92
pines, *see* Datasets
plclust, **390**, 395, *509*
plot, 70, **73**, 80, 129, 130, 193, 213, 322, 345, 419, 433, *509*
plot.gam, 328, 329
plot.survfit, 349, 350, 354, 376
plots
    adding axes, 267
    anatomy of, 81
    axes, 80
    basic, 73
    biplot, 388
    boxplot, 172
    common axes, 84
    contour, 77, 279
    dynamic, 76, 382
    enhancing, 78
    equal scales, 76
    greyscale, 77
    histogram, 168–170
    interacting with, 80
    labels, 128
    legends, 81
    line, 74
    multiple, 78
    multivariate, 76
    perspective, 77
    QQ, 86, **164**
        with envelope, 86
    stem-and-leaf, 170–172
    surface, 77
    symbols, 75
    title, 80
pltree, 395
plug-in rule, 486
pmatch, 44, 115, 131
pmax, 32
pmin, 32
pmvnorm, 164
pnorm, 288
point processes
    binomial, 482
    Poisson, 482
    spatial, 480–484
    SSI, 484
    Strauss, 483
points
    colour, 80
    highlighting, 76
    identifying, 80
points, 70, **75**, 80, *509*
poisson, 224
Poisson process, 482
Poisson regression model, 294

polish, median, 254, 255
poly, 324, **470**, *509*
polygon
    frequency, 177
polygon, 11, 70, *509*
polynomials, 208, 470
    orthogonal, 199, 470
polyroot, 444, *509*
porosity, 469
portmanteau test, 446
post.tree, 421, *510*
postscript, 64, 71, **73**, 80, 82, 88–90, *510*
ppinit, 483, *510*
pplik, 483, *510*
ppoints, 165
ppreg, 332, *510*
ppregion, 483, *510*
prcomp, 383
predict, **193**, 208, 209, 228, 274, 340, 424, *510*
predict.factanal, 412
predict.gam, 209, 324, 470
predict.lm, 324
predict.lme, 309
predict.tree, 424
predictive rule, 486
principal component analysis, 382–385, 405
    loadings, 384
    scores, 384
principal coordinate analysis, 385
princomp, **383**, 384, 409, 507, *510*, 511
print, 20, 25, 117, 132, 136, 193, 228, 259, 274, 287, 340, 345, 419, *510*
print.summary.lm, 63
print.trellis, 90, **105**, *510*
printgraph, 72
printing, 62
    compact, 127
    digits, 63
    precision, 63
prmat, 476, *510*
probability distributions, 163
probability forecasts, 495
probit analysis, 233
proc.time, 148, 149
prod, 31
profile, 237, 274, **276**, 283, 287, *510*
programming, object-oriented, 133
proj, 302
projection pursuit, 385
    regression, 331–334
promax rotation, 410
prompt, 146, 147, 520
prompt.screen, 79
proportional hazards, 350, 356–363
    tests of, 362

prune.misclass, 428, *510*
prune.tree, **425**, 428, *510*
ps.options, 64, 73, 88
pseudo-likelihood, 483
pseudo-random number generator, 166
Psim, 483, *510*
pty, 78
pyrethroid, 230

q, **5**, 9, 129, *510*
qda, 397, 487, *511*
QQ plots, 165
qqline, 13, **165**, *511*
qqnorm, 13, 70, **165**, 166, *511*
qqplot, 70, **164**, *511*
qr, **56**, 60, 217, *511*
qr.coef, 57, 511
qr.fitted, 57, 511
qr.Q, 511
qr.R, 511
qr.resid, 217, 511
qr.X, 511
quadratic discriminant analysis, 397, 486
    debiased, 487
    predictive, 486
quadrature spectrum, 456
quantile, 53, **164**, 168, *511*
quantiles, 163–166, 168
quasi, 226–228, 230
quasi-likelihood, 226
quine, *see* Datasets
quitting, 5, 9

Rabbit, *see* Datasets
random effects, 297, 304, 314
random numbers, 166
random variables, generating, 166
range, 32
rank, 53
raov, **297**, 298, *511*
Rayleigh distribution, 352
rbind, 28, **36**, 60, *511*
rbiwt, 259
rcond, 60
Re, 30
read.table, 28, 36, **37**, 61, *511*
Recall, 149
recursion, 149
reference manuals, 2
regression, *see also* linear models and non-
        linear models
    diagnostics, 204–208
    L_1, 257
    logistic, 293, 330
    M-estimation, 257
    non-linear, 267–286
    non-negative coefficients, 284

polynomial, 208, 267
projection-pursuit, 331–334
resistant, 261
robust, 256
trees, 413, 418
weighted, 192
with autocorrelated errors, 463–467
regular sequences, 32
regularization, 339
relative efficiency, 248
REML, 299, 307, 315
remove, 47, 155, 161
reorder.factor, 103
rep, 33, *511*
repeat, 116, 158
repeated measures, 304, 312
repeating sequences, 33
replace, 49
replacement function, 22
replications, 212, 214
resid, 13, **193**, 228, 229, 287, *511*
residuals, 204
    deviance, 228, 362
    from random effect models, 299
    jackknifed, 205
    martingale, 361
    Pearson, 228
    Schoenfeld, 362
    score, 362
    standardized, 204
    studentized, 204
    working, 228
residuals, **193**, 274, 362
resistant methods, 247
response transformation models, 334–337
restart, 132
return, 126
rev, 32
Ripley's K function, 480
rlm, **260**, 263, *511*
rm, 11, **47**, *511*
rms.curv, 282
rmv, 73
rmvnorm, 164
rnorm, 11
robust methods, 247
rock, *see* Datasets
rotate, 409, *511*
rotation of loadings, 405, 409
    oblimin, 410
    promax, 410
    varimax, 409
round, 31, 63
rounding, 31, 63
row, 51
row.names, 27
RowPermutation, 59

Rows, 94
rpois, 129
rreg, **259**, 260, *511*
rts, 431, *511*
rug, 181, **336**, 378, *512*
rule
    Bayes, 397, 416, 486
    Doane's, 168
    recycling, 30
    Scott's, 177
    Sturges', 181

s, **327**, *512*
S+SPATIALSTATS, 467
sabl, 449
sammon, **386**, 402, *512*
Sammon mapping, 386
sample, 166, *512*
sapply, 118, **121**, 158, 159, *512*
scalars, 21
scale, 102, 119, *512*
scale.a, 252, 253
scale.tau, 252, 253, *512*
scaling, multi-dimensional, 385
scan, 36, **38**, *512*
scatter.smooth, 70, **326**, 361
scatterplot, 74
    matrix, 76
    smoothers, 324–327
schur, 60
screen, 79, *512*
screens, 78
screeplot, 384
se.contrast, 211
search, 16, **44**, *512*
search path, 44
seasonality, 449–455
    decompositions, 449
    trading days, 455
segments, 70, **88**, *512*
semat, 476, *512*
seq, 12, **33**, *512*
sequential spatial inhibition *(SSI)*, 484
Session options, 64
set.seed, 166, *512*
shingle, 91, 103, 109
shkap, *see* Datasets
shoes, *see* Datasets
show.settings, 89
shuttle, *see* Datasets
sign, 31
signif, 63
similarity coefficient, 389
simple matching coefficient, 389
simulation, 166
sin, 31
singular value decomposition, 56, 60

sinh, 31
sink, 62, *512*
Sitka, *see* Datasets
skip-layer connections, 338
slice.index, 51, 123
slm, 467
smooth.spline, 326
smoothers, 324–327
    kernel, 326
    loess, 326
    spline, 326
    super-smoother supsmu, 327
software installation, 497
solution locus, 269
solve, **55**, 60, 455, *512*
solve.qr, 512
solve.upper, 55, 501, 512
sort, 32, **52**, *512*
sort.list, 53, *513*
sorting, 32, **52**
    partial, 52
source, **61**, 62, 140, 155, 159, *513*
spatial, 469
spatial correlation, 472
spec.ar, 448, *513*
spec.pgram, **439**, 441, 458, *513*
spec.taper, 441, *513*
spectral analysis, 437–442
    multiple series, 456
    via AR processes, 448
spectral density, 437
spectrum, 437
    co-, 456
    quadrature, 456
spectrum, **439**, 441, 448, *513*
sphercov, 478
spin, 70, 72, **77**, *513*
spline, 12, 321, 326
splines, 321, 324, 378
    interpolating, 326
    smoothing, 326
split, 459, *513*
split-plot experiments, 300, 304
split.screen, **78**, 84, *513*
splom, 13, **91**, 92, 94, 97, 98, 106, 112, 382, *513*
sqrt, 31
SSI, 484, *513*
stack.loss, *see* Datasets
stack.x, *see* Datasets
standardized residuals, 204
Stanford heart transplants, *see* Dataset heart
stars, 70
starting
    under Unix, 4
    under Windows, 8
state.name, 43

`state.x77`, *see* Datasets
statistical tables, 164
statistics
    classical, 172–177
`statlib`, 82, 184, 186, 188, 247, 272, 282,
      304, 327, 341, 343, 379, 385, 463,
      474
`stdres`, 205
`stem`, 170, *513*
stem-and-leaf plots, 170–172
`step`, **220**, 222, 235, 237, *513*
`stepAIC`, 222, **237**, 244, 366, 367, *513*
`stepfun`, 80, *513*
`stepwise`, 221
Stirling's formula, 151
`stl`, **449**, 459, *513*
`stop`, 129, *513*
`stormer`, *see* Datasets
strata, 315, 344
`Strauss`, 483, *514*
Strauss process, 483
strings, character, 21
`stripplot`, 92, **99**, 100, 101, *514*
`structure`, 135
studentized residuals, 204
`studres`, 205
Sturges' formula, 168
`subplot`, 79
`subset`, 41, 42
`substitute`, **128**, 243, *514*
`substring`, 43
`sum`, 31
summaries, univariate, 168
`summary`, 20, 168, 193, 228, 244, 259, 263,
      274, 287, 291, 301, 340, 345, 419,
      514
`summary.coxph`, 359
`summary.gam`, 328
`summary.lm`, 211
`supsmu`, 327, *514*
`surf.gls`, 478, *514*
`surf.ls`, 470, *514*
`Surv`, **344**, 368, *514*
`survdiff`, 349, *514*
`survexp`, 379, *514*
`survfit`, **345**, 347, 359, 361, *514*
survival
    expected rates, 379–380
    parametric models, 350–356
    proportional hazards model, 356–361
    residuals, 361
survivor function, 343
    Altshuler estimator, 346
    confidence intervals, 347
    estimators, 345–346
    Fleming-Harrington estimator, 346
    Kaplan-Meier estimator, 346

tests for equality, 349
`survreg`, **351**, 352, *514*
`svd`, **56**, 60, *514*
`sweep`, 119, *514*
`swiss`, *see* Datasets
`switch`, 113–115
symbol size, 75
symbolic differentiation, 271, 288
symbols
    plotting, 75
`symbols`, 70, 75, *514*
`synchronize`, 47, *514*
`sys.parent`, 156
`system`, 147, 148
system dependencies, 2, 498

`T`, 21, 34
`t`, **35**, 54, 60, *515*
t-test, 3, 126, 127, 189
`t.test`, 29, 127, 177, 189
`table`, **26**, 120, *515*
`tan`, 31
`tanh`, 31
tapering, 441
`tapply`, 118, **120**, 121, 124, 125, *515*
`tempfile`, 147
`terms`, 240
tessellation, Dirichlet, 473
tests
    chi-square, 177
    classical, 176
    goodness-of-fit, 177
    Kolmogorov-Smirnov, 177
    permutation, 189
    Peto-Peto, 349
    portmanteau, 446
`text`, 70, **75**, 420, *515*
tiles, 473
time series
    complex demodulation, 468
    differencing, 434
    filtering, 467
    identification, 444
    multiple, 433, 435, 455–458
    seasonality, 449–455
    second-order summaries, 434–442
    spectral analysis, 437–442
    windows on, 434
`title`, 70, **80**, *515*
titles, 74, 80, 107, 111
`topo`, *see* Datasets
`tperm`, *see* Datasets
`tprint`, 146
`trace`, 140, 146, 156
trace of a matrix, 57
`traceback`, 141, 144, 145, *515*
trading days, 455

translation, 148
transpose of a matrix, 35
tree, 25, 419, 424, *515*
tree.control, 424
trees
    classification, 415–418
    minimum-spanning, 387
    pruning, 425
        cost-complexity, 425
        error-rate, 425, 428
    regression, 418
    snipping, 422
trellis, 89
Trellis graphics, 88–108
    3D plots, 107
    aspect ratio, 94, 96, 106
    basic plots, 91–105
    colour key, 97
    fine control, 106
    keys, 96, 98, 100, 107
    layout, 102
    multiple displays per page, 105
    panel functions, 93
    prepanel functions, 106, 165
    QQ plot, 165
    scatterplot matrices, 91
    strip labels, 100
    trellises of plots, 97
trellis.3d.args, 106
trellis.args, 106
trellis.device, 89
trellis.par.get, 92, **94**, 107
trellis.par.set, 92, **94**
trend surfaces, 470–472
    local, 473
triangulation, Delaunay, 473
trmat, 470, *515*
true.file.name, 147
truehist, **170**, *515*
trunc, 31
ts.intersect, 433, 434, *515*
ts.lines, 433
ts.plot, 433, *515*
ts.points, 433
ts.union, 433, 434, *515*
tspar, 431
Ttest, 135–137
ttest, 126–131, 135, 141, 142
two-way layout, 202, 254
twoway, 255, *515*

ucv, 183
ultrametric dissimilarity, 389
unclass, 134
unique, 32, 48
uniqueness, 406
uniroot, **286**, 484, *515*

Unix, vii, 2, 4–8, 11, 19, 36, 38, 41, 43–45,
    62, 64–67, 71–73, 77, 82, 88–90,
    140, 145, 148, 159, 160, 273, 304,
    317, 497, 498, 505, 519, 520
unix, 147, 148
unix.time, 148, 149
unlink, 147
unlist, 24, 64, 127, 128
unpack, 60
update, 17, 42, 193, 196, 203, 218, 220,
    515
upper case, 148
usa, 17
UseMethod, 136
usr, 84

VA lung cancer data, *see* Dataset cancer.vet
var, 6, **32**, 141, 168, *515*
var.test, 177, 194
varcomp, **297**, 298, 303, *515*
variance components, 297, 302
variance matrix, 168, 228
    between-class, 395
    estimated, *see* vcov
    robust estimation, 266
    within-class, 395
varimax rotation, 409
variogram, 477, 478
variogram, 478
vcov, **228**, 274, *515*
vcov.nlminb, 293
vcov.nlreg, 286
vecnorm, 55
vectorization, 150, 158
vectorized calculations, 117
vectors, 21
    character, 42
        string lengths, 42
        substrings, 43
    indexing, 21, 22, **47**
    norm, 55
version 4.0, vii, 2, 9, 66, 67, 69, 72, 75, 89,
    108, 109, 111, 164, 168, 188, 262,
    274, 281, 520
vi, 140

waiting, *see* Dataset faithful
warning, 128, *516*
Weibull distribution, 343
while, 116, 117, 158
wilcox.test, 177
win.colorscheme, 72
win.graph, 71, 77
win.printer, 71, 73
win3, 148
window, 434, *516*
window size, 75

Windows, vii, 2, 3, 8–11, 38, 43–45, 66, 67,
　　　69, 71–73, 77, 89, 108, 140, 145,
　　　147, 148, 304, 317, 343, 497, 498,
　　　517, 520
　clipboard, 38
Winsorizing, 251
wireframe, 92, 96, 106, 470, *516*
working directory, 4, 5, 8, 9
write, **61**, 63, *516*
write.matrix, 61
write.table, 61, *516*
wtloss, *see* Datasets

xor, 35
xyplot, 92, **96**, 304, *516*

Yule–Walker equations, 444, 446, 456